ELECTROMAGNETISM

CONTENTS

PREFACE

Like *Electromagnetic Fields and Waves* by the same authors, this book aims to give the reader a working knowledge of electromagnetism. Those who will use it should refer to a collection of essays by Alfred North Whitehead entitled *The Aims of Education,*† particularly the first essay, which carries the same title. At the outset, Whitehead explains his point: "The whole book is a protest against dead knowledge, that is to say, against inert ideas . . . ideas that are merely received into the mind without being utilized, or tested, or thrown into fresh combinations." It is in this spirit that we have given more than 90 examples and set 332 problems, 27 of which are solved in detail.

The book is intended for courses at the freshman or sophomore level, either as an introduction to the subject for physicists and engineers, or as a last course in electromagnetism for students in other disciplines.

It is assumed that the reader has had a one-term course on differential and integral calculus. No previous knowledge of vectors, multiple integrals, differential equations, or complex numbers is assumed.

The features of our previous book that have been so appreciated will be found here as well: the examples mentioned above, problems drawn from the current literature, a summary at the end of each chapter, and three-dimensional figures, this time drawn in true perspective.

The book is divided into twenty short chapters suitable for self-paced instruction. Following an introductory chapter on vectors, the discussion of electromagnetism ranges from Coulomb's law through plane waves in dielectrics. Electric circuits are discussed at some length in three chapters: 5, 17, and 18. Chapter 16 is entirely devoted to complex numbers and phasors. There is no way of dealing properly with alternating currents without using complex numbers, and, contrary to what is often assumed, complex numbers are not much of

†Free Press, New York, 1976.

an obstacle at this level. Our discussion of relativity occupies only part of one chapter (Chapter 10)—just enough to explain briefly the Lorentz transformation and the transformation of electric and magnetic fields. This naturally leaves a host of questions unanswered, but should be sufficient in the present context.

THE PROBLEMS

The longer problems concern devices and methods of measurement described in the physics and electrical engineering literature of the past few years. They are "programmed" in the sense that they progress by small steps and that the intermediate results are usually given.

These longer problems are meant to give the reader the opportunity to learn how to make approximations and to build models amenable to quantitative analysis. It is of course hoped that they can teach the heuristic process involved in solving problems in the field of electromagnetism. In fact, after working through many such problems, students end up surprised to see that they can deal with real situations.

These problems contain a large amount of peripheral information and should provide some interesting reading. They should also incite the reader to apply his newly acquired knowledge to other fields and should stimulate creativity. Many of them can serve for open-ended experiments.

The easier problems are marked *E,* from Chapter 2 on. They can be solved in just a few lines, but they nonetheless often require quite a bit of thought. Some problems are marked *D*. Those are quite difficult.

A number of problems require curve plotting. This is because curves are so much more meaningful than formulas. It is assumed that the student has access to a programmable calculator, or possibly to a computer. Otherwise the calculations would be tedious.

On the average, about two problems are solved in detail at the end of each chapter.

UNITS AND SYMBOLS

The units and symbols used are those of the *Système International d'Unités,* designated SI in all languages.† The system originated with the proposal made

†See *ASTM/IEEE Standard Metric Practice,* published by the Institute of Electrical and Electronics Engineers, Inc., 345 East 47th Street, New York, N.Y. 10017.

by the Italian engineer Giovanni Giorgi in 1901 that electrical units be based on the meter-kilogram-second system. The Giorgi system grew with the years and came to be called, first the MKS system, and later the MKSA system (A for ampere). Its development was fostered mostly by the International Bureau of Weights and Measures, but also by several other international bodies such as the International Council of Scientific Unions, the International Electrotechnical Commission, and the International Union of Pure and Applied Physics.

Appendix D provides a conversion table for passing from cgs to SI units, and inversely. "Further use of the cgs units of electricity and magnetism is deprecated."†

SUPPLEMENTARY READING

The following books are recommended for supplementary reading:

Electromagnetic Fields and Waves: Second Edition, by the same authors and the same publisher, published in 1970. There are several references to this book in the footnotes.

Standard Handbook for Electrical Engineers, McGraw-Hill, New York, 1968.

Electronics Engineers' Handbook, McGraw-Hill, New York, 1975.

Reference Data for Radio Engineers, Howard Sams and Company, Inc., Indianapolis, 1970.

Electrostatics and Its Applications, A. D. Moore, Editor, John Wiley, New York, 1973.

These books can be found in most physics and engineering libraries. The reader will find the second and third references to be inexhaustible sources of information on practical applications.

ACKNOWLEDGMENTS

I am indebted first of all to my students, who have taught me so much in the course of innumerable discussions. I am also indebted to the persons who assisted me in writing this book: to François Lorrain, who took part in the preparation of the manuscript and who prepared sketches of the three-dimensional objects that appear in the figures in true perspective; to Ronald Liboiron, who also worked on the preparation of the figures; to Jean-Guy Desmarais and

†*Ibid.*, p. 11.

Guy Bélanger, both of whom revised the complete text; and particularly to Lucie Lecomte, Nancy Renz, and Alice Chénard, who typed or retyped over a thousand pages of manuscript.

I am also grateful to Robert Mann, who took such great care in editing both this book and the previous one; and to Pearl C. Vapnek, of W. H. Freeman and Company, for her meticulous work on the proofs.

Finally, I wish to thank collectively all those persons who wrote to me in connection with *Electromagnetic Fields and Waves,* even though they all received a prompt answer and were duly thanked individually. Their feedback was invaluable.

Dale Corson's duties as President, and later as Chancellor, of Cornell University have unfortunately prevented him from working on this offshoot of *Electromagnetic Fields and Waves,* of which he is co-author.

PAUL LORRAIN

Montreal, 1978

LIST OF SYMBOLS

SPACE, TIME, MECHANICS

Element of length	dl, ds, dr
Total length	l, L, s, r
Element of area	da
Total area	S
Element of volume	$d\tau$
Total volume	τ
Solid angle	Ω
Normal to a surface	$\mathbf{n}$
Wavelength	λ
Radian length	$\lambdabar = \lambda/2\pi$
Time	t
Period	$T = 1/f$
Frequency	$f = 1/T$
Angular frequency, angular velocity	$\omega = 2\pi f$
Velocity	v
Mass	m
Mass density	ρ
Momentum	$\mathbf{p}$
Moment of inertia	I
Force	$\mathbf{F}$
Torque	$\mathbf{T}$
Pressure	p
Energy	W
Power	P

ELECTRICITY AND MAGNETISM

Quantity of electricity	Q
Volume charge density	ρ
Surface charge density	σ
Linear charge density	λ
Electric potential	V
Induced electromotance	$\mathcal{V}$
Electric field intensity	$\mathbf{E}$
Electric displacement	$\mathbf{D}$
Permittivity of vacuum	ϵ_0

Relative permittivity	ϵ_r
Permittivity	$\epsilon = \epsilon_r \epsilon_0 = D/E$
Electric dipole moment	$\mathbf{p}$
Electric polarization, electric dipole moment per unit volume	$\mathbf{P}$
Electric susceptibility	χ_e
Mobility	$\mathcal{M}$
Electric current	I
Volume current density	$\mathbf{J}$
Surface current density	$\boldsymbol{\alpha}$
Vector potential	$\mathbf{A}$
Magnetic induction	$\mathbf{B}$
Magnetic field intensity	$\mathbf{H}$
Magnetic flux	Φ
Magnetic flux linkage	Λ
Permeability of vacuum	μ_0
Relative permeability	μ_r
Permeability	$\mu = \mu_r \mu_0 = B/H$
Magnetic dipole moment per unit volume	$\mathbf{M}$
Magnetic susceptibility	χ_m
Magnetic dipole moment	$\mathbf{m}$
Resistance	R
Capacitance	C
Self-inductance	L
Mutual inductance	M
Impedance	Z
Admittance	Y
Resistivity	ρ
Conductivity	σ
Poynting vector	$\mathcal{S}$

MATHEMATICAL SYMBOLS

Approximately equal to	$\approx$
Proportional to	$\propto$
Exponential of x	e^x, $\exp x$
Average	$\bar{B}$
Real part of z	$\mathrm{Re}\, z$
Modulus of z	$\lvert z \rvert$
Decadic log of x	$\log x$
Natural log of x	$\ln x$
Arc tangent x	$\mathrm{arc\,tan}\, x$
Complex conjugate of z	z^*
Vector	$\mathbf{E}$
Gradient	$\boldsymbol{\nabla}$
Divergence	$\boldsymbol{\nabla}\cdot$
Curl	$\boldsymbol{\nabla}\times$
Laplacian	$\boldsymbol{\nabla}^2$
Unit vectors in Cartesian coordinates	$\mathbf{i}, \mathbf{j}, \mathbf{k}$
Unit vector along $\mathbf{r}$	$\mathbf{r}_1$
Field point	x, y, z
Source point	x', y', z'

ELECTROMAGNETISM

CHAPTER 1

VECTORS

Electric and magnetic phenomena are described in terms of the *fields* of electric charges and currents. For example, one expresses the force between two electric charges as the product of the magnitude of one of the charges and the field of the other.

The object of this first chapter is to describe the mathematical methods used to deal with fields. All the material in this chapter is essential for a proper understanding of what follows.

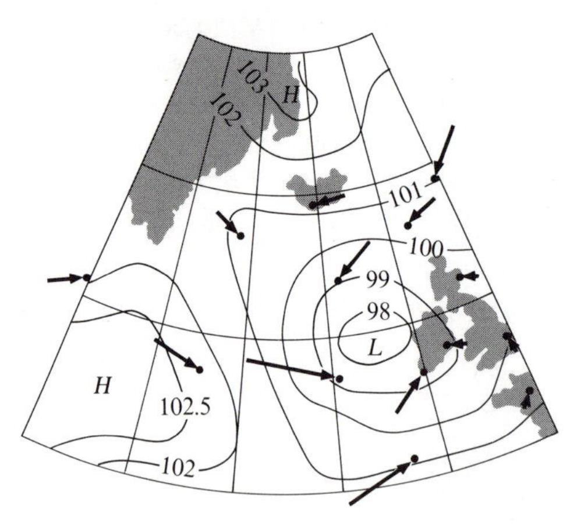

Figure 1-1 Pressure and wind-velocity fields over the North Atlantic on November 1, 1967, at 6 hours, Greenwich Mean Time. The curved lines are *isobars*, or lines of equal pressure. Pressures are given in kilopascals. High-pressure areas are denoted by H, and low-pressure areas by L. In this case an arrow indicates the direction and velocity of the wind at its tip; arrow lengths are proportional to wind velocities (the longest arrow in this figure represents a wind velocity of 25 meters per second). Wind vectors are given only at a few points, where actual measurements were made.

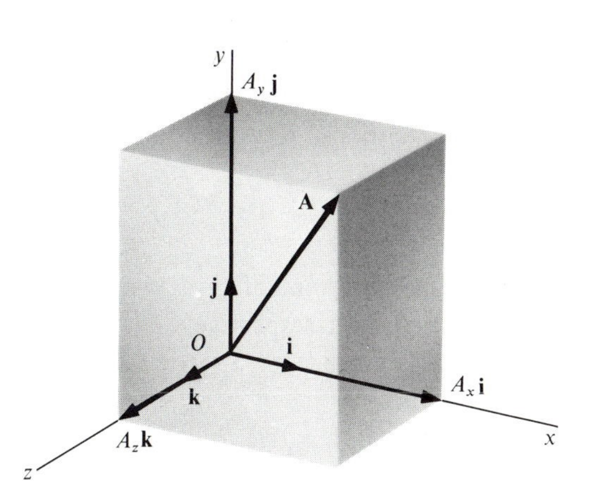

Figure 1-2 A vector **A** and the three vectors $A_x\mathbf{i}$, $A_y\mathbf{j}$, $A_z\mathbf{k}$, which, when placed end-to-end, are equivalent to **A**.

Mathematically, a field is a function that describes a physical quantity at all points in space. In *scalar fields*, this quantity is specified by a single number for each point. Pressure, temperature, and electric potential are examples of scalar quantities that can vary from one point to another in space. For *vector fields*, both a number and a direction are required. Wind velocity, gravitational force, and electric field intensity are examples of such vector quantities. Both types of field are illustrated in Fig. 1-1.

Vector quantities will be indicated by **boldface** type; *italic* type will indicate either a scalar quantity or the magnitude of a vector quantity.†

We shall follow the usual custom of using *right-hand Cartesian coordinate systems*, as in Fig. 1-2; the positive z direction is the direction of advance of a right-hand screw rotated in the sense that turns the positive x-axis into the positive y-axis through the 90° angle.

1.1 VECTORS

A vector can be specified by its *components* along any three mutually perpendicular axes. In the Cartesian coordinate system of Fig. 1-2, for example, the components of the vector **A** are A_x, A_y, A_z.

† In a handwritten text, it is convenient to identify a vector by means of an arrow over the symbol, for example, $\vec{A}$.

The vector **A** can be uniquely expressed in terms of its components through the use of *unit vectors* **i**, **j**, **k**, which are defined as vectors of unit magnitude in the positive x, y, z directions, respectively:

$$\mathbf{A} = A_x\mathbf{i} + A_y\mathbf{j} + A_z\mathbf{k}. \tag{1-1}$$

The vector **A** is thus the sum of three vectors of magnitude A_x, A_y, A_z, parallel to the x-, y-, z-axes, respectively.

The *magnitude of* **A** is

$$A = (A_x^2 + A_y^2 + A_z^2)^{1/2}. \tag{1-2}$$

The *sum of two vectors* is obtained by adding their components:

$$\mathbf{A} + \mathbf{B} = (A_x + B_x)\mathbf{i} + (A_y + B_y)\mathbf{j} + (A_z + B_z)\mathbf{k}. \tag{1-3}$$

Subtraction is simply addition with one of the vectors changed in sign:

$$\mathbf{A} - \mathbf{B} = \mathbf{A} + (-\mathbf{B}) = (A_x - B_x)\mathbf{i} + (A_y - B_y)\mathbf{j} + (A_z - B_z)\mathbf{k}. \tag{1-4}$$

1.2 SCALAR PRODUCT

The *scalar*, or *dot product*, is the scalar quantity obtained on multiplying the magnitude of the first vector by the magnitude of the second and by the cosine of the angle between the two vectors. In Fig. 1-3, for example,

$$\mathbf{A} \cdot \mathbf{B} = AB \cos(\varphi - \theta). \tag{1-5}$$

It follows from this definition that the usual commutative and distributive rules of ordinary arithmetic multiplication apply to the scalar product:

$$\mathbf{A} \cdot \mathbf{B} = \mathbf{B} \cdot \mathbf{A}, \tag{1-6}$$

and, for any three vectors,

$$\mathbf{A} \cdot (\mathbf{B} + \mathbf{C}) = \mathbf{A} \cdot \mathbf{B} + \mathbf{A} \cdot \mathbf{C}. \tag{1-7}$$

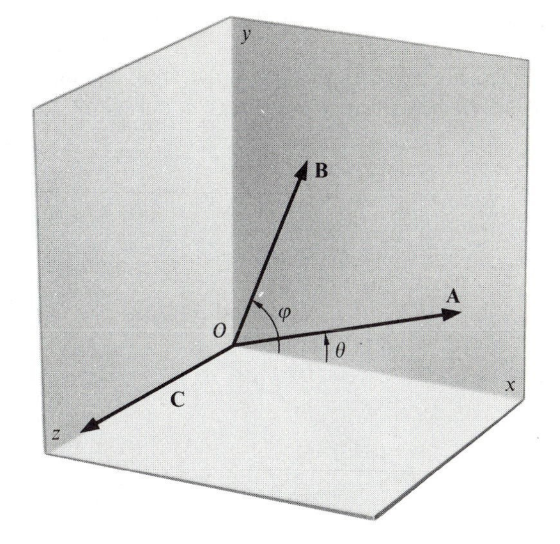

Figure 1-3 Two vectors **A** and **B** in the xy-plane. Their scalar product is $AB \cos(\varphi - \theta)$. The vector **C** is their vector product $\mathbf{A} \times \mathbf{B}$.

The latter property will be verified graphically in Prob. 1-3. It also follows that

$$\mathbf{i} \cdot \mathbf{i} = 1, \ \mathbf{j} \cdot \mathbf{j} = 1, \mathbf{k} \cdot \mathbf{k} = 1, \tag{1-8}$$

$$\mathbf{j} \cdot \mathbf{k} = 0, \mathbf{k} \cdot \mathbf{i} = 0, \ \mathbf{i} \cdot \mathbf{j} = 0. \tag{1-9}$$

Then,

$$\mathbf{A} \cdot \mathbf{B} = (A_x\mathbf{i} + A_y\mathbf{j} + A_z\mathbf{k}) \cdot (B_x\mathbf{i} + B_y\mathbf{j} + B_z\mathbf{k}), \tag{1-10}$$

$$= A_xB_x + A_yB_y + A_zB_z. \tag{1-11}$$

It is easy to check that this result is correct for two vectors in a plane, as in Fig. 1-3:

$$\mathbf{A} \cdot \mathbf{B} = AB \cos(\varphi - \theta) = AB \cos \varphi \cos \theta + AB \sin \varphi \sin \theta, \tag{1-12}$$

$$= A_xB_x + A_yB_y. \tag{1-13}$$

1.2.1 EXAMPLE: WORK DONE BY A FORCE

A simple physical example of the scalar product is the *work done by a force* **F** acting through a displacement **s**: $W = \mathbf{F} \cdot \mathbf{s}$, as in Fig. 1-4.

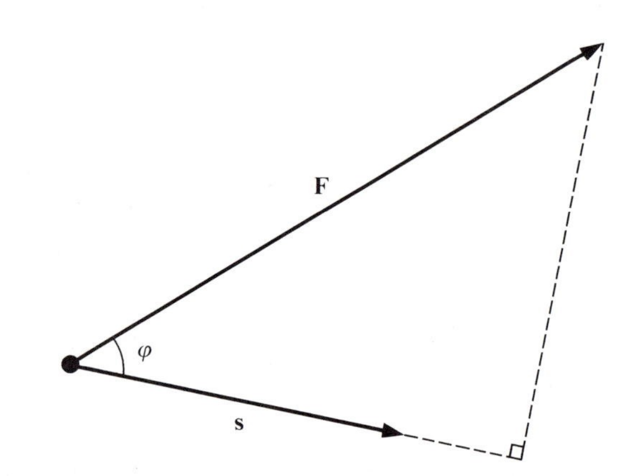

Figure 1-4 The work done by a force **F** whose point of application moves by a distance **s** is $(F \cos \varphi)s$, or $\mathbf{F} \cdot \mathbf{s}$.

1.3 VECTOR PRODUCT

The *vector*, or *cross product*, of two vectors is a vector whose direction is perpendicular to the plane containing the two initial vectors and whose magnitude is the product of the magnitudes of those vectors and the sine of the angle between them. We indicate the vector product thus:

$$\mathbf{A} \times \mathbf{B} = \mathbf{C}. \tag{1-14}$$

The magnitude of **C** is

$$C = |AB \sin (\varphi - \theta)|, \tag{1-15}$$

with φ and θ defined as in Fig. 1-3. The direction of **C** is given by the right-hand screw rule: it is the direction of advance of a right-hand screw whose axis, held perpendicular to the plane of **A** and **B**, is rotated in the sense that rotates the first-named vector (**A**) into the second-named (**B**) through the smaller angle.

The commutative rule is *not* followed for the vector product, since inverting the order of **A** and **B** inverts the direction of **C**:

$$\mathbf{A} \times \mathbf{B} = -(\mathbf{B} \times \mathbf{A}). \tag{1-16}$$

The distributive rule, however, is followed for any three vectors:

$$\mathbf{A} \times (\mathbf{B} + \mathbf{C}) = (\mathbf{A} \times \mathbf{B}) + (\mathbf{A} \times \mathbf{C}). \tag{1-17}$$

This will be shown in Prob. 1-7.

From the definition of the vector product it follows that

$$\mathbf{i} \times \mathbf{i} = 0, \quad \mathbf{j} \times \mathbf{j} = 0, \quad \mathbf{k} \times \mathbf{k} = 0, \tag{1-18}$$

and, for the usual right-handed coordinate systems, such as that of Fig. 1-2,

$$\mathbf{i} \times \mathbf{j} = \mathbf{k}, \quad \mathbf{j} \times \mathbf{k} = \mathbf{i}, \quad \mathbf{k} \times \mathbf{i} = \mathbf{j}, \quad \mathbf{j} \times \mathbf{i} = -\mathbf{k}, \text{ and so on.} \tag{1-19}$$

Writing out the vector product of **A** and **B** in terms of the components,

$$\mathbf{A} \times \mathbf{B} = (A_x\mathbf{i} + A_y\mathbf{j} + A_z\mathbf{k}) \times (B_x\mathbf{i} + B_y\mathbf{j} + B_z\mathbf{k}), \tag{1-20}$$

$$= (A_yB_z - A_zB_y)\mathbf{i} + (A_zB_x - A_xB_z)\mathbf{j} + (A_xB_y - A_yB_x)\mathbf{k}, \tag{1-21}$$

$$= \begin{vmatrix} \mathbf{i} & \mathbf{j} & \mathbf{k} \\ A_x & A_y & A_z \\ B_x & B_y & B_z \end{vmatrix}. \tag{1-22}$$

We can check this result for the two vectors of Fig. 1-3 by expanding $\sin(\varphi - \theta)$ and noting that the vector product is in the positive z direction.

1.3.1 EXAMPLES: TORQUE, AREA OF A PARALLELOGRAM

A good physical example of the vector product is the *torque* **T** produced by a force **F** acting with a moment arm **r** about a point O, as in Fig. 1-5, where $\mathbf{T} = \mathbf{r} \times \mathbf{F}$.

A second example is the *area of a parallelogram*, as in Fig. 1-6, where the area $\mathbf{S} = \mathbf{A} \times \mathbf{B}$. The area is thus represented by a vector perpendicular to the surface.

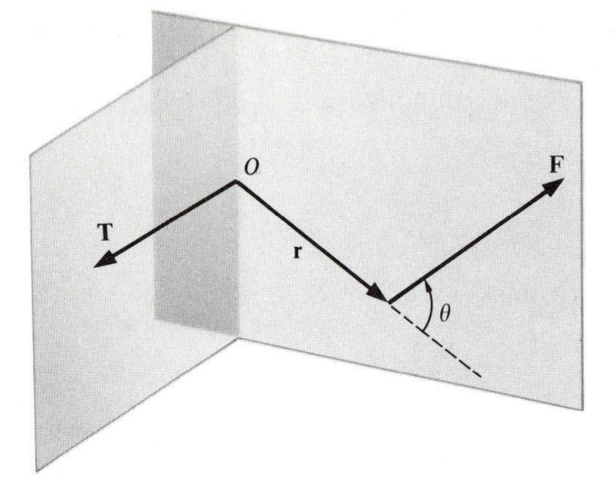

Figure 1-5 An example of vector multiplication. The torque **T** of the force **F** about the point O is $\mathbf{r} \times \mathbf{F}$. This vector has a magnitude of $rF \sin \theta$ and is oriented as shown.

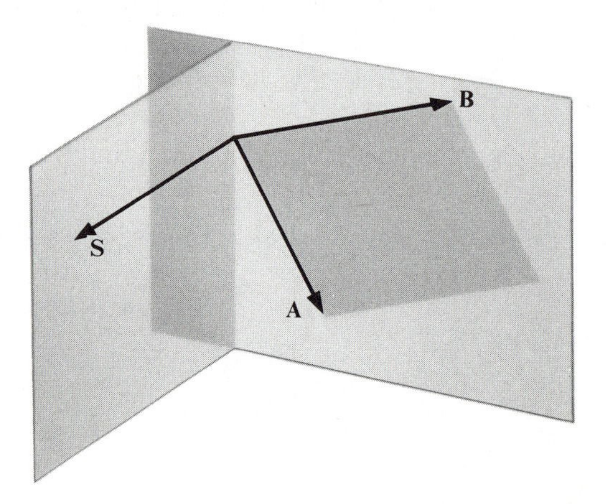

Figure 1-6 The area of the parallelogram is $\mathbf{A} \times \mathbf{B} = \mathbf{S}$. The vector **S** is normal to the parallelogram.

1.4 *THE TIME DERIVATIVE*

We shall often be concerned with the rates of change of scalar and vector quantities with both time and space coordinates, and thus with the time and space derivatives.

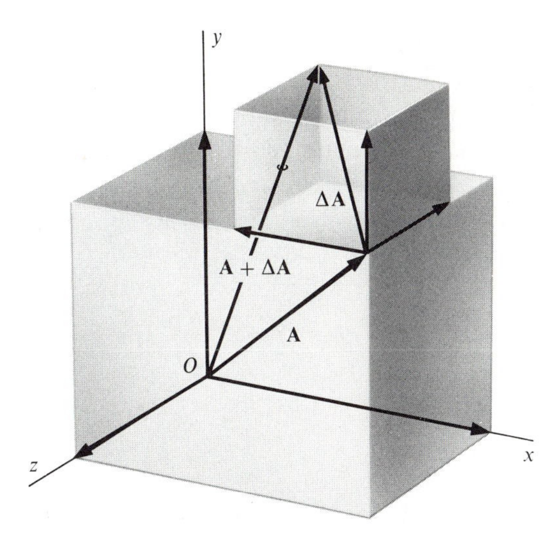

Figure 1-7 A vector **A**, its increment Δ**A**, and their components.

The time derivative of a vector quantity is straightforward. In a time Δt, a vector **A**, as in Fig. 1-7, may change by $\Delta\mathbf{A}$, which in general represents a change both in magnitude and in direction. On dividing $\Delta\mathbf{A}$ by Δt and taking the limit in the usual way, we arrive at the definition of the time derivative $d\mathbf{A}/dt$:

$$\frac{d\mathbf{A}}{dt} = \lim_{\Delta t \to 0} \frac{\mathbf{A}(t + \Delta t) - \mathbf{A}(t)}{\Delta t}, \tag{1-23}$$

$$= \lim_{\Delta t \to 0} \frac{\Delta A_x \mathbf{i} + \Delta A_y \mathbf{j} + \Delta A_z \mathbf{k}}{\Delta t}, \tag{1-24}$$

$$= \frac{dA_x}{dt}\mathbf{i} + \frac{dA_y}{dt}\mathbf{j} + \frac{dA_z}{dt}\mathbf{k}. \tag{1-25}$$

The time derivative of a vector is thus equal to the vector sum of the time derivatives of its components.

1.4.1 *EXAMPLES: POSITION, VELOCITY, AND ACCELERATION*

The time derivative of the *position* **r** of a point is its *velocity* **v**, and the time derivative of **v** is the *acceleration* **a**. See Prob. 1-10.

1.5 THE GRADIENT

Let us consider a scalar quantity that is a continuous and differentiable function of the coordinates and has the value f at a certain point in space, as in Fig. 1-8. For example, f could be the electrostatic potential V. We wish to know how f changes over the distance **dl** measured from that point. Now

$$df = \frac{\partial f}{\partial x} dx + \frac{\partial f}{\partial y} dy = \mathbf{A} \cdot \mathbf{dl}, \tag{1-26}$$

where

$$\mathbf{A} = \frac{\partial f}{\partial x} \mathbf{i} + \frac{\partial f}{\partial y} \mathbf{j}, \quad \text{and } \mathbf{dl} = dx\, \mathbf{i} + dy\, \mathbf{j}. \tag{1-27}$$

The vector **A**, whose components are the rates of change of f with distance along the coordinate axes, is called the *gradient* of the scalar quantity f. The operation on the scalar f defined by the term gradient is indicated by the symbol $\boldsymbol{\nabla}$, called "del". Thus

$$\mathbf{A} = \boldsymbol{\nabla} f. \tag{1-28}$$

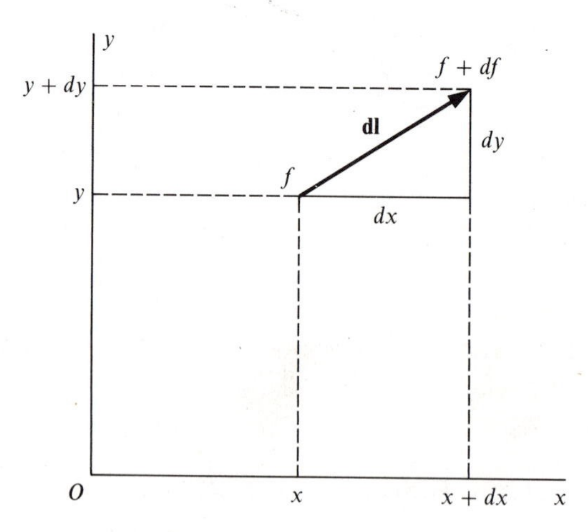

Figure 1-8 The quantity f, a function of position, changes from f to $f + df$ over the distance **dl**.

In three dimensions,

$$\boldsymbol{\nabla} \equiv \mathbf{i}\frac{\partial}{\partial x} + \mathbf{j}\frac{\partial}{\partial y} + \mathbf{k}\frac{\partial}{\partial z}. \tag{1-29}$$

The partial differentiations indicated are carried out on whatever scalar quantity stands to the right of the $\boldsymbol{\nabla}$ symbol. Thus

$$\boldsymbol{\nabla} f = \frac{\partial f}{\partial x}\mathbf{i} + \frac{\partial f}{\partial y}\mathbf{j} + \frac{\partial f}{\partial z}\mathbf{k}, \tag{1-30}$$

and

$$|\boldsymbol{\nabla} f| = \left[\left(\frac{\partial f}{\partial x}\right)^2 + \left(\frac{\partial f}{\partial y}\right)^2 + \left(\frac{\partial f}{\partial z}\right)^2\right]^{1/2}. \tag{1-31}$$

Thus,

$$df = \boldsymbol{\nabla} f \cdot \mathbf{dl} = |\boldsymbol{\nabla} f|\, |\mathbf{dl}| \cos\theta, \tag{1-32}$$

where θ is the angle between the vectors $\boldsymbol{\nabla} f$ and $\mathbf{dl}$.

We now ask what direction one should choose for $\mathbf{dl}$ in order that df be maximum. The answer is: the direction in which $\theta = 0$, that is, the direction of $\boldsymbol{\nabla} f$. The gradient of f is thus a vector whose magnitude and direction are those of the maximum space rate of change of f.

The gradient of a scalar function at a given point is a vector having the following properties:

1) Its components are the rates of change of the function along the directions of the coordinate axes.
2) Its magnitude is the maximum rate of change of the function with distance.
3) Its direction is that of the maximum rate of change of the function.
4) It points toward *larger* values of the function.

The gradient is a vector point-function derived from a scalar point-function.

1.5.1 EXAMPLES: TOPOGRAPHIC MAP, ELECTRIC FIELD INTENSITY

Fig. 1-9 shows the gradient of the elevation on a *topographic map*.

In the next chapter we shall see that, in an electrostatic field, the *electric field intensity* $\mathbf{E}$ is equal to minus the gradient of the electric potential V: $\mathbf{E} = -\nabla V$.

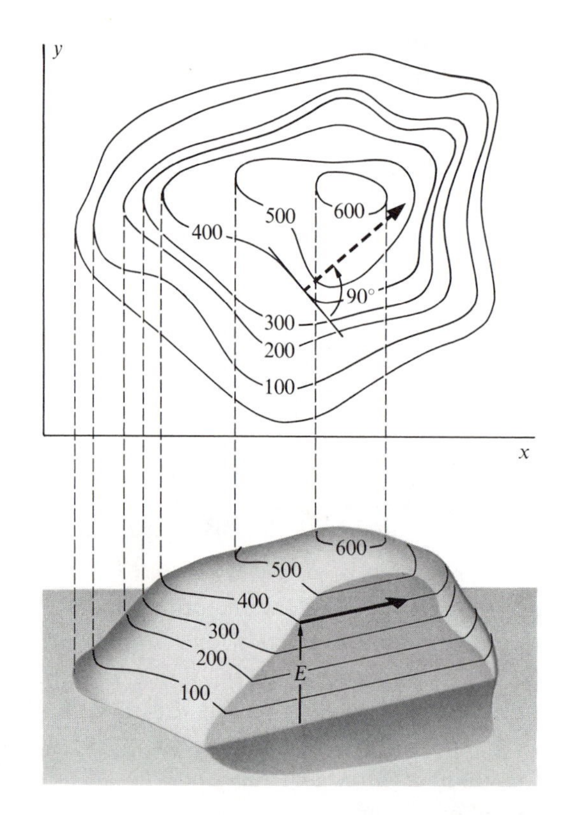

Figure 1-9 Topographic map of a hill. The numbers shown give the elevation E in meters. The gradient of E is the slope of the hill at the point considered, and it points toward an *in*crease in elevation: $\nabla E = (\partial E/\partial x)\mathbf{i} + (\partial E/\partial y)\mathbf{j}$. The arrow shows ∇E at one point where the elevation is 400 meters.

1.6 THE SURFACE INTEGRAL

Consider the x-y plane. To locate a point in this plane, one requires two coordinates, x and y. Now consider a given spherical surface, as in Fig. 1-10.

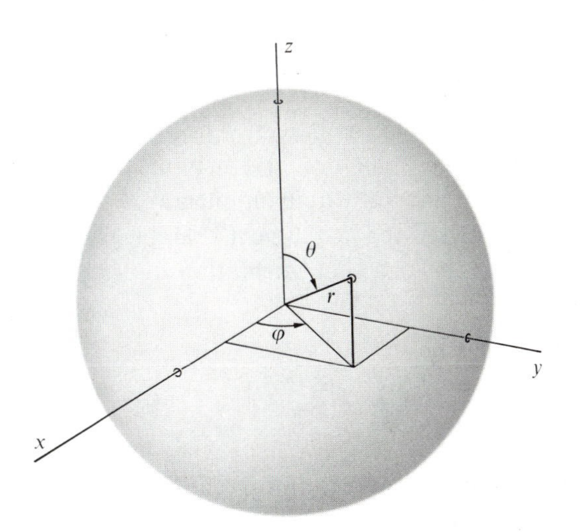

Figure 1-10 Spherical surface of radius r. The two coordinates θ and φ suffice to determine the position of a point P on the surface. How do these coordinates compare with the latitude and the longitude at the surface of the earth?

To fix a point we again need two coordinates, θ and φ. Of course we could use the three coordinates x, y, z. But those three coordinates are redundant, on the given sphere, since $x^2 + y^2 + z^2 = r^2$. So we only need two coordinates, θ and φ.

This is a general rule: one always needs to specify two coordinates to identify a point on a given surface.

If we require the area of the given surface, we must integrate the element of area, say $dx\,dy$, over *both* variables. We then have a *surface integral*.

More generally, an integral evaluated over two variables (that are not necessarily space coordinates) is called a *double integral*.

Suppose the surface carries a charge. Then, to find the total charge, we must integrate the surface charge density, multiplied by the element of area, over *both* variables.

So, a surface integral is a double integral evaluated over the two coordinates that are required to fix a point on a given surface.

If the surface is situated in the x-y plane, the surface integral is of the form

$$\int_{y=c}^{y=d} \left\{ \int_{x=a}^{x=b} f(x,y)\,dx \right\} dy,$$

where $f(x,y)$ is a function of x and y, and where the limits a and b on x may be functions of y. The limits c and d are constants. We first evaluate the integral between the braces. This gives either a constant or a function of y. Then this constant, or function, is integrated over y. The double integral is therefore an integral within an integral.

As we shall see in the next example, one arrives at the same result by inverting the order of integration and writing the double integral in the form

$$\int_{x=a'}^{x=b'} \left\{ \int_{y=c'}^{y=d'} f(x,y)\, dy \right\} dx,$$

where the limits of integration are different: the limits c' and d' can now be functions of x, while a' and b' are constants.

We have shown braces in the above expressions so as to indicate the manner in which a double integral is evaluated, but these braces are never used in practice.

1.6.1 *EXAMPLE: AREA OF A CIRCLE*

It is always advisable to test one's skill at understanding and using new concepts by applying them to extremely simple cases. So, let us calculate the *area of a circle* by means of a surface integral.

So as to simplify the calculation, we shall find the area of the top right-hand sector of the circle of Fig. 1-11, and then multiply by 4.

The element of area dA is $dx\, dy$. We proceed as follows. We first slide dA from left to right, to generate the shaded strip in the figure. Then we slide the strip from $y = 0$ to $y = R$, while adjusting its length correctly. This generates the 90-degree sector.

Let us first find an integral for the strip. The strip has a height dy. It starts at $x = 0$ and ends at the edge of the circle, where

$$x^2 + y^2 = R^2. \tag{1-33}$$

Thus, at the right-hand end of the strip,

$$x = (R^2 - y^2)^{1/2}, \tag{1-34}$$

and the strip has an area

$$\left\{ \int_0^{(R^2-y^2)^{1/2}} dx \right\} dy.$$

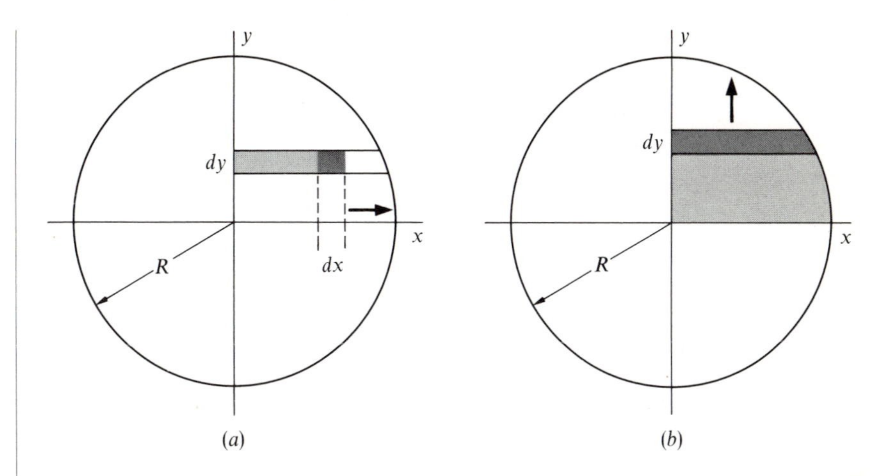

Figure 1-11 (*a*) The infinitesimal element of area $dx\,dy$ sweeps from the y-axis to the periphery of the circle to generate the shaded strip. (*b*) The strip now sweeps from the x-axis to the top of the circle to generate the top-right quadrant.

The total area A of the circle is given by

$$\frac{A}{4} = \int_{y=0}^{y=R} \left\{ \int_0^{(R^2-y^2)^{1/2}} dx \right\} dy, \tag{1-35}$$

$$= \int_0^R (R^2 - y^2)^{1/2}\,dy = R\int_0^R \left(1 - \frac{y^2}{R^2}\right)^{1/2} dy. \tag{1-36}$$

Setting $y/R = \sin\theta$, then

$$\left(1 - \frac{y^2}{R^2}\right)^{1/2} = \cos\theta, \qquad dy = R\cos\theta\,d\theta. \tag{1-37}$$

Also, when $y = 0, \theta = 0$, and when $y = R, \theta = \pi/2$. Then

$$\frac{A}{4} = R\int_0^{\pi/2} R\cos^2\theta\,d\theta = \frac{\pi}{4}R^2, \tag{1-38}$$

$$A = \pi R^2, \tag{1-39}$$

as expected.

If we used a vertical strip, it would extend from

$$y = 0 \quad \text{to} \quad y = (R^2 - x^2)^{1/2}. \tag{1-40}$$

Then we would have that

$$\frac{A}{4} = \int_{x=0}^{x=R} \int_{y=0}^{y=(R^2-x^2)^{1/2}} dy\, dx, \tag{1-41}$$

$$= \int_0^R (R^2 - x^2)^{1/2}\, dx. \tag{1-42}$$

This integral is similar to that of Eq. 1-36 and the area of the circle is again πR^2.

1.6.2 *EXAMPLE: ELECTRICALLY CHARGED DISK*

If we have an *electrically charged disk* with a surface charge density σ given by, say,

$$\sigma = Ky^2, \tag{1-43}$$

then, using a vertical strip and integrating over the complete circle, the total charge is

$$Q = \int_{x=-R}^{x=+R} \left\{ \int_{y=-(R^2-x^2)^{1/2}}^{y=+(R^2-x^2)^{1/2}} Ky^2\, dy \right\} dx, \tag{1-44}$$

$$= \int_{-R}^{+R} \left\{ \left[K\frac{y^3}{3} \right]_{-(R^2-x^2)^{1/2}}^{+(R^2-x^2)^{1/2}} \right\} dx, \tag{1-45}$$

$$= \int_{-R}^{+R} \frac{2K}{3} (R^2 - x^2)^{3/2}\, dx. \tag{1-46}$$

Using a table of integrals,† we find that

$$Q = \frac{\pi}{4} KR^4. \tag{1-47}$$

1.7 THE VOLUME INTEGRAL

In a *volume integral*, there are three variables, say x, y, z in Cartesian coordinates, or r, θ, φ in polar coordinates (Fig. 1-10). In the more general case where the three variables can be of any nature, say velocities, or voltages, etc., one has a *triple integral*.

Here again, one integrates with respect to each variable in succession, starting with the innermost integral.

† See, for example, Dwight, *Table of Integrals and Other Mathematical Data*, No. 350.03.

1.7.1 EXAMPLE: VOLUME OF A SPHERE

To calculate the *volume of a sphere*, we first find the element of volume. Referring to Fig. 1-12, this is

$$(r\,d\theta)(r\sin\theta\,d\varphi)(dr).$$

Then

$$V = \int_{r=0}^{r=R}\int_{\theta=0}^{\theta=\pi}\int_{\varphi=0}^{\varphi=2\pi} r^2 \sin\theta\,d\varphi\,d\theta\,dr. \tag{1-48}$$

Let us integrate, first with respect to φ, then with respect to θ, and finally with respect to r. This means that we first rotate the element of volume about the vertical axis to generate a ring. Then we slide the ring from $\theta = 0$ to $\theta = \pi$, while keeping the radius r and dr constant. This generates a spherical shell of radius r and thickness dr. Then we vary r from 0 to R to generate the sphere. So

$$V = \int_{r=0}^{r=R}\int_{\theta=0}^{\theta=\pi} 2\pi r^2 \sin\theta\,d\theta\,dr, \tag{1-49}$$

$$= 2\pi \int_{r=0}^{r=R} r^2 \left\{\int_{\theta=0}^{\theta=\pi} \sin\theta\,d\theta\right\} dr, \tag{1-50}$$

$$= 2\pi \int_0^R 2r^2\,dr = \frac{4}{3}\pi R^3. \tag{1-51}$$

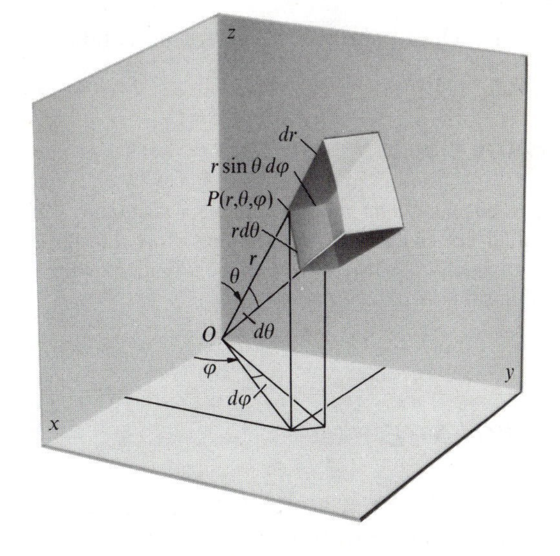

Figure 1-12 Element of volume in spherical coordinates.

This is a particularly simple case because the limits of integration are all constants. Indeed, the triple integral of Eq. 1-48 is really the product of three integrals.

In Prob. 1-19 we shall repeat the calculation in Cartesian coordinates. The limits of integration will then be functions, instead of being constants.

1.8 FLUX

It is often necessary to calculate the flux of a vector through a surface. For example, one might wish to calculate the magnetic flux through the iron core of an electromagnet. The flux of a vector **A** through an infinitesimal surface **da** is

$$d\Phi = \mathbf{A} \cdot \mathbf{da}, \tag{1-52}$$

where the vector **da** is normal to the surface. The flux $d\Phi$ is the component of the vector **A** normal to the surface, multiplied by da. For a finite surface S we find the total flux by integrating $\mathbf{A} \cdot \mathbf{da}$ over the entire surface:

$$\Phi = \int_S \mathbf{A} \cdot \mathbf{da}. \tag{1-53}$$

For a closed surface the vector **da** is taken to point *outward.*

1.8.1 *EXAMPLE: FLUID FLOW*

Let us consider *fluid flow.* We define a vector $\rho\mathbf{v}$, ρ being the fluid density and $\mathbf{v}$ the fluid velocity at a point. The flux of $\rho\mathbf{v}$ through any closed surface is the net rate at which mass leaves the volume bounded by the surface. In an incompressible fluid this flux is zero.

1.9 DIVERGENCE

The outward flux Φ of a vector **A** through a closed surface can be calculated either from the above equation or as follows. Let us consider an infinitesimal volume dx, dy, dz and the vector **A**, as in Fig. 1-13, whose

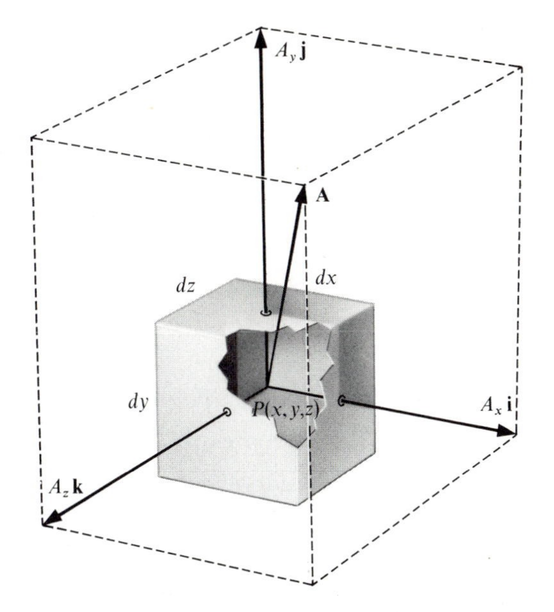

Figure 1-13 Element of volume dx, dy, dz around a point P, where the vector $\mathbf{A}$ has the value illustrated by the arrow.

components A_x, A_y, A_z are functions of the coordinates x, y, z. We consider an infinitesimal volume and first-order variations of $\mathbf{A}$.

The value of A_x at the center of the right-hand face can be taken to be the average value over that face. Through the right-hand face of the volume element, the outgoing flux is

$$d\Phi_R = \left(A_x + \frac{\partial A_x}{\partial x}\frac{dx}{2}\right) dy\, dz, \tag{1-54}$$

since the normal component of $\mathbf{A}$ at the right-hand face is the x-component of $\mathbf{A}$ at that face.

At the left-hand face,

$$d\Phi_L = -\left(A_x - \frac{\partial A_x}{\partial x}\frac{dx}{2}\right) dy\, dz. \tag{1-55}$$

The minus sign before the parenthesis is necessary here because, $A_x\mathbf{i}$ being inward at this face and $\mathbf{da}$ being outward, the cosine of the angle between the two vectors is -1. The net outward flux through the two faces is then

$$d\Phi_R + d\Phi_L = \frac{\partial A_x}{\partial x}\, dx\, dy\, dz = \frac{\partial A_x}{\partial x}\, d\tau, \tag{1-56}$$

where $d\tau$ is the volume of the infinitesimal element.

If we calculate the net flux through the other pairs of faces in the same manner, we find the total outward flux for the element of volume $d\tau$ to be

$$d\Phi_{\text{tot}} = \left(\frac{\partial A_x}{\partial x} + \frac{\partial A_y}{\partial y} + \frac{\partial A_z}{\partial z}\right) d\tau. \tag{1-57}$$

Suppose now that we have two adjoining infinitesimal volume elements and that we add the flux through the bounding surface of the first volume to the flux through the bounding surface of the second. At the common face the fluxes are equal in magnitude but opposite in sign, and they cancel as in Fig. 1-14. The flux from the first volume plus that from the second is the flux through the bounding surface of the combined volumes.

To extend this calculation to a finite volume, we sum the individual fluxes for each of the infinitesimal volume elements in the finite volume, and the total outward flux is

$$\Phi_{\text{tot}} = \int_\tau \left(\frac{\partial A_x}{\partial x} + \frac{\partial A_y}{\partial y} + \frac{\partial A_z}{\partial z}\right) d\tau. \tag{1-58}$$

At any given point in the volume, the quantity

$$\frac{\partial A_x}{\partial x} + \frac{\partial A_y}{\partial y} + \frac{\partial A_z}{\partial z}$$

is thus the *outgoing* flux per unit volume. We call this the *divergence* of the vector $\mathbf{A}$ at the point.

The divergence of a vector point-function is a scalar point-function. According to the rules for the scalar product,

$$\boldsymbol{\nabla} \cdot \mathbf{A} = \frac{\partial A_x}{\partial x} + \frac{\partial A_y}{\partial y} + \frac{\partial A_z}{\partial z}, \tag{1-59}$$

where the operator $\boldsymbol{\nabla}$ is defined as in Eq. 1-29. The operator $\boldsymbol{\nabla}$ is not a vector, of course, but it is convenient to use the notation of the scalar product to indicate the operation that is carried out.

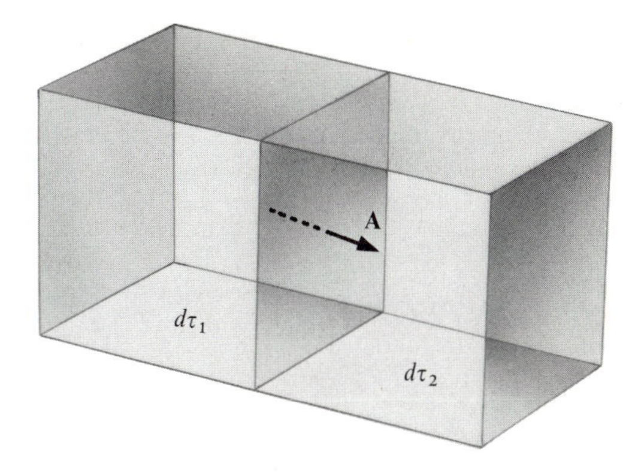

Figure 1-14 At the common face, the flux of **A** *out* of $d\tau_1$ is equal in magnitude, but opposite in sign, to the flux of **A** *out* of $d\tau_2$.

1.10 *THE DIVERGENCE THEOREM*

The total outward flux of Eq. 1-58 is also equal to the surface integral of the normal outward component of **A**. Thus

$$\int_S \mathbf{A} \cdot d\mathbf{a} = \int_\tau \left(\frac{\partial A_x}{\partial x} + \frac{\partial A_y}{\partial y} + \frac{\partial A_z}{\partial z} \right) d\tau = \int_\tau \nabla \cdot \mathbf{A} \, d\tau. \tag{1-60}$$

This is the *divergence theorem.* Note that the integral on the left involves only the values of **A** on the surface S, whereas the two integrals on the right involve the values of **A** throughout the volume τ enclosed by S. The divergence theorem is a generalization of the fundamental theorem of the calculus (see Prob. 1-20).

We can now redefine the divergence of the vector **A** as follows. If the volume τ is allowed to shrink sufficiently, so that $\nabla \cdot \mathbf{A}$ does not vary appreciably over it, then

$$\int_S \mathbf{A} \cdot d\mathbf{a} = (\nabla \cdot \mathbf{A}), \qquad (\tau \to 0), \tag{1-61}$$

and

$$\nabla \cdot \mathbf{A} = \lim_{\tau \to 0} \frac{1}{\tau} \int_S \mathbf{A} \cdot d\mathbf{a}. \tag{1-62}$$

The divergence is thus the outward flux per unit volume, as the volume τ approaches zero.

1.10.1 EXAMPLES: INCOMPRESSIBLE FLUID, EXPLOSION

In an *incompressible fluid*, $\nabla \cdot (\rho \mathbf{v})$ is everywhere equal to zero, since the outward mass flux per unit volume is zero.

Within an *explosion*, $\nabla \cdot (\rho \mathbf{v})$ is positive.

1.11 THE LINE INTEGRAL

The integral

$$\int_a^b \mathbf{A} \cdot \mathbf{dl},$$

evaluated from the point a to the point b on some specified curve, is a *line integral*. Each element of length $\mathbf{dl}$ on the curve is multiplied by the local value of $\mathbf{A}$ according to the rule for the scalar product. These products are then summed to obtain the value of the integral.

A vector field $\mathbf{A}$ is said to be *conservative* if the line integral of $\mathbf{A} \cdot \mathbf{dl}$ around any closed curve is zero:

$$\oint \mathbf{A} \cdot \mathbf{dl} = 0. \tag{1-63}$$

The circle on the integral sign indicates that the path of integration is closed. We shall find in Sec. 2.5 that an electrostatic field is conservative.

1.11.1 EXAMPLE: WORK DONE BY A FORCE

The *work done by a force* $\mathbf{F}$ acting from a to b along some specified path is

$$W = \int_a^b \mathbf{F} \cdot \mathbf{dl}, \tag{1-64}$$

where both $\mathbf{F}$ and $\mathbf{dl}$ must be known functions of the coordinates if the integral is to be evaluated analytically.

Let us calculate the work done by a force $\mathbf{F}$ that is in the y direction and has a magnitude proportional to y, as it moves around the circular path from a to b in Fig. 1-15. Since

$$\mathbf{F} = ky\,\mathbf{j} \qquad \text{and} \qquad \mathbf{dl} = dx\,\mathbf{i} + dy\,\mathbf{j}, \tag{1-65}$$

$$W = \int_a^b \mathbf{F} \cdot \mathbf{dl} = \int_0^r ky\,dy = kr^2/2. \tag{1-66}$$

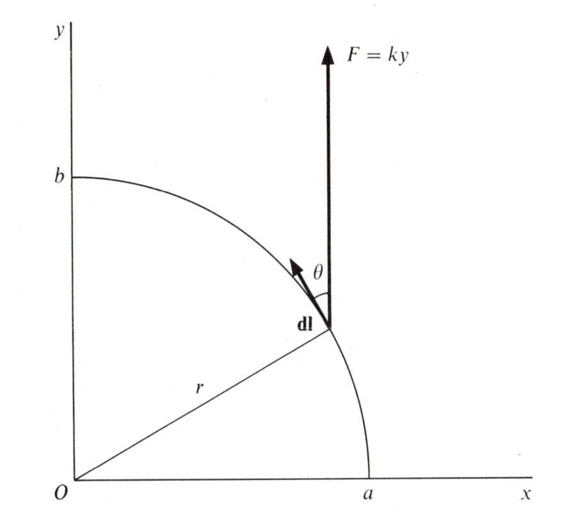

Figure 1-15 The force **F** is proportional to y, and its point of application moves from a to b. The work done is given by the line integral of $\mathbf{F} \cdot \mathbf{dl}$ over the circular path of radius r.

1.12 THE CURL

Let us calculate the value of the line integral of $\mathbf{A} \cdot \mathbf{dl}$ in the more general case where it is not zero.

For an infinitesimal element of path $\mathbf{dl}$ in the xy-plane, and from the definition of the scalar product,

$$\mathbf{A} \cdot \mathbf{dl} = A_x\, dx + A_y\, dy. \tag{1-67}$$

Thus, for any closed path in the xy-plane and for any $\mathbf{A}$,

$$\oint \mathbf{A} \cdot \mathbf{dl} = \oint A_x\, dx + \oint A_y\, dy. \tag{1-68}$$

Now consider the infinitesimal path in Fig. 1-16. There are two contributions to the first integral on the right-hand side of the above equation, one at $y - (dy/2)$ and one at $y + (dy/2)$:

$$\oint A_x\, dx = \left(A_x - \frac{\partial A_x}{\partial y}\frac{dy}{2}\right) dx - \left(A_x + \frac{\partial A_x}{\partial y}\frac{dy}{2}\right) dx. \tag{1-69}$$

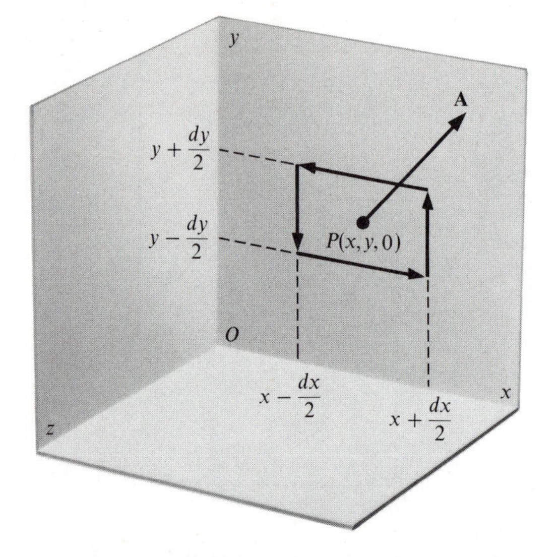

Figure 1-16 Closed, rectangular path in the xy-plane, centered on the point $P(x, y, 0)$, where the vector **A** has the components A_x, A_y. The integration around the path is performed in the direction of the arrows.

There is a minus sign before the second term because the path element at $y + (dy/2)$ is in the negative x direction. Therefore

$$\oint A_x \, dx = -\frac{\partial A_x}{\partial y} \, dy \, dx. \tag{1-70}$$

Similarly,

$$\oint A_y \, dy = \frac{\partial A_y}{\partial x} \, dx \, dy, \tag{1-71}$$

and

$$\oint \mathbf{A} \cdot \mathbf{dl} = \left(\frac{\partial A_y}{\partial x} - \frac{\partial A_x}{\partial y} \right) dx \, dy \tag{1-72}$$

for the infinitesimal path of Fig. 1-16.

If we set

$$g_3 = \frac{\partial A_y}{\partial x} - \frac{\partial A_x}{\partial y}, \tag{1-73}$$

then

$$\oint \mathbf{A} \cdot \mathbf{dl} = g_3 \, da, \tag{1-74}$$

where $da = dx\,dy$ is the area enclosed on the xy-plane by the infinitesimal path. Note that the above equation is correct only if the line integral is evaluated in the positive direction in the xy-plane, that is, in the direction in which one would have to turn a right-hand screw to make it advance in the positive direction of the z-axis. This is known as the *right-hand screw rule.*

Let us now consider g_3 and the other two symmetrical quantities as the components of a vector

$$\left(\frac{\partial A_z}{\partial y} - \frac{\partial A_y}{\partial z}\right)\mathbf{i} + \left(\frac{\partial A_x}{\partial z} - \frac{\partial A_z}{\partial x}\right)\mathbf{j} + \left(\frac{\partial A_y}{\partial x} - \frac{\partial A_x}{\partial y}\right)\mathbf{k},$$

which may be written as

$$\boldsymbol{\nabla} \times \mathbf{A} = \begin{vmatrix} \mathbf{i} & \mathbf{j} & \mathbf{k} \\ \dfrac{\partial}{\partial x} & \dfrac{\partial}{\partial y} & \dfrac{\partial}{\partial z} \\ A_x & A_y & A_z \end{vmatrix}. \tag{1-75}$$

We shall call this vector the *curl* of **A**. The quantity g_3 is then its z component.

If we consider the element of area as a vector **da** pointing in the direction of advance of a right-hand screw turned in the direction chosen for the line integral, then $\mathbf{da} = da\,\mathbf{k}$ and

$$\oint \mathbf{A} \cdot \mathbf{dl} = (\boldsymbol{\nabla} \times \mathbf{A}) \cdot \mathbf{da}. \tag{1-76}$$

This means that the line integral of $\mathbf{A} \cdot \mathbf{dl}$ around the edge of an element of area **da** is equal to the scalar product of the curl of **A** by this element of area, as long as we observe the above sign convention. This is a general result that applies to any element of area **da**, whatever its orientation.

Thus

$$(\nabla \times \mathbf{A})_n = \lim_{S \to 0} \frac{1}{S} \oint \mathbf{A} \cdot \mathbf{dl}: \tag{1-77}$$

the component of the curl of a vector normal to a surface S at a given point is equal to the line integral of the vector around the boundary of the surface, divided by the area of the surface when this area approaches zero around the point.

1.12.1 *EXAMPLE: WATER VELOCITY IN A STREAM*

Let us consider a *stream* in which the velocity $\mathbf{v}$ is proportional to the distance from the bottom. We set the z-axis parallel to the direction of flow, and the x-axis perpendicular to the stream bottom, as in Fig. 1-17. Then

$$v_x = 0, \qquad v_y = 0, \qquad v_z = cx. \tag{1-78}$$

We shall calculate the curl from Eq. 1-77. For $(\nabla \times \mathbf{v})_x$ we choose a path parallel to the yz-plane. In evaluating

$$\oint \mathbf{v} \cdot \mathbf{dl}$$

around such a path we note that the contributions are equal and opposite on the parts parallel to the z-axis, hence $(\nabla \times \mathbf{v})_x = 0$. Likewise, $(\nabla \times \mathbf{v})_z = 0$.

For the y-component we choose a path parallel to the xz-plane and evaluate the integral around it in the sense that would advance a right-hand screw in the positive y direction. On the parts of the path parallel to the x-axis, $\mathbf{v} \cdot \mathbf{dl} = 0$ since $\mathbf{v}$ and $\mathbf{dl}$ are perpendicular. On the bottom part of the path, at a distance x from the yz-plane,

$$\int_z^{z+\Delta z} \mathbf{v} \cdot \mathbf{dl} = cx\,\Delta z, \tag{1-79}$$

whereas at $(x + \Delta x)$

$$\int_z^{z+\Delta z} \mathbf{v} \cdot \mathbf{dl} = -c(x + \Delta x)\,\Delta z. \tag{1-80}$$

For the whole path,

$$\oint \mathbf{v} \cdot \mathbf{dl} = -c\,\Delta x\,\Delta z, \tag{1-81}$$

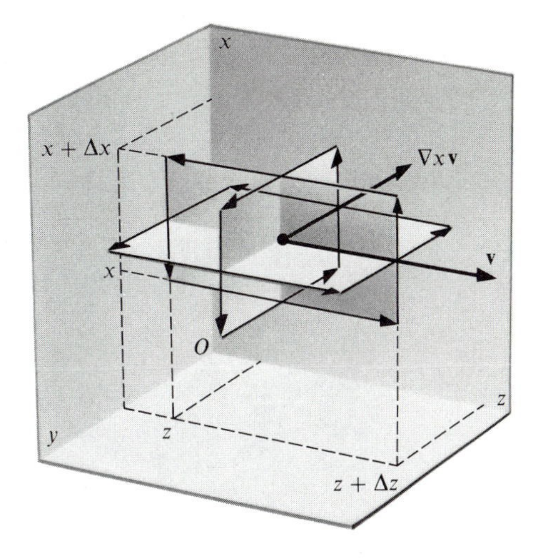

Figure 1-17 The velocity **v** in a viscous fluid is assumed to be in the direction of the z-axis and proportional to the distance x from the bottom. Then $\nabla \times \mathbf{v} = -c\mathbf{j}$.

and the y-component of the curl is

$$(\nabla \times \mathbf{v})_y = \lim_{S \to 0} \frac{\oint \mathbf{v} \cdot \mathbf{dl}}{S} = \frac{-c\,\Delta x\,\Delta z}{\Delta x\,\Delta z} = -c. \tag{1-82}$$

Calculating $\nabla \times \mathbf{v}$ directly from Eq. 1-75,

$$\nabla \times \mathbf{v} = \begin{vmatrix} \mathbf{i} & \mathbf{j} & \mathbf{k} \\ \dfrac{\partial}{\partial x} & \dfrac{\partial}{\partial y} & \dfrac{\partial}{\partial z} \\ 0 & 0 & cx \end{vmatrix} = -c\mathbf{j}, \tag{1-83}$$

which is the same result as above.

1.13 STOKES'S THEOREM

Equation 1-76 is true only for a path so small that $\nabla \times \mathbf{A}$ can be considered constant over the surface $\mathbf{da}$ bounded by the path. What if the path C is so large that this condition is not met? The equation can be extended readily to arbitrary paths. We divide the surface—*any* surface bounded by the

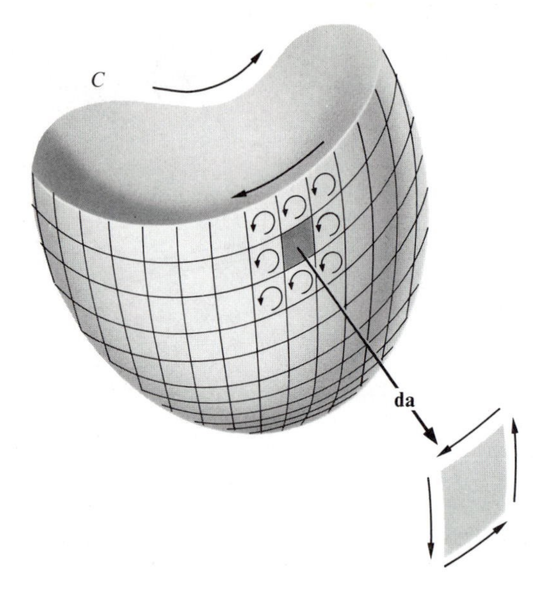

Figure 1-18 An arbitrary surface bounded by the curve C. The sum of the line integrals around the curvilinear squares shown is equal to the line integral around C.

path of integration in question—into elements of area $\mathbf{da}_1$, $\mathbf{da}_2$, and so forth, as in Fig. 1-18. For any one of these small areas,

$$\oint \mathbf{A} \cdot \mathbf{dl}_i = (\nabla \times \mathbf{A}) \cdot \mathbf{da}_i. \tag{1-84}$$

We add the left-hand sides of these equations for all the **da**'s, and then we add all the right-hand sides. The sum of the left-hand sides is the line integral around the external boundary, since there are always two equal and opposite contributions to the sum along every common side between adjacent **da**'s. The sum of the right-hand side is merely the integral of $(\nabla \times \mathbf{A}) \cdot \mathbf{da}$ over the finite surface. Thus, for *any* surface S bounded by a closed curve C,

$$\oint_C \mathbf{A} \cdot \mathbf{dl} = \int_S (\nabla \times \mathbf{A}) \cdot \mathbf{da}. \tag{1-85}$$

This is *Stokes's theorem.*

Figure 1-18 illustrates the sign convention.

1.13.1 EXAMPLE: CONSERVATIVE VECTOR FIELDS

Stokes's theorem can help us to understand *conservative vector fields*. Under what condition is a vector field conservative? In other words, under what condition is the line integral of $\mathbf{A} \cdot \mathbf{dl}$ around an arbitrary path equal to zero? From Stokes's theorem, the line integral of $\mathbf{A} \cdot \mathbf{dl}$ around an arbitrary closed path is zero if and only if $\nabla \times \mathbf{A} = 0$ everywhere. This condition is met if $\mathbf{A} = \nabla f$. Then

$$A_x = \frac{\partial f}{\partial x}, \qquad A_y = \frac{\partial f}{\partial y}, \qquad A_z = \frac{\partial f}{\partial z}, \tag{1-86}$$

and

$$(\nabla \times \mathbf{A})_x = \frac{\partial A_z}{\partial y} - \frac{\partial A_y}{\partial z} = \frac{\partial^2 f}{\partial y \, \partial z} - \frac{\partial^2 f}{\partial z \, \partial y} \equiv 0, \tag{1-87}$$

and so on for the other components of the curl.

The field of $\mathbf{A}$ is therefore conservative, if $\mathbf{A}$ can be expressed as the gradient of some scalar function f.

1.14 THE LAPLACIAN

The divergence of the gradient is of great importance in electromagnetism. Since

$$\nabla f = \frac{\partial f}{\partial x}\mathbf{i} + \frac{\partial f}{\partial y}\mathbf{j} + \frac{\partial f}{\partial z}\mathbf{k}, \tag{1-88}$$

then

$$\nabla \cdot \nabla f = \nabla^2 f = \frac{\partial^2 f}{\partial x^2} + \frac{\partial^2 f}{\partial y^2} + \frac{\partial^2 f}{\partial z^2}. \tag{1-89}$$

The divergence of the gradient is the sum of the second derivatives with respect to the rectangular coordinates. The quantity $\nabla \cdot \nabla f$ is abbreviated to $\nabla^2 f$, and is called the *Laplacian* of f. The operator ∇^2 is called the *Laplace operator*.

1.14.1 EXAMPLE: THE LAPLACIAN OF THE ELECTRIC POTENTIAL

We shall see in Sec. 3.4 that the *Laplacian of the electric potential* is proportional to the space charge density.

1.15 SUMMARY

In Cartesian coordinates a *vector* quantity is written in the form

$$\mathbf{A} = A_x \mathbf{i} + A_y \mathbf{j} + A_z \mathbf{k}, \tag{1-1}$$

where **i**, **j**, **k** are *unit vectors* directed along the x, y, z axes respectively.

The *magnitude* of the vector **A** is the scalar

$$A = (A_x^2 + A_y^2 + A_z^2)^{1/2}. \tag{1-2}$$

Vectors can be added and subtracted:

$$\mathbf{A} + \mathbf{B} = (A_x + B_x)\mathbf{i} + (A_y + B_y)\mathbf{j} + (A_z + B_z)\mathbf{k}, \tag{1-3}$$

$$\mathbf{A} - \mathbf{B} = (A_x - B_x)\mathbf{i} + (A_y - B_y)\mathbf{j} + (A_z - B_z)\mathbf{k}. \tag{1-4}$$

The scalar product (Fig. 1-3) is

$$\mathbf{A} \cdot \mathbf{B} = AB \cos(\varphi - \theta), \tag{1-5}$$

$$= A_x B_x + A_y B_y + A_z B_z. \tag{1-11}$$

It is commutative,

$$\mathbf{A} \cdot \mathbf{B} = \mathbf{B} \cdot \mathbf{A}, \tag{1-6}$$

and distributive,

$$\mathbf{A} \cdot (\mathbf{B} + \mathbf{C}) = \mathbf{A} \cdot \mathbf{B} + \mathbf{A} \cdot \mathbf{C}. \tag{1-7}$$

The *vector product* (Fig. 1-3) is

$$\mathbf{A} \times \mathbf{B} = \mathbf{C} \tag{1-14}$$

with

$$C = |AB \sin (\varphi - \theta)|. \tag{1-15}$$

Another equivalent definition of the vector product is

$$\mathbf{A} \times \mathbf{B} = \begin{vmatrix} \mathbf{i} & \mathbf{j} & \mathbf{k} \\ A_x & A_y & A_z \\ B_x & B_y & B_z \end{vmatrix}. \tag{1-22}$$

The vector product follows the distributive rule

$$\mathbf{A} \times (\mathbf{B} + \mathbf{C}) = (\mathbf{A} \times \mathbf{B}) + (\mathbf{A} \times \mathbf{C}), \tag{1-17}$$

but *not* the commutative rule:

$$\mathbf{A} \times \mathbf{B} = -(\mathbf{B} \times \mathbf{A}). \tag{1-16}$$

The *time derivative* of **A** is

$$\frac{d\mathbf{A}}{dt} = \frac{dA_x}{dt}\mathbf{i} + \frac{dA_y}{dt}\mathbf{j} + \frac{dA_z}{dt}\mathbf{k}. \tag{1-25}$$

The *del operator* is defined as follows:

$$\mathbf{\nabla} = \mathbf{i}\frac{\partial}{\partial x} + \mathbf{j}\frac{\partial}{\partial y} + \mathbf{j}\frac{\partial}{\partial z}, \tag{1-29}$$

and the *gradient* of a scalar function f is

$$\mathbf{\nabla} f = \frac{\partial f}{\partial x}\mathbf{i} + \frac{\partial f}{\partial y}\mathbf{j} + \frac{\partial f}{\partial z}\mathbf{k}. \tag{1-30}$$

The gradient gives the maximum rate of change of f with distance at the point considered, and it points toward larger values of f.

The *surface integral* is a common type of *double integral.* It is used for integrating over two coordinates, say x and y. It is composed of an integral within an integral. The inner integral is evaluated first.

The *volume integral* is one form of *triple integral.* It is used for integrating functions of three space coordinates, say x, y, z. In this case, we have an integral that is within an integral that is within an integral. The innermost integral is evaluated first, leaving a double integral, and so on.

The *flux* Φ of a vector $\mathbf{A}$ through a surface S is

$$\Phi = \int_S \mathbf{A} \cdot \mathbf{da}. \tag{1-53}$$

For a closed surface the vector **da** points outward.

The *divergence* of $\mathbf{A}$,

$$\boldsymbol{\nabla} \cdot \mathbf{A} = \frac{\partial A_x}{\partial x} + \frac{\partial A_y}{\partial y} + \frac{\partial A_z}{\partial z}, \tag{1-59}$$

is the outward flux of $\mathbf{A}$ per unit volume at the point considered.

The *divergence theorem* states that

$$\int_\tau \boldsymbol{\nabla} \cdot \mathbf{A}\, d\tau = \int_S \mathbf{A} \cdot \mathbf{da}, \tag{1-60}$$

where τ is the volume bounded by the surface S.

The *line integral*

$$\int_a^b \mathbf{A} \cdot \mathbf{dl}$$

over a specified curve is the sum of the terms $\mathbf{A} \cdot \mathbf{dl}$ for each element **dl** of the curve between the points a and b.

For a closed curve C that bounds a surface S, we have *Stokes's theorem*:

$$\oint_C \mathbf{A} \cdot \mathbf{dl} = \int_S (\boldsymbol{\nabla} \times \mathbf{A}) \cdot \mathbf{da}, \tag{1-85}$$

where

$$\boldsymbol{\nabla} \times \mathbf{A} = \begin{vmatrix} \mathbf{i} & \mathbf{j} & \mathbf{k} \\ \dfrac{\partial}{\partial x} & \dfrac{\partial}{\partial y} & \dfrac{\partial}{\partial z} \\ A_x & A_y & A_z \end{vmatrix} \tag{1-75}$$

is the *curl* of the vector function $\mathbf{A}$.

The *Laplacian* is the divergence of the gradient:

$$\nabla \cdot \nabla f = \nabla^2 f = \frac{\partial^2 f}{\partial x^2} + \frac{\partial^2 f}{\partial y^2} + \frac{\partial^2 f}{\partial z^2}. \qquad (1\text{-}89)$$

PROBLEMS

1-1 Show that the two vectors $\mathbf{A} = 9\mathbf{i} + \mathbf{j} - 6\mathbf{k}$ and $\mathbf{B} = 4\mathbf{i} - 6\mathbf{j} + 5\mathbf{k}$ are perpendicular to each other.

1-2 Show that the angle between the two vectors $\mathbf{A} = 2\mathbf{i} + 3\mathbf{j} + \mathbf{k}$ and $\mathbf{B} = \mathbf{i} - 6\mathbf{j} + \mathbf{k}$ is 130.5°.

1-3 The vectors **A**, **B**, **C** are coplanar. Show graphically that

$$\mathbf{A} \cdot (\mathbf{B} + \mathbf{C}) = \mathbf{A} \cdot \mathbf{B} + \mathbf{A} \cdot \mathbf{C}.$$

1-4 If **A** and **B** are adjacent sides of a parallelogram, $\mathbf{C} = \mathbf{A} + \mathbf{B}$ and $\mathbf{D} = \mathbf{A} - \mathbf{B}$ are the diagonals, and θ is the angle between **A** and **B**, show that $(C^2 + D^2) = 2(A^2 + B^2)$ and that $(C^2 - D^2) = 4AB \cos \theta$.

1-5 Consider two unit vectors **a** and **b**, as shown in Fig. 1-19. Find the trigonometric relations for the sine and cosine of the sum and difference of two angles from the values of $\mathbf{a} \cdot \mathbf{b}$ and $\mathbf{a} \times \mathbf{b}$.

Solution: First,

$$\mathbf{a} = \cos \alpha\, \mathbf{i} + \sin \alpha\, \mathbf{j}, \qquad (1)$$

$$\mathbf{b} = \cos \beta\, \mathbf{i} + \sin \beta\, \mathbf{j}. \qquad (2)$$

Now we can write the product $\mathbf{a} \cdot \mathbf{b}$ in two different ways. From Eq. 1-11,

$$\mathbf{a} \cdot \mathbf{b} = \cos \alpha \cos \beta + \sin \alpha \sin \beta, \qquad (3)$$

and, from Eq. 1-5,

$$\mathbf{a} \cdot \mathbf{b} = \cos (\alpha - \beta). \qquad (4)$$

Thus

$$\cos (\alpha - \beta) = \cos \alpha \cos \beta + \sin \alpha \sin \beta. \qquad (5)$$

If we set $\beta' = -\beta$,

$$\cos (\alpha + \beta') = \cos \alpha \cos \beta' - \sin \alpha \sin \beta'. \qquad (6)$$

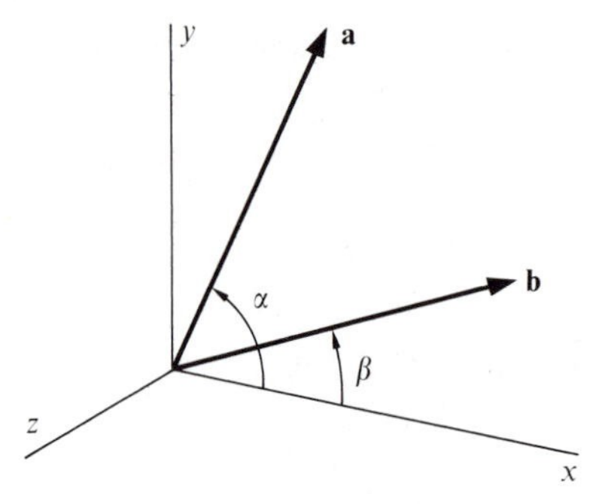

Figure 1-19 Two unit vectors, **a** and **b**, situated in the xy-plane. See Prob. 1-5.

Similarly, from Eq. 1-22,

$$\mathbf{a} \times \mathbf{b} = \begin{vmatrix} \mathbf{i} & \mathbf{j} & \mathbf{k} \\ \cos \alpha & \sin \alpha & 0 \\ \cos \beta & \sin \beta & 0 \end{vmatrix}, \tag{7}$$

$$= (\cos \alpha \sin \beta - \cos \beta \sin \alpha)\mathbf{k} \tag{8}$$

and, from Eqs. 1-14 and 1-15,

$$\mathbf{a} \times \mathbf{b} = -\sin (\alpha - \beta)\mathbf{k}. \tag{9}$$

Thus

$$\sin (\alpha - \beta) = \sin \alpha \cos \beta - \cos \alpha \sin \beta. \tag{10}$$

Setting again $\beta' = -\beta$,

$$\sin (\alpha + \beta') = \sin \alpha \cos \beta' + \cos \alpha \sin \beta'. \tag{11}$$

1-6 Show that the magnitude of $(\mathbf{A} \times \mathbf{B}) \cdot \mathbf{C}$ is the volume of parallelepiped whose edges are **A**, **B**, **C**, and show that $(\mathbf{A} \times \mathbf{B}) \cdot \mathbf{C} = \mathbf{A} \cdot (\mathbf{B} \times \mathbf{C})$.

1-7 Show that $\mathbf{A} \times (\mathbf{B} + \mathbf{C}) = \mathbf{A} \times \mathbf{B} + \mathbf{A} \times \mathbf{C}$.

1-8 Show that $\mathbf{a} \times (\mathbf{b} \times \mathbf{c}) = \mathbf{b}(\mathbf{a} \cdot \mathbf{c}) - \mathbf{c}(\mathbf{a} \cdot \mathbf{b})$.

1-9 If $\mathbf{r} \cdot (d\mathbf{r}/dt) = 0$, show that $r =$ constant.

1-10 A gun fires a bullet at a velocity of 500 meters per second and at an angle of 30° with the horizontal.

Find the position vector **r**, the velocity vector **v**, and the acceleration vector **a** of the bullet, t seconds after the gun is fired. Sketch the trajectory and show the three vectors for some time t.

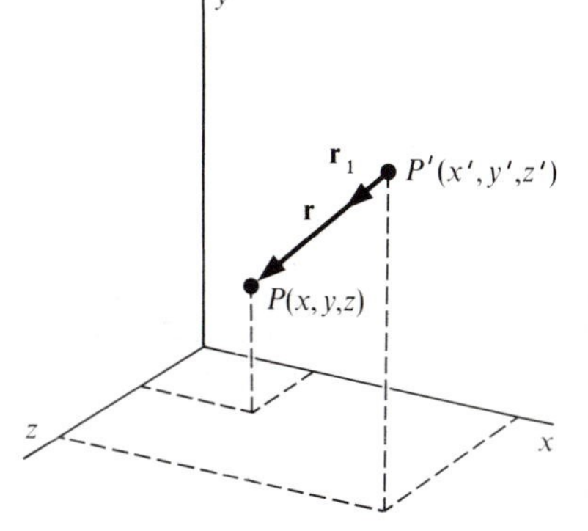

Figure 1-20 Two points P' and P in space, with the vector $\mathbf{r}$ and the unit vector $\mathbf{r}_1$. We shall often use such pairs of points, with P' at the source, and P at the point where the field is calculated. For example, one might calculate the electric field intensity at P with the element of charge at P'. See Prob. 1-13.

1-11 Let $\mathbf{r}$ be the radius vector from the origin of coordinates to any point, and let $\mathbf{A}$ be a constant vector. Show that $\boldsymbol{\nabla}(\mathbf{A} \cdot \mathbf{r}) = \mathbf{A}$.

1-12 Show that $(\mathbf{A} \cdot \boldsymbol{\nabla})\mathbf{r} = \mathbf{A}$.

1-13 The vector $\mathbf{r}$ is directed from $P'(x',y',z')$ to $P(x,y,z)$, as in Fig. 1-20.

a) If the point P is fixed and the point P' is allowed to move, show that the gradient of $(1/r)$ under these conditions is

$$\boldsymbol{\nabla}'\left(\frac{1}{r}\right) = \frac{\mathbf{r}_1}{r^2},$$

where $\mathbf{r}_1$ is the unit vector along $\mathbf{r}$. Show that this is the maximum rate of change of $1/r$.

b) Show similarly that, if P' is fixed and P is allowed to move,

$$\boldsymbol{\nabla}\left(\frac{1}{r}\right) = -\frac{\mathbf{r}_1}{r^2}.$$

1-14 The components of a vector $\mathbf{A}$ are

$$A_x = y\frac{\partial f}{\partial z} - z\frac{\partial f}{\partial y}, \qquad A_y = z\frac{\partial f}{\partial x} - x\frac{\partial f}{\partial z}, \qquad A_z = x\frac{\partial f}{\partial y} - y\frac{\partial f}{\partial x},$$

where f is a function of x, y, z. Show that

$$\mathbf{A} = \mathbf{r} \times \boldsymbol{\nabla} f, \qquad \mathbf{A} \cdot \mathbf{r} = 0, \qquad \text{and} \qquad \mathbf{A} \cdot \boldsymbol{\nabla} f = 0.$$

1-15 a) Show that $\boldsymbol{\nabla} \cdot \mathbf{r} = 3$.

b) What is the flux of $\mathbf{r}$ through a spherical surface of radius a?

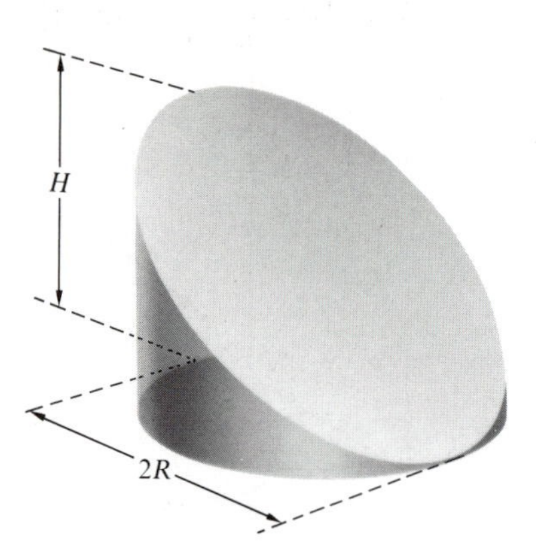

Figure 1-21 Truncated cylinder. See Prob. 1-18.

1-16 Show that

$$\nabla \cdot (f\mathbf{A}) = f\nabla \cdot \mathbf{A} + \mathbf{A} \cdot \nabla f,$$

where f is a scalar function and $\mathbf{A}$ is a vector function.

1-17 The vector $\mathbf{A} = 3x\mathbf{i} + y\mathbf{j} + 2z\mathbf{k}$, and $f = x^2 + y^2 + z^2$.

a) Show that $\nabla \cdot (f\mathbf{A})$ at the point (2,2,2) is 120, by first calculating $f\mathbf{A}$ and then calculating its divergence.

b) Calculate $\nabla \cdot (f\mathbf{A})$ by first finding ∇f and $\nabla \cdot \mathbf{A}$, and then using the identity of Prob. 1-16 above.

c) If x, y, z are measured in meters, what are the units of $\nabla \cdot (f\mathbf{A})$?

1-18 Calculate the volume of the portion of cylinder shown in Fig. 1-21.

Set the xy-plane on the base of the cylinder and the origin on the axis. Write the element of volume of height h as $h\,dx\,dy$, with

$$h = \frac{H}{2}\left(1 - \frac{x}{R}\right).$$

1-19 Calculate the volume of a sphere of radius R in Cartesian coordinates.

1-20 Consider a function $f(x)$. The fundamental theorem of the calculus states that

$$\int_a^b \frac{df(x)}{dx}\,dx = f(b) - f(a).$$

Let $\mathbf{A} = f(x)\mathbf{i}$, and consider a cylindrical volume τ parallel to the x-axis, extending from $x = a$ to $x = b$, as in Fig. 1-22.

Show that the divergence theorem, applied to the field $\mathbf{A}$ and the volume τ, yields the fundamental theorem of the calculus in the form stated above.

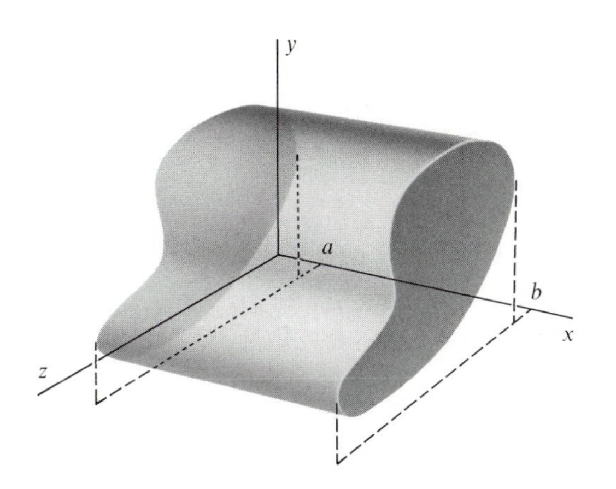

Figure 1-22 Cylindrical volume parallel to the x-axis, extending from $x = a$ to $x = b$. See Prob. 1-20.

1-21 The gravitational forces exerted by the sun on the planets are always directed toward the sun and depend only on the distance r. This type of field is called a *central force field.*

Find the potential energy at a distance r from a center of attraction when the force varies as $1/r^2$. Set the potential energy equal to zero at infinity.

1-22 Show that the gravitational field is conservative. Disregard the curvature of the earth and assume that g is a constant.

What if one takes into account the curvature of the earth and the fact that g decreases with altitude? Is the gravitational field still conservative?

1-23 Let P and Q be any two paths in space that have the same end points a and b.

Show that, if $\mathbf{A}$ is a conservative field,

$$\underset{\text{over } P}{\int_a^b \mathbf{A} \cdot d\mathbf{l}} = \underset{\text{over } Q}{\int_a^b \mathbf{A} \cdot d\mathbf{l}}.$$

1-24 Let $\mathbf{A}$ be a conservative field and P_0 a fixed point in space. Let

$$f(x,y,z) = \int_{P_0}^{(x,y,z)} \mathbf{A} \cdot d\mathbf{l},$$

where the line integral is calculated along any path joining P_0 to (x,y,z).

Show that $\mathbf{A} = \boldsymbol{\nabla} f$.

1-25 The azimuthal force exerted on an electron in a certain betatron (Prob. 11-9) is proportional to $r^{0.4}$.

Show that the force is non-conservative.

1-26 A vector field is defined by $\mathbf{A} = f(r)\mathbf{r}$.

Show that $\boldsymbol{\nabla} \times \mathbf{A}$ is equal to zero.

1-27 Show that $\boldsymbol{\nabla} \times (f\mathbf{A}) = (\boldsymbol{\nabla} f) \times \mathbf{A} + f(\boldsymbol{\nabla} \times \mathbf{A})$, where f is scalar and $\mathbf{A}$ is a vector.

1-28 Show that $\boldsymbol{\nabla} \cdot (\mathbf{A} \times \mathbf{D}) = \mathbf{D} \cdot (\boldsymbol{\nabla} \times \mathbf{A}) - \mathbf{A} \cdot (\boldsymbol{\nabla} \times \mathbf{D})$, where $\mathbf{A}$ and $\mathbf{D}$ are any two vectors.

1-29 Show that $\boldsymbol{\nabla} \cdot \boldsymbol{\nabla} \times \mathbf{A} = 0$.

1-30 One of the four Maxwell equations states that $\boldsymbol{\nabla} \times \mathbf{E} = -\partial \mathbf{B}/\partial t$, where $\mathbf{E}$ and $\mathbf{B}$ are respectively the electric field intensity in volts per meter and the magnetic induction in teslas at a point.

Show that the line integral of $\mathbf{E} \cdot d\mathbf{l}$ is 20.0 microvolts over a square 100 millimeters on the side when $\mathbf{B}$ is $2.00 \times 10^{-3}t$ tesla in the direction normal to the square, where t is the time in seconds.

1-31 In Sec. 1-14 we defined the Laplacian of a scalar point-function f. It is also useful to define the Laplacian of a vector point-function $\mathbf{A}$:

$$\boldsymbol{\nabla}^2\mathbf{A} = \boldsymbol{\nabla}^2 A_x\mathbf{i} + \boldsymbol{\nabla}^2 A_y\mathbf{j} + \boldsymbol{\nabla}^2 A_z\mathbf{k}.$$

Show that $\boldsymbol{\nabla}^2(\boldsymbol{\nabla} f) = \boldsymbol{\nabla}(\boldsymbol{\nabla}^2 f)$.

1-32 Show that

$$\boldsymbol{\nabla} \times \boldsymbol{\nabla} \times \mathbf{A} = \boldsymbol{\nabla}(\boldsymbol{\nabla} \cdot \mathbf{A}) - \boldsymbol{\nabla}^2\mathbf{A}. \tag{1}$$

We shall need this result in Chapter 20.

Solution: Let us find the x-components of these three quantities. Then we can find the y- and z-components by symmetry.

Since

$$\boldsymbol{\nabla} \times \boldsymbol{\nabla} \times \mathbf{A} = \begin{vmatrix} \mathbf{i} & \mathbf{j} & \mathbf{k} \\ \dfrac{\partial}{\partial x} & \dfrac{\partial}{\partial y} & \dfrac{\partial}{\partial z} \\ \dfrac{\partial A_z}{\partial y} - \dfrac{\partial A_y}{\partial z} & \dfrac{\partial A_x}{\partial z} - \dfrac{\partial A_z}{\partial x} & \dfrac{\partial A_y}{\partial x} - \dfrac{\partial A_x}{\partial y} \end{vmatrix}, \tag{2}$$

then its x-component is

$$\frac{\partial}{\partial y}\left(\frac{\partial A_y}{\partial x} - \frac{\partial A_x}{\partial y}\right) - \frac{\partial}{\partial z}\left(\frac{\partial A_x}{\partial z} - \frac{\partial A_z}{\partial x}\right).$$

The x-component of $\boldsymbol{\nabla}(\boldsymbol{\nabla} \cdot \mathbf{A})$ is

$$\frac{\partial}{\partial x}\left(\frac{\partial A_x}{\partial x} + \frac{\partial A_y}{\partial y} + \frac{\partial A_z}{\partial z}\right), \tag{3}$$

while that of $\nabla^2 \mathbf{A}$ is

$$\frac{\partial^2 A_x}{\partial x^2} + \frac{\partial^2 A_x}{\partial y^2} + \frac{\partial^2 A_x}{\partial z^2}. \tag{4}$$

Then we should have that

$$\frac{\partial^2 A_y}{\partial y\, \partial x} - \frac{\partial^2 A_x}{\partial y^2} - \frac{\partial^2 A_x}{\partial z^2} + \frac{\partial^2 A_z}{\partial z\, \partial x}$$

$$= \frac{\partial^2 A_x}{\partial x^2} + \frac{\partial^2 A_y}{\partial x\, \partial y} + \frac{\partial^2 A_z}{\partial x\, \partial z} - \frac{\partial^2 A_x}{\partial x^2} - \frac{\partial^2 A_x}{\partial y^2} - \frac{\partial^2 A_x}{\partial z^2}, \tag{5}$$

which is correct.

The y-component of the given equation can be found from Eq. 5 by replacing x by y, y by z, and z by x. This gives another identity, and similarly for the z-component.

CHAPTER 2

FIELDS OF STATIONARY ELECTRIC CHARGES: I

Coulomb's Law, Electric Field Intensity **E**, *Electric Potential V*

We begin our study of electric and magnetic phenomena by investigating the fields of stationary electric charges.

We shall start with Coulomb's law and deduce from it, in this chapter, the electric field intensity and the electric potential for any distribution of stationary electric charges.

2.1 *COULOMB'S LAW*

It is found experimentally that the force between two stationary electric *point* charges Q_a and Q_b (a) acts along the line joining the two charges, (b) is proportional to the product $Q_a Q_b$, and (c) is inversely proportional to the square of the distance r separating the charges.

If the charges are extended, the situation is more complicated in that the "distance between the charges" has no definite meaning. Moreover, the presence of Q_b can modify the charge distribution within Q_a, and vice versa, leading to a complicated variation of force with distance.

We thus have *Coulomb's law* for stationary point charges:

$$\mathbf{F}_{ab} = \frac{1}{4\pi\epsilon_0} \frac{Q_a Q_b}{r^2} \mathbf{r}_1, \tag{2-1}$$

where $\mathbf{F}_{ab}$ is the force exerted *by* Q_a *on* Q_b, and $\mathbf{r}_1$ is a unit vector pointing in the direction *from* Q_a *to* Q_b, as in Fig. 2-1. The force is repulsive if Q_a and Q_b are of the same sign; it is attractive if they are of different signs. As usual, F is measured in newtons, Q in *coulombs*, and r in meters. The constant ϵ_0

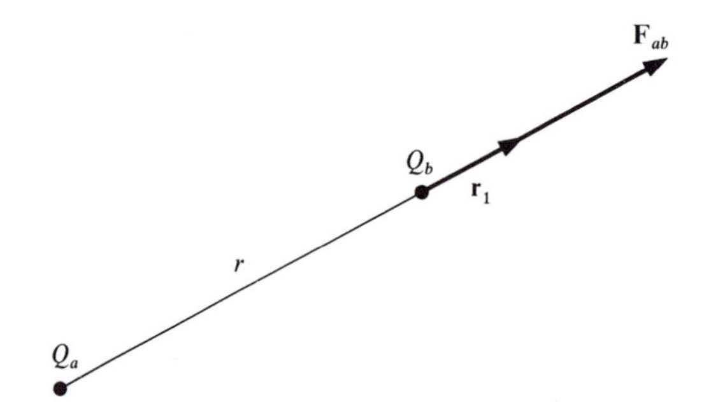

Figure 2-1 Charges Q_a and Q_b separated by a distance r. The force exerted on Q_b by Q_a is $\mathbf{F}_{ab}$ and is in the direction of $\mathbf{r}_1$ along the line joining the two charges.

is the *permittivity of free space*:

$$\epsilon_0 = 8.854\,187\,82 \times 10^{-12} \text{ farad/meter.}^\dagger \tag{2-2}$$

Coulomb's law applies to a pair of point charges in a vacuum. It also applies in dielectrics and conductors if $\mathbf{F}_{ab}$ is the direct force between Q_a and Q_b, irrespective of the forces arising from other charges within the medium.

Substituting the value of ϵ_0 into Coulomb's law,

$$\mathbf{F}_{ab} \approx 9 \times 10^9 \frac{Q_a Q_b}{r^2} \mathbf{r}_1, \tag{2-3}$$

where the factor of 9 is accurate to about one part in 1000; it should really be 8.987 551 79.

The Coulomb forces in nature are enormous when compared with the gravitational forces. The gravitational force between two masses m_a and m_b separated by a distance r is

$$F = 6.672\,0 \times 10^{-11} m_a m_b / r^2. \tag{2-4}$$

---

† From Eq. 2-1, ϵ_0 is expressed in coulombs squared per newton-square meter. However, a coulomb squared per newton-meter, or a coulomb squared per joule, is a farad, as we shall see in Sec. 4.3.

The gravitational force on a proton at the surface of the sun (mass = 2.0×10^{30} kilograms, radius = 7.0×10^{8} meters) is equal to the electric force between a proton and one *microgram* of electrons, separated by a distance equal to the sun's radius.

It is remarkable indeed that we should not be conscious of these enormous forces in everyday life. The positive and negative charges carried respectively by the proton and the electron are very nearly, if not exactly, the same. Experiments have shown that they differ, if at all, by at most one part in 10^{22}. Ordinary matter is thus neutral, and the enormous Coulomb forces prevent the accumulation of any appreciable quantity of charge of either sign.

2.2 *THE ELECTRIC FIELD INTENSITY* E

We think of the force between the point charges Q_a and Q_b in Coulomb's law as the product of Q_a and the *field* of Q_b, or vice versa. We define the *electric field intensity* **E** to be the force per unit charge exerted on a test charge in the field. Thus the electric field intensity due to the point charge Q_a is

$$\mathbf{E}_a = \frac{\mathbf{F}_{ab}}{Q_b} = \frac{Q_a}{4\pi\epsilon_0 r^2}\mathbf{r}_1. \tag{2-5}$$

The electric field intensity is measured in volts per meter.

The electric field intensity due to the point charge Q_a is the same, whether the test charge Q_b is in the field or not, even if Q_b is large compared to Q_a.

2.3 *THE PRINCIPLE OF SUPERPOSITION*

If the electric field is produced by more than one charge, each one produces its own field, and the resultant **E** is simply the vector sum of the individual **E**'s. This is the *principle of superposition.*

2.4 *THE FIELD OF AN EXTENDED CHARGE DISTRIBUTION*

For an extended charge distribution,

$$\mathbf{E} = \frac{1}{4\pi\epsilon_0}\int_{\tau'} \frac{\rho \mathbf{r}_1}{r^2}\, d\tau'. \tag{2-6}$$

The meaning of the various terms under the integral sign is illustrated in Fig. 2-2: ρ is the electric charge density at the source point (x',y',z'); $\mathbf{r}_1$ is a unit vector pointing from the source point to the field point (x,y,z) where $\mathbf{E}$ is calculated; r is the distance between these two points; $d\tau'$ is the element of volume $dx'\,dy'\,dz'$. If there exist surface distributions of charge, then we must add a surface integral, with the volume charge density ρ replaced by the surface charge density σ and the volume τ' replaced by the surface S'.

The electric charge density inside macroscopic bodies is of course not a smooth function of x', y', z'. However, nuclei and electrons are so small and so closely packed, that one can safely use an average electric charge density ρ, as above, to calculate macroscopic fields.

When the electric field is produced by a charge distribution that is disturbed by the introduction of a finite test charge Q', $\mathbf{E}$ is the force per unit charge as the magnitude of the test charge Q' tends to zero:

$$\mathbf{E} = \lim_{Q' \to 0} \frac{\mathbf{F}}{Q'}. \tag{2-7}$$

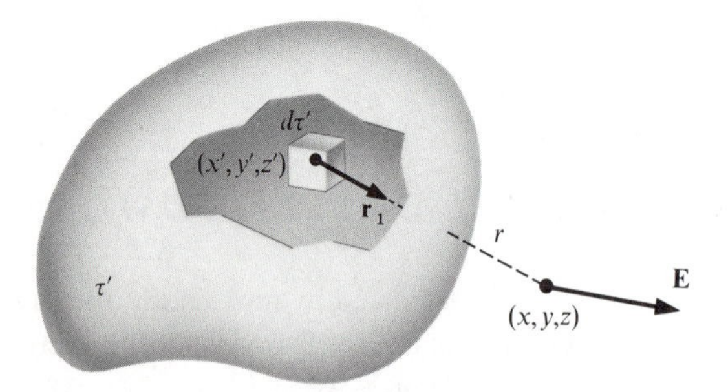

Figure 2-2 Source point (x',y',z') in a charge distribution and field point (x,y,z).

2.5 THE ELECTRIC POTENTIAL V

Consider a test point charge Q' that can be moved about in an electric field. The work W required to move it at a constant speed from a point P_1 to a point P_2 along a given path is

$$W = -\int_{P_1}^{P_2} \mathbf{E}Q' \cdot \mathbf{dl}. \tag{2-8}$$

The negative sign is required to obtain the work done *against* the field. Here again, we assume that Q' is so small that the charge distributions are not appreciably disturbed by its presence.

If the path is closed, the total work done is

$$W = -\oint \mathbf{E}Q' \cdot \mathbf{dl}. \tag{2-9}$$

Let us evaluate this integral. To simplify matters, we first consider the electric field produced by a single point charge Q. Then

$$\oint \mathbf{E}Q' \cdot \mathbf{dl} = \frac{QQ'}{4\pi\epsilon_0} \oint \frac{(\mathbf{r}_1 \cdot \mathbf{dl})}{r^2}. \tag{2-10}$$

Figure 2-3 shows that the term under the integral on the right is simply dr/r^2 or $-d(1/r)$. The sum of the increments of $(1/r)$ over a closed path is zero, since r has the same value at the beginning and at the end of the path. Then the line integral is zero, and the net work done in moving a point charge Q' around any closed path in the field of a point charge Q, which is fixed, is zero.

If the electric field is produced, not by a single point charge Q, but by some fixed charge distribution, then the line integrals corresponding to each individual charge of the distribution are all zero. Thus, for any distribution of fixed charges,

$$\oint \mathbf{E} \cdot \mathbf{dl} = 0. \tag{2-11}$$

An electrostatic field is therefore conservative (Sec. 1.11). It can be shown that this important property follows from the sole fact that the Coulomb force is a central force (see Prob. 1-21).

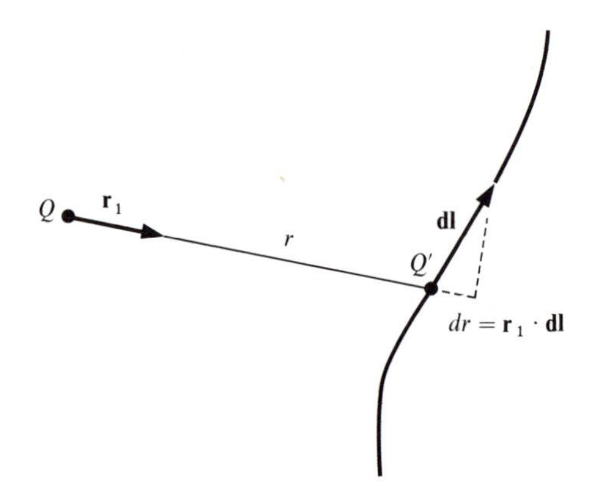

Figure 2-3 The product $\mathbf{r}_1 \cdot \mathbf{dl}$ is simply dr.

Then, from Stokes's theorem (Sec. 1.13), at all points in space,

$$\nabla \times \mathbf{E} = 0, \tag{2-12}$$

and we can write that

$$\mathbf{E} = -\nabla V, \tag{2-13}$$

where V is a scalar point function, since $\nabla \times \nabla V \equiv 0$.

We can thus describe an electrostatic field completely by means of the function $V(x, y, z)$, which is called the *electric potential.* The negative sign is required in order that the electric field intensity $\mathbf{E}$ point toward a *decrease* in potential, according to the usual convention.

It is important to note that V is not uniquely defined; we can add to it any quantity that is independent of the coordinates without affecting $\mathbf{E}$ in any way.

As shown in Prob. 1-23, the work done in moving a test charge at a constant speed from a point P_1 to a point P_2 is independent of the path.

We must remember that we are dealing here with electro*statics.* If there were moving charges present, $\nabla \times \mathbf{E}$ would not necessarily be zero, and ∇V would then describe only part of the electric field intensity $\mathbf{E}$. We shall investigate these more complicated phenomena later on, in Chapter 11.

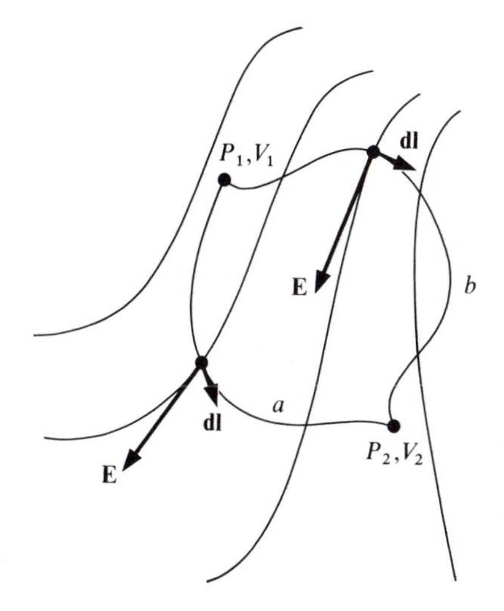

Figure 2-4 Two paths in a conservative electric field **E**, with endpoints P_1, at potential V_1, and P_2, at potential V_2. The integral of $\mathbf{E} \cdot \mathbf{dl}$ along path a is the same as along path b and is $V_1 - V_2$.

According to Eq. 2-13,

$$\mathbf{E} \cdot \mathbf{dl} = -\boldsymbol{\nabla} V \cdot \mathbf{dl} = -dV. \tag{2-14}$$

Then, for any two points P_1 and P_2 as in Fig. 2–4,

$$V_2 - V_1 = -\int_1^2 \mathbf{E} \cdot \mathbf{dl} = \int_2^1 \mathbf{E} \cdot \mathbf{dl}. \tag{2-15}$$

Note that the electric field intensity $\mathbf{E}(x, y, z)$ determines only the *difference* between the potentials at two different points. When we wish to speak of the electric potential at a given point, we must therefore arbitrarily define the potential in a given region of space to be zero. It is usually convenient to choose the potential at infinity to be zero. Then the potential V at the point 2 is

$$V = \int_2^\infty \mathbf{E} \cdot \mathbf{dl}. \tag{2-16}$$

The work W required to bring a charge Q' from a point at which the potential is defined to be zero to the point considered is VQ'. Thus V is W/Q' and can be defined to be the work per unit charge. The potential V is expressed in joules per coulomb, or in volts.

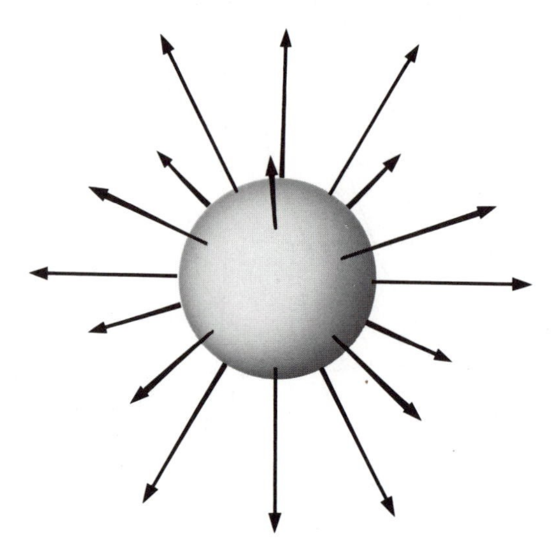

Figure 2-5 Equipotential surface and lines of force near a point charge.

When the field is produced by a single point charge Q, the potential at a distance r is

$$V = \int_r^\infty \frac{Q}{4\pi\epsilon_0} \frac{dr}{r^2} = \frac{Q}{4\pi\epsilon_0 r}. \tag{2-17}$$

It will be observed that the sign of the potential V is the same as that of Q.

The principle of superposition applies to the electric potential V as well as to the electric field intensity **E**. The potential V at a point P due to a charge distribution of density ρ is therefore

$$V = \frac{1}{4\pi\epsilon_0} \int_{\tau'} \frac{\rho \, d\tau'}{r}, \tag{2-18}$$

where r is the distance between the point P and the element of charge $\rho \, d\tau'$.

The points in space that are at a given potential define an *equipotential surface*. For example, an equipotential surface about a point charge is a concentric sphere as in Fig. 2-5. We can see from Eq. 2-13 that **E** is everywhere normal to the equipotential surfaces (Sec. 1.5).

If we join end-to-end infinitesimal vectors representing **E**, we get a curve in space—called a *line of force*—that is everywhere normal to the equipotential surfaces, as in Fig. 2-5. The vector **E** is everywhere tangent to a line of force.

2.5.1 EXAMPLE: THE ELECTRIC DIPOLE

The *electric dipole* shown in Fig. 2-6 is one type of charge distribution that is encountered frequently. We shall return to it in Chapter 6.

The electric dipole consists of two charges, one positive and the other negative, of the same magnitude, separated by a distance s. We shall calculate V and **E** at a distance r that is large compared to s.

At P,

$$V = \frac{Q}{4\pi\epsilon_0}\left(\frac{1}{r_b} - \frac{1}{r_a}\right), \tag{2-19}$$

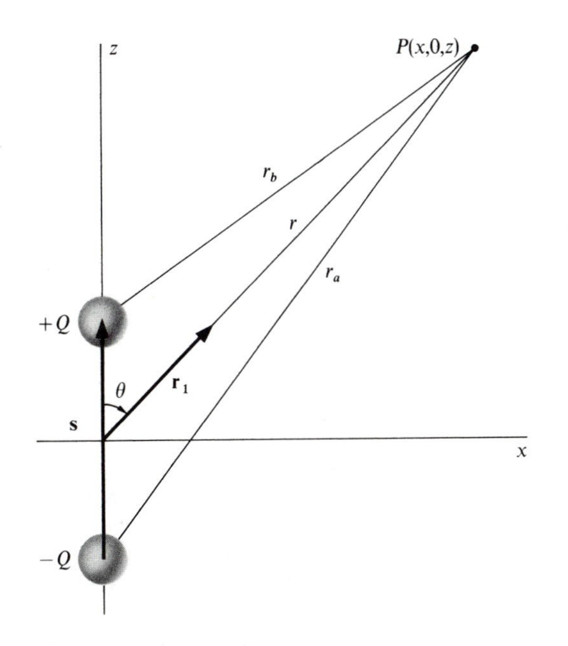

Figure 2-6 The two charges $+Q$ and $-Q$ form a dipole. The electric potential at P is the sum of the potentials due to the individual charges. The vector **s** points from $-Q$ to $+Q$, and $\mathbf{r}_1$ is a unit vector pointing from the origin to the point of observation P.

where

$$r_a^2 = r^2 + \left(\frac{s}{2}\right)^2 + rs\cos\theta, \tag{2-20}$$

$$r_a = r\left[1 + \left(\frac{s}{2r}\right)^2 + \frac{s}{r}\cos\theta\right]^{1/2}, \tag{2-21}$$

$$\approx r\left(1 + \frac{s}{r}\cos\theta\right)^{1/2}, \tag{2-22}$$

$$\frac{r}{r_a} \approx \frac{1}{\left(1 + \frac{s}{r}\cos\theta\right)^{1/2}}, \tag{2-23}$$

$$\approx 1 - \frac{s}{2r}\cos\theta.^\dagger \tag{2-24}$$

Similarly,

$$\frac{r}{r_b} \approx 1 + \frac{s}{2r}\cos\theta, \tag{2-25}$$

and

$$V = \frac{Qs}{4\pi\epsilon_0 r^2}\cos\theta. \tag{2-26}$$

This expression is valid for $r^3 \gg s^3$.

It is interesting to note that the potential due to a dipole falls off as $1/r^2$, whereas the potential from a single point charge varies only as $1/r$. This comes from the

† Remember that

$$(1 + a)^n = 1 + na + \frac{n(n-1)}{2!}a^2 + \frac{n(n-1)(n-2)}{3!}a^3 + \cdots$$

If n is a positive integer, the series stops when the coefficient is zero. Try, for example, $n = 1$ and $n = 2$. If n is not a positive integer, the series converges for $a^2 < 1$.

This series is very often used when $a \ll 1$. Then

$$(1 + a)^n \approx 1 + na.$$

For example,

$$(1 + a)^2 \approx 1 + 2a, \qquad (1 + a)^{-1/2} \approx 1 - \frac{a}{2}, \text{ etc.}$$

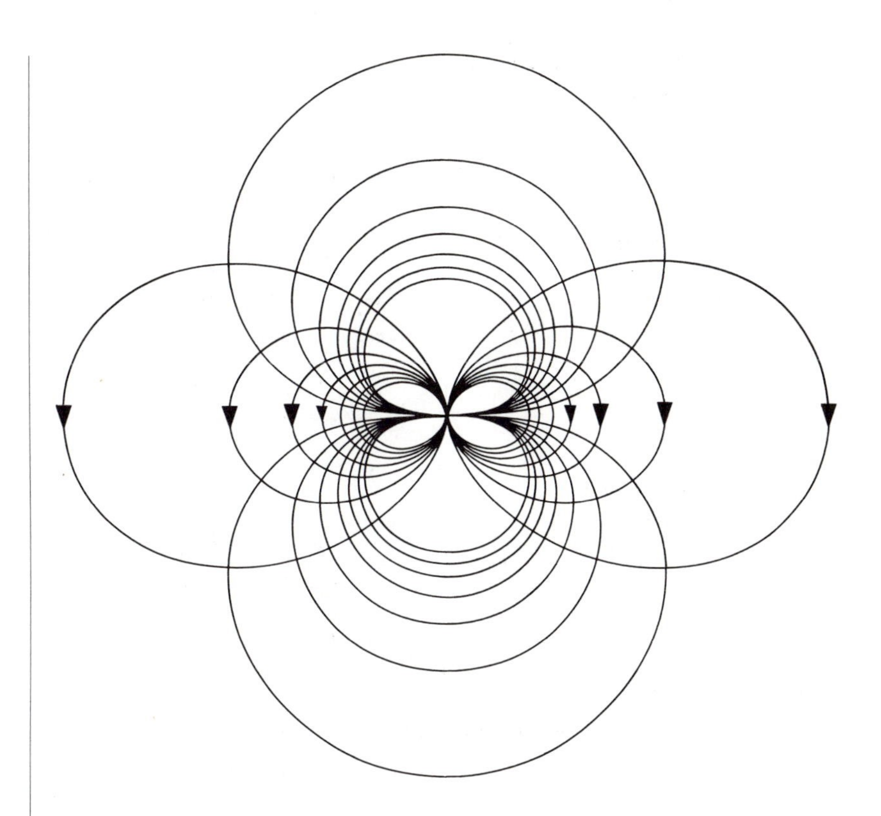

Figure 2-7 Lines of force (arrows) and equipotential lines for the dipole of Fig. 2-6. The dipole is vertical, at the center of the figure, with the positive charge close to and above the negative charge. In the central region the lines come too close together to be shown.

fact that the charges of a dipole appear close together for an observer some distance away, and that their fields cancel more and more as the distance r increases.

We define the *dipole moment* $\mathbf{p} = Q\mathbf{s}$ as a vector whose magnitude is Qs and that is directed *from* the negative *to* the positive charge. Then

$$V = \frac{\mathbf{p} \cdot \mathbf{r}_1}{4\pi\epsilon_0 r^2}. \tag{2-27}$$

Figure 2-7 shows equipotential lines for an electric dipole. Equipotential surfaces are generated by rotating equipotential lines around the vertical axis.

Let us now calculate **E** at the point $P(x,0,z)$ in the plane $y = 0$ as in Fig. 2-6. We have that

$$V = \frac{p}{4\pi\epsilon_0}\frac{\cos\theta}{r^2} = \frac{p}{4\pi\epsilon_0}\frac{z}{r^3} \tag{2-28}$$

and, at $P(x,0,z)$,

$$E_x = -\frac{\partial V}{\partial x} = \frac{3p}{4\pi\epsilon_0}\frac{xz}{r^5} = \frac{3p}{4\pi\epsilon_0}\frac{\sin\theta\cos\theta}{r^3}, \tag{2-29}$$

$$E_y = -\frac{\partial V}{\partial y} = \frac{3p}{4\pi\epsilon_0}\frac{yz}{r^5} = 0, \tag{2-30}$$

since $y = 0$, and

$$E_z = -\frac{\partial V}{\partial z} = -\frac{p}{4\pi\epsilon_0}\left(\frac{1}{r^3} - \frac{3z^2}{r^5}\right), \tag{2-31}$$

$$= \frac{p}{4\pi\epsilon_0}\frac{3\cos^2\theta - 1}{r^3}. \tag{2-32}$$

Figure 2-7 shows lines of force for the electric dipole.

2.6 *SUMMARY*

It is found empirically that the force exerted *by* a point charge Q_a *on* a point charge Q_b is

$$\mathbf{F}_{ab} = \frac{1}{4\pi\epsilon_0}\frac{Q_aQ_b}{r^2}\mathbf{r}_1, \tag{2-1}$$

where r is the distance between the charges and $\mathbf{r}_1$ is a unit vector pointing from Q_a to Q_b. This is *Coulomb's law*. We consider the force $\mathbf{F}_{ab}$ as being the product of Q_b by the *electric field intensity* due to Q_a,

$$\mathbf{E}_a = \frac{Q_a}{4\pi\epsilon_0 r^2}\mathbf{r}_1, \tag{2-5}$$

or vice versa.

According to the *principle of superposition*, two or more electric field intensities acting at a given point add vectorially. For an extended charge distribution,

$$\mathbf{E} = \frac{1}{4\pi\epsilon_0} \int_{\tau'} \frac{\rho \mathbf{r}_1}{r^2} \, d\tau'. \tag{2-6}$$

The electrostatic field is conservative,

$$\nabla \times \mathbf{E} = 0, \tag{2-12}$$

hence

$$\mathbf{E} = -\nabla V, \tag{2-13}$$

where

$$V = \frac{1}{4\pi\epsilon_0} \int_{\tau'} \frac{\rho \, d\tau'}{r} \tag{2-18}$$

is the *electric potential*; $\rho \, d\tau'$ is the element of charge contained within the element of volume $d\tau'$, and r is the distance between this element and the point where V is calculated.

PROBLEMS

2-1E† *COULOMB'S LAW*

a) Calculate the electric field intensity that would be just sufficient to balance the gravitational force of the earth on an electron.

b) If this electric field were produced by a second electron located below the first one, what would be the distance between the two electrons?

The charge on an electron is -1.6×10^{-19} coulomb and its mass is 9.1×10^{-31} kilogram.

2-2E *SEPARATION OF PHOSPHATE FROM QUARTZ*

Crushed Florida phosphate ore consists of particles of quartz mixed with particles of phosphate rock. If the mixture is vibrated, the quartz becomes charged

† The letter *E* indicates that the problem is relatively easy.

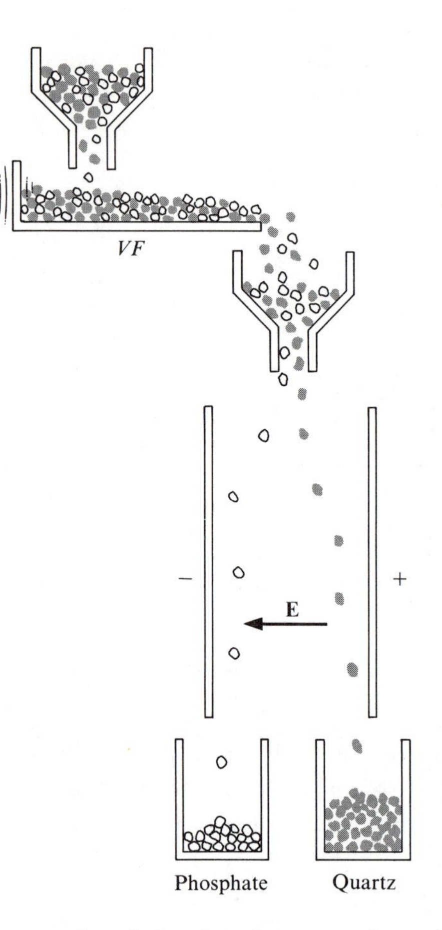

Figure 2-8 Electrostatic separation of phosphate from quartz in crushed phosphate ore. The vibrating feeder VF charges the phosphate particles positively, and the quartz particles negatively. See Prob. 2-2.

negatively and the phosphate positively. The phosphate can then be separated out as in Fig. 2-8.

Over what minimum distance must the particles fall if they must be separated by at least 100 millimeters?

Set $E = 5 \times 10^5$ volts per meter and the specific charge equal to 10^{-5} coulomb per kilogram.

2-3E *ELECTRIC FIELD INTENSITY*

A charge $+Q$ is situated at $x = a$, $y = 0$, and a charge $-Q$ at $x = -a$, $y = 0$. Calculate the electric field intensity at the point $x = 0$, $y = a$.

2-4 *ELECTRIC FIELD INTENSITY*

A circular disk of charge has a radius R and carries a surface charge density σ. Find the electric field intensity E at a point P on the axis at a distance a. What is the value of E when $a \ll R$?

You can solve this problem by calculating the field at P due to a ring of charge of radius r and width dr, and then integrating from $r = 0$ to $r = R$.

2-5E *CATHODE-RAY TUBE*

The electron beam in the cathode-ray tube of an oscilloscope is deflected vertically and horizontally by two pairs of parallel deflecting plates that are maintained at appropriate voltages. As the electrons pass between one pair, they are accelerated by the electric field, and their kinetic energy increases. Thus, under steady-state conditions, we achieve an increase in the kinetic energy of the electrons without any expenditure of power in the deflecting plates, as long as the beam does not touch the plates.

Could this phenomenon be used as the basis of a perpetual motion machine?

2-6E *CATHODE-RAY TUBE*

In a certain cathode-ray tube, the electrons are accelerated under a difference of potential of 5 kilovolts. After being accelerated, they travel horizontally over a distance of 200 millimeters.

Calculate the downward deflection over this distance caused by the gravitational force.

An electron carries a charge of -1.6×10^{-19} coulomb and has a mass of 9.1×10^{-31} kilogram.

2-7E *MACROSCOPIC PARTICLE GUN*

It is possible to obtain a beam of fine particles in the following manner. Figure 2-9 shows a parallel-plate capacitor with a hole in the upper plate. If a particle of dust, say, is introduced in the space between the plates, it sooner or later comes into contact with one of the plates and acquires a charge of the same polarity. It then flies over to the opposite plate, and the process repeats itself. The particle oscillates back and forth, until it is lost either at the edges or through the central hole. One can thus obtain a beam emerging from the hole by admitting particles steadily into the capacitor. In order to achieve high velocities, the gun must, of course, operate in a vacuum.

Beams of macroscopic particles are used for studying the impact of micrometeorites.

Now it has been found that a spherical particle of radius R lying on a charged plate acquires a charge

$$Q = 1.65 \times 4\pi\epsilon_r\epsilon_0 R^2 E_0 \text{ coulomb},$$

where E_0 is the electric field intensity in the absence of the particle, and ϵ_r is the relative permittivity of the particle (Sec. 6.7). We assume that the particle is at least slightly conducting.

Assuming that the plates are 10 millimeters apart and that a voltage difference of 15 kilovolts is applied between them, find the velocity of a spherical particle 1

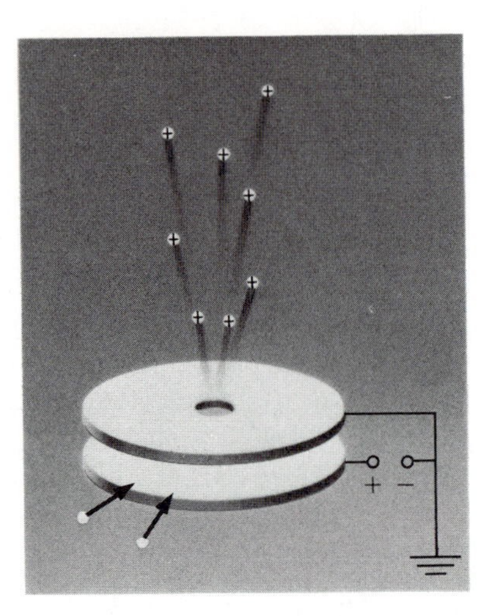

Figure 2-9 Device for producing a beam of fine particles. See Prob. 2-7.

micrometer in diameter as it emerges from the hole in the upper plate. Assume that the particle has the density of water and that $\epsilon_r = 2$.

2-8E *ELECTROSTATIC SPRAYING*

When painting is done with an ordinary spray gun, part of the paint escapes deposition. The fraction lost depends on the shape of the surface, on drafts, etc., and can be as high as 80%. The use of the ordinary spray gun in large-scale industrial processes would therefore result in intolerable waste and pollution.

The efficiency of spray painting can be increased to nearly 100%, and the pollution reduced by a large factor, by charging the droplets of paint, electrically, and applying a voltage difference between the gun and the object to be coated.

It is found that, in such devices, the droplets carry a specific charge of roughly one coulomb per kilogram.

Assuming that the electric field intensity in the region between the gun and the part is at least 10 kilovolts per meter, what is the minimum ratio of the electric force to the gravitational force?

2-9 *THE RUTHERFORD EXPERIMENT*

In 1906, in the course of a historic experiment that demonstrated the small size of the atomic nucleus, Rutherford observed that an alpha particle ($Q = 2 \times 1.6 \times 10^{-19}$ coulomb) with a kinetic energy of 7.68×10^6 electron-volts making a head-on collision with a gold nucleus ($Q = 79 \times 1.6 \times 10^{-19}$ coulomb) is repelled backwards.

The *electron-volt* is the kinetic energy acquired by a particle carrying one electronic charge when it is accelerated through a difference of potential of one volt:

$$1 \text{ electron-volt} = 1.6 \times 10^{-19} \text{ joule.}$$

a) What is the distance of closest approach at which the electrostatic potential energy is equal to the initial kinetic energy? Express your result in femtometers (10^{-15} meter).

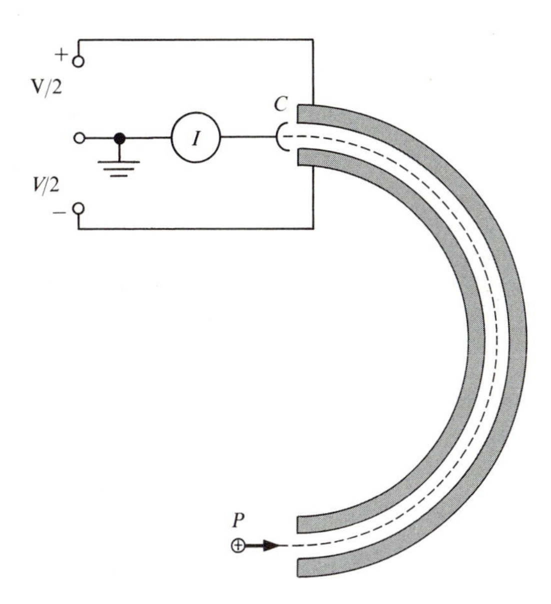

Figure 2-10 Cylindrical electrostatic analyser. A particle P can reach the collector C only if it is positive and if it satisfies the equation given in Prob. 2-11. Otherwise, it stops on one of the cylindrical electrodes. For a given charge-to-mass ratio Q/m, one can select different velocities by changing V.

A circle with the letter I represents an ammeter, while a circle with a letter V represents a voltmeter.

b) What is the maximum force of repulsion?

c) What is the maximum acceleration in g's?

The mass of the alpha particle is about 4 times that of a proton, or $4 \times 1.7 \times 10^{-27}$ kilogram.

2-10 *ELECTROSTATIC SEED SORTING*

Normal seeds can be separated from discolored ones and from foreign objects by means of an electrostatic seed-sorting machine that operates as follows. The seeds are observed by a pair of photocells as they fall one by one inside a tube. If the color is not right, voltage is applied to a needle that deposits a charge on the seed. The seeds then fall between a pair of electrically charged plates that deflect the undesired ones into a separate bin.

One such machine can sort dry peas at the rate of 100 per second, or about 2 tons per 24-hour day.

a) If the seeds are dropped at the rate of 100 per second, over what distance must they fall if they must be separated by 20 millimeters when they pass between the photocells? Neglect air resistance.

b) Assuming that the seeds acquire a charge of 1.5×10^{-9} coulomb, that the deflecting plates are parallel and 50 millimeters apart, and that the potential difference between them is 25 kilovolts, how far should the plates extend below the charging needle if the charged seeds must be deflected by 40 millimeters on leaving the plates?

Assume that the charging needle and the top of the deflecting plates are close to the photocell.

2-11E *CYLINDRICAL ELECTROSTATIC ANALYSER*

Figure 2-10 shows a cylindrical electrostatic analyser or velocity selector. It consists of a pair of cylindrical conductors separated by a radial distance of a few millimeters.

Show that, if the radial distance between the cylindrical surfaces of average radius R is a, and if the voltage between them is V, then the particles collected at I have a velocity

$$v = \left(\frac{Q}{m}\frac{VR}{a}\right)^{1/2}.$$

2-12 *PARALLEL-PLATE ANALYSER*

Figure 2-11 shows a parallel-plate analyser. The instrument serves to measure the energy distribution of the charged particles emitted by a source. It is often used for sources of electrons. It has been used, for example, to investigate the energy distribution of electrons in the plasma formed at the focus of a lens illuminated by a powerful laser.

If the source emits particles of various energies, then, for a given value of V and for a given geometry, the collector receives only those particles that have the

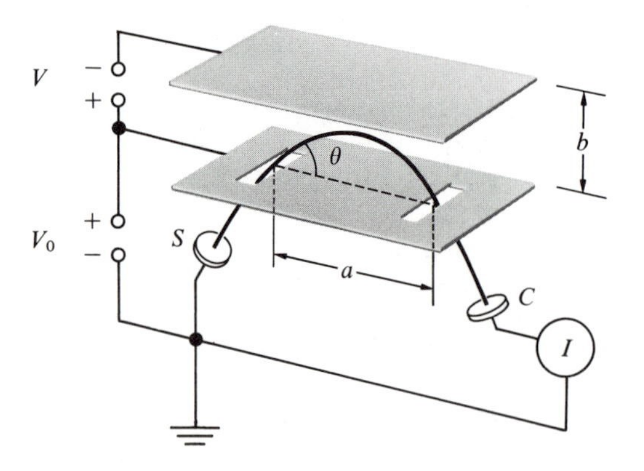

Figure 2-11 Parallel-plate analyser for measuring the energy distribution of particles emerging from a source S. The particles follow a ballistic trajectory in the uniform electric field between the two plates. Particles of the correct energy and, of course, of the correct polarity are collected at C. The other particles hit either one of the two parallel plates. The polarities shown apply to electrons or to negative ions. See Prob. 2-12.

corresponding energy. One therefore observes the energy spectrum of the particles by measuring the collector current as a function of V. As we shall see below, we assume that the particles all carry the same charge.

In the figure, particles of charge Q have a kinetic energy QV_0.

This type of analyser is relatively simple to build. It can also be, at the same time, compact and accurate. For example, in one case, the electrodes measured 40 mm × 115 mm, with $b = 15$ mm and $a = 50$ mm. With slit widths of 0.25 mm, the detector current as a function of V for mono-energetic electrons gave a sharp peak as in Fig. 2-12 having a width at half maximum of only 1%.

a) Find a for a particle of mass m and charge Q.

Solution: We can treat the particle as a projectile of mass m, charge Q, and initial velocity v_0, with

$$\left(\frac{1}{2}\right) m v_0^2 = QV_0, \tag{1}$$

$$v_0 = \left(\frac{2QV_0}{m}\right)^{1/2}. \tag{2}$$

The particle has a vertical acceleration, downwards, equal to QE/m, or

$$\frac{Q(V/b)}{m}.$$

Then the time of flight, from slit to slit, is

$$T = 2\frac{v_0 \sin\theta}{Q(V/b)/m} = \frac{2v_0 mb}{QV}\sin\theta, \tag{3}$$

and the distance a is given by

$$a = (v_0 \cos\theta)T = \frac{2v_0^2 mb}{QV}\sin\theta\cos\theta, \tag{4}$$

$$= \frac{4bV_0}{V}\sin\theta\cos\theta = \frac{2bV_0}{V}\sin 2\theta. \tag{5}$$

Note that, for a given value of V_0, a is independent of both m and Q. It is proportional to V_0. If all the particles carry the same charge Q, a is proportional to their energy QV_0.

b) The optimum value of θ may be defined as that for which the horizontal distance a traveled by the particle is maximum. One can then tolerate a slightly divergent beam at the entrance slit, with an average θ of θ_{opt}.

Find θ_{opt}.

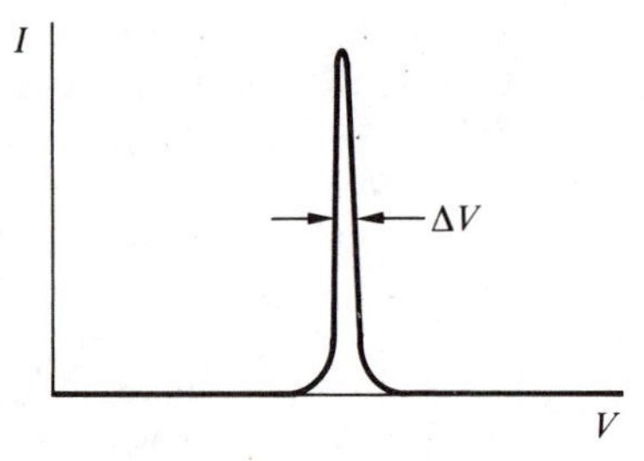

Figure 2-12 The current I at the collector C of Fig. 2-11, as a function of V.

Solution: Since the distance a must be maximum, $da/d\theta$ must be zero. Now $\sin 2\theta$ has a maximum at $\theta = 45°$, so $\theta_{opt} = 45°$.

Note that, since a is maximum at $\theta = 45°$ this is also the condition for maximum resolution.

c) Now find V_0 as a function of V, for given values of a and b, at θ_{opt}.

Solution: From Eq. 5, with $\theta = 45°$,

$$V_0 = \frac{a}{2b} V. \tag{6}$$

d) Find da/dV_0 for $\theta = \theta_{opt}$.

Solution: At $\theta = \theta_{opt}$,

$$\frac{da}{dV_0} = \frac{2b}{V}. \tag{7}$$

e) Find the minimum value of V as a function of V_0 at θ_{opt}.

Solution: In all the above equations, V is always divided by b. To find V we use the fact that V must be sufficient to reduce the vertical velocity to zero:

$$QV > \frac{1}{2} m(v_0 \sin 45°)^2, \tag{8}$$

$$V > \frac{m}{2Q} \frac{2QV_0}{m} \frac{1}{2} = \frac{V_0}{2}. \tag{9}$$

Comparing now with Eq. 6, we see that a/b must be smaller than 4.

This analyser is practical for energies QV_0 less than, say, 10 kiloelectron-volts. It would be inconvenient to use voltages V larger than a few kilovolts. For higher energies one would use the analyser of Prob. 2-11.

2-13 *CYLINDRICAL AND PARALLEL-PLATE ANALYSERS COMPARED*

Show that both the cylindrical analyser of Prob. 2-11 and the parallel-plate analyser of Prob. 2-12 measure the ratio

$$\frac{(1/2)mv^2}{Q},$$

where m is the mass of the particle, v its velocity, and Q its charge.

2-14 *ION THRUSTER*

Synchronous satellites describe circular orbits above the equator at the altitude where their angular velocity is equal to that of the earth. Since these satellites are extremely costly to build and to launch, they must remain operational for several years. However, it is impossible to adjust their initial position and velocity with sufficient accuracy to maintain them fixed with respect to the earth for long periods of time. Moreover, they are perturbed by the moon, and their antennas must remain constantly directed towards the earth.

Synchronous satellites are therefore equipped with *thrusters* whose function is to exert the small forces necessary to keep them properly oriented and on their prescribed orbits. Even then, they keep wandering about their reference position. For example, their altitude can vary by as much as 15 kilometers.

The thrust is $m'v$, where m' is the mass of propellant ejected per unit time and v is the exhaust velocity with respect to the satellite. Clearly, m' should be small. Then v should be large. However, as we shall see, too large a v could lead to an excessive power consumption. So the choice of velocity depends on the maximum allowable value for m' and on the available power.

One way of achieving large values of v is to eject a beam of charged particles, as in Fig. 2-13. The propellant is ionized in an ion source and ejected as a positive ion beam.

By itself, the beam cannot exert a thrust, for the following reason. Initially, the net charge on the satellite is zero. When it ejects positive particles, it acquires a negative charge. The particles are thus held back by the electrostatic force, and, once the satellite is sufficiently charged, they fall back on the satellite.

To achieve a thrust, the ions must be neutralized on leaving the satellite. This is achieved by means of a heated filament, as in Fig. 2-13. The filament emits electrons that are attracted by the positively charged beam.

a) The current I of positive ions in the beam is carried by particles of mass m and charge ne, where e is the electronic charge.

Show that the thrust is

$$F = \left(\frac{2mV}{ne}\right)^{1/2} I.$$

b) What is the value of F for a 0.1 ampere beam of protons with $V = 50$ kilovolts?

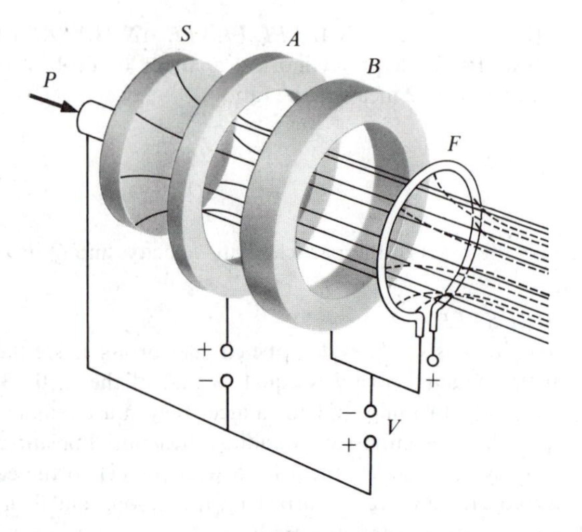

Figure 2-13 Schematic diagram of an ion thruster. The propellant is admitted at P and ionized in S; A is a beam-shaping electrode and B is the accelerating electrode, maintained at a voltage V with respect to S; the heated filament F injects electrons into the ion beam to make it neutral. Electrode B is part of the outer skin of the satellite. See Prob. 2-14.

c) If we call P the power IV spent in accelerating the beam, show that

$$F = (2m'P)^{1/2} = \frac{2P}{v} = \left(\frac{2m}{neV}\right)^{1/2} P.$$

Note that, since $F = (2m'P)^{1/2}$, for given values of m' and P, the thrust is independent of the charge-to-mass ratio of the ions.

Also, since F is $2P/v$, the thrust is *inversely* proportional to v, for a given power expenditure P.

Finally, the last expression shows that, again for a given P, it is preferable to use heavy ions carrying a single charge ($n = 1$) and to use a low accelerating voltage V.

d) If the filament is left unheated and if the beam current is 0.1 ampere, how long does it take the body of the satellite to attain a voltage of 50 kilovolts?

Assume that the satellite is spherical and that it has a radius of 1 meter.

2-15 COLLOID THRUSTER

Before reading this solved problem you should first read Prob. 2-14.

Figure 2-14 shows a so-called colloid thruster. A conducting fluid is pumped into a hollow needle maintained at a high positive voltage with respect to the satellite's outer skin. Microscopic droplets form around the edge of the opening, where the electric field intensity and the surface charge density are extremely high.

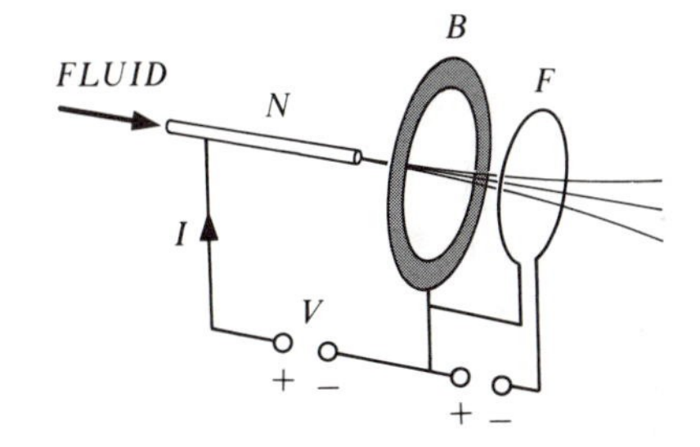

Figure 2-14 Colloid thruster. The hollow needle N ejects a beam of positively charged, high-velocity droplets. Electrode B and filament F are as in Fig. 2-13. See Prob. 2-15.

These droplets, carrying specific charges Q/m as large as 4×10^4 coulombs per kilogram, are accelerated in the electric field and form a high-energy jet. As a rule, several needles are grouped together so as to obtain a larger thrust.

a) Calculate the thrust exerted by a single needle for the specific charge given above, when $V = 10$ kilovolts and $I = 5$ microamperes.

Use the results of Prob. 2-14.

Solution:

$$F = \left(\frac{2m}{QV}\right)^{1/2} P, \tag{1}$$

$$= \left(\frac{2}{4 \times 10^4 \times 10^4}\right)^{1/2} 10^4 \times 5 \times 10^{-6}, \tag{2}$$

$$= 3.5 \text{ micronewtons.} \tag{3}$$

b) Calculate the mass of fluid ejected per second.

Solution: The mass m' ejected per second is m/Q times the charge ejected per second, which is 5×10^{-6} coulomb:

$$m' = 5 \times 10^{-6}/4 \times 10^4 = 1.3 \times 10^{-10} \text{ kilogram/second.} \tag{4}$$

c) The thrust is so small that it is not practical to measure it directly. The method used is illustrated in Fig. 2-15. The thruster is first put into operation. Then switch S is opened and the voltage IR across R is observed on an oscilloscope, as a function of the time. The curve has the shape shown in Fig. 2-16a.

Calculate the thrust as a function of V, I, D, T.

Solution: The thrust is

$$F = m'v = \frac{2}{v}\left(\frac{1}{2}m'v^2\right) = \frac{2}{v} IV, \tag{5}$$

$$= 2IV\frac{T}{D} = 2\frac{V}{D} IT. \tag{6}$$

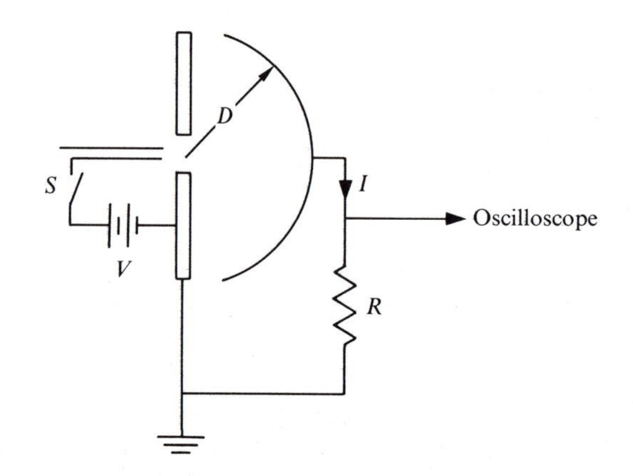

Figure 2-15 Method used for measuring the thrust of the colloid thruster of Fig. 2-14. The current I produced by the charged droplets collected on the hemispherical electrode gives a voltage IR that can be observed as a function of the time on an oscilloscope.

d) If there are several types of droplets, the curve of I as a function of t has several plateaus as in Fig. 2-16b.

How can one calculate the thrust in this case?

Solution: In the last expression for F in Eq. 6, we see that F is $2V/D$ times the area under the curve of Fig. 2-16a. Thus, in the case of Fig. 2-16b,

$$F = 2\frac{V}{D}\int_0^\infty I\,dt. \tag{7}$$

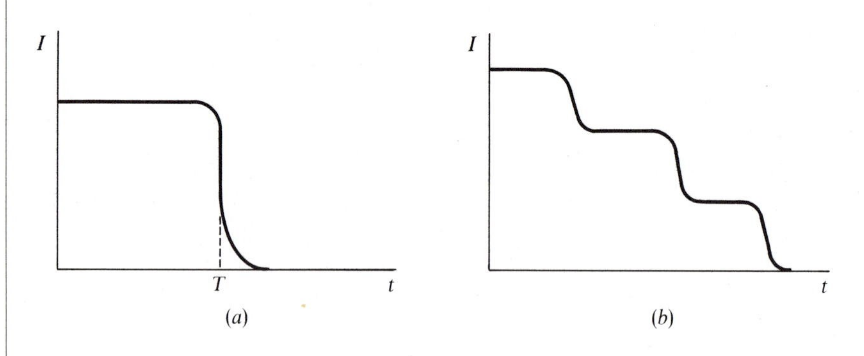

Figure 2-16 (*a*) Current I as a function of the time in the set-up of Fig. 2-15, starting at the instant when switch S is opened, when all the droplets are of the same size and carry equal charges. (*b*) If there are several types of droplets of different sizes or charges, the curve of I as a function of t has one plateau for each type.

CHAPTER 3

FIELDS OF STATIONARY ELECTRIC CHARGES: II

Gauss's Law, Poisson's and Laplace's Equations, Uniqueness Theorem

Now that we have discussed the two fundamental concepts of electric field intensity **E** and electric potential V, we shall study Gauss's law, Poisson's equation, and the uniqueness theorem. This will give us three powerful methods for calculating electrostatic fields.

But first we must study solid angles.

3.1 *SOLID ANGLES*

3.1.1 *THE ANGLE SUBTENDED BY A CURVE AT A POINT*

Consider the curve C and the point P of Fig. 3-1. They are situated in a plane. We wish to find the angle α subtended by C at P. This is

$$\alpha = \int_C \frac{dl'}{r} = \int_C \frac{dl \sin \theta}{r} \text{ radians.} \tag{3-1}$$

The integrals are evaluated over the curve C: for the second one, we multiply each element of length dl by $\sin\theta$ and divide by r; then we sum the results.

The arc c subtends the same angle α at P.

If now the curve is closed, as in Fig. 3-2a, with P inside C, the angle subtended at P by the small circle of radius b is 2πb/b, or 2π radians. The angle subtended by C at *any* inside point P is therefore also 2π radians.

If P is outside C, as in Fig. 3-2b, then the segments C_1 and C_2 subtend angles of the same magnitude but of opposite signs, since $\sin \theta$ is positive on C_1 and negative on C_2. Thus the angle subtended by C at *any* outside point is zero.

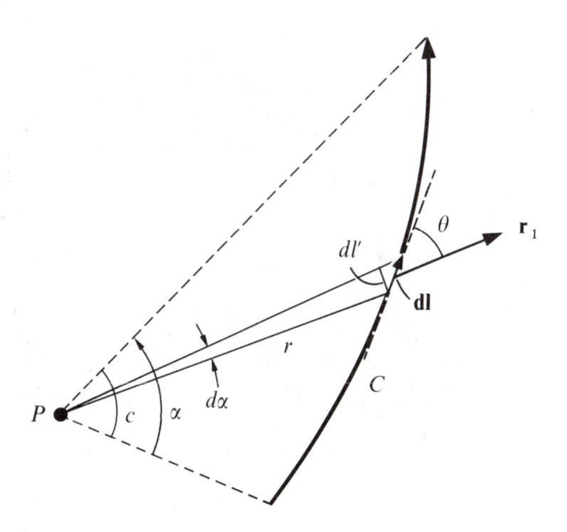

Figure 3-1 A curve C and a point P situated in a plane. The angle subtended by C at P is α. The element dl' is the component of $\mathbf{dl}$ in the direction perpendicular to the radius vector.

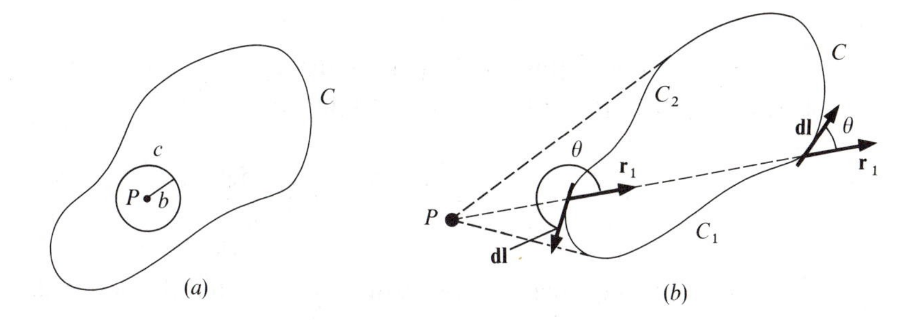

Figure 3-2 (*a*) Closed curve C and a point P situated inside. The angle subtended by C at P is the same as that subtended by the small circle, namely 2π radians.
(*b*) When the point P is outside C, the angle subtended by C at P is zero because segments C_1 on the right and C_2 on the left of the points where the tangents to the curve go through P subtend equal and opposite angles.

3.1.2 *THE SOLID ANGLE SUBTENDED BY A CLOSED SURFACE AT A POINT*

Figure 3-3 shows a closed surface S and a point P situated inside. The small cone intersects an element of area da on S. This small cone defines the *solid angle* subtended by da at P.

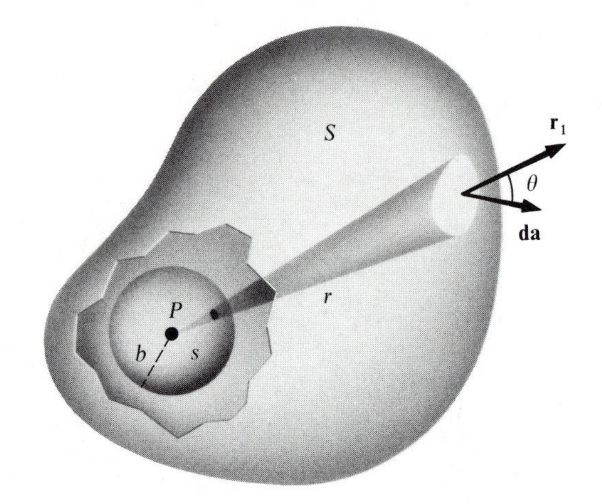

Figure 3-3 A closed surface S and a point P situated inside. The cone defines the solid angle subtended by the element of area **da** at P. The solid angle subtended at P by the complete surface S is the same as that subtended by the small sphere of radius b, namely 4π steradians.

By definition, this solid angle is the area da, projected on a plane perpendicular to the radius vector, and divided by r^2:

$$d\Omega = \frac{da \cos\theta}{r^2} = \frac{\mathbf{r}_1 \cdot \mathbf{da}}{r^2}, \tag{3-2}$$

where θ is the angle between the radius vector $\mathbf{r}$ and the vector $\mathbf{da}$, normal to the surface and pointing outward.

Solid angles are expressed in *steradians.*

The total solid angle subtended by S at P is

$$\Omega = \int \frac{da \cos\theta}{r^2}. \tag{3-3}$$

This solid angle is also the solid angle subtended at P by the small sphere s of radius a. Thus the solid angle subtended by S at *any* inside point is

$$\Omega = 4\pi a^2/a^2 = 4\pi \text{ steradians.} \tag{3-4}$$

Note that a solid angle is dimensionless, like an ordinary angle.

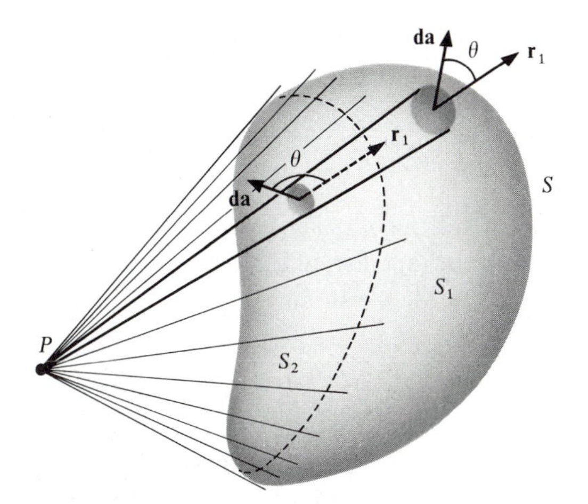

Figure 3-4 Closed surface S and a point P situated outside. The solid angle subtended by S at P is zero.

If the point P is situated outside the surface as in Fig. 3-4, then $\mathbf{r}_1 \cdot \mathbf{da}$ is positive over S_1 and negative over S_2, and the total solid angle subtended at *any* outside point is zero.

The situation remains unchanged if the surface is convoluted in such a way that a line drawn from P cuts the surface at more than one point. The total solid angle subtended by a closed surface is still 4π at an inside point, and zero at an outside point.

3.2 *GAUSS'S LAW*

Gauss's law relates the flux of **E** through a closed surface to the total charge enclosed within the surface.

By using this law one can find the electric field of simple charge distributions with a minimum of effort. Let us suppose that a point charge Q is located at P in Fig. 3-3. We can calculate the flux of **E** through the closed surface as follows. By definition, the flux of **E** through the element of area **da** is

$$\mathbf{E} \cdot \mathbf{da} = \frac{Q}{4\pi\epsilon_0} \frac{\mathbf{r}_1 \cdot \mathbf{da}}{r^2}, \tag{3-5}$$

where $\mathbf{r}_1 \cdot \mathbf{da}$ is the projection of **da** on a plane normal to $\mathbf{r}_1$. Then

$$\mathbf{E} \cdot \mathbf{da} = \frac{Q}{4\pi\epsilon_0}\, d\Omega, \tag{3-6}$$

where $d\Omega$ is the element of solid angle subtended by **da** at the point P.

To find the total flux of **E**, we integrate over the whole surface S. Since the point P is inside the surface, by hypothesis, the integral of $d\Omega$ is 4π and

$$\int_S \mathbf{E} \cdot \mathbf{da} = Q/\epsilon_0. \tag{3-7}$$

If more than one point charge resides within S, the fluxes add algebraically, and the total flux of **E** leaving the volume is equal to the total enclosed charge divided by ϵ_0.

If the charge enclosed by the surface S' is distributed over a finite volume, the total enclosed charge is

$$Q = \int_{\tau'} \rho\, d\tau', \tag{3-8}$$

where ρ is the charge density and τ' is the volume enclosed by the surface S'. Then

$$\int_{S'} \mathbf{E} \cdot \mathbf{da}' = \frac{1}{\epsilon_0} \int_{\tau'} \rho\, d\tau'. \tag{3-9}$$

This is Gauss's law stated in integral form.

Applying the divergence theorem (Sec. 1.10) to the left-hand side,

$$\int_{\tau'} \boldsymbol{\nabla} \cdot \mathbf{E}\, d\tau' = \frac{1}{\epsilon_0} \int_{\tau'} \rho\, d\tau'. \tag{3-10}$$

Since this equation is valid for any closed surface, the integrands must be equal and, at any point in space,

$$\boldsymbol{\nabla} \cdot \mathbf{E} = \rho/\epsilon_0. \tag{3-11}$$

This is Gauss's law stated in differential form. It concerns the *derivatives* of **E** with respect to the coordinates, and not **E** itself.

3.2.1

EXAMPLE: INFINITE SHEET OF CHARGE

Figure 3-5 shows an *infinite sheet of charge*, of surface charge density σ. On each side, $E = \sigma/2\epsilon_0$.

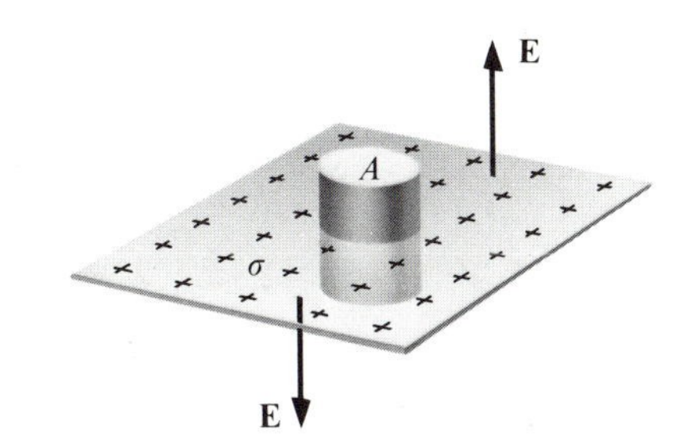

Figure 3-5 Portion of an infinite sheet of charge of density σ coulombs per square meter. The imaginary box of cross-section A encloses a charge σA. The flux of **E** leaving the box is $2EA$. Therefore, $E = \sigma/2\epsilon_0$.

3.2.2

EXAMPLE: SPHERICAL CHARGE

Let us calculate **E**, both inside and outside a *spherical charge* of radius R and uniform density ρ, as in Fig. 3-6. The electric field intensity **E** is a function of the distance r from the center O of the sphere to the point considered.

We shall use the subscript o to indicate that we are dealing with the field outside the charge distribution, and the subscript i inside. The total charge Q is $(4/3)\pi R^3\rho$.

Consider an imaginary sphere of radius $r > R$, concentric with the charged sphere. We know that $\mathbf{E}_o$ must be radial. Then, according to the integral form of Gauss's law,

$$4\pi r^2 E_o = Q/\epsilon_0, \tag{3-12}$$

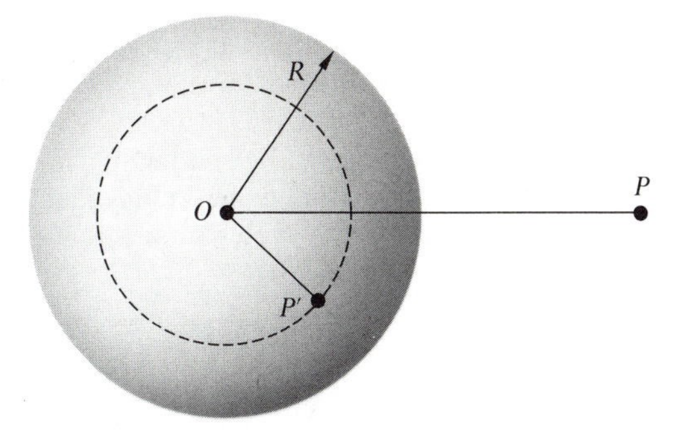

Figure 3-6 Uniform spherical charge. We can calculate the electric field at a point P outside the sphere, as well as at a point P' inside the sphere, by using Gauss's law.

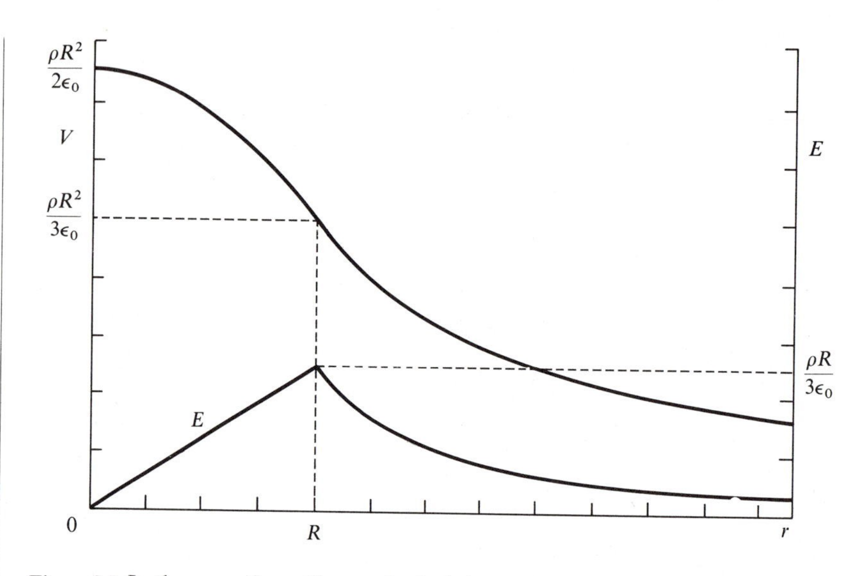

Figure 3-7 In the case of a uniform spherical charge distribution, the electric field intensity E rises linearly from the center to the surface of the sphere and falls off as the inverse square of the distance outside the sphere. On the other hand, the electric potential falls from a maximum at the center in parabolic fashion inside the sphere, and as the inverse first power of the distance outside.

and

$$E_o = \frac{1}{4\pi\epsilon_0} \frac{Q}{r^2}, \tag{3-13}$$

as if the total charge Q were situated at the center of the sphere.

If the charge were not distributed with spherical symmetry, E_o would not be uniform over the imaginary sphere. Gauss's law would then only give the average value of the normal component of $\mathbf{E}_o$ over the sphere.

To calculate $\mathbf{E}_i$ at an internal point, we draw an imaginary sphere of radius r through the point P'. Symmetry requires again that $\mathbf{E}_i$ be radial; thus

$$4\pi r^2 E_i = (4/3)\pi r^3 \rho/\epsilon_0, \tag{3-14}$$

and

$$E_i = \rho r/3\epsilon_0. \tag{3-15}$$

Figure 3-7 shows E and V as functions of the radial distance r. Note that $E_i = E_o$ at $r = R$. This is required by Gauss's law, since the charge contained within a spherical shell of zero thickness at $r = R$ is zero. Note also that $V_i = V_o$ at $r = R$. A discontinuity in V would require an infinite electric field intensity, since $\mathbf{E}$ is $-dV/dr$.

3.3 CONDUCTORS

A conductor can be defined as a material inside which charges can flow freely.

Since we are dealing with electrostatics, we assume that the charges have reached their equilibrium positions and are fixed in space. Then, inside a conductor, there is zero electric field, and all points are at the same potential.

If a conductor is placed in an electric field, charges flow temporarily within it so as to produce a second field that cancels the first one at every point inside the conductor.

Coulomb's law applies within conductors, even though the *net* field is zero. Gauss's law, $\nabla \cdot \mathbf{E} = \rho/\epsilon_0$, is also valid. Then, since $\mathbf{E} = 0$, the charge density ρ must be zero.

As a corollary, any net static charge on a conductor must reside on its surface.

At the surface of a conductor, $\mathbf{E}$ must be normal, for if there were a tangential component of $\mathbf{E}$, charges would flow along the surface, which would be contrary to our hypothesis. Then, according to Gauss's law, just outside the surface, $E = \sigma/\epsilon_0$, as in Fig. 3-8, σ being the surface charge density.

It is paradoxical that one should be able to express the electric field intensity at the surface of a conductor in terms of the local surface charge density σ alone, despite the fact that the field is of course due to *all* the charges, whether they are on the conductor or elsewhere.

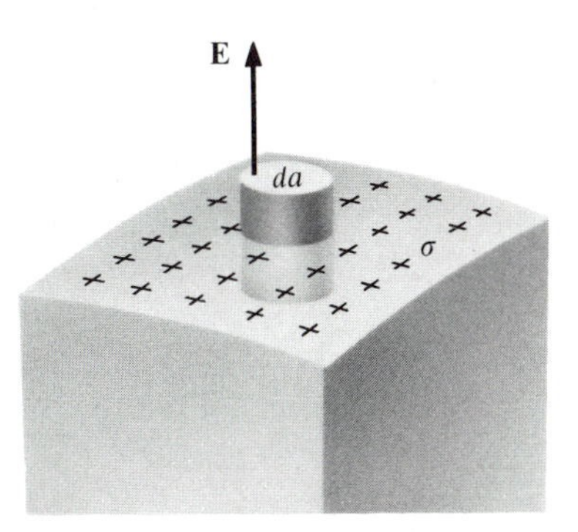

Figure 3-8 Portion of a charged conductor carrying a surface charge density σ. The charge enclosed by the imaginary box is $\sigma\, da$. There is zero field inside the conductor. Then, from Gauss's law, $E = \sigma/\epsilon_0$.

3.3.1

EXAMPLE: HOLLOW CONDUCTOR

A *hollow conductor*, as in Fig. 3-9, has a charge on its inner surface that is equal in magnitude and opposite in sign to any charge that may be enclosed within the hollow. This is readily demonstrated by considering a Gaussian surface that lies within the conductor and that encloses the hollow. Since **E** is everywhere zero on this surface, the total enclosed charge must be zero.

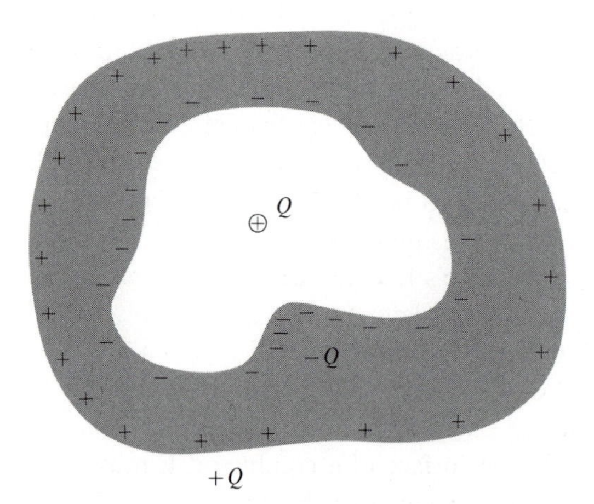

Figure 3-9 Cross-section of a hollow conductor. A charge Q in the hollow induces a total charge $-Q$ on the inner surface. If the conductor has a zero net charge, a charge $+Q$ is induced on the outer surface.

3.3.2

EXAMPLE: ISOLATED CHARGED CONDUCTING PLATE

An *isolated charged conducting plate* of infinite extent must carry equal charge densities σ on its two faces, as in Fig. 3-10a. Outside, $E = 2(\sigma/2\epsilon_0) = \sigma/\epsilon_0$, while, inside the plate, the fields of the two surface distributions cancel and $E = 0$.

3.3.3

EXAMPLE: PAIR OF PARALLEL CONDUCTING PLATES CARRYING CHARGES OF EQUAL MAGNITUDES AND OPPOSITE SIGNS

If one has a *pair of parallel conducting plates*, as in Fig. 3-10b, *carrying charges of equal magnitude and opposite signs*, the charges position themselves as in the figure and $E = 0$ everywhere except in the region between the plates.

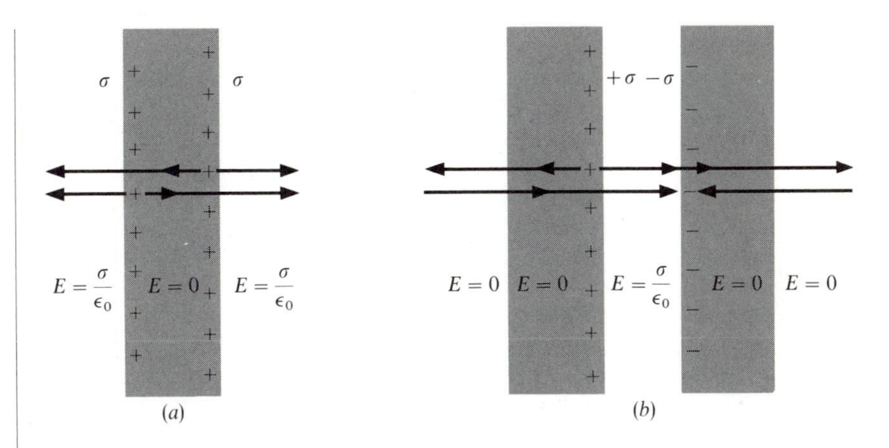

Figure 3-10 (*a*) Charged conducting plate carrying equal surface charge densities σ on each side. The field inside the plate is zero. (*b*) Pair of parallel plates carrying surface charge densities $+\sigma$ and $-\sigma$. The electric field inside the plates and outside is zero.

3.4 POISSON'S EQUATION

According to Gauss's law, Eq. 3-11, $\nabla \cdot \mathbf{E} = \rho/\epsilon_0$. Now, since $\mathbf{E} = -\nabla V$, from Eq. 2-13,

$$\nabla^2 V \equiv \frac{\partial^2 V}{\partial x^2} + \frac{\partial^2 V}{\partial y^2} + \frac{\partial^2 V}{\partial z^2} = -\frac{\rho}{\epsilon_0}. \tag{3-16}$$

This is Poisson's equation.

Poisson's equation often serves as the starting point for calculating electrostatic fields. It states that, within a constant factor, the charge density ρ is equal to the sum of the second derivatives of V with respect to x, y, z.

3.4.1 EXAMPLE: FLAT ION BEAM

Figure 3-11 shows a *flat ion beam* situated between two grounded parallel conducting plates. A complete study of the electric and magnetic fields of the beam would be quite elaborate, but we shall disregard the motion of the ions and limit ourselves to the electric field, in the simple case where ρ is uniform inside the beam

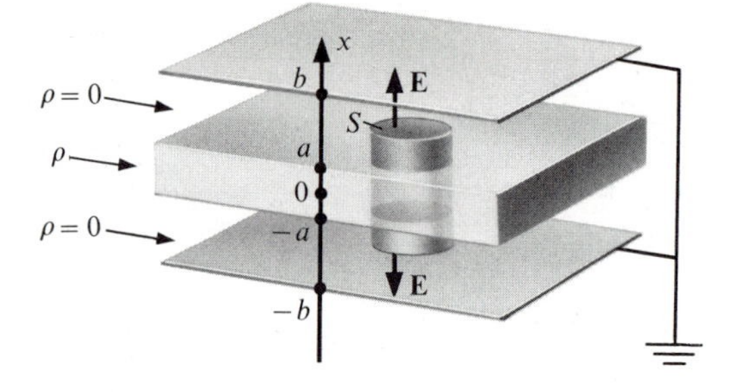

Figure 3-11 Flat ion beam of uniform space charge density ρ, between two grounded conducting plates.

and where edge effects are negligible. We shall use the subscript i on **E** and V inside the beam, and the subscript o outside.

Inside the beam,

$$\nabla^2 V_i = \frac{d^2 V_i}{dx^2} = -\frac{\rho}{\epsilon_0}, \tag{3-17}$$

$$E_i = -\frac{dV_i}{dx} = \frac{\rho x}{\epsilon_0} + A, \tag{3-18}$$

where A is a constant of integration. By symmetry, $\mathbf{E}_i = 0$ at $x = 0$. Therefore $A = 0$; inside the beam,

$$E_i = \rho x/\epsilon_0, \tag{3-19}$$

$$V_i = -\frac{\rho}{2\epsilon_0}x^2 + B, \tag{3-20}$$

where B is another constant of integration.

Now consider a cylindrical volume whose upper and lower faces are outside the beam, at equal distances from the plane $x = 0$, as in Fig. 3-11. The magnitude of $\mathbf{E}_o$ is the same at both faces. Let S be the area of one face. According to Gauss's law,

$$2E_o S = 2aS\rho/\epsilon_0, \tag{3-21}$$

$$E_o = a\rho/\epsilon_0 \qquad x \geq a, \tag{3-22}$$

$$E_o = -a\rho/\epsilon_0 \qquad x \leq -a. \tag{3-23}$$

Therefore, outside the beam,

$$V_o = -\frac{a\rho x}{\epsilon_0} + C \qquad x \geq a. \tag{3-24}$$

$$V_o = \frac{a\rho x}{\epsilon_0} + D \qquad x \leq -a, \tag{3-25}$$

where C and D are again constants of integration. Since $V = 0$ both at $x = b$ and at $x = -b$,

$$C = D = a\rho b/\epsilon_0 \tag{3-26}$$

and

$$V_o = \frac{a\rho}{\epsilon_0}(b - x) \qquad x \geq a, \tag{3-27}$$

$$= \frac{a\rho}{\epsilon_0}(b + x) \qquad x \leq -a. \tag{3-28}$$

Now there can be no discontinuity of V at the surface of the beam, for otherwise $\mathbf{E} = -\nabla V$ would be infinite. Then Eqs. 3-20 and 3-27 must agree at $x = a$:

$$-\frac{\rho}{2\epsilon_0}a^2 + B = \frac{a\rho}{\epsilon_0}(b - a), \tag{3-29}$$

$$B = \frac{a\rho}{\epsilon_0}\left(b - \frac{a}{2}\right). \tag{3-30}$$

Equating the V's from Eqs. 3-20 and 3-28 at $x = -a$ gives the same result.

Finally, inside the beam,

$$V_i = \frac{\rho}{\epsilon_0}\left[-\frac{x^2}{2} + a\left(b - \frac{a}{2}\right)\right], \tag{3-31}$$

$$E_i = \rho x/\epsilon_0. \tag{3-32}$$

Above, for $x \geq a$,

$$V_o = \frac{a\rho}{\epsilon_0}(b - x), \tag{3-33}$$

$$E_o = a\rho/\epsilon_0, \tag{3-34}$$

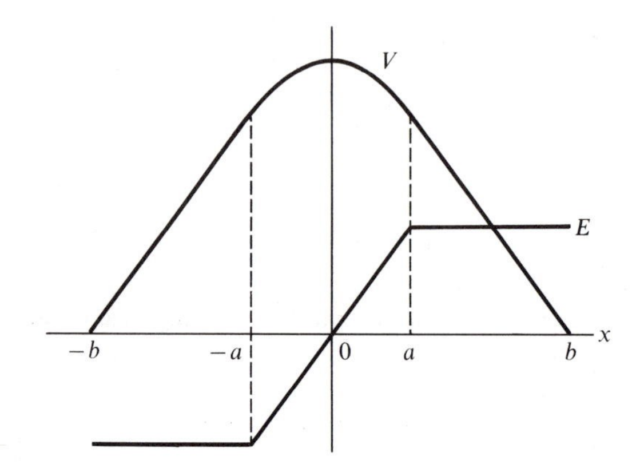

Figure 3-12 The electric field intensity E and potential V as functions of x, between the two plates in Fig. 3-11.

and below, for $x \leq -a$,

$$V_o = \frac{a\rho}{\epsilon_0}(b + x), \tag{3-35}$$

$$E_o = -a\rho/\epsilon_0. \tag{3-36}$$

The curves for V and E as functions of x are shown in Fig. 3-12.

3.5 *LAPLACE'S EQUATION*

When $\rho = 0$, Poisson's equation becomes

$$\nabla^2 V = 0. \tag{3-37}$$

This is *Laplace's equation.* This equation applies not only to electrostatics but also to heat conduction, hydro- and aerodynamics, elasticity, and so on.

3.6 *THE UNIQUENESS THEOREM*

According to the *uniqueness theorem*, Poisson's equation has only one solution $V(x,y,z)$, for a given charge density $\rho(x,y,z)$ and for given boundary conditions.†

† For a proof of this theorem, see our *Electromagnetic Fields and Waves*, p. 142.

This is an important theorem. If, somehow, by intuition or by analogy, one finds a function V that satisfies both Poisson's equation and the given boundary conditions, then it is the correct V.

3.7 IMAGES

The method of images involves the conversion of an electric field into another equivalent electric field that is simpler to calculate.

With this method, one modifies the field in such a way that, *for the region of interest*, both the charge distribution and the boundary conditions are conserved. The method is a remarkable application of the uniqueness theorem. It is best explained by means of an example.

3.7.1 EXAMPLE: POINT CHARGE NEAR AN INFINITE GROUNDED CONDUCTING PLATE

Consider a *point charge Q at a distance D from an infinite conducting plate connected to ground*, as in Fig. 3-13a. This plate may be taken to be at zero potential. It is clear that if we remove the grounded conductor and replace it by a charge $-Q$ at a distance D behind the plane, then every point of the plane will be equidistant from Q and from $-Q$, and will thus be at zero potential. The field to the left of the conducting sheet is therefore unaffected.

The charge $-Q$ is said to be the *image* of the charge Q in the plane.

The potential V at a point $P(x, y)$ as in Fig. 3-13b, is given by

$$4\pi\epsilon_0 V = \frac{Q}{r} - \frac{Q}{r'}, \tag{3-38}$$

where

$$r = (x^2 + y^2)^{1/2} \quad \text{and} \quad r' = [(2D - x)^2 + y^2]^{1/2}. \tag{3-39}$$

The components of the electric field intensity at P are those of $\boldsymbol{\nabla} V$:

$$4\pi\epsilon_0 E_x = -4\pi\epsilon_0 \frac{\partial V}{\partial x} = \frac{Qx}{r^3} + \frac{Q(2D - x)}{r'^3}, \tag{3-40}$$

$$4\pi\epsilon_0 E_y = -4\pi\epsilon_0 \frac{\partial V}{\partial y} = \frac{Qy}{r^3} - \frac{Qy}{r'^3}. \tag{3-41}$$

The lines of force and the equipotentials are shown in Fig. 3-14.

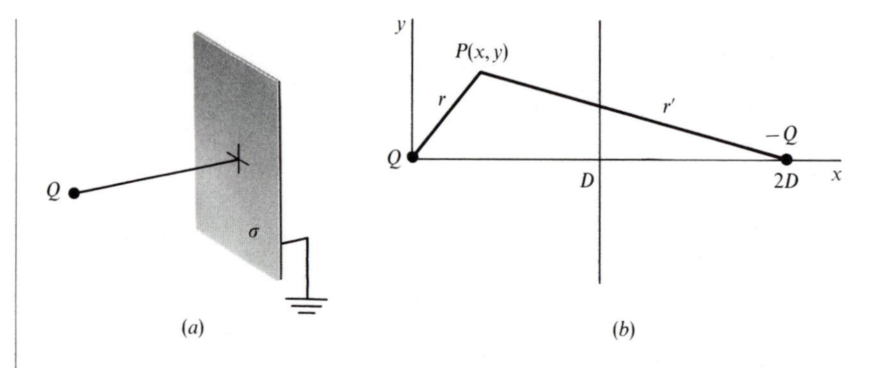

Figure 3-13 (*a*) Point charge Q near a grounded conducting plate. (*b*) The conducting plate has been replaced by the image charge $-Q$ to calculate the field at P.

At the surface of the conducting plate, $x = D$, $r = r'$,

$$E_x = 2QD/4\pi\epsilon_0 r^3, \; E_y = 0, \tag{3-42}$$

and the induced charge density is $-\epsilon_0 E_x$. In this particular case, the surface charge density is negative, and the electric field intensity points to the right at the surface of the conducting plate.

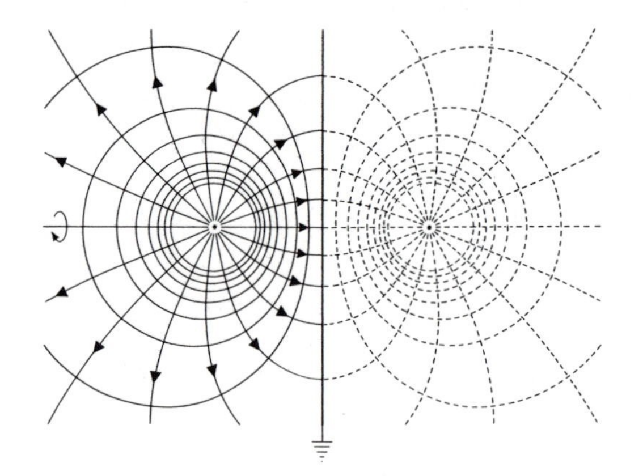

Figure 3-14 Lines of force (identified by arrows) and equipotentials for a point charge near a grounded conducting plate. Equipotentials and lines of force near the charge cannot be shown because they get too close together. Equipotential surfaces are generated by rotating the figure about the axis designated by the curved arrow. The image field to the right of the conducting plate is indicated by broken lines.

It will be observed that image charges are located *outside* the region where we calculate the field. In this case we require the field in the region to the left of the plate, whereas the image is to the right.

Now what is the electrostatic force of attraction between the charge Q and the grounded plate? It is obviously the same as between two charges Q and $-Q$ separated by a distance $2D$, since Q cannot tell whether it is in the presence of a point charge $-Q$ or of a grounded plate. The force on Q is thus $Q^2/4\pi\epsilon_0(2D)^2$. This is always true. The force between a charge and a conductor is always given correctly by the Coulomb force between the charge and its image charges.

3.8 SUMMARY

The *solid angle* subtended by a closed surface S at a point P is 4π steradians if P is situated inside S, and zero if P is situated outside.

Gauss's law follows from Coulomb's law. It can be stated either in integral or in differential form:

$$\int_{S'} \mathbf{E} \cdot \mathbf{da}' = \frac{1}{\epsilon_0} \int_{\tau'} \rho \, d\tau', \tag{3-9}$$

where S' is the surface that encloses τ', or

$$\boldsymbol{\nabla} \cdot \mathbf{E} = \rho/\epsilon_0. \tag{3-11}$$

Poisson's equation,

$$\boldsymbol{\nabla}^2 V = -\rho/\epsilon_0, \tag{3-16}$$

then follows from Eq. 3-11 and from $\mathbf{E} = -\boldsymbol{\nabla} V$.

When the charge density is zero we have *Laplace's equation,*

$$\boldsymbol{\nabla}^2 V = 0. \tag{3-37}$$

According to the *uniqueness theorem*, for a given ρ and for a given set of boundary conditions, there is only one possible potential satisfying Poisson's equation.

The method of *images* is one remarkable application of the uniqueness theorem. For example, the field of a point charge Q in front of an infinite grounded conducting plate is the same as if the plate were replaced by a charge $-Q$ at the position of the image of Q. The method of images gives the correct field only outside the region where the images are situated.

PROBLEMS

3-1E *ANGLE SUBTENDED BY A LINE AT A POINT*

What is the angle θ subtended at a point P by a line of length a situated at a distance b as in Fig. 3-15?

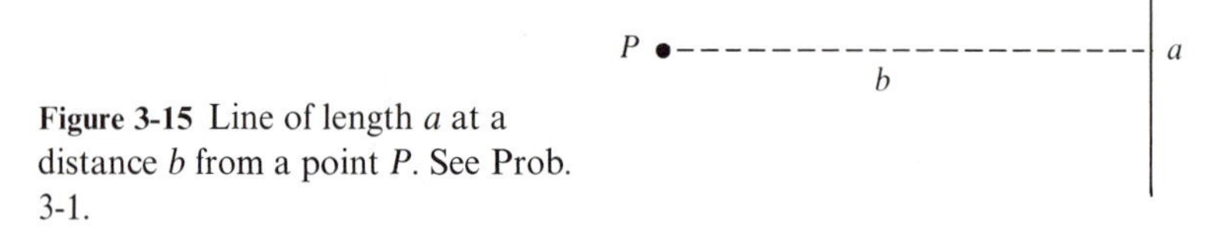

Figure 3-15 Line of length a at a distance b from a point P. See Prob. 3-1.

3-2 *SOLID ANGLE SUBTENDED BY A DISK AT A POINT*

What is the solid angle subtended at a point P by a circular area of radius a at a distance b as in Fig. 3-16?

Hint: Consider first the solid angle subtended at P by a ring of radius r and width dr, and then integrate.

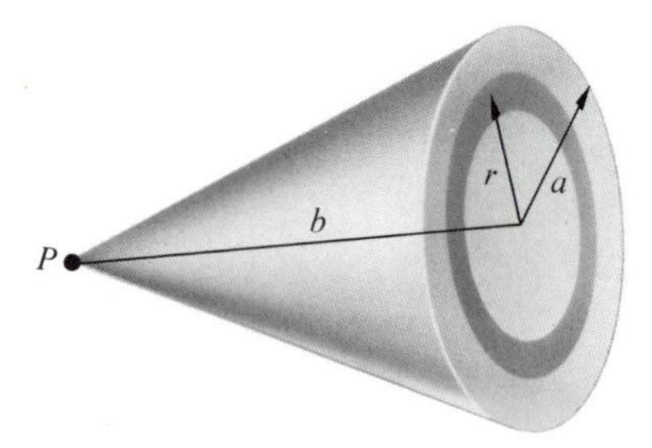

Figure 3-16 Circular area of radius a situated at a distance b from the point P. See Prob. 3-2.

3-3E *GAUSS'S LAW*

Could one use Gauss's law to calculate the electric field intensity near a dipole?

3-4E *SURFACE DENSITY OF ELECTRONS ON A CHARGED BODY*

The maximum electric field intensity that can be maintained in air is 3×10^6 volts per meter.

a) Calculate the surface charge density that will produce this field.

b) Consider a metal with an interatomic spacing of 0.3 nanometer. What is the approximate number of atoms per square meter?

c) How many extra electrons are required, per atom, to give a surface charge density of the above magnitude?

3-5 *THE ELECTRIC FIELD IN A NUCLEUS*

a) Calculate the electric field intensity in volts per meter at the surface of an iodine nucleus (53 protons and 74 neutrons).

b) Find the electric potential at the center of the nucleus.

Assume that the charge density is uniform and that the radius of the nucleus is $1.25 \times 10^{-15} A^{1/3}$ meter, where A is the total number of particles in the nucleus.

3-6E *THE SPACE DERIVATIVES OF E_x, E_y, E_z*

List all the relations between the various space derivatives of the three components of **E**, using Gauss's law and the fact that an electrostatic field is conservative.

3-7 *PHYSICALLY IMPOSSIBLE FIELDS*

An electric field points everywhere in the z direction.

a) What can you conclude about the values of the partial derivatives of E_z with respect to x, to y, and to z, (i) if the volume charge density ρ is zero, (ii) if the volume charge density is not zero?

b) Sketch lines of force for some fields that are possible and for some that are not.

3-8 *PROTON BEAM*

A 1.00-microampere beam of protons is accelerated through a difference of potential of 10,000 volts.

a) Calculate the volume charge density in the beam, once the protons have been accelerated, assuming that the current density is uniform within a diameter of 2.00 millimeters, and zero outside this diameter. (The current density in an ion beam is in fact largest in the center and falls off gradually with increasing radius.)

Solution: The charge per unit length λ is equal to the current I, divided by the velocity v of the protons:

$$\lambda = I/v. \tag{1}$$

Also,

$$(1/2)mv^2 = eV, \tag{2}$$

where m is the proton mass, e is the proton charge, and V is the accelerating voltage. Thus,

$$\lambda = \frac{I}{(2eV/m)^{1/2}} = I\left(\frac{m}{2eV}\right)^{1/2} \tag{3}$$

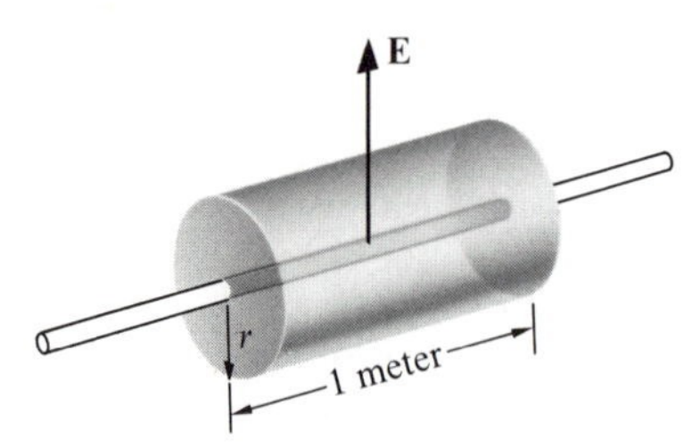

Figure 3-17 Cylindrical surface surrounding the ion beam of Prob. 3-8 for calculating the magnitude of **E** using Gauss's law.

and

$$\rho = \lambda/\pi R^2 = 2.300 \times 10^{-7} \text{ coulomb/meter}^3, \tag{4}$$

where R is the radius of the beam.

b) Calculate the radial electric field intensity, both inside and outside the beam.

Solution: Outside the beam, **E** is radial. Applying Gauss's law to a cylindrical surface as in Fig. 3-17,

$$2\pi r(\epsilon_0 E_o) = \lambda, \tag{5}$$

$$E_o = \lambda/2\pi\epsilon_0 r, \tag{6}$$

$$= \frac{I}{2\pi\epsilon_0}\left(\frac{m}{2eV}\right)^{1/2}\frac{1}{r}, \tag{7}$$

$$= 1.299 \times 10^{-2}/r \text{ volts/meter}, \tag{8}$$

$$= 12.99 \text{ volts/meter at } r = 10^{-3} \text{ meter}. \tag{9}$$

Inside the beam, we find E by applying Gauss's law to a cylindrical surface of radius $r < R$. Then

$$E_i = \pi r^2 \rho/2\pi\epsilon_0 r = \rho r/2\epsilon_0, \tag{10}$$

$$= 1.299 \times 10^4 r \text{ volts/meter}, \tag{11}$$

$$= 12.99 \text{ volts/meter at } r = 10^{-3} \text{ meter}. \tag{12}$$

c) Draw a graph of the radial electric field intensity for values of r ranging from zero to 10 millimeters.

Solution: See Fig. 3-18.

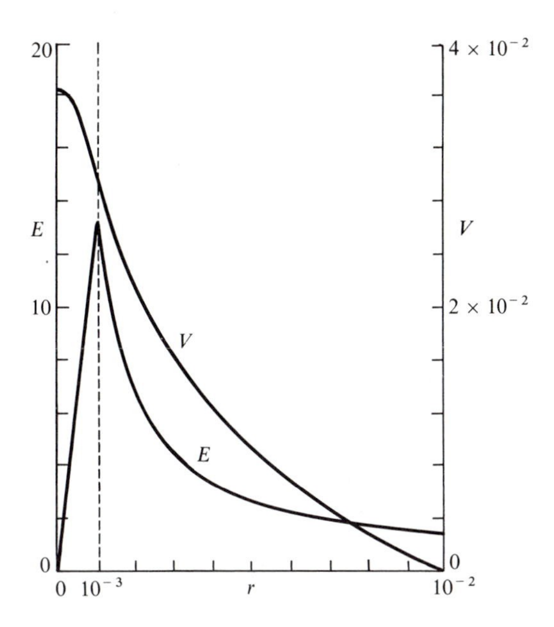

Figure 3-18 Potential V and electric field intensity E as functions of the distance r from the axis for the proton beam of Prob. 3-8. The beam has a radius of 1 millimeter.

d) Now let the beam be situated on the axis of a grounded cylindrical conducting tube with an inside radius of 10.0 millimeters.

Are the above values of E still valid?

Solution: They are still valid: E_o depends solely on λ and r, and E_i on ρ and r. It is the value of V that is affected by the presence of the tube.

e) Draw a graph of V inside the tube.

Solution: Outside the beam, $V = 0$ at $r = 10^{-2}$ meter and beyond. Then

$$V_o = \int_r^{10^{-2}} E_o\, dr = \int_r^{10^{-2}} 1.299 \times 10^{-2}\, dr/r, \tag{13}$$

$$= 1.299 \times 10^{-2}[\ln r]_r^{10^{-2}}, \tag{14}$$

$$= 1.299 \times 10^{-2}[\ln 10^{-2} - \ln r], \tag{15}$$

$$= -(5.982 + 1.299 \ln r)10^{-2} \text{ volt}, \tag{16}$$

$$= 2.991 \times 10^{-2} \text{ volt at } r = 10^{-3} \text{ meter}. \tag{17}$$

Inside the beam,

$$V_i = 2.991 \times 10^{-2} + \int_r^{10^{-3}} 1.299 \times 10^4\, r\, dr, \tag{18}$$

$$= 2.991 \times 10^{-2} + 6.495 \times 10^3(10^{-6} - r^2) \text{ volt}, \tag{19}$$

$$= 36.41 \times 10^{-3} \text{ volt at } r = 0. \tag{20}$$

Figure 3-18 shows V as a function of r.

3-9 *ION BEAM*

An ion beam of uniform density ρ passes between a pair of parallel plates, one at $x = 0$, at zero potential, and one at $x = a$, at the potential V_0. Neglect the motion of the ions, as well as edge effects. The beam completely fills the space between the plates.

Find V and the transverse E as functions of x.

3-10 *A UNIFORM AND A NON-UNIFORM FIELD*

Consider a pair of parallel conducting plates, one at $x = 0$ that is grounded, and one at $x = 0.1$ meter that is maintained at a potential of 100 volts.

a) Express V as a function of x when $\rho = 0$. Show the equipotentials at $V =$ 0, 10, 20, . . . 100 volts. Draw a curve of E as a function of x.

b) Do the same for the case where ρ/ϵ_0 is uniform and equal to 10^4.

3-11 *VACUUM DIODE*

In a vacuum diode, electrons are emitted by a heated cathode and collected by an anode. In the simplest case, the electrodes are plane, parallel, and close together, so edge effects can be neglected.

Since the electrons move at a finite velocity, the space charge density is not zero. It is given by

$$\rho = -\frac{4\epsilon_0 V_0}{9 s^{4/3} x^{2/3}},$$

where V_0 is the voltage at the anode (the voltage at the cathode is zero); s is the distance between the electrodes; and x is the distance of a point to the cathode. So, at $x = 0$, $V = 0$, while, at $x = s$, $V = V_0$.

This peculiar charge distribution comes from the fact that the electrons are accelerated in the electric field, which itself depends on the charge density, and hence on the electron velocity. We assume that the current is limited by the space charge, and not by electron emission at the heated cathode.

a) Use Poisson's equation to find V as a function of x.

b) At any point, the current density J is ρv, where v is the velocity of the electrons at that point.

Use the value of ρv at $x = s$ to find J.

3-12 *IMAGES*

Two charges $+Q$ and $-Q$ are separated by a distance a and are at a distance b from a large conducting sheet as in Fig. 3-19.

Find the x and y components of the force exerted on the right-hand charge.

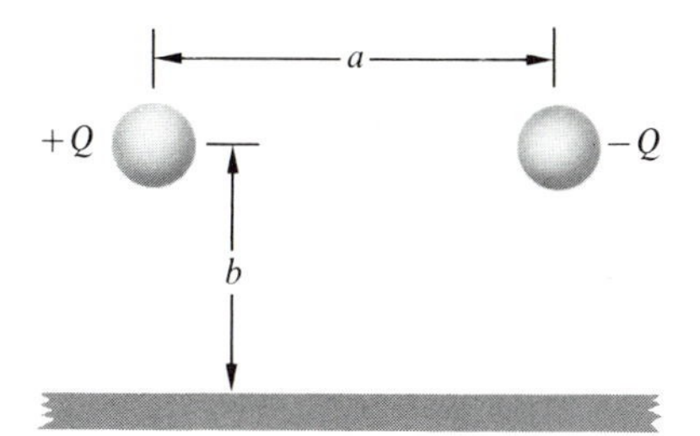

Figure 3-19 Pair of charges near a conducting surface. See Prob. 3-12.

3-13 *IMAGES*

A charge Q is situated near a conducting plate forming a right angle as in Fig. 3-20.
Find the electric field intensity at P.
There are three images of equal magnitudes.

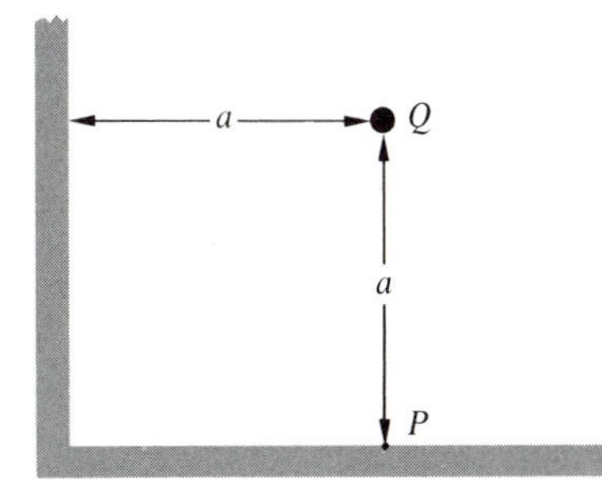

Figure 3-20 Point charge near a conductor forming a right angle. See Prob. 3-13.

3-14 *IMAGES*

A point charge Q is situated at a distance D from an infinite conducting plate connected to ground, as in Fig. 3-13.

a) Find the surface charge density on the conductor as a function of the radius.
b) Show that the total charge on the conductor is $-Q$.

3-15 *ANTENNA IMAGES*

Antennas are often mounted near conducting surfaces. One example is the whip antenna shown in Fig. 18-4.

Figure 3-21 shows schematically three antennas, or portions of antennas, near conducting surfaces. The arrows show the direction of current flow at a given instant.

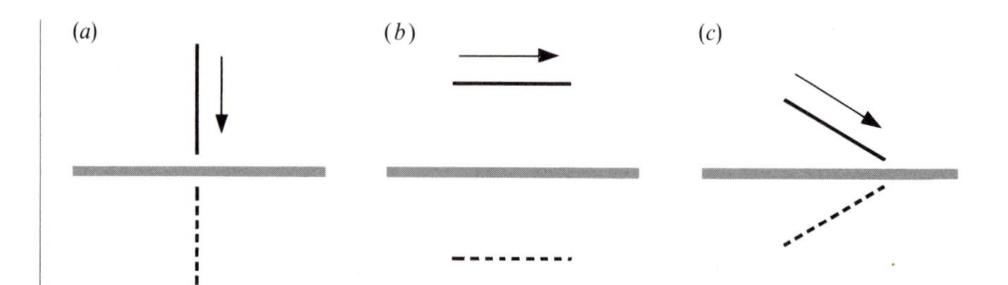

Figure 3-21 Three antennas situated above conducting planes, and their images. The arrows show the direction of the currents in the antennas, at a given instant. See Prob. 3-15.

Draw figures showing the direction of current flow in the images.

Solution: (a) Consider a positive charge moving downward in the vertical antenna. Its image is a negative charge that moves upward. Now a negative charge moving upward gives a downward current. So the current in the image is as shown in Fig. 3-22a.

b) Similarly, a positive charge moving to the right has a negative image also moving to the right. The current in the image is in the opposite direction, as shown in Fig. 3-22b.

c) A positive charge moving to the right and downward has a negative image moving to the right and upward. The image current is in the direction shown in Fig. 3-22c.

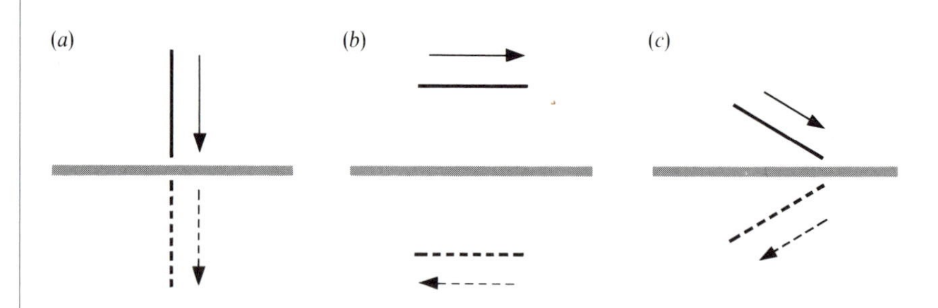

Figure 3-22 Relation between the current directions in the antennas and in their images.

CHAPTER 4

FIELDS OF STATIONARY ELECTRIC CHARGES: III

Capacitance, Energy, and Forces

In this chapter we shall complete our study of the electric fields of stationary electric charges in a vacuum. We shall start with the concept of capacitance. This will lead us to the energy stored in an electric field, and this energy, in turn, will give us the force exerted on a conductor situated in an electric field.

4.1 *CAPACITANCE OF AN ISOLATED CONDUCTOR*

Imagine a conductor situated a long distance away from any other body. As charge is added to it, its potential rises, the magnitude of the change in potential being proportional to the amount of charge added and depending on the geometrical configuration of the conductor.

The ratio

$$C = Q/V \tag{4-1}$$

is called the *capacitance* of the isolated conductor. A capacitance is positive, by definition.

The unit of capacitance is the coulomb per volt or *farad*.

4.1.1 *EXAMPLE: ISOLATED SPHERICAL CONDUCTOR*

An *isolated spherical conductor* of radius R carries a total surface charge Q. Then its capacitance is

$$C = \frac{Q}{V} = \frac{Q}{Q/4\pi\epsilon_0 R} = 4\pi\epsilon_0 R. \tag{4-2}$$

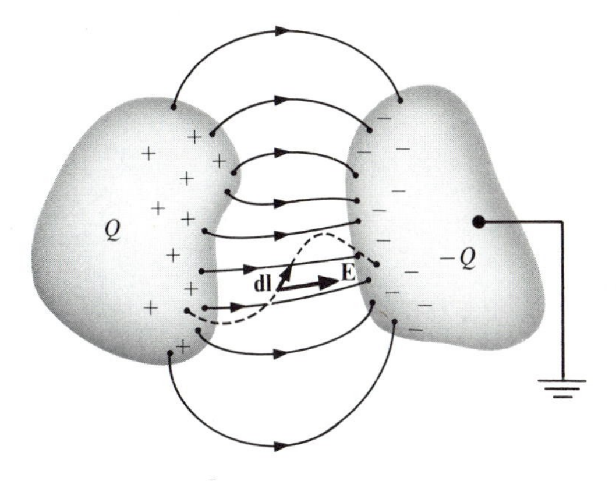

Figure 4-1 Two conductors carrying equal and opposite charges. The integral of $\mathbf{E} \cdot \mathbf{dl}$ from one conductor to the other, along any path, gives the potential difference V. The capacitance between the conductors is Q/V.

4.2 *CAPACITANCE BETWEEN TWO CONDUCTORS*

We have just defined the capacitance of an isolated conductor. We can proceed in a similar way to define the capacitance between two conductors, as in Fig. 4-1.

Initially, both conductors are uncharged. Let us assume that one of the conductors is grounded, so that its potential is kept equal to zero. We then gradually add small charges to the other until it carries a charge Q. In the process, charges of opposite sign are attracted to the grounded conductor until it carries a charge $-Q$. At that point the grounded conductor is still at zero potential, while the other conductor is at a potential V.

The capacitance between the two conductors is again defined as in Eq. 4-1:

$$C = Q/V. \tag{4-3}$$

The capacitance between two conductors is also positive, by definition.

In other words, if we establish a potential difference V between two conductors by giving them charges $+Q$ and $-Q$, then the capacitance between the conductors is Q/V.

4.2.1 *EXAMPLE: PARALLEL-PLATE CAPACITOR*

Figure 4-2 shows a *parallel-plate capacitor* with the spacing s grossly exaggerated. We neglect edge effects. We have a uniform positive charge density σ on the top

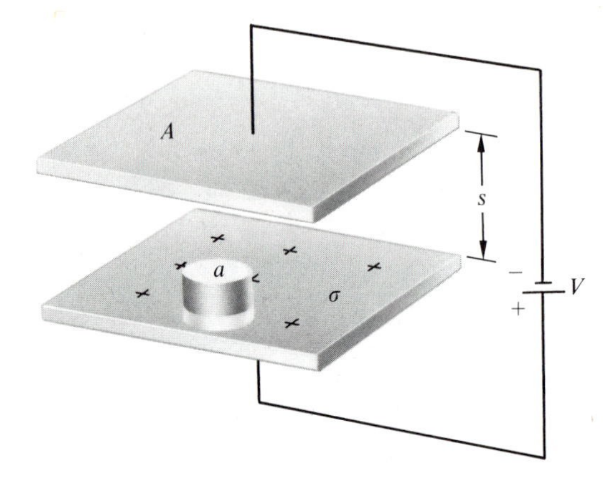

Figure 4-2 Parallel-plate capacitor. The lower end of the small cylindrical figure is situated inside the lower plate where $\mathbf{E} = 0$.

surface of the lower plate and a uniform negative charge density $-\sigma$ on the bottom surface of the upper plate. The electric field intensity $\mathbf{E}$ points upward.

Applying Gauss's law to a cylindrical volume as in the figure, the outward flux of $\mathbf{E}$ is Ea, where a is the cross-section of the cylinder. Since the enclosed charge is σa, then

$$Ea = \sigma a/\epsilon_0, \qquad E = \sigma/\epsilon_0, \tag{4-4}$$

so that

$$V = Es = \sigma s/\epsilon_0. \tag{4-5}$$

Now $Q = \sigma A$, where A is the area of one plate, and

$$C = Q/V = \epsilon_0 A/s. \tag{4-6}$$

This is the capacitance between two parallel conducting plates of area A separated by a distance s.

4.2.2 CAPACITORS CONNECTED IN PARALLEL

Figure 4-3a shows two capacitors C_1 and C_2 connected in parallel and carrying charges Q_1 and Q_2. We wish to find the value of C in Fig. 4-3b that will give the same V for the same charge $Q_1 + Q_2$. This is simple:

$$Q_1 + Q_2 = C_1 V + C_2 V = CV. \tag{4-7}$$

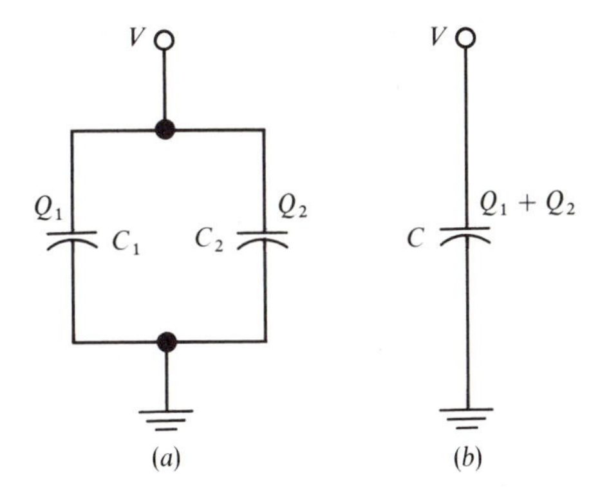

Figure 4-3 The single capacitor C has the same capacitance as the two capacitors C_1 and C_2 connected in parallel.

Then

$$C = C_1 + C_2. \tag{4-8}$$

So, the capacitance of two capacitors connected in parallel is the sum of the two capacitances. If we have several capacitors in parallel, the total capacitance is the sum of the individual capacitances.

4.2.3 *CAPACITORS CONNECTED IN SERIES*

Figure 4-4a shows two capacitors C_1 and C_2 connected in series. We first connect points A and B to ground, temporarily. This makes the voltages V_1 and V_2 zero. The capacitor plates are uncharged. We now disconnect A and B from ground.

We then apply a charge Q to A. We assume that the voltmeters draw zero current. No charge can flow into or out of B. A charge $-Q$ leaves the upper plate of C_2 and goes to the lower plate of C_1. This leaves a charge $+Q$ on the top plate of C_2. Then the lower plate of C_1 takes on a charge $-Q$, as in the figure.

Then

$$V_1 = Q/C_1, \qquad V_2 = Q/C_2. \tag{4-9}$$

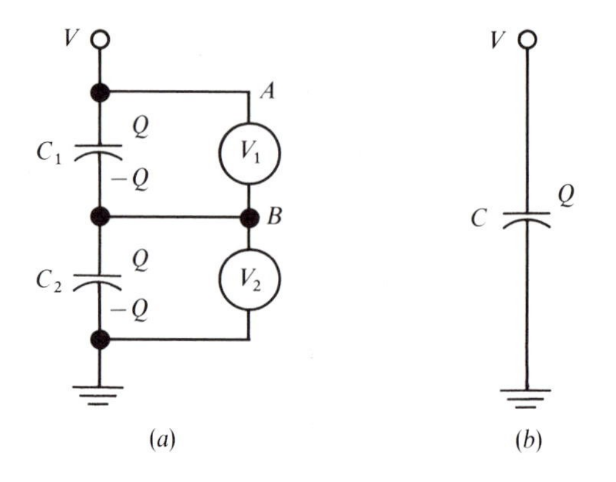

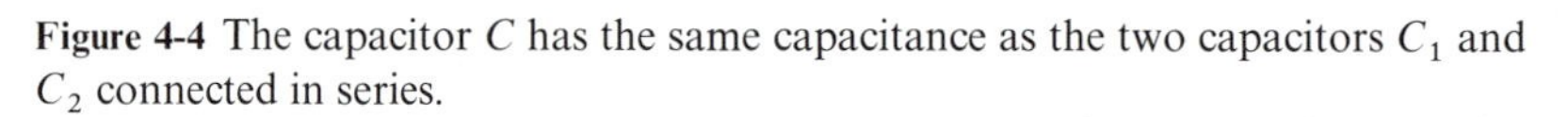
Figure 4-4 The capacitor C has the same capacitance as the two capacitors C_1 and C_2 connected in series.

Now what should be the value of C in Fig. 4-4b to give a voltage V equal to $V_1 + V_2$ with the same charge Q? We must have

$$V = \frac{Q}{C} = \frac{Q}{C_1} + \frac{Q}{C_2}, \tag{4-10}$$

or

$$\frac{1}{C} = \frac{1}{C_1} + \frac{1}{C_2}, \tag{4-11}$$

$$C = \frac{C_1 C_2}{C_1 + C_2}. \tag{4-12}$$

If we have three capacitors in series,

$$\frac{1}{C} = \frac{1}{C_1} + \frac{1}{C_2} + \frac{1}{C_3}, \tag{4-13}$$

$$C = \frac{C_1 C_2 C_3}{C_2 C_3 + C_3 C_1 + C_1 C_2}. \tag{4-14}$$

Now the inverse of the capacitance, $1/C$, is called the *elastance*. So, if capacitors are connected in series, the total elastance is the sum of the individual elastances.

Note that, when capacitors are connected in parallel, the voltages are the same. When capacitors are connected in series, the charges are the same.

4.3 POTENTIAL ENERGY OF A CHARGE DISTRIBUTION

We can find a general expression for the potential energy of a charge distribution by considering the energy expended in charging a pair of conductors, as in Sec. 4.2.

When we add a small charge dQ to the left-hand conductor, we must expend an energy

$$dW = V\,dQ = \frac{Q}{C}\,dQ, \tag{4-15}$$

where Q is the charge already on the conductor. Then, to charge the conductor to a potential $V = Q/C$, we must supply an energy

$$W = \int_0^Q \frac{Q}{C}\,dQ = \frac{Q^2}{2C} = \frac{1}{2}QV = \frac{1}{2}CV^2. \tag{4-16}$$

The reason for the factor of one-half should be clear from the reasoning that we have used to arrive at it. It is that, on the average, the potential V at the time of arrival of a charge dQ is just one-half the final potential.

More generally, if we have a number of charged conducting bodies, the stored energy is the sum of similar terms for each one:

$$W = \frac{1}{2}\sum_i Q_i V_i. \tag{4-17}$$

For a continuous charge distribution of density $\rho(x',y',z')$ occupying a volume τ', we can replace Q_i by $\rho\,d\tau'$ and the summation by an integration:

$$W = \frac{1}{2}\int_{\tau'} V\rho\,d\tau', \tag{4-18}$$

where V and ρ are both functions of the coordinates in the general case.

4.4 ENERGY DENSITY IN AN ELECTRIC FIELD

The energy stored in a parallel-plate capacitor is therefore

$$\frac{1}{2}CV^2 = \frac{1}{2}\epsilon_0 \frac{A}{s} V^2 = \frac{1}{2}\epsilon_0 \left(\frac{V}{s}\right)^2 As, \tag{4-19}$$

$$= \left(\frac{1}{2}\epsilon_0 E^2\right) As, \tag{4-20}$$

where As is the volume occupied by the field. The stored energy may therefore be calculated by associating with each point an energy density $(\epsilon_0/2)E^2$ joules per cubic meter.

This is a general result. The energy associated with a charge distribution, that is, the energy required to assemble it, starting with the charge spread over an infinite volume, is

$$W = \frac{\epsilon_0}{2}\int_\tau E^2\, d\tau, \tag{4-21}$$

where the volume τ extends to all the region where the field under consideration differs from zero.

The energy stored in an electric field can therefore be expressed, either as in Eq. 4-18, or as in Eq. 4-21. Note that the terms under the integral signs in these two equations are unrelated. For example, at a point where $\rho = 0$, $V\rho$ is zero but E^2 is usually not zero.

4.4.1 *EXAMPLE: ISOLATED SPHERICAL CONDUCTOR*

For the *isolated spherical conductor* of radius R and carrying a charge Q,

$$W = \frac{\epsilon_0}{2}\int_R^\infty E^2\, d\tau. \tag{4-22}$$

Using a spherical shell of radius r and thickness dr as element of volume $d\tau$,

$$W = \frac{\epsilon_0}{2}\int_R^\infty \left(\frac{Q}{4\pi\epsilon_0 r^2}\right)^2 4\pi r^2\, dr, \tag{4-23}$$

$$= \frac{Q^2}{8\pi\epsilon_0 R}. \tag{4-24}$$

This energy is $\frac{1}{2}QV$, or $\frac{1}{2}CV^2$, where C is the capacitance of the sphere, as in Sec. 4.1.1.

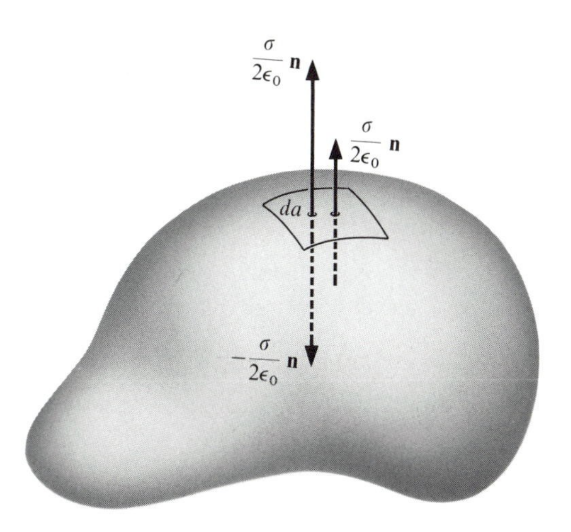

Figure 4-5 The local electric charge density at the surface of a conductor gives rise to oppositely directed electric field intensities $\sigma/2\epsilon_0$ as shown by the two arrows on the left; the other charges on the conductor give rise to the field $\sigma/2\epsilon_0$ shown by the arrow on the right. The net result is σ/ϵ_0 outside, and zero inside. The vector $\mathbf{n}$ is a unit vector normal to the conductor surface, and it points outward.

4.5 *FORCES ON CONDUCTORS*

An element of charge $\sigma\, da$ on the surface of a conductor experiences the electric field of all the other charges in the system and is therefore subject to an electric force. In a static field, this force must be perpendicular to the surface of the conductor, for otherwise the charges would move along the surface. Since the charge $\sigma\, da$ is bound to the conductor by internal forces, the force acting on $\sigma\, da$ is transmitted to the conductor itself.

To calculate the magnitude of this force, we consider a conductor with a surface charge density σ and a field $\mathbf{E}$ at the surface. From Gauss's law, the electric field intensity just outside the conductor is σ/ϵ_0. This field is perpendicular to the surface. Now the force on the element of charge $\sigma\, da$ is *not* $E\sigma\, da$, since the field that acts on $\sigma\, da$ is the field due only to the *other* charges in the system.

Let us first calculate the electric field intensity produced by $\sigma\, da$ itself. We can do this by using Gauss's law. The total flux emerging from $\sigma\, da$ must be $\sigma\, da/\epsilon_0$, half of it inward and half outward, as in Fig. 4-5. Then the electric field intensity due to the *local* surface charge density must be $\sigma/2\epsilon_0$ inward and $\sigma/2\epsilon_0$ outward.

The element of charge $\sigma\, da$ itself therefore produces exactly half the total field at a point outside, arbitrarily close to the surface. This is reasonable, for the nearby charge is more effective than the rest.

Now if $\sigma\, da$ produces half the field, then all the other charges must produce the other half, and the electric field intensity acting on $\sigma\, da$ must be $\sigma/2\epsilon_0$.

The force dF on the element of area da of the conductor is therefore given by the product of its charge $\sigma\, da$ multiplied by the field of all the other charges:

$$dF = \frac{\sigma}{2\epsilon_0}\,\sigma\, da, \tag{4-25}$$

and the force per unit area is

$$\frac{dF}{da} = \frac{\sigma^2}{2\epsilon_0} = \frac{1}{2}\,\epsilon_0 E^2. \tag{4-26}$$

Note that this force is just equal to the energy density of Sec. 4.4 at the surface of the conductor. We can show that this is correct by using the *method of virtual work*. In this method we assume a small change in the system and apply the principle of conservation of energy. Let us imagine that the conducting bodies in the field are disconnected from their power supplies. Then, if the electric forces are allowed to perform mechanical work, they must do so at the expense of the electric energy stored in the field. Imagine that a small area a of a conductor is allowed to be pulled into the field by a small distance x. The mechanical work performed is ax times the the force per unit area. It is also equal to the energy lost by the field, which is ax times the energy density. Thus the force per unit area is equal to the energy density.

If we use the same type of argument for a gas, we find that its energy density is equal to its pressure.

Note that the electric force on a conductor always tends to pull the conductor into the field. In other words, an electric field exerts a negative pressure on a conductor.

In order to visualize electric forces, it is useful to imagine that electric lines of force are real and under *tension*—also, that they arrange themselves in space as if they repelled each other.

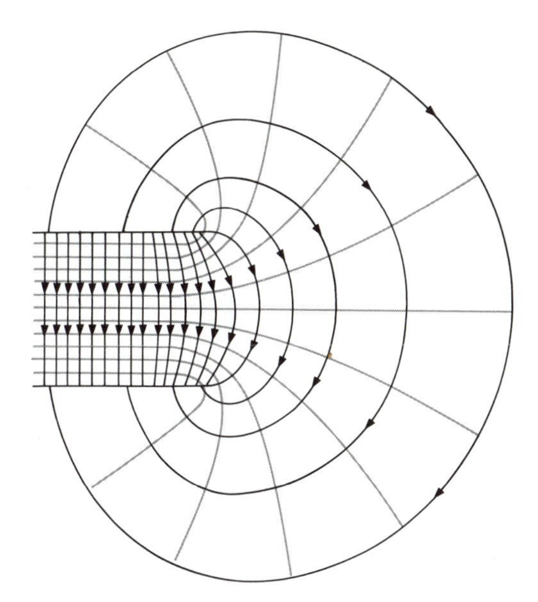

Figure 4-6 Lines of force and equipotentials near the edge of a parallel-plate capacitor.

For example, the lines of force between the plates of a capacitor tend to pull the plates together and bulge out at the edges of the plates, as in Fig. 4-6.

4.5.1 *EXAMPLE: ELECTRIC FORCE FOR* $E = 3 \times 10^6$ *VOLTS PER METER*

The maximum electric field intensity that can be sustained in air at normal temperature and pressure is about 3×10^6 volts per meter. The force is then 40 newtons per square meter.

4.5.2 *EXAMPLE: PARALLEL-PLATE CAPACITOR*

Let us calculate the forces on the plates of a *parallel-plate capacitor* of area S carrying a charge density $+\sigma$ on one plate and $-\sigma$ on the other, as in Fig. 4-7. We assume that the distance between the plates is small compared to their linear extent, in order to make edge effects negligible.

The force per unit area on the plates is equal to the energy density in the field $\epsilon_0 E^2/2$, and the force of attraction between the plates is therefore $\epsilon_0 E^2 S/2$, or $(\sigma^2/2\epsilon_0)S$.

We shall calculate this force in two other ways in Probs. 4-13 and 4-14.

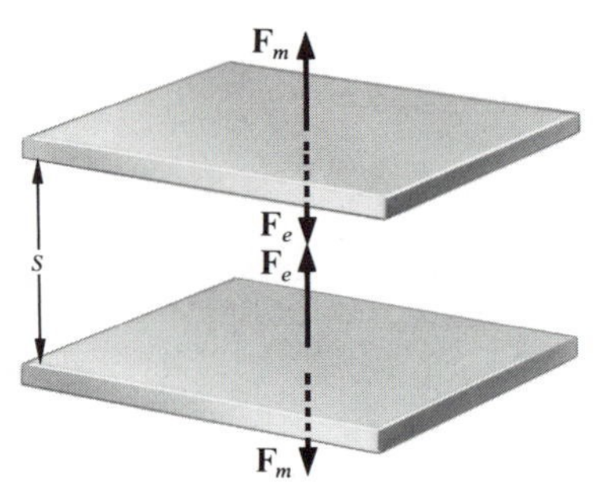

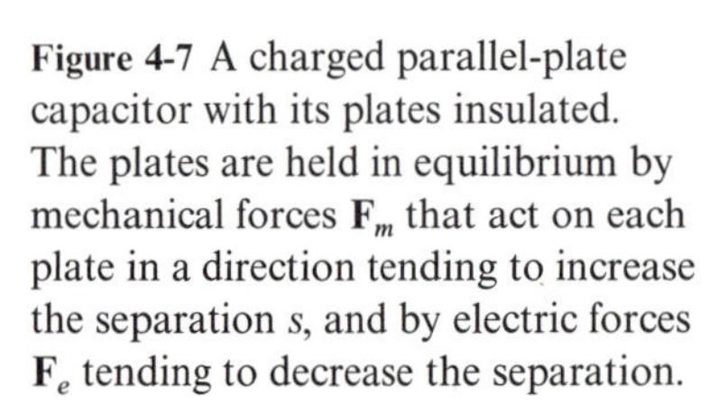

Figure 4-7 A charged parallel-plate capacitor with its plates insulated. The plates are held in equilibrium by mechanical forces $\mathbf{F}_m$ that act on each plate in a direction tending to increase the separation s, and by electric forces $\mathbf{F}_e$ tending to decrease the separation.

4.6 *SUMMARY*

If Q is the net charge carried by an isolated conductor, and if V is its potential, the ratio Q/V is called its *capacitance.*

The *capacitance between two isolated conductors* is Q/V, where Q is the charge transferred from one to the other, and V is the resulting potential difference.

The capacitance of two or more capacitors connected in *parallel* is the sum of the capacitances.

The inverse of a capacitance is an *elastance.*

The elastance of two or more capacitors connected in *series* is the sum of the elastances.

The *potential energy* associated with a charge distribution can be written either as

$$W = \frac{1}{2}\int_{\tau'} V\rho \, d\tau' \tag{4-18}$$

or as

$$W = \frac{\epsilon_0}{2}\int_{\tau} E^2 \, d\tau. \tag{4-21}$$

In the first integral, τ' must be chosen to include all the charge distribution, and in the second, must include all regions of space where E is nonvanishing. The assignment of an *energy density* $\epsilon_0 E^2/2$ to every point in space leads to the correct potential energy for the whole charge distribution.

Finally, the *force per unit area* on a charged conductor is equal to $\sigma^2/2\epsilon_0$, or to the energy density in the electric field, $\epsilon_0 E^2/2$.

Figure 4-8 Parallel-plate capacitor with seven plates. See Prob. 4-3.

PROBLEMS

4-1E Show that ϵ_0 is expressed in farads per meter.

4-2E *THE EARTH'S ELECTRIC FIELD*

a) The earth has a radius of 6.4×10^6 meters. What is its capacitance?

b) The earth carries a negative charge that gives a field of about 100 volts per meter at the surface.

Calculate the total charge.

c) Calculate the potential at the surface of the earth.

4-3E *PARALLEL-PLATE CAPACITOR*

Show that the capacitance of a parallel-plate capacitor having N plates, as in Fig. 4-8, is

$$C = 8.85(N - 1)A/t \text{ picofarads},$$

where N is the number of plates, A is the area of one side of one plate, and t is the distance between the plates. One picofarad is 10^{-12} farad.

4-4E *PARALLEL-PLATE CAPACITOR*

Can you suggest reasonable dimensions for a 1 picofarad (10^{-12} farad) air-insulated parallel-plate capacitor that is to operate at a potential difference of 500 volts?

4-5 *PARALLEL-PLATE CAPACITOR*

A parallel-plate capacitor has plates of area S separated by a distance s.

How is the capacitance affected by the introduction of an insulated sheet of metal of thickness s', parallel to the plates?

4-6 *CYLINDRICAL CAPACITOR*

Calculate the capacitance per unit length C' between two coaxial cylinders of radii R_1 and R_2 as in Fig. 4-9.

Calculate C' when $R_2 = 2R_1$.

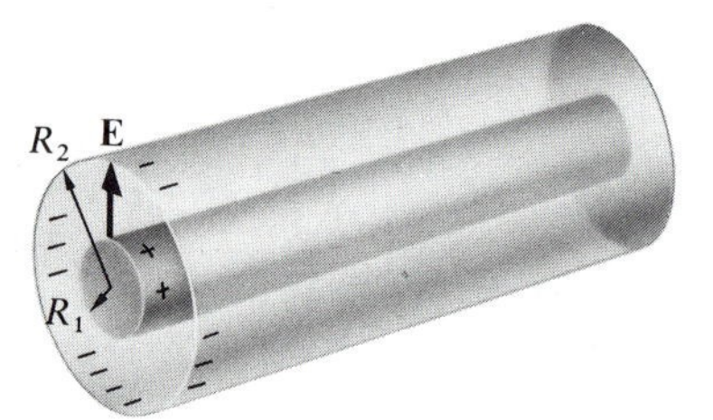

Figure 4-9 Cylindrical capacitor. See Prob. 4-6.

Solution: Let us assume that the inner conductor is positive and that it carries a charge of λ coulombs per meter of length. Then the capacitance per unit length is λ divided by the voltage V of the inner conductor with respect to the outer one, where

$$V = \int_{R_1}^{R_2} E\,dr = \int_{R_1}^{R_2} \frac{\lambda}{2\pi\epsilon_0 r}\,dr, \tag{1}$$

$$= \frac{\lambda}{2\pi\epsilon_0} \ln \frac{R_2}{R_1}. \tag{2}$$

Thus

$$C' = \frac{2\pi\epsilon_0}{\ln (R_2/R_1)}. \tag{3}$$

For $R_2 = 2R_1$,

$$C' = 2\pi\epsilon_0/\ln 2 = 8.026 \times 10^{-11} \text{ farad/meter}. \tag{4}$$

4-7 DROPLET GENERATOR

There are a number of devices that utilize charged droplets of liquid. Electrostatic spray guns (Prob. 2-8), colloid thrusters for spacecraft (Prob. 2-15), and ink-jet printers (Prob. 4-17) are examples. As we shall see, drop size is controlled by the specific charge (coulombs per kilogram) carried by the fluid.

a) When a charged droplet splits into two parts, electrostatic repulsion drives the droplets apart and the electrostatic energy decreases.

Calculate the loss of electrostatic energy when a droplet of radius R carrying a charge Q splits into two equal-sized droplets of charge $Q/2$ and radius R'.

Assume that the droplets are repelled to a distance that is large compared to R'. Neglect evaporation.

Solution: Since

$$2 \times \frac{4}{3}\pi R'^3 = \frac{4}{3}\pi R^3, \tag{1}$$

then

$$R' = R/2^{1/3}. \tag{2}$$

Initially, the charge Q is at the potential of the surface of the drop of radius R and the electrostatic energy is

$$\frac{QV}{2} = \frac{1}{2}\frac{Q^2}{4\pi\epsilon_0 R} = \frac{Q^2}{8\pi\epsilon_0 R}. \tag{3}$$

After splitting, the electrostatic energy is

$$2\frac{(Q/2)^2 2^{1/3}}{8\pi\epsilon_0 R} = \frac{1}{2^{2/3}}\frac{Q^2}{8\pi\epsilon_0 R}, \tag{4}$$

and the electrostatic energy gained is

$$\Delta W = \frac{Q^2}{8\pi\epsilon_0 R}\left(\frac{1}{2^{2/3}} - 1\right) = -0.3700\frac{Q^2}{8\pi\epsilon_0 R}. \tag{5}$$

b) The energy associated with surface tension is equal to a constant that is a characteristic of the liquid, multiplied by the surface area. The constant is called the *surface tension* of the liquid. The surface tension of water is 7.275×10^{-2} joule per square meter.

When a droplet splits, the surface area increases, and the energy associated with the surface tension increases proportionately.

Calculate the gain in this energy when a droplet of water of radius R splits into two equal-sized droplets.

Solution: The increase in surface energy is

$$\Delta W = (2 \times 4\pi R'^2 - 4\pi R^2)T = \left(\frac{2}{2^{2/3}} - 1\right)4\pi R^2 T, \tag{6}$$

$$= 0.2599 \times 4\pi R^2 T = 0.2376R^2, \tag{7}$$

where T is the surface tension of water.

c) A droplet of water has a radius of one micrometer and carries a specific charge of one coulomb per kilogram.

Will the droplet split?

Solution: The droplet will split if the total ΔW is negative. The charge Q is numerically equal to the mass, in this case:

$$Q = 1000 \times \frac{4}{3}\pi R^3 \text{ coulombs.} \tag{8}$$

Thus

$$\Delta W = -0.3700 \frac{[(4000\pi/3)R^3]^2}{8\pi\epsilon_0 R} + 0.2376R^2, \tag{9}$$

$$= -2.919 \times 10^{16} R^5 + 0.2376R^2. \tag{10}$$

For $R = 10^{-6}$,

$$\Delta W = -2.919 \times 10^{-14} + 0.2376 \times 10^{-12} \text{ joule.} \tag{11}$$

The droplet will not split.

d) What is the radius of the largest droplet of water that is stable with this specific charge?

Solution: For a specific charge of one coulomb per kilogram, a droplet of water will not split spontaneously if its radius is such that

$$-2.919 \times 10^{16} R^5 + 0.2376R^2 > 0,$$

or for radii smaller than

$$\left(\frac{0.2376}{2.919 \times 10^{16}}\right)^{1/3} = 2.012 \times 10^{-6} \text{ meter.} \tag{12}$$

4-8E *ELECTROSTATIC ENERGY*

a) Two capacitors are connected in series as in Fig. 4-10a. Calculate the ratio of the stored energies.

b) Calculate this ratio for capacitors connected in parallel as in Fig. 4-10b.

4-9E *ELECTROSTATIC ENERGY*

Two capacitors of capacitances C_1 and C_2 have charges Q_1 and Q_2, respectively.

a) Calculate the amount of energy dissipated when they are connected in parallel.

b) How is this energy dissipated?

4-10 *PROTON BOMB*

a) Calculate the electric potential energy of a sphere of radius R carrying a total charge Q uniformly distributed throughout its volume.

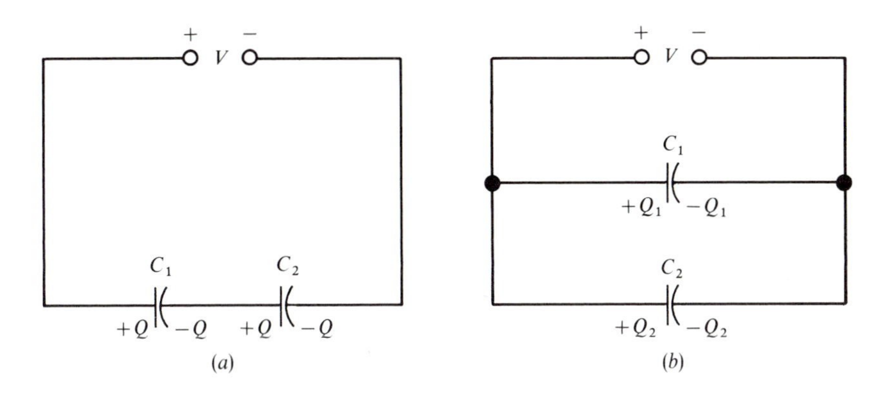

Figure 4-10 Two capacitors: (*a*) connected in series, and (*b*) connected in parallel. See Prob. 4-8.

b) Now calculate the gravitational potential energy of a sphere of radius R' of total mass M.

c) The moon has a mass of 7.33×10^{22} kilograms and a radius of 1.74×10^{6} meters. Calculate its gravitational potential energy.

d) Imagine that you can assemble a sphere of protons with a density equal to that of water. What would be the radius of this sphere if its electric potential energy were equal to the gravitational potential energy of the moon?

4-11 *ELECTROSTATIC MOTOR*

Can you suggest a rough design for an electrostatic motor?

Draw a sketch, explain its operation, specify voltages and currents, and make a rough estimate of what its power would be.

4-12 *ELECTROSTATIC PRESSURE*

A light spherical balloon is made of conducting material. It is suggested that it could be kept spherical simply by connecting it to a high-voltage source. The balloon has a diameter of 100 millimeters, and the maximum breakdown voltage in air is 3×10^{6} volts per meter.

a) What must be the voltage of the source if the electric force is to be as large as possible?

b) What gas pressure inside the balloon would produce the same effect?

4-13 *PARALLEL-PLATE CAPACITOR*

Calculate the force of attraction F between the plates of a parallel-plate capacitor. Assume that the capacitor is connected to a battery supplying a constant voltage V.

Use the method of virtual work, assuming a small increase ds in the spacing s between the plates, and set

$$dW_B + dW_m = dW_e,$$

where dW_B is the work done *by* the battery, $dW_m = Fds$ is the mechanical work done *on* the system, and dW_e is the *in*crease in the energy stored in the electric field.

You should find that

$$F = \frac{1}{2}\epsilon_0 \frac{V^2}{s^2} S = \frac{1}{2}\epsilon_0 E^2 S.$$

Note that one half of the energy supplied by the battery appears as mechanical work, and one half as an increase in the energy stored in the electric field. This is a general rule.

4-14 *PARALLEL-PLATE CAPACITOR*

Calculate the force of attraction between the plates of a parallel-plate capacitor, as in the preceding problem, assuming now that the capacitor is charged and disconnected from the battery.

In this case,

$$dW_m = dW_e.$$

You should find the same result.

4-15 *OSCILLATING PARALLEL-PLATE CAPACITOR*

Figure 4-11 shows a parallel-plate capacitor whose upper plate of mass m is supported by a spring insulated from ground. A voltage V is applied to the upper plate. There is zero tension in the spring when the length of the air gap is x_0.

Set $m = 0.1$ kilogram, $x_0 = 0.01$ meter, $k = 150$ newtons per meter, $V = 100$ volts, and $A = 0.01$ square meter. Assume that the mass of the spring is negligible, and neglect edge effects.

a) Set the various potential energies of the complete system equal to zero when the tension in the spring is zero, at $x = x_0$. Show that the potential energy is then

$$W = mg(x - x_0) + \frac{1}{2}k(x - x_0)^2 - \frac{1}{2}\epsilon_0 AV^2\left(\frac{1}{x} - \frac{1}{x_0}\right).$$

Draw a curve of W as a function of x between $x = 10^{-6}$ and $x = 10^{-2}$ meter. Use a logarithmic scale for the x-axis.

b) Show that, at equilibrium,

$$mg + k(x - x_0) + \frac{1}{2}\epsilon_0 AV^2/x^2 = 0.$$

You can arrive at this result in two different ways.

There is stable equilibrium at $x_{eq} = 3.46 \times 10^{-3}$ meter and unstable equilibrium at $x = 2.93 \times 10^{-5}$ meter.

c) If the system is brought to the point of stable equilibrium, and then disturbed slightly, it oscillates. For small oscillations, the restoring force F is proportional to the

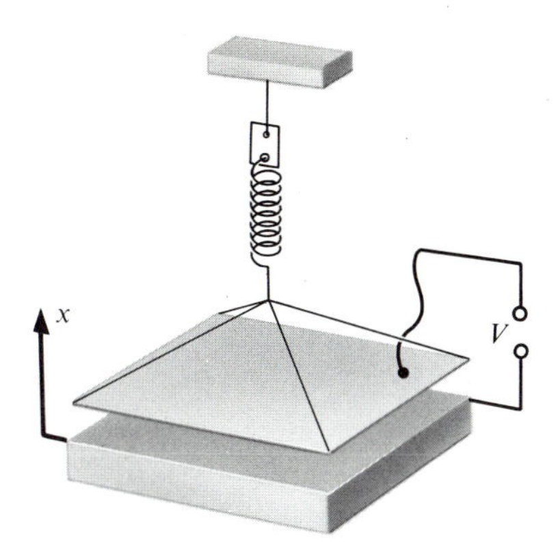

Figure 4-11 Parallel-plate capacitor, with the lower plate fixed in position and the upper plate suspended by a spring insulated from ground. See Prob. 4-15.

displacement and we have a simple harmonic motion:

$$F = m\left(\frac{d^2x}{dt^2}\right)_{eq} = -\left(\frac{dW}{dx}\right)_{eq} = -K(x - x_{eq}),$$

where the derivatives must be evaluated in the neighborhood of $x = x_{eq}$.

The circular frequency is $\omega = (K/m)^{1/2}$, where

$$K = \left(\frac{d^2W}{dx^2}\right)_{eq} = -\left(\frac{dF}{dx}\right)_{eq}.$$

Show that

$$\omega = \left[\frac{k - (\epsilon_0 AV^2/x_{eq}^3)}{m}\right]^{1/2}.$$

This gives a frequency of 6.16 hertz.

4-16 *HIGH-VOLTAGE GENERATOR*

Imagine the following simple-minded high-voltage generator. A parallel-plate capacitor has one fixed plate that is permanently connected to ground, and one plate that is movable. When the plates are close together at the distance s, the capacitor is charged by a battery to a voltage V. Then the movable plate is disconnected from the battery and moved out to a distance ns. The voltage on this plate then increases to nV, if we neglect edge effects. Once the voltage has been raised to nV, the plate is discharged through a load resistance.

a) Verify that there is conservation of energy.

b) Can you suggest a rough design for such a high-voltage generator with a more convenient geometry?

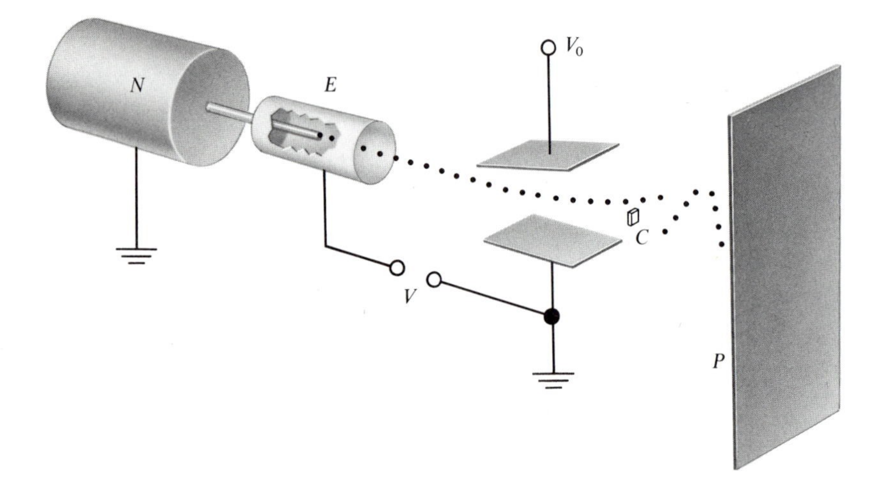

Figure 4-12 Ink-jet printer. A nozzle N forms a fine jet that breaks up into droplets inside electrode E. A variable voltage V applied to the electrode induces a known charge on each droplet in succession. The droplets are deflected in the constant electric field between the deflecting plates and impinge on the paper form P. Uncharged droplets are collected at C. See Prob. 4-17.

4-17 *INK-JET PRINTER*

The ink-jet printer is one solution to the problem of printing data at high speed. It operates more or less like an oscilloscope, except that it utilizes microscopic droplets of ink, instead of electrons. See Prob. 4-7.

Its principle of operation is shown in Fig. 4-12. A nozzle produces a fine jet of conducting ink that separates into droplets inside a cylindrical electrode. The potential on the electrode is either positive or zero. At a given instant, the charge carried by a droplet that breaks off from the jet depends on the charge induced on the jet and hence on V.

The droplets are then deflected as in an oscilloscope, except that here the deflecting field is constant, each droplet being deflected according to its charge. Uncharged droplets are collected in an ink sump. The jet operates continuously, and only a small fraction of the droplets is actually used.

The deflection is vertical and the jet moves horizontally.

The pressure in the nozzle is modulated at frequencies up to 0.5 megahertz, producing up to 5×10^5 droplets per second, with radii of the order of 10 micrometers.

a) Let the radius of the jet be R_1 and the inside radius of electrode E be R_2. What is the charge per unit length on the jet, neglecting end effects?

b) The jet breaks up into larger droplets of radius $2R_1$. What length of jet is required to form one droplet?

c) Calculate the charge Q per droplet.

d) Calculate the specific charge in coulombs per kilogram for $R_1 = 20$ micrometers, $R_2 = 5$ millimeters, $V = 100$ volts. Set the ink density equal to that of water (1000 kilograms per cubic meter).

e) Calculate the velocity of the droplets if they are produced at the rate of 10^5 per second and if they are separated by a distance of 100 micrometers.

f) Calculate the vertical deflection and the vertical velocity of a droplet upon leaving the deflecting plates, assuming a uniform transverse field of 10^5 volts per meter extending over a horizontal distance of 40 millimeters.

CHAPTER 5

DIRECT CURRENTS IN ELECTRIC CIRCUITS

Until now we have limited ourselves to the study of the fields of stationary charges. We shall now consider the flow of electric current, and we shall develop methods for calculating the currents in complex arrangements of conductors.

We shall not be concerned as yet with magnetic fields.

5.1 *CONDUCTION OF ELECTRICITY*

We have seen in Sec. 3.3 that, under static conditions, **E** is zero inside a conductor. However, if an electric field is maintained by an external source—for example, when one connects a length of copper wire between the terminals of a battery—then charge carriers drift in the field and there is an electric current.

Within the conductor, the *current density* vector **J** points in the direction of flow of positive charge carriers; for negative carriers, **J** points in the direction opposite to the flow, as in Fig. 5-1. The magnitude of **J** is the quantity of charge flowing through an infinitesimal surface perpendicular to the direction of flow, per unit area and per unit time. Current density is expressed in coulombs per square meter per second, or in amperes per square meter. In *good conductors*, such as copper, the charge carriers are the conduction electrons, and we have the situation shown in Fig. 5-1b. *Semiconductors* contain either or both of two types of mobile charges, namely conduction electrons and holes. A *hole* is a vacancy left by the release of an electron by an atom. A hole can migrate in the material as if it were a positive particle.

For ordinary conductors, the current density **J** is proportional to the electric field intensity **E**:

$$\mathbf{J} = \sigma \mathbf{E}, \tag{5-1}$$

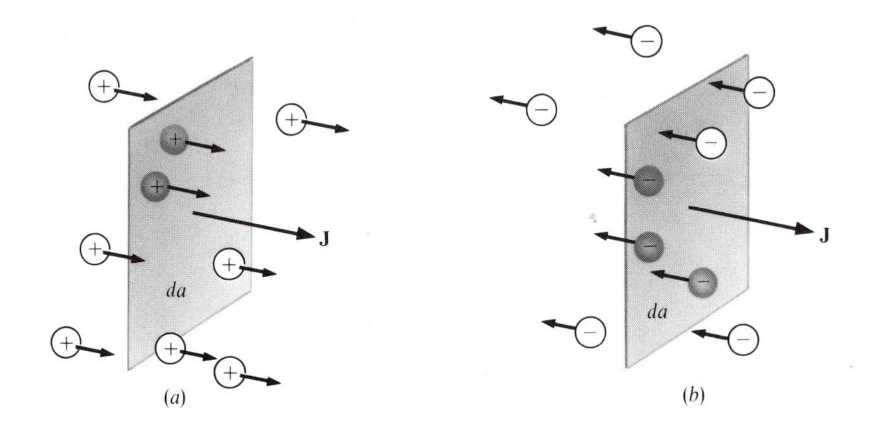

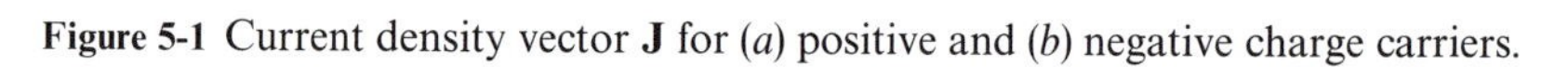
Figure 5-1 Current density vector **J** for (*a*) positive and (*b*) negative charge carriers.

Table 5-1

Conductor	Conductivity σ (siemens per meter)
Aluminum	3.54×10^7
Brass (65.8 Cu, 34.2 Zn)	1.59×10^7
Chromium	3.8×10^7
Copper	5.80×10^7
Gold	4.50×10^7
Graphite	1.0×10^5
Magnetic iron	1.0×10^7
Mumetal (75 Ni, 2 Cr, 5 Cu, 18 Fe)	0.16×10^7
Nickel	1.3×10^7
Sea water	≈ 5
Silver	6.15×10^7
Tin	0.870×10^7
Zinc	1.86×10^7

where σ is the *conductivity* of the material. Conductivity is expressed in amperes per volt per meter, or in *siemens*† per meter. Table 5-1 gives the conductivities of a number of common materials.

The drift velocity v of conduction electrons is easily calculated. It is surprisingly low. If n is the number of conduction electrons per cubic meter and if e is the magnitude of the electron charge, then

$$J = nev. \tag{5-2}$$

In copper, there is one conduction electron per atom, and 8.5×10^{28} atoms per cubic meter, so $n = 8.5 \times 10^{28}$ electrons per cubic meter. For a current of one ampere in a wire having a cross-section of one square millimeter, $J = 10^6$ amperes per square meter and

$$v = \frac{10^6}{8.5 \times 10^{28} \times 1.6 \times 10^{-19}} \approx 7.4 \times 10^{-5} \text{ meter/second,} \tag{5-3}$$

or about 0.26 meter per *hour*.‡

5.1.1 CONSERVATION OF CHARGE

Now consider a volume τ bounded by a surface S inside conducting material. The integral of $\mathbf{J} \cdot \mathbf{da}$ over the surface is the charge flowing out of S in one second, or the charge lost by the volume τ in one second. Then

$$\int_S \mathbf{J} \cdot \mathbf{da} = -\frac{d}{dt}\int_\tau \rho \, d\tau. \tag{5-4}$$

This is the *law of conservation of charge.*

Using the divergence theorem (Sec. 1.10) on the left-hand side,

$$\int_\tau \boldsymbol{\nabla} \cdot \mathbf{J} \, d\tau = -\int_\tau \frac{\partial \rho}{\partial t} \, d\tau. \tag{5-5}$$

† After Ernst Werner von Siemens (1816–1892). The word therefore takes a terminal *s* in the singular: one siemens. The siemens was formerly called a "mho."

‡ Then how is it that one can, say, turn on a light kilometers away, instantaneously? The moment one closes the switch, a guided electromagnetic wave is launched along the wire. This wave travels at the velocity of light and establishes the electric field in and around the wire.

Since this equation is valid for any volume,

$$\boldsymbol{\nabla} \cdot \mathbf{J} = -\frac{\partial \rho}{\partial t}. \tag{5-6}$$

This is again the law of conservation of charge, stated in differential form.

5.1.2 CHARGE DENSITY IN A HOMOGENEOUS CONDUCTOR

Under steady-state conditions, the time derivative in Eq. 5-6 is zero. Then

$$\boldsymbol{\nabla} \cdot \mathbf{J} = \boldsymbol{\nabla} \cdot \sigma \mathbf{E} = 0. \tag{5-7}$$

If now we have a homogeneous conductor with σ independent of the coordinates, $\boldsymbol{\nabla} \cdot \mathbf{E}$ is zero. Thus, from Eq. 3-11, the net charge density inside a *homogeneous* conductor carrying a current under steady-state conditions is zero.

5.2 OHM'S LAW

Figure 5-2 shows an element of conductor of length l and cross-section A. If a difference of potential V is maintained between its two ends, then

$$J = I/A = \sigma E = \sigma V/l, \tag{5-8}$$

and

$$I = \frac{\sigma V A}{l} = \frac{V}{l/\sigma A} = \frac{V}{R}, \tag{5-9}$$

where

$$R = l/\sigma A \tag{5-10}$$

is the *resistance* of the element. Resistances are expressed in volts per ampere or in *ohms*.

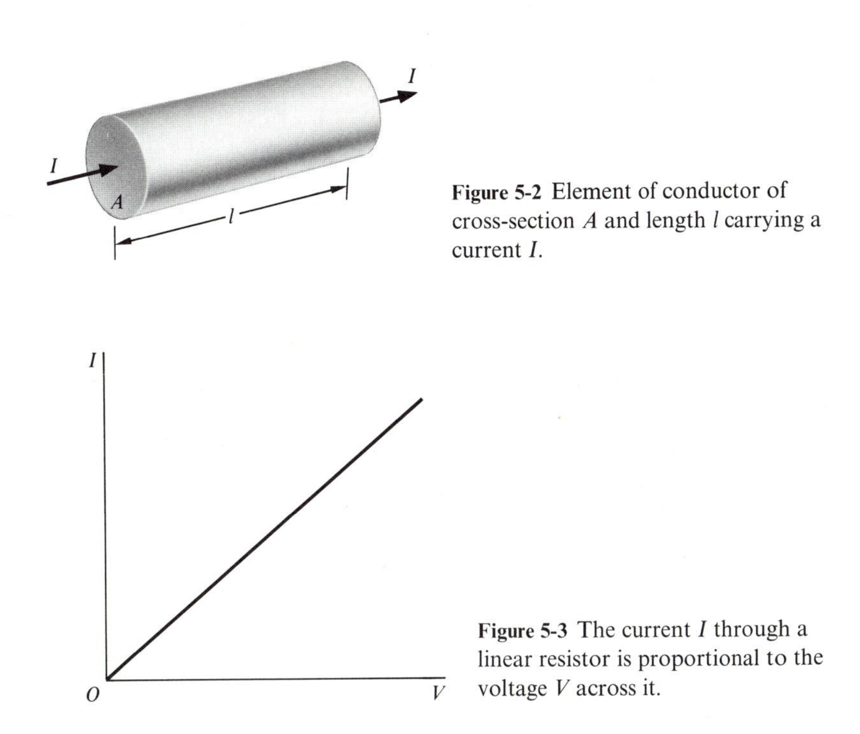

Figure 5-2 Element of conductor of cross-section A and length l carrying a current I.

Figure 5-3 The current I through a linear resistor is proportional to the voltage V across it.

Equation 5-9 shows that the current through a resistor is proportional to the voltage across it, as in Fig. 5-3. This is *Ohm's law*. A resistor satisfying this law is said to be *ohmic*, or *linear*. Equation 5-1 is another, more general form of Ohm's law.

The current and the voltage are measured as in Fig. 5-4. It is assumed that the current flowing through the voltmeter V is negligible compared to that through R. That is usually the case.

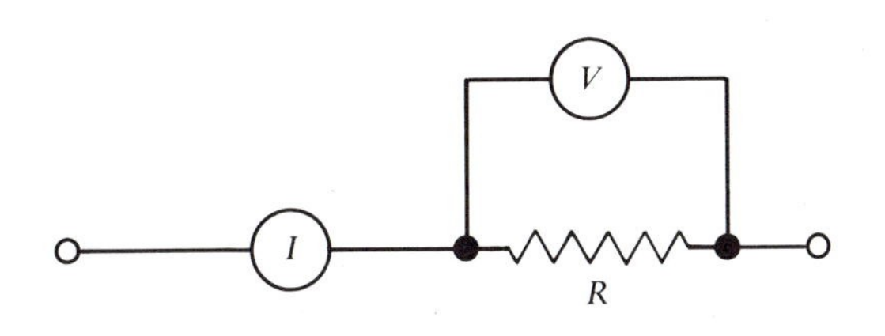

Figure 5-4 Measurement of the current I through a resistor R, and of the voltage V across it. *A circle marked I represents an ammeter, and a circle marked V, a voltmeter.*

5.2.1 THE JOULE EFFECT

When a current flows through a conductor, the charge carriers gain kinetic energy by moving in the electric field. Their velocity never becomes large, however, because they can travel only a short distance before colliding with an atom or a molecule. The kinetic energy gained by the carriers thus serves to raise the temperature of the conducting medium. This phenomenon is called the *Joule effect*.

The energy that is dissipated as heat in this way is easily calculated. Consider a cubic meter of conductor with **E** parallel to one edge. The current density J is the charge flowing through the cube in one second. The voltage difference across the cube is E. Then the kinetic energy gained by the carriers and lost to the conductor as heat in one second is JE, or σE^2, or J^2/σ, from Eq. 5-1. So the heat energy produced per cubic meter in one second is σE^2 or J^2/σ.

If one has a resistance R carrying a current I, then the voltage across it is IR and the kinetic energy gained by the charge carriers in one second is $I(IR)$. Then the power dissipated as heat is I^2R.

5.3 NON-LINEAR RESISTORS

Although Ohm's law applies to ordinary resistors, it is by no means general.

5.3.1 EXAMPLE: INCANDESCENT LAMPS

Figure 5-5 shows the current as a function of voltage for a 60-watt *incandescent lamp*. The resistance V/I is about 20 ohms at a few volts, and about 250 ohms at 100 volts. The reason is that, in the interval, the temperature of the tungsten filament has changed from 300 to 2400 kelvins.

If the temperature of the filament were maintained constant by external means, say by immersing it in a constant-temperature oil bath, then it would follow Ohm's law.

5.3.2 EXAMPLE: VOLTAGE-DEPENDENT RESISTORS

Voltage-dependent resistors utilize ceramic semiconductors for which $V = CI^B$, where C and B are constants, and B is less than unity. This equation applies at room temperature. Figure 5-6 shows the current-voltage curve for one particular type with $C = 100$ and $B = 0.2$.

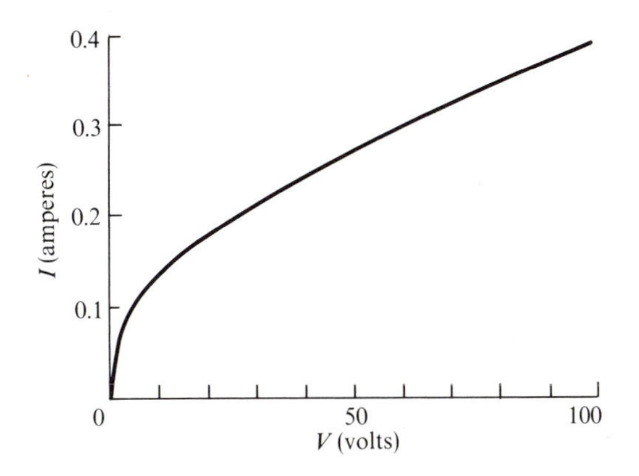

Figure 5-5 Current–voltage relationship for a 60-watt incandescent lamp.

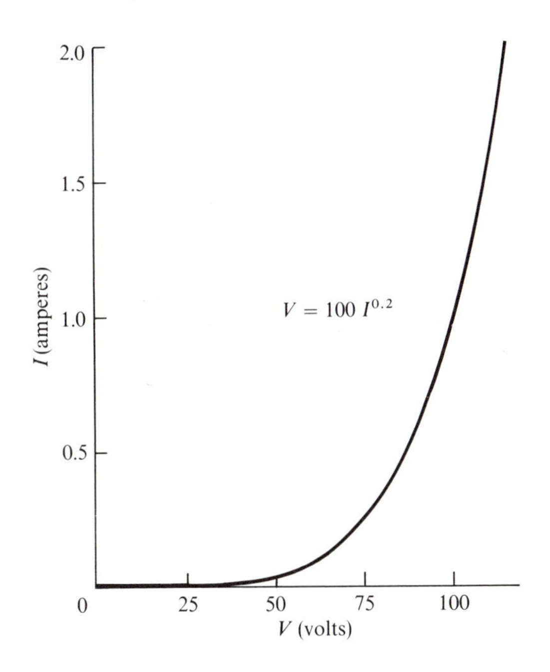

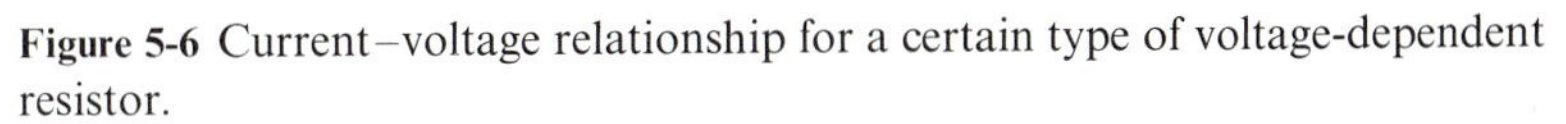

Figure 5-6 Current–voltage relationship for a certain type of voltage-dependent resistor.

Voltage-dependent resistors are used on telephone lines for short-circuiting lightning to ground, for suppressing sparks, and so on, since their resistance decreases at large voltages.

We shall be concerned henceforth only with linear resistors.

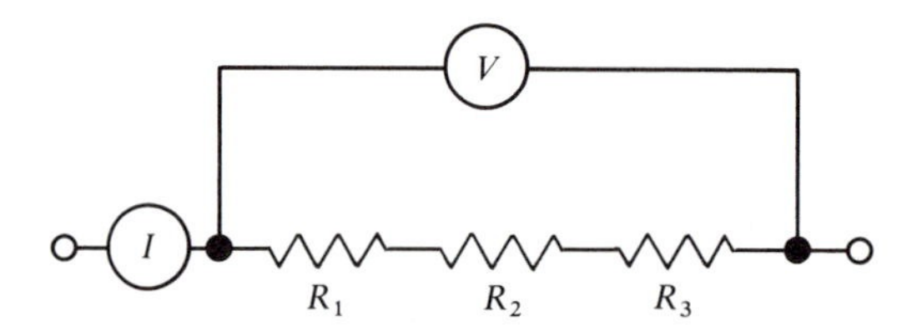

Figure 5-7 Three resistors connected in series.

5.4 RESISTORS CONNECTED IN SERIES

Figure 5-7 shows three resistors connected in *series*. Since

$$V = IR_1 + IR_2 + IR_3 = I(R_1 + R_2 + R_3) = IR_s, \qquad (5\text{-}11)$$

the resistors have a total resistance

$$R_s = R_1 + R_2 + R_3. \qquad (5\text{-}12)$$

The resistance of a set of resistors connected in series is the sum of the individual resistances.

5.5 RESISTORS CONNECTED IN PARALLEL

In Fig. 5-8 the resistors are connected in *parrallel*. Then

$$I = \frac{V}{R_1} + \frac{V}{R_2} + \frac{V}{R_3} = V\left(\frac{1}{R_1} + \frac{1}{R_2} + \frac{1}{R_3}\right) = \frac{V}{R_p}, \qquad (5\text{-}13)$$

and the resistance R_p is given by

$$\frac{1}{R_p} = \frac{1}{R_1} + \frac{1}{R_2} + \frac{1}{R_3}, \qquad (5\text{-}14)$$

$$R_p = \frac{R_1 R_2 R_3}{R_2 R_3 + R_3 R_1 + R_1 R_2}. \qquad (5\text{-}15)$$

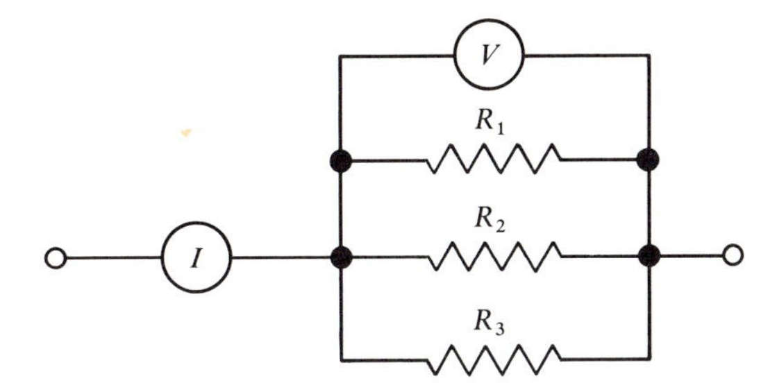

Figure 5-8 Three resistors connected in parallel.

It is useful to remember that, for two resistances in parallel,

$$R_p = \frac{R_1 R_2}{R_1 + R_2}. \tag{5-16}$$

It is often convenient to use *conductance*

$$G = 1/R, \tag{5-17}$$

instead of resistance. Then

$$G_p = G_1 + G_2 + G_3. \tag{5-18}$$

The conductance of a set of resistors connected in parallel is the sum of the individual conductances.

Conductances are expressed in amperes per volt or in siemens. We have already used the siemens in Sec. 5.1.

5.6 *THE PRINCIPLE OF SUPERPOSITION*

The principle of superposition which we studied in Sec. 2.3 applies to electric circuits comprising sources and *linear* resistors: if a circuit comprises two or more sources, each source acts independently of the others and, at any point in the circuit, the current is the algebraic sum of the currents produced by the individual sources.

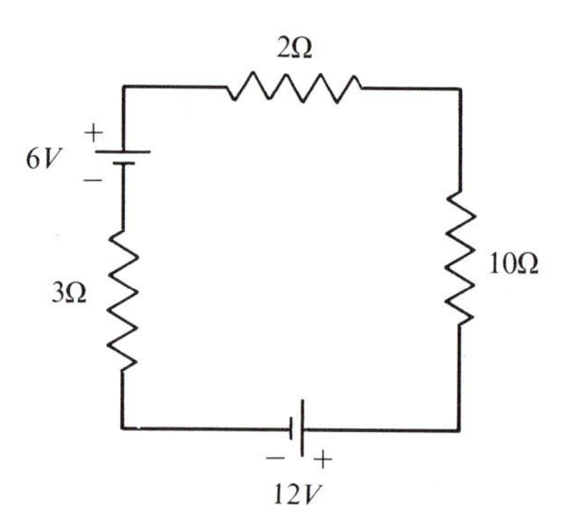

Figure 5-9 The current flowing through the circuit is the algebraic sum of the current due to the 6-volt battery, plus that due to the 12-volt battery.

5.6.1 *EXAMPLE: SIMPLE CIRCUIT WITH TWO SOURCES*

Figure 5-9 shows a simple circuit comprising three resistors and two sources, all connected in series. The total resistance is 15 ohms. The 12-volt source produces a counterclockwise current of $12/15 = 0.8$ ampere, while the 6-volt source produces a clockwise current of 0.4 ampere. The net current is 0.4 ampere, counterclockwise.

The power dissipated as heat in the 2-ohm resistor is $(0.4)^2 \times 2 = 0.32$ watt, and the total power dissipated in the circuit is $(0.4)^2 \times 15 = 2.4$ watts.

5.7 *KIRCHOFF'S LAWS*

Let us consider a general circuit composed of resistors and voltage sources, for example, that of Fig. 5-10.

Junctions such as *A, B, C, D, E* are called *nodes*; connections between nodes, such as *AB, BD, DC . . .*, are called *branches*; closed circuits, such as *ABDE*, or *ABCDE*, or *BCD*, are called *meshes*.

Kirchoff's laws provide a method for calculating the branch currents when the values of the resistances and of the voltages supplied by the sources are known.

Kirchoff's current law states that, at any given node, the sum of the incoming currents is equal to the sum of the outgoing currents. This is simply because charge does not accumulate at the nodes. For example, at *A*,

$$I_1 = I_6 + I_7. \tag{5-19}$$

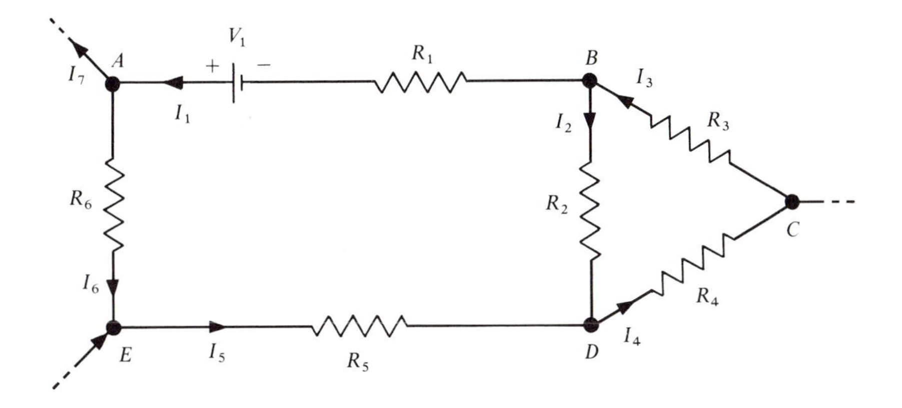

Figure 5-10 Part of a complex circuit.

According to *Kirchoff's voltage law*, the voltage rise, or the voltage drop, around any given mesh is zero. For example, the voltage rise from A to B, to D, to E, and back to A, is clearly zero, so

$$-V_1 + I_1R_1 - I_2R_2 + I_5R_5 + I_6R_6 = 0. \tag{5-20}$$

To find the branch currents, one writes down as many equations of the above two types as there are branches, and one solves the resulting system of simultaneous equations.

The directions chosen for the currents are arbitrary. If the calculations give a positive value for, say, I_1, then that current flows in the direction of the arrow. If the calculations give a negative value, then the current flows in the opposite direction.

5.7.1 *EXAMPLE: THE BRANCH CURRENTS IN A TWO-MESH CIRCUIT*

For the circuit of Fig. 5-11, we have three unknowns, I_1, I_2, I_3. Hence we need three equations. At node A,

$$I_1 = I_2 + I_3. \tag{5-21}$$

For the left-hand mesh, starting at B, clockwise,

$$V - I_1R - I_3R = 0. \tag{5-22}$$

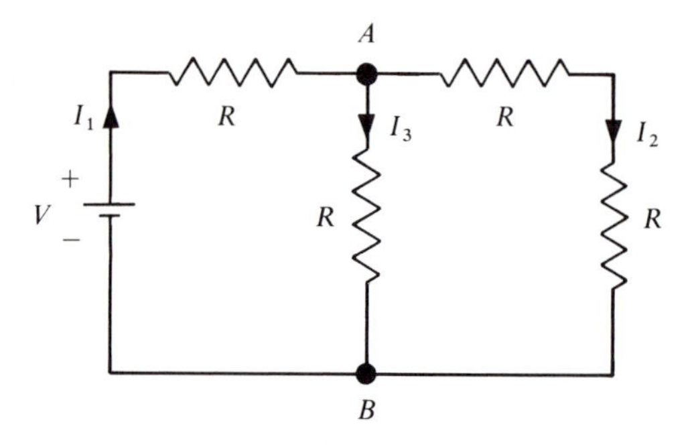

Figure 5-11 One can calculate I_1, I_2, I_3, given R and V.

Similarly, for the right-hand mesh, starting at B, clockwise,

$$I_3R - I_2R - I_2R = 0. \tag{5-23}$$

Solving, we find that

$$I_1 = 3V/5R, \qquad I_2 = V/5R, \qquad I_3 = 2V/5R. \tag{5-24}$$

5.8 *SUBSTITUTION THEOREM*

Let us return to Fig. 5-10 and to Eq. 5-20. Consider the resistance R_2. The voltage drop across it is I_2R_2. Let us assume that I_2 is really in the direction of the arrow. Then the upper end of R_2 is positive. Now let us substitute for R_2 a voltage source as in Fig. 5-12. Equation 5-20 is unaffected. Similarly, the corresponding equation for the mesh BCD is unaffected. Then the currents I_1, I_2, I_3, and so forth, are unaffected.

Thus, if a current I flows through a resistance R, one can replace the resistance by a voltage source IR of the same polarity without affecting any of the currents in the circuit.

Inversely, if we have a voltage source V passing a current I, *with the current entering the source at the positive terminal*, one can replace the source by a resistance V/I without affecting any of the currents in the circuit.

This is the *substitution theorem*.

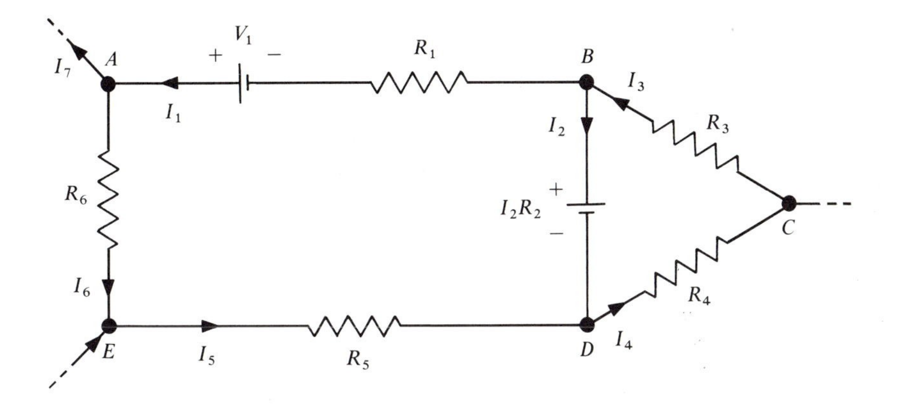

Figure 5-12 Resistor R_2 of Fig. 5-10 has been replaced by a battery.

5.8.1 *EXAMPLE: RESISTOR REPLACED BY A BATTERY*

In Fig. 5-13a,

$$I_1 = I_2 + I_3, \tag{5-25}$$

$$I_1R + I_2R = V, \tag{5-26}$$

$$-I_2R + I_3R = 0, \tag{5-27}$$

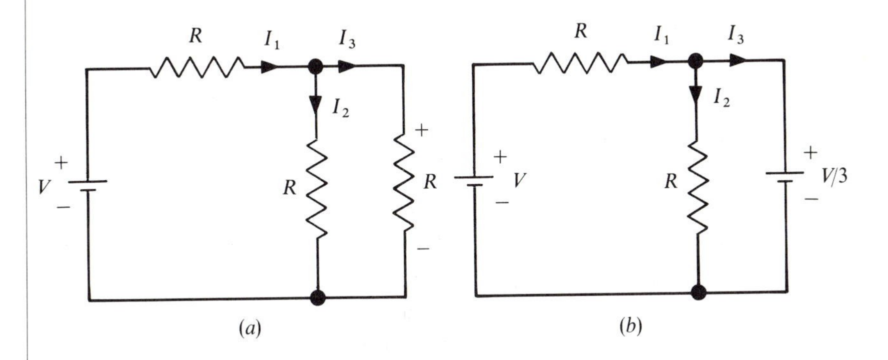

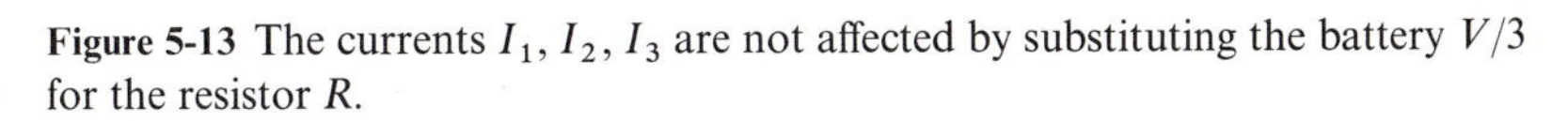
Figure 5-13 The currents I_1, I_2, I_3 are not affected by substituting the battery $V/3$ for the resistor R.

giving

$$I_1 = 2V/3R, \qquad I_2 = V/3R, \qquad I_3 = V/3R. \tag{5-28}$$

Using now the substitution theorem, we can replace the resistor on the right by a source $I_3R = V/3$, giving the circuit of Fig. 5-13b. Then

$$I_1 = I_2 + I_3, \tag{5-29}$$

$$I_1R + I_2R = V, \tag{5-30}$$

$$-I_2R + (V/3) = 0. \tag{5-31}$$

Solving, we find the same values for I_1, I_2, I_3 as before.

Inversely, one can substitute the circuit of Fig. 5-13a for that of Fig. 5-13b without affecting I_1, I_2, I_3.

5.8.2 *EXAMPLE: BATTERY REPLACED BY A RESISTOR*

In Sec. 5.6.1 we found that the current in Fig. 5-9 is counterclockwise. Since a current of 0.4 ampere flows into the positive terminal of the 6-volt source, we can replace this source by a resistance of $6/0.4 = 15$ ohms without affecting the current flowing in the circuit. Let us check. We now have a 12-volt source and a total resistance of $15 + 10 + 3 + 2 = 30$ ohms. Then the current is $12/30 = 0.4$ ampere, as previously.

5.9 *MESH CURRENTS*

If we use Kirchoff's laws as such, the number of unknown currents is equal to the number of branches. It is possible to reduce the number of unknowns in the following way. Instead of using branch currents as in Fig. 5-10, one uses *mesh currents* as in Fig. 5-14, since there are fewer mesh currents than branch currents. Kirchoff's current law is then automatically satisfied, and we need to apply only the voltage law.

The advantage of using mesh currents comes from the fact that the time required to solve a set of simultaneous equations decreases rapidly as the number of equations is reduced.

In the end, one must find the branch currents, but that is a simple matter, once the mesh currents are known.

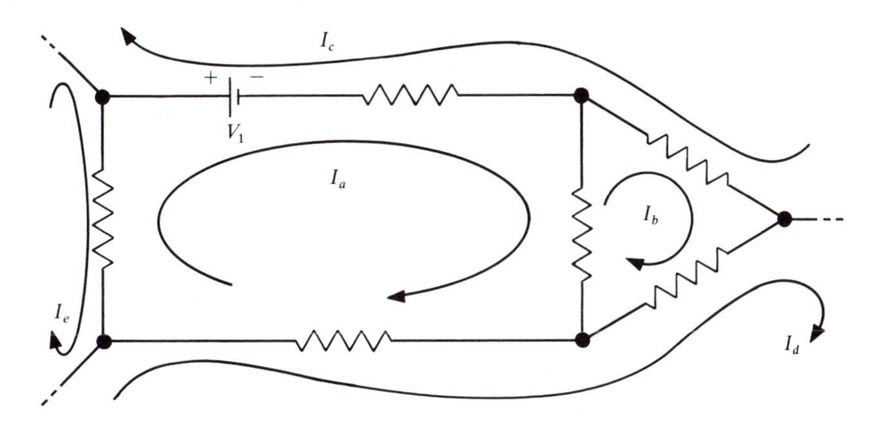

Figure 5-14 Mesh currents in the circuit of Fig. 5-10.

5.9.1 *EXAMPLE: THE MESH CURRENTS IN A TWO-MESH CIRCUIT*

In the circuit of Fig. 5-15 we have two meshes. Starting at B in both cases and proceeding in the clockwise direction,

$$V - I_a R - (I_a - I_b)R = 0, \tag{5-32}$$

$$(I_a - I_b)R - I_b 2R = 0, \tag{5-33}$$

and

$$I_a = 3V/5R, \qquad I_b = V/5R. \tag{5-34}$$

Note that we have only two unknowns, I_a and I_b, instead of the three we had previously.

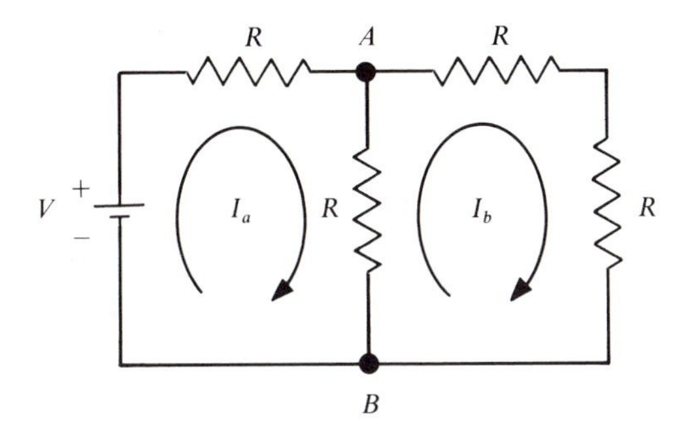

Figure 5-15 Mesh currents in the circuit of Fig. 5-11.

5.10 DELTA-STAR TRANSFORMATIONS

Figure 15-6 shows three nodes *A, B, C*, *delta-connected* in part *a* and *star-connected* in part *b*. The nodes *A, B, C* form part of a larger circuit, and can be connected together either as in Fig. 5-16a or as in Fig. 5-16b. We shall see that, if certain relations are satisfied between the three resistances on the left and the three on the right, one circuit may be substituted for the other without disturbing the mesh currents. In fact, if one had two boxes, one containing a delta circuit, and the other containing a star circuit satisfying Eqs. 5-38 to 5-40, with only the terminals *A, B, C* showing on each box, there would be no way of telling which box contained the delta and which contained the star. The two circuits are thus said to be *equivalent*.

This transformation is often used for simplifying the calculation of mesh currents.

We could find R_A, R_B, R_C in terms of R_a, R_b, R_c, and inversely, by assuming the same mesh currents I_A, I_B, I_C in the two circuits, as in the figure, and then making the voltages $V_A - V_B$, $V_B - V_C$, $V_C - V_A$ in Fig. 5-16a equal to the corresponding voltages in Fig. 5-16b. This is the object of Prob.

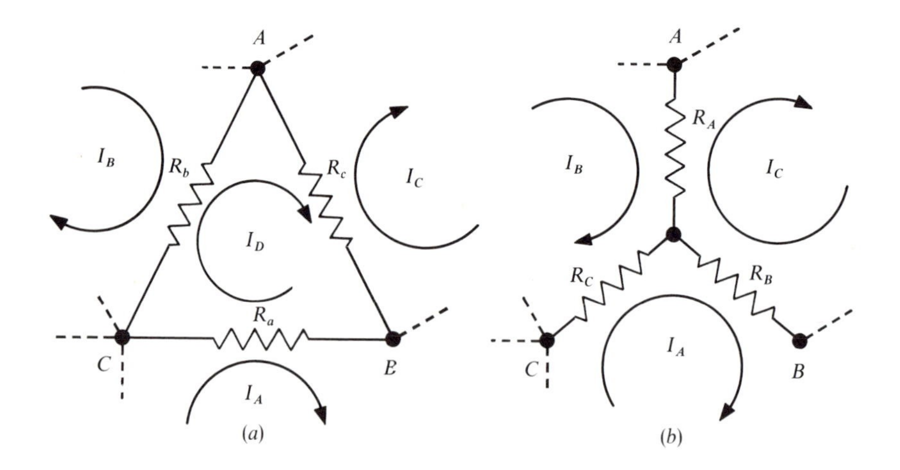

Figure 5-16 (*a*) Three-node circuit having the shape of a capital delta. (*b*) Four-node circuit having the shape of a star. Nodes *A, B, C* are part of a more extensive circuit. It is shown that, if either Eqs. 5-38 to 5-40 or Eqs. 5-44 to 5-46 are satisfied, the above two circuits are completely equivalent.

5-19. We shall use here another derivation that is somewhat less convincing, but shorter.

We disconnect the delta and the star from the rest of the circuit. If the circuits are equivalent, then the resistance between A and B must be the same in both:

$$R_{AB} = \frac{R_c(R_a + R_b)}{R_a + R_b + R_c} = R_A + R_B. \tag{5-35}$$

Similarly,

$$R_{BC} = \frac{R_a(R_b + R_c)}{R_a + R_b + R_c} = R_B + R_C, \tag{5-36}$$

$$R_{CA} = \frac{R_b(R_c + R_a)}{R_a + R_b + R_c} = R_C + R_A. \tag{5-37}$$

Note that, in the above equations, R_A is associated with R_bR_c, R_B with R_cR_a, and R_C with R_aR_b. Thus, simply by inspection, we have that

$$R_A = \frac{R_bR_c}{R_a + R_b + R_c}, \tag{5-38}$$

$$R_B = \frac{R_cR_a}{R_a + R_b + R_c}, \tag{5-39}$$

$$R_C = \frac{R_aR_b}{R_a + R_b + R_c}. \tag{5-40}$$

Now we also need the inverse relationship. We could deduce it from the above three equations, but that would be rather long, and not very instructive. Instead, let us use the conductance between node A and the nodes B and C short-circuited together. This gives

$$G_b + G_c = \frac{G_A(G_B + G_C)}{G_A + G_B + G_C}. \tag{5-41}$$

Similarly,

$$G_c + G_a = \frac{G_B(G_C + G_A)}{G_A + G_B + G_C}, \tag{5-42}$$

$$G_a + G_b = \frac{G_C(G_A + G_B)}{G_A + G_B + G_C}. \tag{5-43}$$

Again, by inspection,

$$G_a = \frac{G_B G_C}{G_A + G_B + G_C}, \tag{5-44}$$

$$G_b = \frac{G_C G_A}{G_A + G_B + G_C}, \tag{5-45}$$

$$G_c = \frac{G_A G_B}{G_A + G_B + G_C}. \tag{5-46}$$

5.10.1 *EXAMPLE: TRANSFORMATION OF A SYMMETRICAL DELTA AND OF A SYMMETRICAL STAR*

In Fig. 5-16a, suppose R_a, R_b, R_c are all 100-ohm resistances. Then, from Eqs. 5-38 to 5-40, the equivalent star-connected circuit is made up of three resistances of $(100 \times 100)/300 = 33$ ohms each.

If, instead, we had a star-connected circuit with resistances R_A, R_B, R_C of 100 ohms each, then the equivalent delta-connected circuit would have three identical resistances with conductances given by Eqs. 5-44 to 5-46: $10^{-4}/(3/100) = 10^{-2}/3$ siemens. The resistances in the delta would thus each have a resistance of 300 ohms.

5.11 VOLTAGE AND CURRENT SOURCES

The ideal voltage source supplies a constant voltage that is independent of the current drawn, and hence of the load resistance. A good electronically voltage-stabilized power supply is close to ideal, up to a specified current, beyond which the voltage falls as in Fig. 5-17.

Similarly, the ideal current source supplies a constant current that is independent of the voltage across the load. Again, current-stabilized

Figure 5-17 Output voltage as a function of output current, for a voltage source.

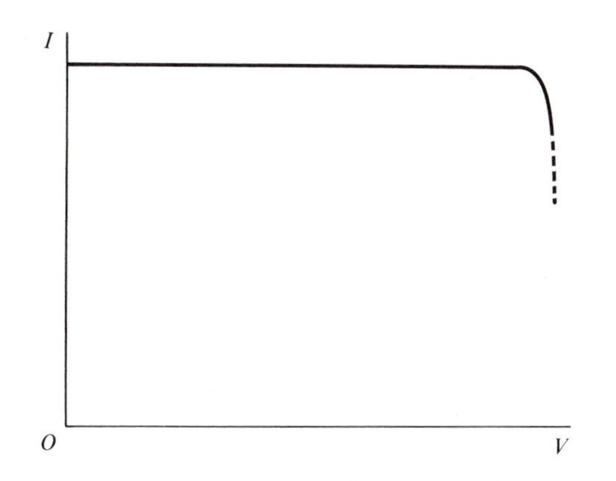

Figure 5-18 Output current as a function of output voltage, for a current source.

supplies are nearly ideal, but only up to a specified voltage beyond which the current decreases. This is shown in Fig. 5-18.

5.12 THÉVENIN'S THEOREM

It is found experimentally that the voltage supplied by a source decreases when the load current increases. This is true, even for voltage-stabilized sources: the plateau in Fig. 5-17 is not really flat but slopes down slightly.

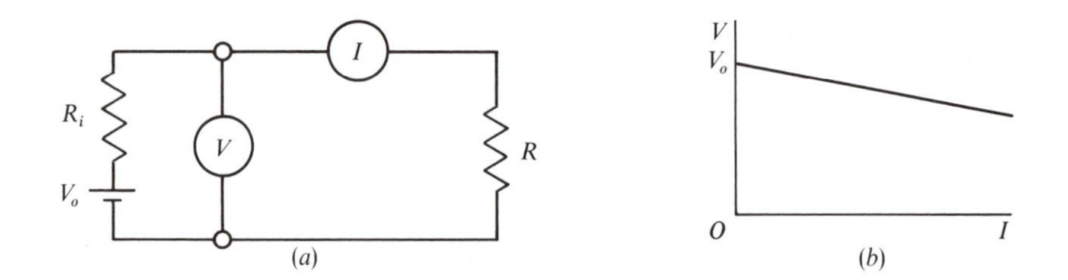

Figure 5-19 (*a*) Any real source is equivalent to an ideal source V_0 in series with a resistance R_i. The source is shown here connected to a load resistance R drawing a current I. The voltage V between the terminals is $V_0 - IR_i$, or IR. (*b*) The terminal voltage V as a function of the load current I. The slope is $-R_i$.

Thévenin's theorem states that any source—a flashlight battery, an oscillator, a voltage- or current-stabilized power supply, etc.—is equivalent to an ideal voltage source V_0 in series with a resistance R_i, as in Fig. 5-19, where R_i is called the *output resistance*, or the *source resistance*. An ideal voltage source, by definition, has a zero R_i.

In Fig. 5-19, the current I is $V_0/(R_i + R)$, while the voltage V is IR. The voltage V is smaller than V_0 by the factor $R/(R_i + R)$, because of the voltage drop on R_i. With R disconnected ($R \to \infty$), V is equal to V_0.

The internal resistance R_i is given by the slope of the curve of V as a function of I. If one increases I by decreasing R, then V decreases and R_i is equal to $|\Delta V/\Delta I|$. For one particular electronically *voltage*-stabilized power supply, the output voltage drops by one millivolt when the output current increases from zero to 100 milliamperes. Thus R_i is 0.01 ohm. A small battery charger has an internal resistance that is of the order of one ohm: if the load current increases by one ampere, the output voltage drops by about one volt.

A *current*-stabilized power supply has a large V_0 and a large R_i. Its output current is V_0/R_i, as long as $R \ll R_i$. Thus its output voltage $(V_0/R_i)R$ is proportional to the load resistance R. If the load resistor is disconnected, then R is infinite and the output voltage rises to some limiting value that can be much smaller than V_0 and that is fixed internally. In one particular current-stabilized supply, V_0 is 50 kilovolts, R_i is 100 kilohms, and the limiting voltage is 50 volts.

Thus, if V_0 and R_i are known, one can predict the values of V and I for a given load resistance R. Thévenin's theorem therefore provides us with a simple model for describing the operation of a real source.

Thévenin's theorem is also useful for calculating currents flowing through circuits, as we shall see in the following examples.

5.12.1

EXAMPLE: THÉVENIN'S THEOREM APPLIED TO A SIMPLE CIRCUIT

Let us see how Thévenin's theorem applies to the circuit of Fig. 5-20a, where the part of the circuit to the left of the terminals is considered as a source feeding the load resistance R. Here V_1 is a voltage-stabilized source with an internal resistance that is negligible compared to R_1.

The Thévenin equivalent of this source is shown in Fig. 5-20b. The voltage V_0 is the voltage between the terminals of the circuit of Fig. 5-20a when the resistor R is removed. Also, the resistance R_i is the resistance measured at the terminals when R is removed, and when V_1 is reduced to zero. This is simply R_1 and R_2 in parallel. The current $V_0/(R_i + R)$ in Fig. 5-20b is the same current one would find directly from Fig. 5-20a using mesh currents as in Sec. 5.9.

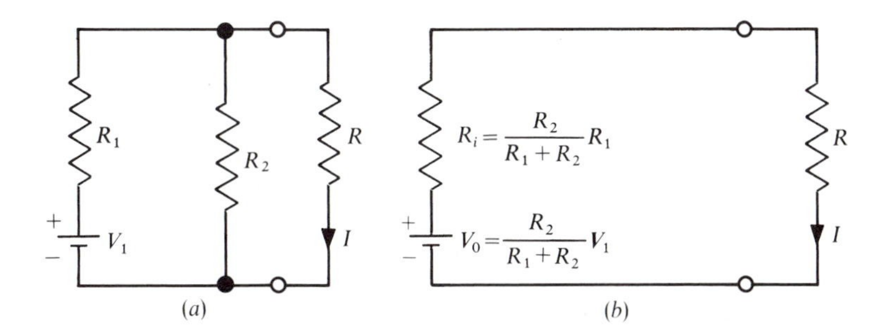

Figure 5-20 The circuit to the left of the terminals in (*a*) gives the same current I in the load resistance R as the circuit to the left of the terminals in (*b*).

5.12.2

EXAMPLE AND REVIEW: THE WHEATSTONE BRIDGE

Figure 5-21a shows a *Wheatstone bridge*, similar to that of Prob. 5-8, except that we have shown the resistances R_s and R_d of the source and of the detector, respectively, and that R_d is not infinite here.

We wish to find the current in the detector, when the resistance of the upper left-hand branch is $R(1 + x)$, as a function of the current I in that branch.

When $x = 0$, the bridge is balanced and the current through the detector is zero (Prob. 5-8).

We could solve this problem by using either branch or mesh currents. Instead, we shall solve it in a more roundabout way that will use most of the tricks that we have learned in this chapter.

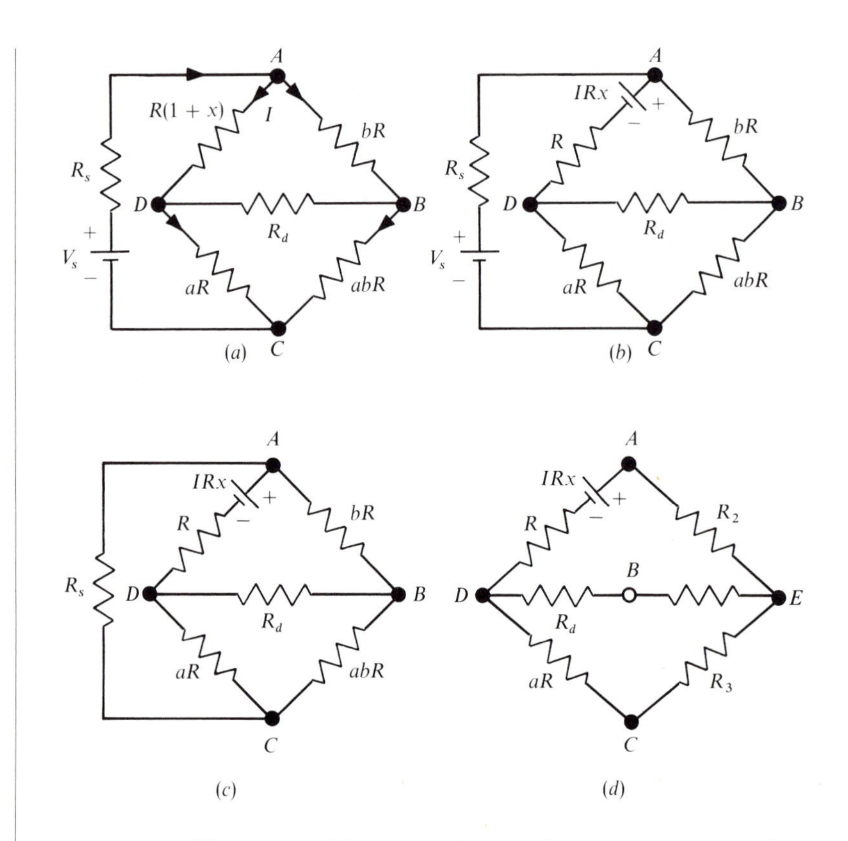

Figure 5-21 (*a*) Wheatstone bridge circuit, showing the internal resistance of the source R_s and the resistance of the detector R_d. When $x = 0$, the bridge is balanced and no current flows through R_d. (*b*, *c*, *d*) Successive transformations of the circuit that are used in calculating the current through the detector as a function of x.

a) Let the current through the resistance $R(1 + x)$ be I. Then, using the *substitution theorem*, we can replace the resistance Rx by a source IRx and we have the circuit of Fig. 5-21b.

b) We now have two sources, V_s and IRx. According to the *principle of superposition*, each one acts independently of the other. The currents due to V_s are now those for the balanced bridge. Thus the current through the detector is due only to the source IRx, and we may disregard V_s. We now have circuit *c*.

c) Resistances R_s, bR, abR form a delta. If we use the *delta-star transformation*, we have circuit *d*. We now have a fifth node, *E*.

Using the transformation rules,

$$R_2 = \frac{(bR)R_s}{bR + R_s + abR}, \qquad R_3 = \frac{(abR)R_s}{abR + R_s + bR}, \tag{5-47}$$

$$= BR, \qquad = aBR, \tag{5-48}$$

where B is R_2/R.

d) We can now use *Thévenin's theorem* to find the current through the dectector We consider R_d as the load resistance, and all the rest of the circuit as a source.

The open-circuit voltage V_0 is that between D and B when the branch DB is cut. The current flowing around the outer branches is

$$\frac{IRx}{R + R_2 + R_3 + aR},$$

and

$$V_0 = \frac{IRx}{R + R_2 + R_3 + aR}(aR + R_3), \tag{5-49}$$

$$= \frac{IRx}{R + BR + aBR + aR}(aR + aBR), \tag{5-50}$$

$$= IRx\frac{a(1 + B)}{(1 + a)(1 + B)}, \tag{5-51}$$

$$= \frac{a}{1 + a} IRx. \tag{5-52}$$

Note that this voltage is independent of R_s, of b, and of the quantity B of Eq. 5-48.

We now require the resistance of $R + R_i$ of Sec. 5.12, which is the resistance measured at a cut in the branch DB, when the source IRx is replaced by a short-circuit. We can find this resistance from Fig. 5-21d, but it is simpler to return to Fig. 5-21b. We imagine a source inserted in the branch DB, with both V_s and IRx replaced by short-circuits. The bridge is balanced. Then a source in the branch DB gives zero current in the branch AC, so we may disregard the resistance R_s. So

$$R_d + R_i = R_d + \frac{(R + bR)(aR + abR)}{R + bR + aR + abR}, \tag{5-53}$$

$$= R_d + \frac{a(1 + b)(1 + b)}{(1 + b)(1 + a)}R, \tag{5-54}$$

$$= R_d + \frac{a(1 + b)}{1 + a}R \tag{5-55}$$

and, finally,

$$I_d = \frac{V_0}{R_d + R_i} = \frac{aIRx/(1 + a)}{R_d + [a(1 + b)R/(1 + a)]}, \tag{5-56}$$

$$= \frac{Ix}{(1 + b) + \left(1 + \frac{1}{a}\right)(R_d/R)}. \tag{5-57}$$

If we wish to maximize the ratio I_d/I, then R_d/R should be small, b should be small, and a large. These last two conditions are not at all critical and one can say, as a rule of thumb, that both a and b should be approximately equal to unity.

5.13 SOLVING A SIMPLE DIFFERENTIAL EQUATION

In the next section we shall have to solve the following equation:

$$a\frac{dx}{dt} + bx = c, \tag{5-58}$$

where a, b, c are known constants, and x is an unknown function of t.

This is a *differential equation.* Its *general solution* is composed of two parts: (a) the *particular solution,* obtained by conserving only the last term on the left,

$$x_1 = c/b, \tag{5-59}$$

plus (b) the *complementary function,* which is the solution for $c = 0$, or

$$x_2 = Ae^{-(b/a)t}, \tag{5-60}$$

where A is a *constant of integration.*

Thus

$$x = x_1 + x_2 = \frac{c}{b} + Ae^{-(b/a)t}. \tag{5-61}$$

The numerical value of the *constant of integration* A is obtained, as a rule, from the known value of x at $t = 0$.

It is easy to check this solution by substituting it in the original equation (Eq. 5-58). A uniqueness theorem states that this solution is the only possible one.

5.14 TRANSIENTS IN RC CIRCUITS

When a source is connected to a circuit comprising resistors and capacitors, *transient* currents flow until the capacitors are charged. Such currents are large, at first, and then decay exponentially with time.

5.14.1 EXAMPLE: SIMPLE RC CIRCUIT

Figure 5-22 shows a capacitor C, initially uncharged, that is connected at time $t = 0$ to a voltage source V through a resistor R.

We can find the voltage across the capacitor by using Kirchoff's voltage law. Adding up the voltage rises around the loop, starting at the lower left-hand corner, clockwise,

$$V - IR - \frac{Q}{C} = 0. \tag{5-62}$$

Since I is dQ/dt,

$$R\frac{dQ}{dt} + \frac{1}{C}Q = V \tag{5-63}$$

and, from the previous section,

$$Q = CV + Ae^{-t/RC}. \tag{5-64}$$

Since, by hypothesis, $Q = 0$ at $t = 0$, $A = -CV$ and

$$Q = CV(1 - e^{-t/RC}). \tag{5-65}$$

After a time $t \gg RC$, $Q \approx CV$.

The voltage across the capacitor is

$$V_C = V(1 - e^{-t/RC}). \tag{5-66}$$

Figure 5-23 shows V_C/V as a function of time for $R = 1$ megohm, $C = 1$ microfarad, $RC = 1$ second.

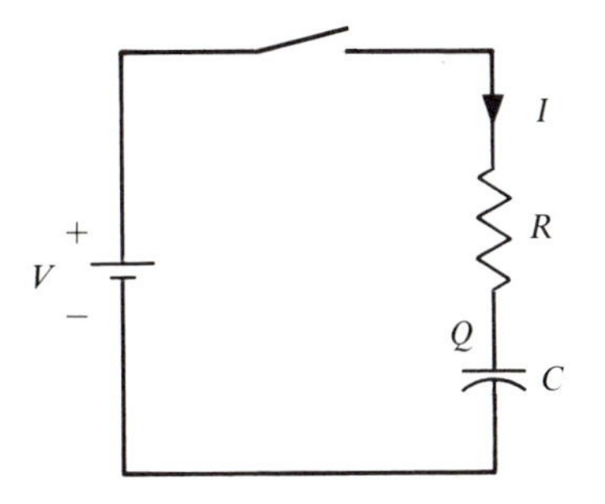

Figure 5-22 An *RC* circuit.

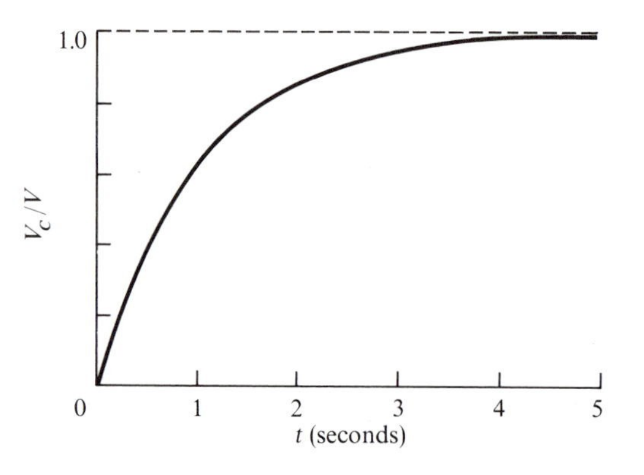

Figure 5-23 The voltage V_C across the capacitor C in Fig. 5-22 as a function of the time. It is assumed that $V_C = 0$ at $t = 0$ and that $RC = 1$.

The product RC is called the *time constant* of the circuit. For $t = RC$, $V_C/V = 1 - 1/e \approx 2/3$.

It will be shown in Prob. 5-23 that, if C is discharged through R, $V_C = Ve^{-t/RC}$.

5.15 SUMMARY

The *current density vector* **J** points in the direction of flow of positive charge and is expressed in amperes per square meter.

For any surface S bounding a volume τ,

$$\int_S \mathbf{J} \cdot \mathbf{da} = -\frac{d}{dt} \int_\tau \rho \, d\tau, \tag{5-4}$$

which is the *law of conservation of charge.*

The net charge density inside a homogeneous conductor carrying a current under steady-state conditions is zero.

For ordinary conductors, *Ohm's law* applies:

$$\mathbf{J} = \sigma \mathbf{E}, \tag{5-1}$$

where **J** is the current density, σ is the conductivity, and **E** is the electric field intensity. The electric charge density inside a uniform current-carrying conductor is zero under steady-state conditions.

Ohm's law can also be stated as

$$I = V/R, \tag{5-9}$$

where I is the current flowing through a resistance R, across which a voltage V is maintained.

The power dissipated as heat by the *Joule effect* per cubic meter is J^2/σ, and the power dissipated in a resistor R carrying a current I is I^2R.

Resistors for which I is proportional to V are said to be *linear*, or *ohmic*. Other resistors are said to be *non-linear*.

The resistance of a set of *resistors connected in series* is equal to the sum of their individual resistances.

The conductance of a set of *resistors connected in parallel* is equal to the sum of their individual conductances.

The *principle of superposition* states that, as long as all the resistors in a circuit are linear, each source acts independently of the others and produces its own set of currents.

Kirchoff's current law states that the algebraic sum of the currents entering a node, in an electric circuit, is equal to zero. His *voltage law* states that the sum of the voltage drops around a mesh is equal to zero. Currents in the various branches of a circuit are usually calculated by using mesh currents and the voltage law.

According to the *substitution theorem*, one can replace a resistance R in a circuit by a voltage source IR of the correct polarity without altering the currents flowing through the circuit or, inversely, if a current enters a voltage source V at its positive terminal, one can replace the source by a resistance V/I without changing the currents.

One can transform the *delta-connected circuit* of Fig. 5-16a into the *star-connected circuit* of Fig. 5-16b by means of the following equations:

$$R_A = \frac{R_b R_c}{R_a + R_b + R_c}, \tag{5-38}$$

$$R_B = \frac{R_c R_a}{R_a + R_b + R_c}, \tag{5-39}$$

$$R_C = \frac{R_a R_b}{R_a + R_b + R_c}. \tag{5-40}$$

Also,

$$G_a = \frac{G_B G_C}{G_A + G_B + G_C}, \tag{5-44}$$

$$G_b = \frac{G_C G_A}{G_A + G_B + G_C}, \tag{5-45}$$

$$G_c = \frac{G_A G_B}{G_A + G_B + G_C}. \tag{5-46}$$

A *voltage source* supplies a voltage that is nearly independent of the current through the load; a *current source* supplies a current that is nearly independent of the voltage across the load.

Thévenin's theorem states that any source may be represented by an ideal voltage source, in series with a resistance called its *output resistance.*

When a source is connected to a circuit comprising capacitors, *transient currents* flow temporarily in the circuit until all the capacitors are charged.

PROBLEMS

5-1E *CONDUCTION IN A NON-UNIFORM MEDIUM*

Two plane parallel copper electrodes are separated by a plate of thickness s whose conductivity σ varies linearly from σ_0, near the positive electrode, to $\sigma_0 + a$ near the negative electrode. Neglect edge effects. The current density is J.

Find the electric field intensity in the conducting plate, as a function of the distance x from the positive electrode.

5-2E *RESISTIVE FILM*

A square film of Nichrome (an alloy of nickel and chromium) is deposited on a sheet of glass, and copper electrodes are then deposited on two opposite edges.

Show that the resistance between the electrodes depends only on the thickness of the film and on its conductivity, as long as the film is square.

This resistance is given in *ohms per square*. Nichrome films range approximately from 40 to 400 ohms per square.

5-3E *RESISTOJET*

Figure 5-24 shows the principle of operation of the resistojet, which is used as a thruster for correcting either the trajectory or the attitude of a satellite. See also Probs. 2-14, 2-15, and 10-11.

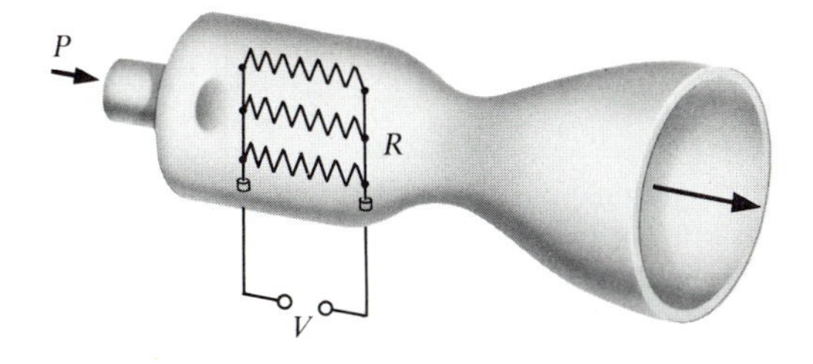

Figure 5-24 Schematic diagram of a resistojet. The propellant gas P is admitted on the left, heated to a high temperature by the resistor, and exhausted through the nozzle. See Prob. 5-3.

Assuming complete conversion of the electric energy into kinetic energy, calculate the thrust for an input power of 3 kilowatts and a flow of 0.6 gram of hydrogen per second.

5-4E *JOULE LOSSES*

A resistor has a resistance of 100 kilohms and a power rating of one-quarter watt. What is the maximum voltage that can be applied across it?

5-5E *VOLTAGE DIVIDER*

Figure 5-25 shows a voltage divider or attenuator. A source supplies a voltage V_i to the input, and a high-resistance load (not shown) $R \gg R_2$ is connected across R_2. The voltage on R_2 and R is V_o.

Show that

$$\frac{V_o}{V_i} = \frac{R_2}{R_1 + R_2}.$$

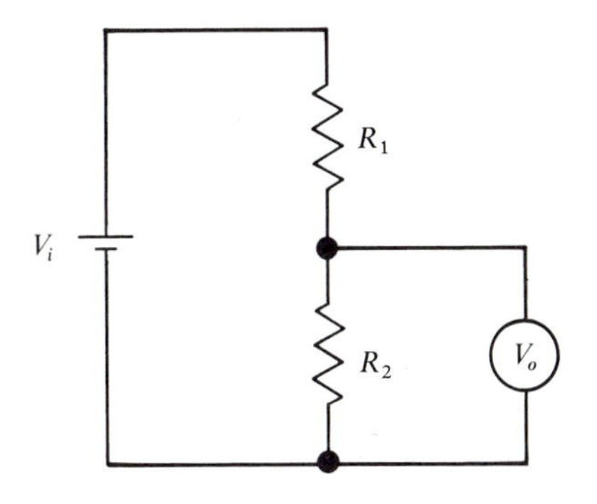

Figure 5-25 Voltage divider: V_o is $R_2/(R_1 + R_2)$ times V_i. See Prob. 5-5.

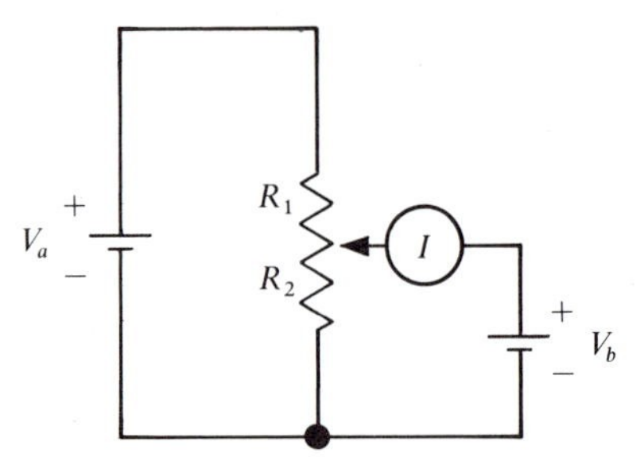

Figure 5-26 Potentiometer circuit. The voltage V_a is known, and V_b is unknown. When the sliding contact is adjusted for zero current, V_b/V_a is equal to $R_2/(R_1 + R_2)$. See Prob. 5-6.

5-6E *POTENTIOMETER*

Figure 5-26 shows a potentiometer circuit. Show that, when $I = 0$,

$$V_b = \frac{R_2}{R_1 + R_2} V_a.$$

This circuit is extensively used for measuring the voltage supplied by a source V_b, without drawing any current from it. The circuit is used in strip-chart recorders. In that case the current I is amplified to actuate the motor that displaces the pen, and simultaneously moves the tap on the resistance in a direction to decrease I. This is one type of *servomechanism.*

5-7E *SIMPLE CIRCUIT*

Show that, in Fig. 5-27,

$$V' = \frac{R_2 - R_1}{R_2 + R_1} V$$

if the voltmeter draws a negligible amount of current.

We shall use this result in Prob. 15-5.

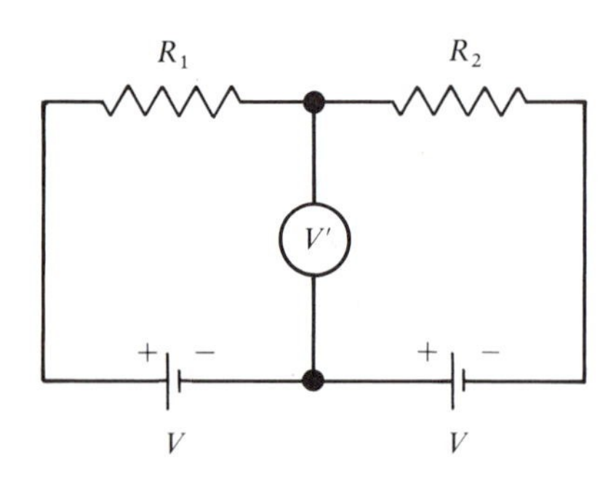

Figure 5-27 See Prob. 5-7.

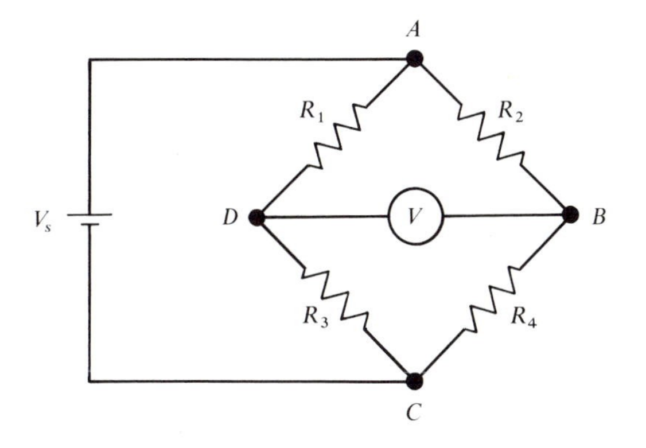

Figure 5-28 Wheatstone bridge. See Prob. 5-8.

5-8

WHEATSTONE BRIDGE

Figure 5-28 shows a circuit known as a Wheatstone bridge. As we shall see, the voltage V is zero when

$$\frac{R_1}{R_3} = \frac{R_2}{R_4}. \tag{1}$$

The bridge is then said to be *balanced*. As a rule, the bridge is used *un*balanced, the voltage V being a measure of the unknown resistance, say R_1.

The circuit is so widely used today that it is almost impossible to list all its applications. Some of the better known are the following.

1. If one of the resistors is a temperature-sensitive resistor, or *thermistor*, the bridge serves as a *thermometer*, V being a measure of the temperature.

2. If one of the resistors is a temperature-sensitive wire, heated by the bridge current and immersed in a gas, V is a measure of the gas velocity, or of its turbulence. One then has a *hot-wire anemometer*.

3. In the *Pirani vacuum gauge* the heated wire is in a partial vacuum and is cooled more or less by the residual gas, according to the pressure.

4. In certain *gas analyzers*, the heated wire is exposed to the unknown gas. Then V is a measure of the thermal conductivity of the gas. Such analyzers are used, in particular, for carbon dioxide. The thermal conductivity of carbon dioxide is roughly one half that of air (16.6, against 26.1 milliwatts per meter-kelvin).

5. In *flammable-gas detectors* the wire is heated sufficiently to ignite a sample of the gas contained in a small cell. The heat generated by the combustion increases the temperature and changes the resistance of the wire. The deflection of the voltmeter pointer is a measure of the flammability.

6. If one of the resistors is a fine wire whose resistance changes when elongated, the bridge serves as a *strain gauge* for measuring microscopic displacements or deformations.

7. If the strain gauge is attached to a diaphragm, one has a *pressure gauge.*

a) Show that, when $V = 0$, Eq. 1 is satisfied.

Solution: Let us suppose that node C is grounded.

The voltage V being zero, there is zero current in the branch DB. The current in the left-hand side is $V_s/(R_1 + R_3)$ and the voltage at D is

$$V_D = \frac{V_s}{R_1 + R_3} R_3. \tag{2}$$

Similarly, at B,

$$V_B = \frac{V_s}{R_2 + R_4} R_4. \tag{3}$$

Since $V_D = V_B$,

$$\frac{R_1 + R_3}{R_3} = \frac{R_2 + R_4}{R_4}, \tag{4}$$

and, subtracting 1 on both sides,

$$\frac{R_1}{R_3} = \frac{R_2}{R_4}. \tag{5}$$

b) Suppose now that R_1 increases by the factor $1 + x$ as in Sec. 5.12.2. We require V/V_s as a function of x, where x can vary from -0.2 to $+0.2$. The voltmeter V draws a negligible current.

Set

$$R_3 = aR_1, \qquad R_2 = bR_1, \qquad R_4 = abR_1, \tag{6}$$

again as in Sec. 5.12.2. With $x = 0$, the bridge is balanced.

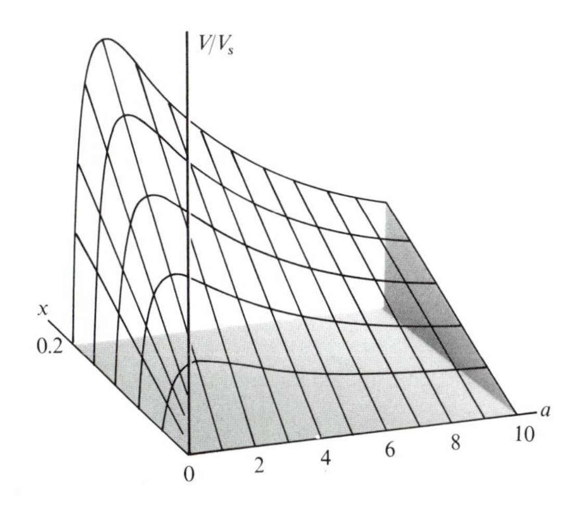

Figure 5-29 The ratio V/V_s for the Wheatstone bridge of Fig. 5-28, as a function of x and a.

Find V/V_s as a function of x, a, b.
Note that V/V_s turns out to be independent of b.

Solution: We now have that

$$V = V_B - V_D = \left(\frac{R_4}{R_2 + R_4} - \frac{R_3}{R_1(1 + x) + R_3}\right) V_s, \tag{7}$$

$$\frac{V}{V_s} = \frac{R_4/R_2}{1 + (R_4/R_2)} - \frac{R_3/R_1}{(1 + x) + (R_3/R_1)}, \tag{8}$$

$$= \frac{a}{1 + a} - \frac{a}{1 + x + a}, \tag{9}$$

$$= \frac{ax}{(1 + a)(1 + a + x)}. \tag{10}$$

Figure 5-29 shows V/V_s as a function of both x and a.

c) For what value of a is V/V_s maximum, for a given value of x? This is the condition for maximum sensitivity. Figure 5-29 shows that this condition is $a \approx 1$.

Solution: V/V_s is maximum when $d(V/V_s)/da = 0$. Thus we set

$$\frac{(1 + a)(1 + a + x)x - ax[(1 + a + x) + (1 + a)]}{(1 + a)^2(1 + a + x)^2} = 0, \tag{11}$$

$$(1 + a)(1 + a + x) = a(2 + 2a + x), \tag{12}$$

$$1 + a + x + a + a^2 + ax = 2a + 2a^2 + ax, \tag{13}$$

$$a = (1 + x)^{1/2}. \tag{14}$$

Since x is small, the condition for maximum sensitivity is $a \approx 1$, as we had guessed.

d) Draw curves of V/V_s as a function of x, for $x = -0.2$ to $x = +0.2$, and for $a = 0.3$, 1, and 3.

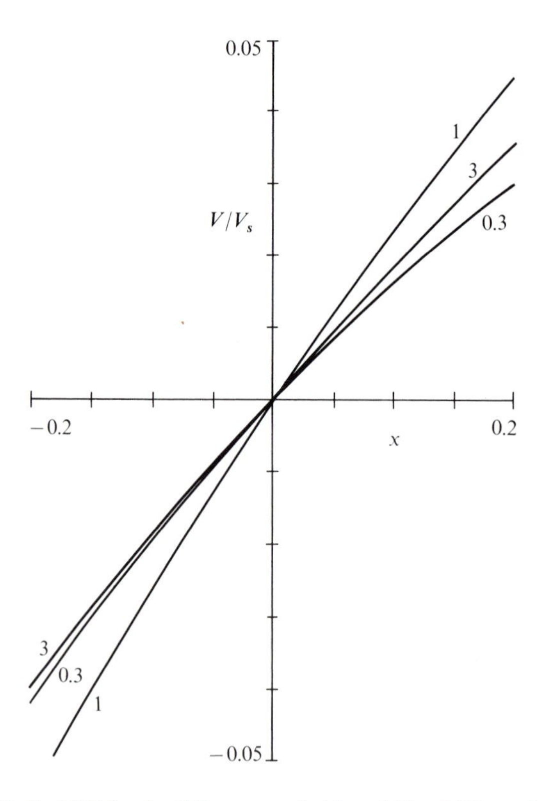

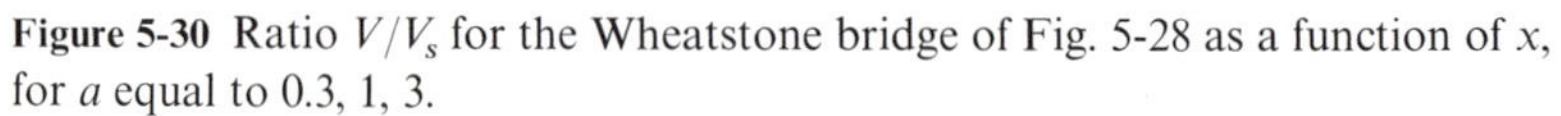
Figure 5-30 Ratio V/V_s for the Wheatstone bridge of Fig. 5-28 as a function of x, for a equal to 0.3, 1, 3.

Note the slight non-linearity of the curves. The non-linearity decreases with increasing a.

Solution: See Fig. 5-30.

5-9 *AMPLIFIER*

Figure 5-31a shows one element of an *analog computer*. Its function is to amplify the input voltage by the factor $-R_2/R_1$:

$$V_o = -\frac{R_2}{R_1} V_i.$$

The triangular figure is an *operational amplifier* whose gain is $-A$. Such amplifiers have very high gains, of the order of 10^4 to 10^9, and draw a negligible amount of current at their input terminals.

The accuracy of the gain is limited only by the stability of the ratio R_2/R_1. The drift in R_2/R_1 due to aging, temperature changes, and so forth, can be smaller than the drift in A by orders of magnitude.

a) You can show in the following way that the above equation is correct. First, you need an expression for the voltage V_{iA} at the input to the amplifier. Since the amplifier draws essentially zero current at its input terminals, you can use the circuit of Fig. 5-31b. Then, setting $V_o = -AV_{iA}$, you can show that

$$\frac{V_o}{V_i} = -\frac{R_2}{R_1 + [(R_1 + R_2)/A]} \approx -\frac{R_2}{R_1}.$$

The gain is $-R_2/R_1$ if $A \gg 1$ and if $A \gg R_2/R_1$.

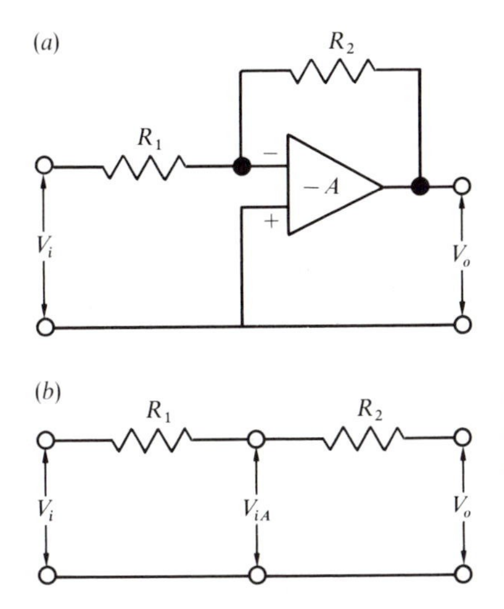

Figure 5-31 (*a*) Multiplying circuit. The triangle represents an amplifier whose gain is $-A$. The circuit has a gain of $-R_2/R_1$, as long as $A \gg 1$ and $A \gg R_2/R_1$. (*b*) Equivalent circuit for calculating the gain. See Prob. 5-9.

b) What is the minimum value of A if $R_1 = 1000$ ohms, $R_2 = 2000$ ohms, and if V_o must be equal to $-(R_2/R_1)V_i$ within one part in 1000?

5-10E $G = e$!

You are given a large number of 6-ohm resistors. How can they be connected to give a circuit whose conductance is equal to e siemens, where e is the base of the natural logarithms?

Hint: Write out the series for e.

5-11E *TETRAHEDRON*

Six equal resistors R form the edges of a tetrahedron, as in Fig. 5-32.

a) A battery is connected between A and B.

Show that, by symmetry, points C and D will be at the same potential. That being so, there is zero current in branch CD, and we can remove it.

b) What is the resistance between A and B?

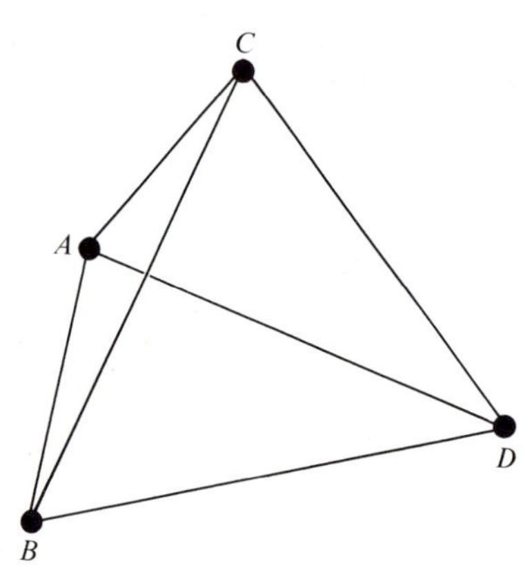

Figure 5-32 See Prob. 5-11.

5-12E *CUBE*

Twelve equal resistances R form the edges of a cube, as in Fig. 5-33.

a) A battery is connected between A and G.

Can you find one set of three points that are at the same potential?

Can you find another set?

b) Now imagine that you have a sheet of copper connecting the first set, and another sheet of copper connecting the second set. This does not affect the currents in the resistors.

You can now show that

$$R_{AG} = (5/6)R.$$

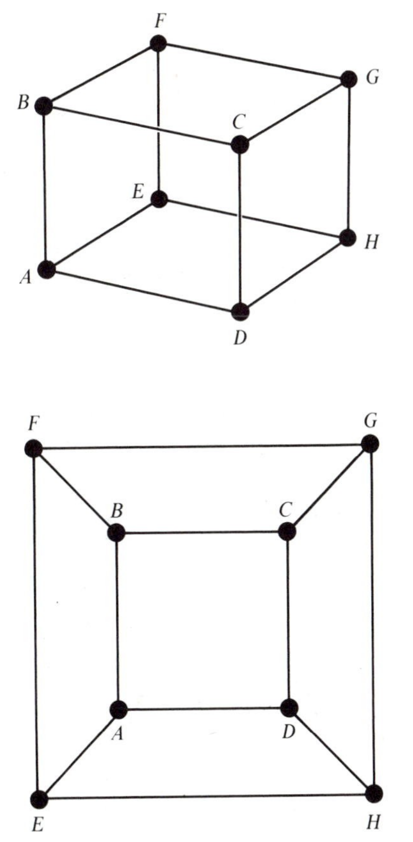

Figure 5-33 See Prob. 5-12.

Figure 5-34 See Prob. 5-13.

5-13E *CUBE*

Twelve equal resistances R form the edges of a cube, as in Fig. 5-33. To find the resistance between A and C, we first flatten the cube as in Fig. 5-34.

a) A battery is connected between A and C.

Which four points are at the same potential?

Which two branches can be removed without affecting the currents?

b) Now show that

$$R_{AC} = (3/4)R.$$

5-14 *CUBE*

First, read Probs. 5-11 to 5-13.

Now show that, in Fig. 5-33,

$$R_{AD} = (1.4/2.4)R.$$

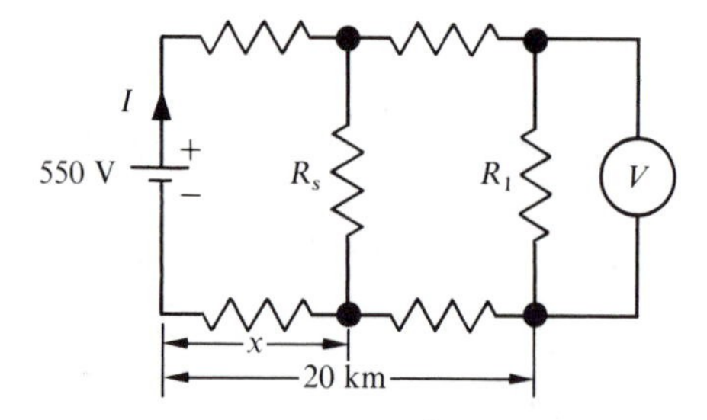

Figure 5-35 A 550-volt source feeds a load resistance R_l through a 20-kilometer line. The position of a fault R_s can be found from measurements of the current I, with and without a short-circuit across R_l. See Prob. 5-15.

5-15D† *LINE FAULT LOCATION*

In Fig. 5-35, a 550-volt source feeds a fixed load resistance R_l through a pair of copper ($\sigma = 5.8 \times 10^7$ siemens per meter) wires 20 kilometers long. The wires are 3 millimeters in diameter.

Suddenly, during a storm, the current I supplied by the source increases and the voltage V across R_l falls. It is concluded that a fault has developed somewhere along the line, and that there is a shunt resistance R_s at some distance x from the source.

If the load resistance R_l is disconnected, I is found to be 3.78 amperes. When the line terminals at the load are short-circuited, I is 7.20 amperes.

What is the distance x?

5-16E *UNIFORM RESISTIVE NET*

Figure 5-36 shows a two-dimensional network composed of equal resistances. The outside nodes are either connected to batteries in some arbitrary way, or left unconnected.

a) Use Kirchoff's current law to show that the potential at any inside node is equal to the average potential at the four neighboring nodes. For example,

$$V_0 = \frac{V_A + V_B + V_C + V_D}{4}.$$

b) What is the rule for a similar three-dimensional network?

5-17 *POTENTIAL DIVIDER*

Find the ratio V_o/V_i in Fig. 5-37.

---

† Problems marked *D* are relatively difficult.

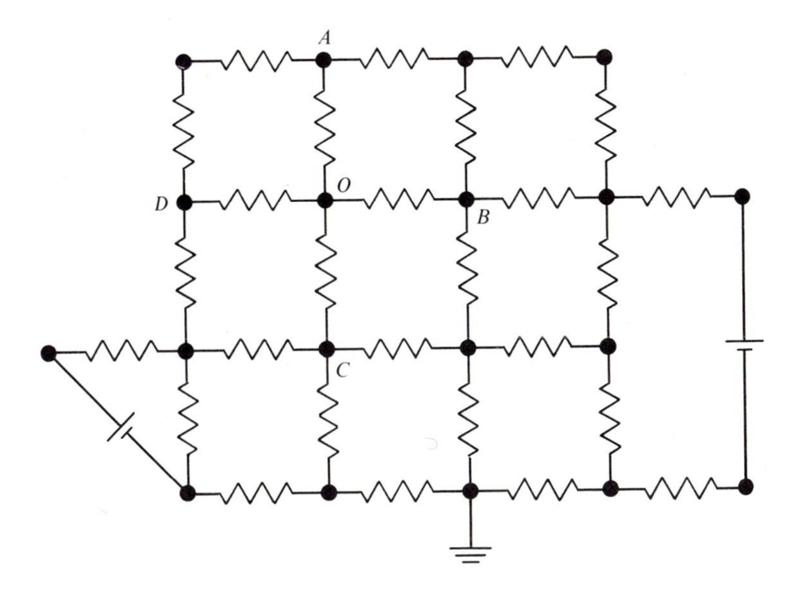

Figure 5-36 Two-dimensional network composed of equal resistances R. Arbitrary voltages are applied at the periphery. It is shown in Prob. 5-16 that the voltage on any inside node is equal to the average value of the voltages on the four neighboring nodes.

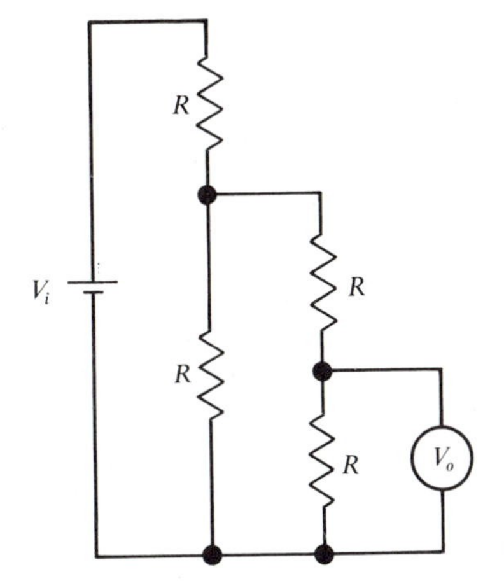

Figure 5-37 See Prob. 5-17.

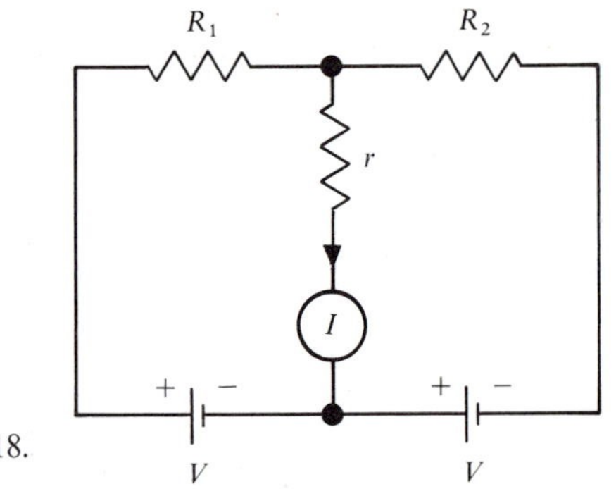

Figure 5-38 See Prob. 5-18.

5-18 *SIMPLE CIRCUIT WITH TWO SOURCES*

Show that, in Fig. 5-38,

$$I = \frac{R_2 - R_1}{R_1 R_2 + r(R_1 + R_2)} V.$$

5-19 *DELTA-STAR TRANSFORMATION*

Find the equations for either the delta-star or the star-delta transformation by assuming mesh currents, as in Fig. 5-16, and making the voltages $V_A - V_B$, $V_B - V_C$, $V_C - V_A$ in part *a* the same as those in part *b*.

Hint: Find one equation of the form

$$(\quad . \quad)I_A + (\quad . \quad)I_B + (\quad . \quad)I_C = 0.$$

Then, since it must be valid, whatever the values of the mesh currents, the parentheses must all be identically equal to zero. This should give you one of the equations of one set; the other two equations can be found by symmetry.

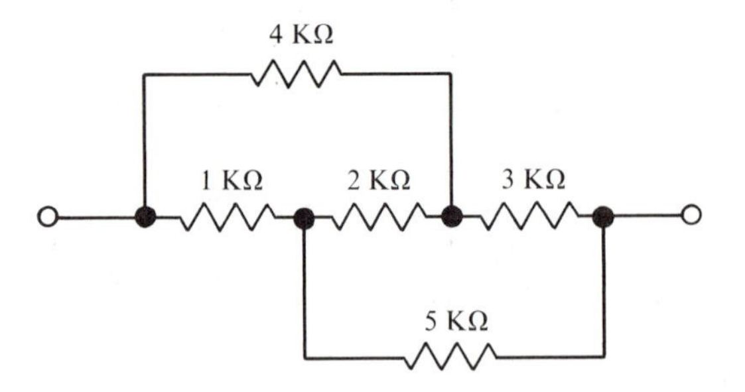

Figure 5-39 See Prob. 5-20. The symbol Ω stands for ohm.

5-20 *DELTA-STAR TRANSFORMATION*

Find the resistance of the circuit shown in Fig. 5-39.

5-21E *OUTPUT RESISTANCE OF A BRIDGE CIRCUIT*

Show that the output resistance of the bridge circuit shown in Fig. 5-40, as seen at the voltmeter V, is R.

Assume that the source V_s has a zero output resistance.

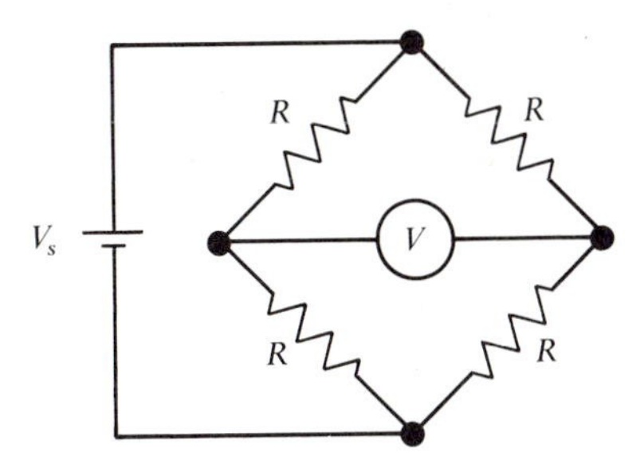

Figure 5-40 See Prob. 5-21.

5-22E *OUTPUT RESISTANCE OF AN AUTOMOBILE BATTERY*

It is found that the voltage at the terminals of a defective automobile battery drops from 12.5 to 11.5 volts when the headlights are turned on.

What is the approximate value of the output resistance?

5-23E *DISCHARGING A CAPACITOR THROUGH A RESISTOR*

A resistance of 1 megohm is connected across a 1 microfarad capacitor initially charged to 100 volts.

Calculate the voltage across the capacitor as a function of the time.

5-24E *RAMP GENERATOR*

A constant-current source feeds a current I to a capacitor C. At $t = 0$ the capacitor voltage is zero.

Find the voltage across C as a function of the time.

5-25 *CHARGING A CAPACITOR THROUGH A RESISTOR*

A source supplying a voltage V charges a capacitor C through a resistance R.

Calculate the energy supplied by the source, that dissipated by the resistor, and that stored in the capacitor, after an infinite time.

You should find that one half of the energy is dissipated in R, and that the other half is stored in C.

5-26 *RC TRANSIENT*

In the circuit of Fig. 5-41, the capacitor is initially uncharged.

a) At $t = 0$ the switch is closed. Find the voltage V as a function of t.

b) After a long time the switch is opened. Find again V as a function of t, setting $t = 0$ at the moment the switch is opened.

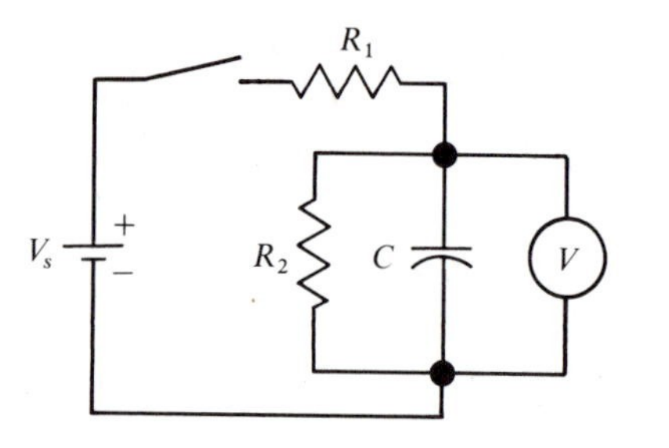

Figure 5-41 See Prob. 5-26.

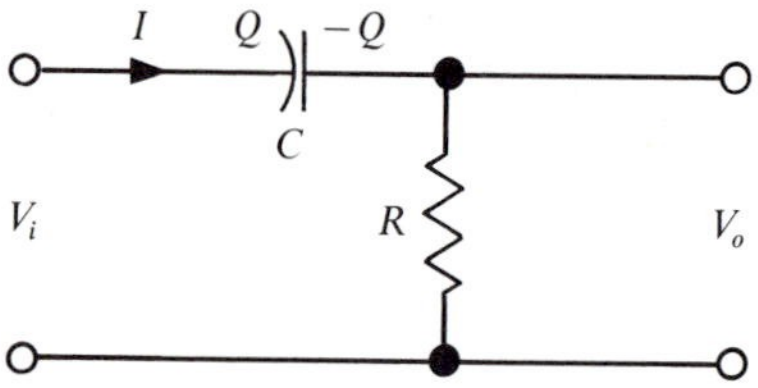

Figure 5-42 *RC* differentiating circuit. See Prob. 5-27.

5-27E *RC DIFFERENTIATING CIRCUIT*

Figure 5-42 shows an *RC* differentiating circuit. The load resistance connected at V_o is large compared to R.

Show that, if the voltage drop across R is negligible compared to that across C,

$$V_o \approx RC\frac{dV_i}{dt}.$$

Note that $V_o \ll V_i$. See Prob. 5-29.

5-28E *DIFFERENTIATING A SQUARE WAVE*

A square wave as in Fig. 5-43 is applied to the *RC* differentiating circuit of Fig. 5-42.

Sketch a curve of the output voltage as a function of the time.

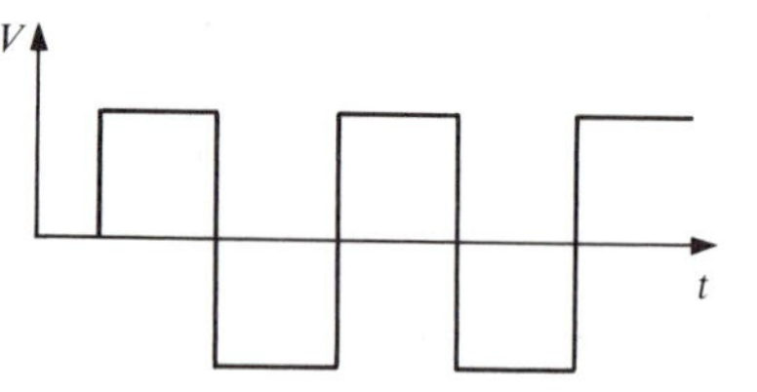

Figure 5-43 Ideal square wave. In practice, the straight lines are portions of exponential curves. See Prob. 5-28.

5-29 *DIFFERENTIATING CIRCUIT*

The circuit of Fig. 5-42 is simple and inexpensive, but $V_o \ll V_i$.

Figure 5-44a shows a much superior, but more complex, differentiating circuit. The triangle represents an operational amplifier as in Prob. 5-9.

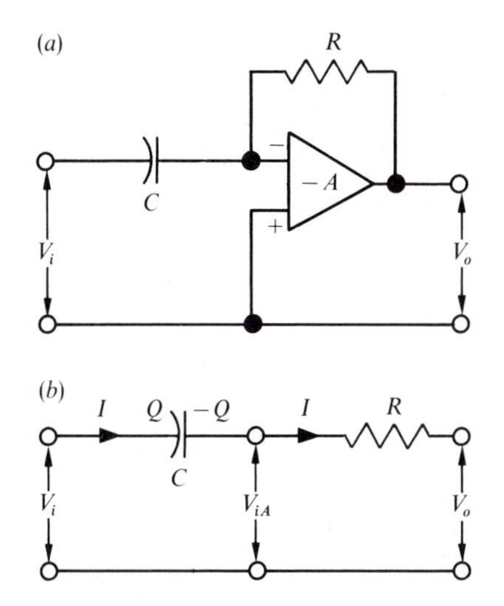

Figure 5-44 (*a*) Differentiating circuit using an amplifier. (*b*) Equivalent circuit used to calculate V_o. See Prob. 5-29.

Show that

$$V_o = -RC\frac{dV_i}{dt},$$

as long as $A \gg 1$ and $|V_o|/RC \gg |dV_o/dt|/A$. Note that RC can be much larger than unity, so that V_o need *not* be much smaller than V_i.

Solution: From Fig. 5-44b,

$$V_i = \frac{Q}{C} + V_{iA}, \qquad V_o = -AV_{iA}, \tag{1}$$

$$\frac{dV_i}{dt} = \frac{1}{C}\frac{dQ}{dt} + \frac{dV_{iA}}{dt} = \frac{1}{C}I - \frac{1}{A}\frac{dV_o}{dt}. \tag{2}$$

Now

$$I = \frac{V_{iA} - V_o}{R} = -\frac{1}{R}\left(1 + \frac{1}{A}\right)V_o, \tag{3}$$

so that

$$\frac{dV_i}{dt} = -\frac{1}{RC}\left(1 + \frac{1}{A}\right)V_0 - \frac{1}{A}\frac{dV_0}{dt}, \tag{4}$$

$$= -\frac{1}{RC}V_o - \frac{1}{A}\left(\frac{V_o}{RC} + \frac{dV_o}{dt}\right), \tag{5}$$

$$\approx -V_o/RC, \tag{6}$$

if $A \gg 1$ and if $|V_o|/RC \gg |dV_o/dt|/A$. Thus, with this approximation,

$$V_o = -RC\frac{dV_i}{dt}. \tag{7}$$

5-30 *RC INTEGRATING CIRCUIT*

Figure 5-45 shows an RC integrating circuit. The current through the load connected at V_o is negligible compared to dQ/dt at C.

Show that, as long as the voltage across C is small compared to that across R,

$$V_o = \frac{1}{RC}\int_0^t V_i\,dt.$$

As in Prob. 5-27, $V_o \ll V_i$. It is assumed that $V_o = 0$ at $t = 0$.

5-31 E *INTEGRATING CIRCUIT*

A square wave as in Fig. 5-43 is applied to the RC integrating circuit of Fig. 5-45. Sketch a curve of the output voltage as a function of the time.

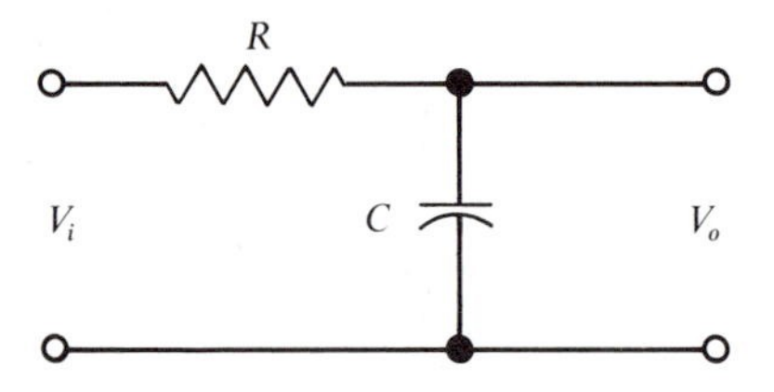

Figure 5-45 RC integrating circuit. See Prob. 5-30.

5-32 *INTEGRATING CIRCUIT*

The integrating circuit shown in Fig. 5-46a performs integrations without the limitation $V_o \ll V_i$ that applies to the circuit of Fig. 5-45. The triangle represents an amplifier as in Probs. 5-9 and 5-29.

Analog computers make extensive use of the circuits of Figs. 5-31, 5-44, and 5-46. Show that

$$V_o = -\frac{1}{RC}\int_0^t V_i\,dt,$$

if $A \gg 1$ and if $|V_o|/RC \ll A|dV_o/dt|$.

Use the circuit of Fig. 5-46b and set

$$\frac{d}{dt}V_o = \frac{d}{dt}\left(V_{iA} - \frac{Q}{C}\right)$$

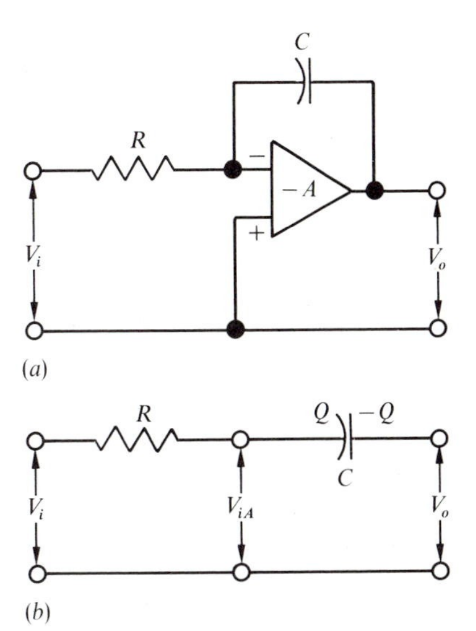

Figure 5-46 (*a*) Integrating circuit using an amplifier. (*b*) Equivalent circuit for calculating V_o. See Prob. 5-32.

5-33 *PULSE-COUNTING CIRCUIT*

The circuit shown in Fig. 5-47 can be used either for counting pulses or for measuring a capacitance, either C_1 or C_2, the other one being known.

The pulse generator G produces f positive square pulses of amplitude V_p per second and

$$C_1 \ll C_2, \qquad fC_1 \ll C_2.$$

Diodes D_1 and D_2 pass currents only in the direction of the arrow.

During a pulse, no current flows through diode D_1, and capacitors C_1 and C_2 charge in series.

Between pulses, the terminals of G are effectively shorted and capacitor C_1 discharges through diode D_1. Diode D_2 is then non-conducting.

a) Sketch a curve of V as a function of the time, for $V \ll V_p$, with $V = 0$ at $t = 0$.

b) What is the value of V after a time t?

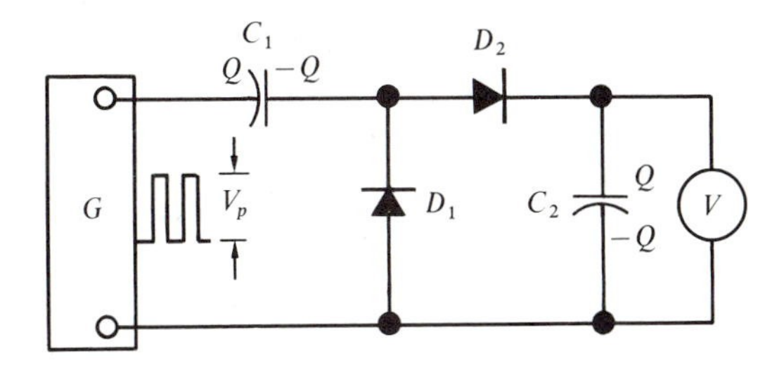

Figure 5-47 This circuit can be used either for measuring the repetition frequency of the current pulses originating in G or for measuring either C_1 or C_2. See Prob. 5-33.

CHAPTER 6

DIELECTRICS: I

Electric Polarization **P**, *Bound Charges, Gauss's Law, Electric Displacement* **D**

Dielectrics differ from conductors in that they have no free charges that can move through the material under the influence of an electric field. In dielectrics, all the electrons are bound; the only possible motion in an electric field is a minute displacement of positive and negative charges in opposite directions. The displacement is usually small compared to atomic dimensions.

A dielectric in which this charge displacement has taken place is said to be *polarized*, and its molecules are said to possess *induced dipole moments.* These dipoles produce their own field, which adds to that of the external charges. The dipole field and the externally applied electric field can be comparable in magnitude.

In addition to displacing the positive and negative charges, an applied electric field can also orient molecules that possess permanent dipole moments. Such molecules experience a torque that tends to align them with the field, but collisions arising from thermal agitation of the molecules tend to destroy the alignment. An equilibrium polarization is thus established in which there is, on the average, a net alignment.

This is, in fact, a simplified view of dielectric behavior, because many solids, such as sodium chloride, for example, are not made up of molecules but of individual ions.

6.1 *THE ELECTRIC POLARIZATION* P

The *electric polarization* **P** is the dipole moment per unit volume at a given point. If **p** is the average electric dipole moment per molecule, and if N is the number of molecules per unit volume, then

$$\mathbf{P} = N\mathbf{p}. \tag{6-1}$$

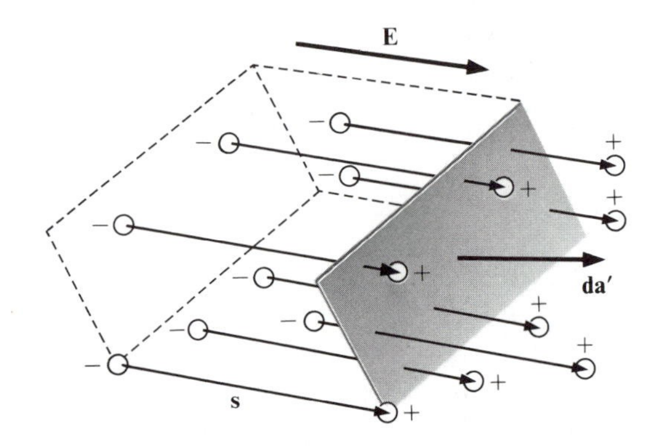

Figure 6-1 Under the action of an electric field **E**, which is the resultant of an external field and of the field of the dipoles within the dielectric, positive and negative charges in the molecules are separated by an average distance **s**. In the process a net charge $dQ = NQ\mathbf{s} \cdot \mathbf{da}'$ crosses the surface $\mathbf{da}'$, N being the number of molecules per unit volume and Q the positive charge in a molecule. The vector $\mathbf{da}'$ is perpendicular to the shaded surface. The circles indicate the centers of charge for the positive and for the negative charges in one molecule.

6.2 *THE BOUND CHARGE DENSITIES* ρ_b *AND* σ_b

We shall now show that the displacement of charge within the dielectric gives rise to net volume and surface charge densities

$$\rho_b = -\nabla \cdot \mathbf{P} \qquad \text{and} \qquad \sigma_b = \mathbf{P} \cdot \mathbf{n}_1, \tag{6-2}$$

where $\mathbf{n}_1$ is the normal to the surface, pointing outward.

Let us first consider the surface density σ_b. We imagine a small element of surface $\mathbf{da}'$ *inside* the dielectric, as in Fig. 6-1. Under the action of the field, positive and negative charges within the molecules separate by an average distance **s**. Positive charges cross the surface by moving in the direction of the field; negative charges cross it by moving in the opposite direction. For the purpose of our calculation, we consider the positive charges to be in the form of point charges Q and the negative charges in the form of point charges $-Q$. Furthermore, we consider the negative charges to be fixed and the positive charges to move a distance **s**. The amount of charge dQ that crosses $\mathbf{da}'$ is then just the total amount of *positive* charge within the imaginary parallelepiped shown in Fig. 6-1. The volume of this parallelepiped is

$$d\tau' = \mathbf{s} \cdot \mathbf{da}', \tag{6-3}$$

and

$$dQ = NQ\mathbf{s} \cdot \mathbf{da}', \tag{6-4}$$

where N is the number of molecules per unit volume and $Q\mathbf{s}$ is the dipole moment $\mathbf{p}$ of a molecule. Then

$$dQ = \mathbf{P} \cdot \mathbf{da}'. \tag{6-5}$$

If $\mathbf{da}'$ is on the surface of the dielectric material, dQ accumulates there as a surface distribution of density

$$\sigma_b = \frac{dQ}{da'} = \mathbf{P} \cdot \mathbf{n}_1, \tag{6-6}$$

where $\mathbf{n}_1$ is a unit vector, normal to the surface and pointing *outward.* The bound surface charge density σ_b is thus equal to the normal component of $\mathbf{P}$ at the surface.

We can show similarly that $-\boldsymbol{\nabla} \cdot \mathbf{P}$ represents a volume density of charge. The net charge that flows out of a volume τ' across an element $\mathbf{da}'$ of its surface is $\mathbf{P} \cdot \mathbf{da}'$, as we found above. The net charge that flows out of the surface S' bounding τ' is thus

$$Q = \int_{S'} \mathbf{P} \cdot \mathbf{da}', \tag{6-7}$$

and the net charge that remains within the volume τ' must be $-Q$. If ρ_b is the volume density of the charge remaining within this volume, then

$$\int_{\tau'} \rho_b \, d\tau' = -Q = -\int_{S'} \mathbf{P} \cdot \mathbf{da}' = -\int_{\tau'} (\boldsymbol{\nabla} \cdot \mathbf{P}) \, d\tau'. \tag{6-8}$$

Since this equation must be true for all τ', the integrands must be equal at every point, and the bound volume charge density is

$$\rho_b = -\boldsymbol{\nabla} \cdot \mathbf{P}. \tag{6-9}$$

We refer to ρ_b and σ_b as *bound* charge densities, as distinguished from the *free* charge densities ρ_f and σ_f. Bound charges are those that accumulate through the displacements that occur on a molecular scale in the polarization process. The other charges are called free charges. The conduction electrons in a conductor and the electrons injected into a dielectric with a high-energy electron beam are examples of free charges.

6.2.1 EXAMPLE: SHEET OF DIELECTRIC

If a *sheet of dielectric* is polarized uniformly in the direction normal to its surface, then the surface density of bound charge σ_b is P, as in Fig. 6-2. Also, $\rho_b = -\nabla \cdot \mathbf{P} = 0$, since $\mathbf{P}$ is independent of the coordinates.

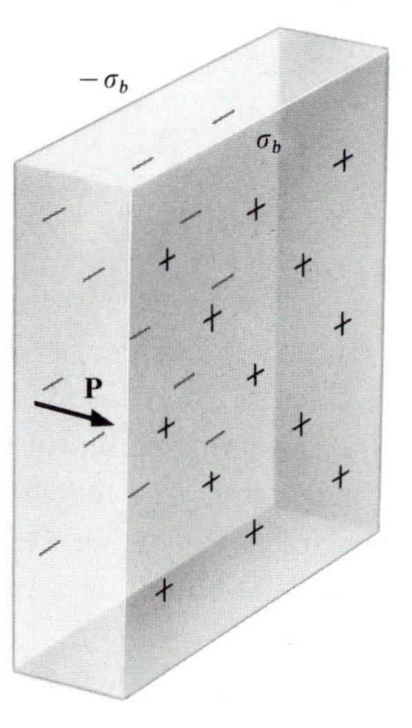

Figure 6-2 Uniformly polarized sheet of dielectric.

6.3 THE ELECTRIC FIELD OUTSIDE AND INSIDE A DIELECTRIC

Figure 6-3 shows a block of polarized dielectric material in which $\mathbf{P}$ is a function of position. We must find the electric field intensity that the dipoles in the dielectric produce at some point A, either inside or outside. Once obtained, this $\mathbf{E}$ can then be added to that produced by other charges to give the total $\mathbf{E}$.

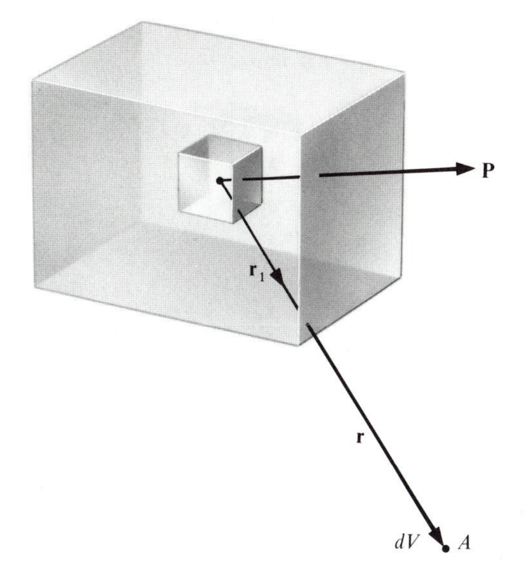

Figure 6-3 Block of dielectric with a dipole moment **P** per unit volume. The dipoles within the element of volume shown inside the block give rise to an electric potential dV at the point A.

Coulomb's law applies to any net accumulation of charge, regardless of other matter that may be present. Thus the field produced by the dielectric, both inside and outside, is the same as if the charge densities ρ_b and σ_b were situated in a vacuum.

6.3.1 *EXAMPLE: SHEET OF DIELECTRIC*

In the polarized sheet of dielectric of Fig. 6-2, $E = 0$ outside and $E = \sigma_b/\epsilon_0$ pointing to the left inside, exactly as if the sheet were replaced by its surface charges.

6.4 *THE DIVERGENCE OF* E *IN DIELECTRICS: GAUSS'S LAW*

We have just seen that the electric field produced by the dipoles of a polarized dielectric can be calculated from the bound charge densities as if they were situated in a vacuum. Let us investigate the implications of this fact on Gauss's law (Sec. 3.2).

Gauss's law relates the flux of **E** through a closed surface to the charge Q enclosed within that surface:

$$\int_S \mathbf{E} \cdot d\mathbf{a} = \int_\tau \nabla \cdot \mathbf{E}\, d\tau = Q/\epsilon_0. \tag{6-10}$$

For dielectrics, Q includes bound as well as free charges:

$$Q = \int_\tau (\rho_f + \rho_b)\, d\tau. \tag{6-11}$$

If we substitute this value of Q into Eq. 6-10 and equate the integrands of the volume integrals, then

$$\boxed{\nabla \cdot \mathbf{E} = \frac{\rho_f + \rho_b}{\epsilon_0}.} \tag{6-12}$$

This is *Gauss's law* in its more general form. It is one of Maxwell's four fundamental equations of electromagnetism. We shall find the other three in Secs. 8.2, 11.4, and 19.2. This one follows from (a) Coulomb's law, (b) the concept of electric field intensity, (c) the principle of superposition, and (d) the fact that the field of the dipoles situated in a polarized medium can be calculated from ρ_b and σ_b.

Note that we have implicitly assumed the existence of the space derivatives of **E**. These derivatives do not exist at the interface between two media; in such cases one must revert to the integral of $\mathbf{E} \cdot d\mathbf{a}$ over a closed surface, which is equal to the total enclosed charge over ϵ_0, according to Eq. 6-10.

6.5 *THE ELECTRIC DISPLACEMENT* D

Since $\rho_b = -\nabla \cdot \mathbf{P}$, from Eq. 6-9, then

$$\nabla \cdot \mathbf{E} = \frac{1}{\epsilon_0}(\rho_f - \nabla \cdot \mathbf{P}), \tag{6-13}$$

or

$$\nabla \cdot (\epsilon_0 \mathbf{E} + \mathbf{P}) = \rho_f. \tag{6-14}$$

The vector $\epsilon_0 \mathbf{E} + \mathbf{P}$ is therefore such that its divergence depends only on the free charge density ρ_f. This vector is called the *electric displacement* and is designated by $\mathbf{D}$:

$$\mathbf{D} = \epsilon_0 \mathbf{E} + \mathbf{P}. \tag{6-15}$$

Thus

$$\boldsymbol{\nabla} \cdot \mathbf{D} = \rho_f. \tag{6-16}$$

In integral form, Gauss's law for $\mathbf{D}$ becomes

$$\int_S \mathbf{D} \cdot \mathbf{da} = \int_\tau \rho_f \, d\tau, \tag{6-17}$$

and the flux of the electric displacement $\mathbf{D}$ through a closed surface is equal to the free charge enclosed by the surface.

Note that the *divergence* of $\mathbf{D}$, as well as the *surface integral* of $\mathbf{D}$, are both unaffected by the bound charges.

6.6 THE ELECTRIC SUSCEPTIBILITY χ_e

In most dielectrics the molecular charge separation is directly proportional to, and in the same direction as, $\mathbf{E}$. Dielectrics that show this simple dependence of polarization on field are said to be *linear* and *isotropic*. In practice, most dielectrics are *homogenous*, as well as linear and isotropic.

The dipole moment per unit volume is then

$$\mathbf{P} = N\mathbf{p} = \chi_e \epsilon_0 \mathbf{E}, \tag{6-18}$$

where N is again the number of molecules per unit volume, and χ_e is a dimensionless constant known as the *electric susceptibility* of the dielectric.

6.7 THE RELATIVE PERMITTIVITY ϵ_r

For a linear and isotropic dielectric we have, from Eqs. 6-15 and 6-18 that

$$\mathbf{D} = \epsilon_0 (1 + \chi_e) \mathbf{E} = \epsilon_0 \epsilon_r \mathbf{E} = \epsilon \mathbf{E}, \tag{6-19}$$

Table 6-1 *Relative Permittivities of Dielectrics Near 25°C.*

Type	Frequency (hertz)		
	100	10^6	10^{10}
Bakelite	5.50	4.45	3.55
Butyl rubber	2.43	2.40	2.38
Fused silica	3.78	3.78	3.78
Lucite	3.20	2.63	2.57
Neoprene	6.70	6.26	4.0
Polystyrene	2.56	2.56	2.54
Steatite	6.55	6.53	6.51
Styrofoam	1.03	1.03	1.03
Teflon	2.1	2.1	2.08
Water	81.	78.2	34.

where

$$\epsilon_r = 1 + \chi_e = \epsilon/\epsilon_0 \tag{6-20}$$

is a dimensionless constant larger than unity and is known as the *relative permittivity* of the material. In a vacuum, $\chi_e = 0$, $\epsilon_r = 1$. The relative permittivity was formerly called the *dielectric constant*. The quantity ϵ is called the *permittivity*.

The relative permittivity of dielectrics lies typically between 2 and 7. However, some non-linear dielectrics have relative permittivities as high as 10^5.

Table 6-1 gives the relative permittivities of some common dielectrics. It will be observed from the table that ϵ_r is frequency-dependent. We shall return to this phenomenon briefly in Sec. 7.7.

6.8 RELATION BETWEEN THE FREE CHARGE DENSITIES AND THE BOUND CHARGE DENSITIES

6.8.1 THE VOLUME CHARGE DENSITIES ρ_f AND ρ_b

In a linear and isotropic dielectric, from Eqs. 6-15 and 6-19,

$$\mathbf{D} = \epsilon_0 \mathbf{E} + \mathbf{P} = \epsilon_r \epsilon_0 \mathbf{E}, \tag{6-21}$$

and thus

$$\mathbf{P} = (\epsilon_r - 1)\epsilon_0 \mathbf{E} = \frac{\epsilon_r - 1}{\epsilon_r} \mathbf{D}. \tag{6-22}$$

Taking now the divergence of both sides and using Eqs. 6-9 and 6-16,

$$\rho_b = -\frac{\epsilon_r - 1}{\epsilon_r} \rho_f. \tag{6-23}$$

Note that ρ_b and ρ_f have opposite signs, so that the total charge density is smaller than ρ_f:

$$\rho_f + \rho_b = \rho_f / \epsilon_r. \tag{6-24}$$

Also, if ρ_f if zero in a linear and isotropic dielectric, which is nearly always the case, then ρ_b is also zero.

6.8.2 THE SURFACE CHARGE DENSITIES σ_f AND σ_b

At the interface between a dielectric and a conductor, there is a bound surface charge density σ_b on the dielectric, and a free surface charge density σ_f on the conductor, as in Fig. 6-4. For steady-state conditions, there is zero electric field inside the conductor and, inside the dielectric, from Gauss's law,

$$\epsilon_0 E = \sigma_f + \sigma_b, \qquad D = \epsilon_r \epsilon_0 E = \sigma_f. \tag{6-25}$$

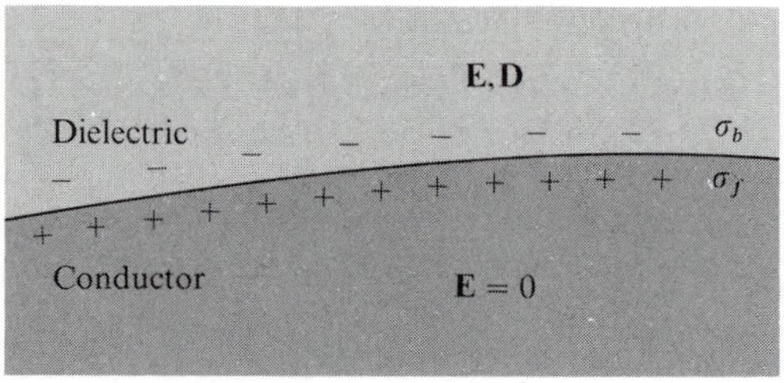

Figure 6-4 Interface between a dielectric and a conductor, with a positive charge density on the surface of the conductor.

Thus

$$\sigma_f + \sigma_b = \sigma_f/\epsilon_r. \tag{6-26}$$

This equation is similar to Eq. 6-24.

6.8.3 *EXAMPLE: DIELECTRIC-INSULATED PARALLEL-PLATE CAPACITOR*

Figure 6-5 shows a dielectric-insulated parallel-plate capacitor with the spacing s exaggerated for clarity. We assume that s is small, compared to the linear extent of the plates, in order that we may neglect the fringing field at the edges.

Then $\mathbf{P}$ is uniform, $\boldsymbol{\nabla} \cdot \mathbf{P} = 0$, and $\rho_b = 0$, which is consistent with the fact that $\rho_f = 0$.

The surface-charge densities $+\sigma_f$ on the lower plate and $-\sigma_f$ on the upper plate produce a uniform electric field directed upward. The polarization in the dielectric gives a bound surface charge density $-\sigma_b$ on the lower surface of the dielectric and $+\sigma_b$ on the upper surface. These bound charges produce a uniform electric field directed downward that cancels part of the field of the charges situated on the plates. Since the net field between the plates must remain equal to V/s, the free charge densities $+\sigma_f$ and $-\sigma_f$ on the plates must be larger than when the dielectric is absent. The presence of the dielectric thus has the effect of increasing the charges on the plates for a given value of V, and hence of increasing the capacitance.

To calculate the capacitance, we apply Gauss's law for the displacement $\mathbf{D}$ to a cylinder as in Fig. 6-5. Then the only flux of $\mathbf{D}$ through the Gaussian surface is through the top, and D is numerically equal to σ_f. Then, $E = \sigma_f/\epsilon_r\epsilon_0$, the potential difference between the plates is $\sigma_f s/\epsilon_r\epsilon_0$, and the capacitance

$$C = \frac{\sigma_f A}{\sigma_f s/\epsilon_r\epsilon_0} = \epsilon_r\epsilon_0 A/s, \tag{6-27}$$

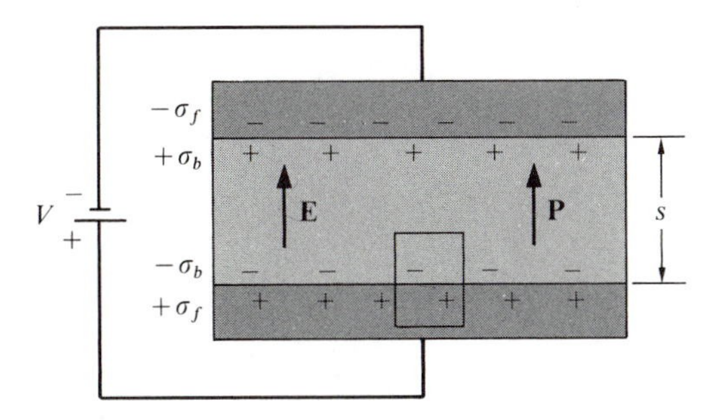

Figure 6-5 Dielectric-insulated parallel-plate capacitor. The small rectangle is the cross-section of an imaginary cylinder used for calculating **D**.

where A is the area of one plate. The capacitance is therefore increased by a factor of ϵ_r through the presence of the dielectric.

The measurement of the capacitance of a suitable capacitor with and without a dielectric provides a convenient method for measuring a relative permittivity ϵ_r.

6.9 *POISSON'S AND LAPLACE'S EQUATIONS*

In homogeneous, linear, and isotropic dielectrics, **D** is proportional to **E**, and ϵ_r is independent of the coordinates. Then, from Eqs. 6-12 and 6-24,

$$\nabla \cdot \mathbf{E} = \frac{\rho_f + \rho_b}{\epsilon_0} = \rho_f/\epsilon_r\epsilon_0. \tag{6-28}$$

Also, since **E** is $-\nabla V$,

$$\nabla^2 V = -\frac{\rho_f + \rho_b}{\epsilon_0} = -\rho_f/\epsilon_r\epsilon_0. \tag{6-29}$$

This is *Poisson's equation* for dielectrics. The expressions involving ϵ_r are valid only in linear and isotropic dielectrics.

Laplace's equation is again.

$$\nabla^2 V = 0, \tag{6-30}$$

when both ρ_f and ρ_b are zero.

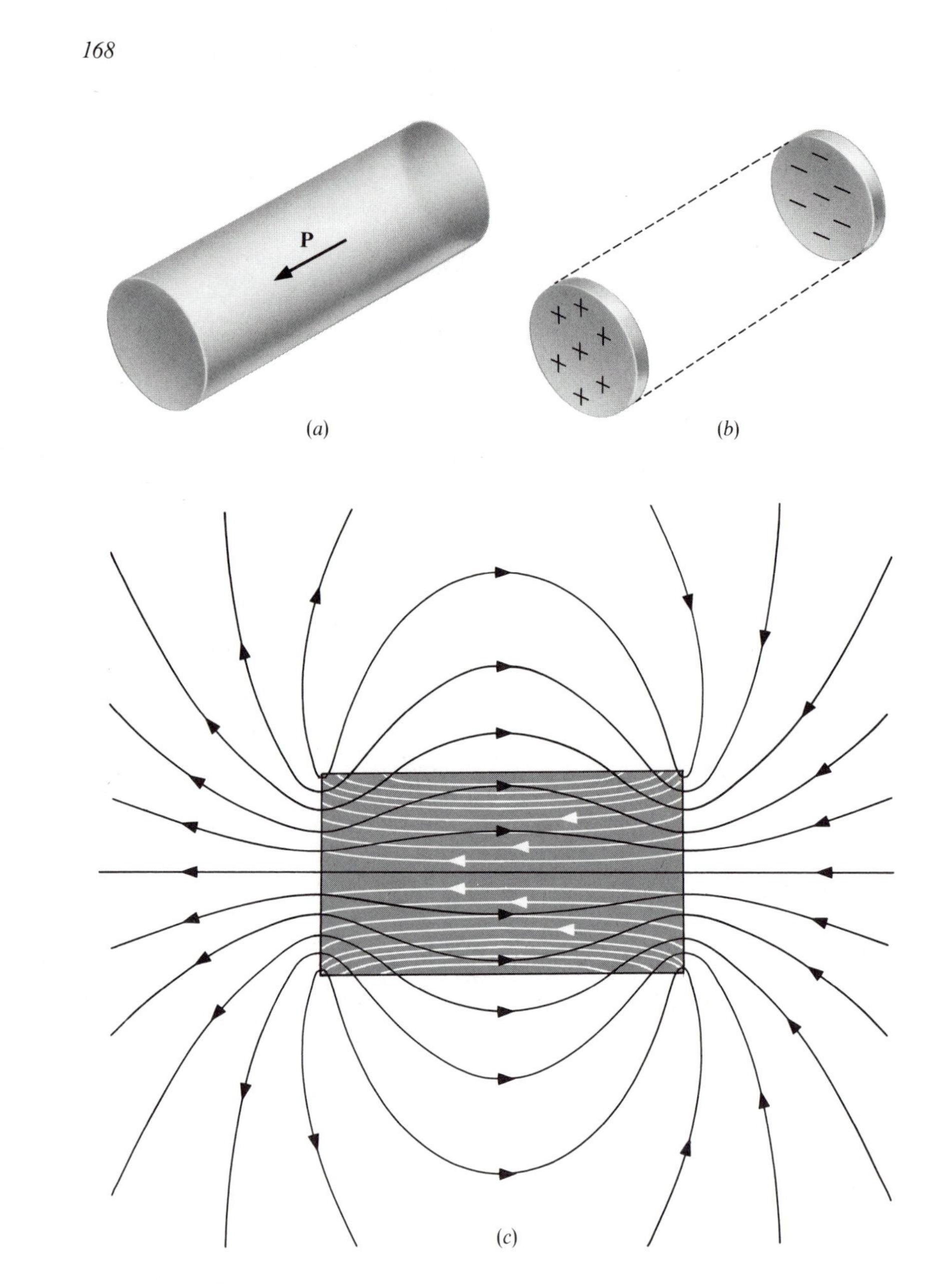

Figure 6-6 (*a*) Bar electret polarized uniformly parallel to its axis. (*b*) The **E** field of the bar electret is the same as that of a pair of circular plates carrying uniform surface charge densities of opposite polarities. (*c*) Lines of **E** (black). Lines of **D** are shown in white inside the electret; outside, they follow the lines of **E**.

6.10 ELECTRETS

An *electret* is the electric equivalent of a permanent magnet. In most dielectrics the polarization disappears immediately when the electric field is removed, but some dielectrics retain their polarization for a very long time. Some polymers have extrapolated lifetimes of a few hundred years at room temperature.

One way of charging a dielectric is to place it in a strong electric field at high temperature. The bound charge density on the surfaces then builds up slowly as the molecules orient themselves. Free charges are also deposited on the surfaces when sparking occurs between the electrodes and the dielectric. The free and the bound charges have opposite signs. The sample is then cooled to room temperature without removing the electric field.

The examples in Secs. 6.2.1 and 6.3.1 apply to a sheet electret with zero free charge density.

Figure 6-6 shows the field of a uniformly polarized bar electret.

6.11 SUMMARY

When a dielectric material is placed in an electric field, positive and negative charges within the molecules are displaced, one with respect to the other, and the material becomes *polarized*. The induced dipole moment per unit volume $\mathbf{P}$ is called the *electric polarization*.

This produces real accumulations of charge that we can use to calculate V and $\mathbf{E}$ both inside and outside the dielectric:

$$\rho_b = -\nabla \cdot \mathbf{P}, \qquad \sigma_b = \mathbf{P} \cdot \mathbf{n}_1, \tag{6-2}$$

where ρ_b is the volume density and σ_b is the surface density of bound charge. The unit vector $\mathbf{n}_1$ is normal to the surface of the dielectric and points *out*ward.

Gauss's law can be expressed in a form that is valid for dielectrics:

$$\boxed{\nabla \cdot \mathbf{E} = \frac{\rho_f + \rho_b}{\epsilon_0}.} \tag{6-12}$$

This is one of Maxwell's four fundamental equations of electromagnetism.

In linear and isotropic dielectrics, the dipole moment per molecule $\mathbf{p}$ is proportional to $\mathbf{E}$, and

$$\mathbf{P} = \epsilon_0 \chi_e \mathbf{E}, \tag{6-18}$$

where χ_e is a dimensionless constant called the *electric susceptibility*.

The *electric displacement*

$$\mathbf{D} = \epsilon_0 \mathbf{E} + \mathbf{P}, \tag{6-15}$$

$$= \epsilon_0(1 + \chi_e)\mathbf{E} = \epsilon_0 \epsilon_r \mathbf{E} = \epsilon \mathbf{E}, \tag{6-19}$$

where ϵ_r is the *relative permittivity* and ϵ is the *permittivity*. Also,

$$\nabla \cdot \mathbf{D} = \rho_f, \tag{6-16}$$

or

$$\int_S \mathbf{D} \cdot \mathbf{da} = \int_\tau \rho_f \, d\tau. \tag{6-17}$$

Inside a dielectric,

$$\rho_f + \rho_b = \rho_f / \epsilon_r, \tag{6-24}$$

while, at the interface between a dielectric and a conductor,

$$\sigma_f + \sigma_b = \sigma_f / \epsilon_r. \tag{6-26}$$

Poisson's equation for dielectrics is

$$\nabla^2 V = -\frac{\rho_f + \rho_b}{\epsilon_0} = -\rho_f / \epsilon_r \epsilon_0, \tag{6-29}$$

and *Laplace's equation* is again

$$\nabla^2 V = 0, \tag{6-30}$$

when ρ_f and ρ_b are both zero.

As usual, expressions involving ϵ_r are valid only in linear and isotropic media.

An *electret* is a piece of dielectric material that is permanently polarized.

PROBLEMS

6-1E *THE DIPOLE MOMENT* **p**

A sample of diamond has a density of 3.5×10^3 kilograms per cubic meter and a polarization of 10^{-7} coulomb per square meter.

a) Compute the average dipole moment per atom.

b) Find the average separation between centers of positive and negative charge. Carbon has a nucleus with a charge $+6e$, surrounded by 6 electrons.

6-2E *THE VOLUME AND SURFACE BOUND CHARGE DENSITIES* ρ_b *AND* σ_b

Consider a block of dielectric with bound charge densities ρ_b and σ_b. Show mathematically that

$$\int_\tau \rho_b \, d\tau + \int_S \sigma_b \, da = 0,$$

where the τ is the volume of the dielectric and S is its surface. In other words, the total net bound charge is zero.

6-3 *BOUND CHARGE DENSITY AT AN INTERFACE*

Show that the bound charge density at the interface between two dielectrics 1 and 2 that is crossed by an electric field is $(\mathbf{P}_1 - \mathbf{P}_2) \cdot \mathbf{n}$. The polarization in 1 is $\mathbf{P}_1$ and is directed *into* the interface; the polarization in 2 is $\mathbf{P}_2$ and points *away* from the interface. The unit vector $\mathbf{n}$ is normal to the interface and points in the direction from 1 to 2.

6-4E *COAXIAL LINE*

Show that $\nabla \cdot \mathbf{E} = 0$ in the dielectric of a coaxial line.

Hint: Apply the divergence theorem to a portion of the dielectric.

6-5 *COAXIAL LINE*

The capacitance per unit length C' of a dielectric-insulated coaxial line is equal to that for an air-insulated line as in Prob. 4-6, multiplied by ϵ_r:

$$C' = \frac{2\pi\epsilon_r\epsilon_0}{\ln(R_2/R_1)}.$$

Table 6-2

B & S Gauge	Diameter in Millimeters
20	0.812
22	0.644
24	0.511
26	0.405
28	0.321
30	0.255
32	0.202
34	0.160

For a certain application, one requires a coaxial cable whose outside diameter must not exceed about 10 millimeters. Then $R_2 \leq 5$ millimeters. Its capacitance per unit length must be as low as possible. Then R_2 must be 5 millimeters, and R_1 as small as possible. The cable will be operated at up to 500 volts.

The insulating material will be Teflon. Teflon has a relative permittivity of 2.1. It can operate reliably in such a cable at electric field intensities as high as 5×10^6 volts per meter. (The maximum electric field intensity before breakdown is called the *dielectric strength* of a dielectric. In a uniform field the dielectric strength is larger for thinner specimens. A film of Teflon 0.1 millimeter thick has a dielectric strength of 10^8 volts per meter.)

a) What is the minimum allowable diameter for the inner conductor?

Remember that you can solve a transcendental equation by trial and error.

b) Table 6-2 shows the diameters of some commonly available copper wires. Finer wires would be impractical because they would tend to break at the connectors fixed to the ends of the cable.

What gauge number do you suggest?

c) What will be the capacitance per meter with the wire size you have chosen?

6-6 *COAXIAL LINE*

A coaxial line is composed of an internal conductor of radius R_1 and an external conductor of radius R_2, separated by a dielectric. The dielectric should normally fill all the space between the conductors. However, due to an error, the outer radius of the dielectric is $R < R_2$, leaving an air space.

a) Assuming that the dielectric is well centered, what is the capacitance when $R_1 = 1$ mm, $R_2 = 5$ mm, $R = 4.5$ mm, and $\epsilon_r = 2.56$ (polystyrene)?

Solution: The outer surface of the dielectric is an equipotential surface. Thus, we can imagine that it is covered with a thin conducting layer. This does not disturb the field. Then we may consider that we have two cylindrical capacitors, one inside the other.

Between R_1 and R we have a capacitance per unit length of

$$\frac{2\pi\epsilon_r\epsilon_0}{\ln(R/R_1)},$$

from Prob. 6-5. Between R and R_2, the capacitance per unit length is

$$\frac{2\pi\epsilon_0}{\ln(R_2/R)}.$$

These two capacitors are in series, and the capacitance per unit length between the inner and the outer conductors is given by

$$\frac{1}{C'} = \frac{\ln(R/R_1)}{2\pi\epsilon_r\epsilon_0} + \frac{\ln(R_2/R)}{2\pi\epsilon_0}, \tag{1}$$

$$= \frac{1}{2\pi\epsilon_0}\left[\frac{1}{\epsilon_r}\ln(R/R_1) + \ln(R_2/R)\right], \tag{2}$$

$$C' = \frac{2\pi\epsilon_0}{\dfrac{1}{\epsilon_r}\ln(R/R_1) + \ln(R_2/R)}, \tag{3}$$

$$= \frac{2\pi \times 8.85 \times 10^{-12}}{\dfrac{1}{2.56}\ln 4.5 + \ln\dfrac{1}{0.9}} = 80.2 \text{ picofarads/meter.} \tag{4}$$

b) Calculate the percent change in capacitance due to the air film.

Solution: Without the air film, the capacitance would be

$$C' = \frac{2\pi \times 2.56 \times 8.85 \times 10^{-12}}{\ln 5} = 88.4 \text{ picofarads/meter.} \tag{5}$$

The air film therefore decreases the capacitance by about 10%.

6-7 CHARGED WIRE EMBEDDED IN DIELECTRIC: THE FREE AND BOUND CHARGES

A conducting wire carrying a charge λ per unit length is embedded along the axis of a circular cylinder of dielectric. The radius of the wire is a; the radius of the cylinder is b.

a) Show that the bound charge on the outer surface of the dielectric is equal to the bound charge on the inner surface, except for sign.

b) Show that the net charge along the axis is λ/ϵ_r per unit length.

6-8 PARALLEL-PLATE CAPACITOR

The space between the plates of a parallel-plate capacitor with a plate separation s and a surface area A is partially filled with a dielectric plate of thickness $t \leq s$ and area A.

a) Show that

$$C = \frac{\epsilon_0 A}{s - t + (t/\epsilon_r)}.$$

b) Call C_0 the capacitance for $t = 0$.
Draw a curve of C/C_0 as a function of t/s for $\epsilon_r = 3$.

6-9 SPHERE OF DIELECTRIC WITH A POINT CHARGE AT ITS CENTER

A sphere of dielectric of radius R contains a point charge Q at its center. Set $R = 20$ millimeters, $Q = 10^{-9}$ coulomb, $\epsilon_r = 3$.

a) Draw graphs of D, E, V as functions of the distance r from the center, out to $r = 100$ millimeters.

Note the discontinuity in E at the surface.

b) Now calculate σ_b at the surface from the value of the polarization P just inside the surface.

c) Now use Gauss's law to explain the discontinuity in E.

6-10 CHARGED DIELECTRIC SPHERE

A dielectric sphere of radius R contains a uniform density of free charge ρ_f. Show that the potential at the center is

$$\frac{(2\epsilon_r + 1)\rho_f R^2}{6\epsilon_r\epsilon_0}.$$

6-11 MEASURING SURFACE CHARGE DENSITIES ON DIELECTRICS

Figure 6-7a shows a schematic diagram of an instrument that has been used to measure charge densities at the surface of dielectrics. See also Prob. 17-9,

The probe P is supported a few millimeters above the surface of the sample and can be displaced horizontally and vertically along three orthogonal directions.

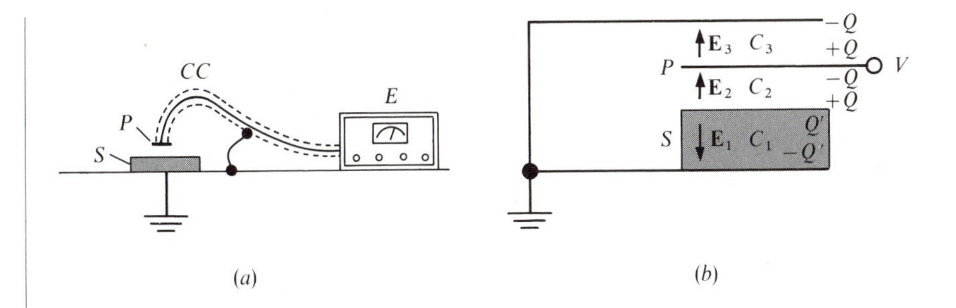

Figure 6-7 (*a*) Instrument for measuring surface charge densities on dielectrics. A conducting probe P is held close to the surface of the sample S. The probe is connected to an electrometer E through a coaxial cable CC. (*b*) Equivalent circuit: C_1 is the capacitance between the top surface of the sample and ground, C_2 is that between the probe and the sample surface, and C_3 is the capacitance of the coaxial cable, plus the input capacitance of the electrometer. See Prob. 6-11.

As we shall see, the surface charge on the sample induces a voltage on P. This voltage is read on the electrometer.

An *electrometer* is a voltmeter that has an extremely high resistance. For example, one type has a resistance of 10^{14} ohms. For comparison, common digital voltmeters have resistances of the order of 10^6 or 10^7 ohms. A good analog voltmeter draws about 100 microamperes. If it is rated at 10 volts, its resistance is $10/10^{-4} = 10^5$ ohms. An electrometer draws essentially zero current and is therefore suitable for measuring the voltage on a small capacitor, which is what we have here.

Figure 6-7b shows the equivalent circuit: we have, effectively, three capacitors connected in series, and the electrometer serves to measure the voltage V.

a) Find the surface charge density σ in terms of C_1, C_2, C_3, V.

Solution: Let the area of the probe be A. Then the charge we are concerned with, at the surface of the dielectric, is $A\sigma$. We neglect the fringing field at the edge of the probe. Part of the field of $A\sigma$ goes to the probe, and part goes to ground. Let

$$A\sigma = Q + Q', \qquad \text{where } Q = A\epsilon_0 E_2, \qquad Q' = A\epsilon_r\epsilon_0 E_1, \tag{1}$$

as in Fig. 6-7b. The figure also shows the charges induced on the conductors. Since the probe and its lead to the electrometer are exceedingly well insulated, the net charge induced on them is zero.

Thus $V = Q/C_3$, and we must find σ, C_3 being known.

Now we have two expressions for the voltage V between the probe and ground:

$$V = \frac{Q}{C_3} = \frac{Q'}{C_1} - \frac{Q}{C_2}. \tag{2}$$

So

$$Q' = Q\left(\frac{C_1}{C_2} + \frac{C_1}{C_3}\right). \tag{3}$$

Substituting in Eq. 1,

$$A\sigma = Q\left(1 + \frac{C_1}{C_2} + \frac{C_1}{C_3}\right). \tag{4}$$

Since $Q = C_3 V$,

$$\sigma = \left(C_1 + C_3 + \frac{C_1 C_3}{C_2}\right)\frac{V}{A}. \tag{5}$$

b) Express σ in terms of the thickness t_1 of the dielectric, the distance t_2 between the probe and the dielectric, and ϵ_r.

Solution:

$$C_1 = \epsilon_r \epsilon_0 A/t_1, \qquad C_2 = \epsilon_0 A/t_2, \tag{6}$$

and

$$\sigma = \left(\frac{\epsilon_r \epsilon_0}{t_1} + \frac{C_3}{A} + \frac{\epsilon_r C_3}{A}\frac{t_2}{t_1}\right)V. \tag{7}$$

6-12E *VARIABLE CAPACITOR UTILIZING A PRINTED-CIRCUIT BOARD*

A printed-circuit board is a sheet of plastic, about one millimeter thick, one side of which is covered with a thin coating of copper. Part of the copper can be etched away, leaving conducting paths that serve to interconnect resistors, capacitors, and so forth. The terminals of these components are soldered directly to the copper.

Figure 6-8 shows a variable capacitor in which a sliding conducting plate lies under a printed-circuit board that has been etched to give a prescribed variation of capacitance with position z.

At the position z, the capacitance is

$$C = \epsilon_r \epsilon_0 \frac{1}{t}\int_0^z y\, dz,$$

where t is the thickness of the plastic sheet of relative permittivity ϵ_r.

Set $\epsilon_r = 3$, $t = 1$ millimeter, and find $y(z)$ giving the following values of C, with C expressed in farads and x in meters:

a) $10^{-9}z$, b) $10^{-8}z^2$.

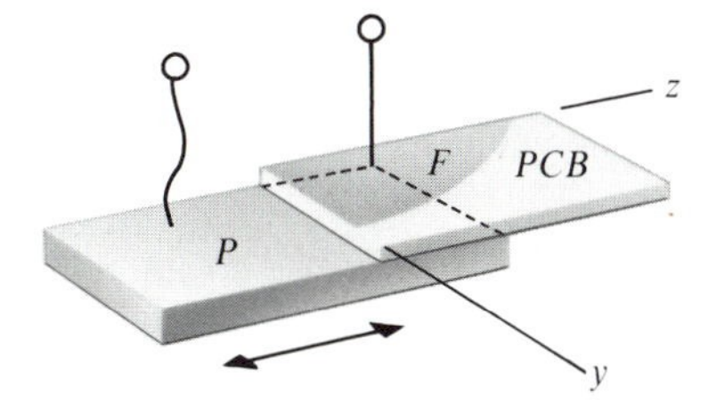

Figure 6-8 Variable capacitor made with a printed-circuit board *PCB* and a conducting plate *P*. The capacitance between the terminals depends on the position of the plate *P* and on the shape of the copper foil *F*. See Prob. 6-12.

In both cases, draw a curve of $y(z)$ from $z = 0$ to $z = 100$ millimeters, showing the unetched region on the circuit board.

6-13E *EQUIPOTENTIAL SURFACES*

A liquid dielectric is situated in an electric field. A thin conducting sheet carrying zero net charge is introduced into the dielectric so as to coincide with an equipotential surface.

a) Is the electric field disturbed?

b) What is the value of σ_f on the surfaces of the sheet at a point where the electric displacement is D?

6-14E *NON-HOMOGENEOUS DIELECTRICS*

Show that a non-homogeneous dielectric can have a volume density of bound charge in the absence of a free charge density.

6-15 *FIELD OF A SHEET OF ELECTRONS TRAPPED IN LUCITE*

When a block of insulating material such as Lucite is bombarded with high-energy electrons, the electrons penetrate into the material and remain trapped inside. In one particular instance, a 0.1 microampere beam bombarded an area of 25 square centimeters of Lucite ($\epsilon_r = 3.2$) for 1 second, and essentially all the electrons were trapped about 6 millimeters below the surface in a region about 2 millimeters thick. The block was 12 millimeters thick.

In the following calculation, neglect edge effects and assume a uniform density for the trapped electrons. Assume also that both faces of the Lucite are in contact with grounded conducting plates.

Show that

a) in the region where the electrons are trapped,

$\rho_f = -2.000 \times 10^{-2}$ coulomb/meter^3,

$\rho_b = 1.375 \times 10^{-2}$ coulomb/meter^3;

b) in the neutral region,

$D_n = -2.000 \times 10^{-5}$ coulomb/meter^2,

$P_n = -1.375 \times 10^{-5}$ coulomb/meter^2,

$E_n = -7.062 \times 10^5$ volts/meter,

$V_n = 7.062 \times 10^5 x - 4{,}237$ volts;

c) at the surfaces,

$\sigma_b = -1.375 \times 10^{-5}$ coulomb/meter^2;

d) in the charged region,

$$\frac{dD_c}{dx} = -2.000 \times 10^{-2} \text{ coulomb/meter}^3,$$

$$D_c = -2.000 \times 10^{-2} x \text{ coulomb/meter}^2,$$

$$\frac{dE_c}{dx} = -7.062 \times 10^8 \text{ volts/meter}^2,$$

$$E_c = -7.068 \times 10^8 x \text{ volts/meter},$$

$$\frac{d^2 V_c}{dx^2} = 7.062 \times 10^8 \text{ volts/meter}^2,$$

$$V_c = 3.531 \times 10^8 x^2 - 3884 \text{ volts}.$$

e) Draw curves of D, E, V as functions of the distance x to the midplane, from $x = -6$ to $x = 6$ millimeters.

f) Show that the stored energy is 1.88×10^{-4} joule.

g) Is there any danger that the block will explode?

6-16 *SHEET ELECTRET*

A sheet electret is polarized in the direction normal to its surface.

Draw a figure showing the polarization **P**, the surface charge densities $+\sigma_b$ and $-\sigma_b$, as well as the electric field intensity **E** and the electric displacement **D**, both inside and outside the electret.

6-17 *RELATION BETWEEN R AND C FOR ANY PAIR OF ELECTRODES*

When a parallel-plate capacitor is filled with a dielectric whose relative permittivity is ϵ_r, its capacitance is C. If the dielectric is slightly conducting, its resistance is R.

Show that $RC = \epsilon_r \epsilon_0 / \sigma$, where σ is the conductivity.

This rule applies to any pair of electrodes submerged in a medium, as long as the conductivity of the electrodes is much larger than that of the medium.

CHAPTER 7

DIELECTRICS: II

Continuity Conditions at an Interface, Energy Density and Forces, Displacement Current

This chapter will complete our study of dielectrics.

We shall first discuss the important continuity conditions that apply at the interface between two media. These conditions concern the potential V, the tangential component of **E**, and the normal component of **D**.

Then we must return to the energy stored in an electrostatic field. This stored energy will give us the forces exerted on conductors immersed in non-conducting liquids. It will also give us the forces acting within the dielectrics themselves.

If the electric field is a function of the time, then the motion of charge within the molecules of the dielectric gives a polarization current. The displacement current is equal to the polarization current plus another current that exists whenever the electric field varies with time, even in a vacuum.

7.1 CONTINUITY CONDITIONS AT THE INTERFACE BETWEEN TWO MEDIA

The quantities V, **E**, and **D** must satisfy certain boundary conditions.

7.1.1 THE POTENTIAL V

At the boundary between two media, V must be continuous, for a discontinuity would imply an infinitely large electric field intensity, which is physically impossible.

The potential is normally set equal to zero at infinity if the charge distribution is of finite extent. The potential is constant throughout any conductor, as long as the electric charges are at rest.

7.1.2 THE NORMAL COMPONENT OF **D**

Consider a short Gaussian cylinder drawn about a boundary surface, as in Fig. 7-1. The end faces of the cylinder are parallel to the boundary and arbitrarily close to it. The boundary carries a free surface charge density σ_f. If the area S is small, D and σ_f do not vary significantly over it and, according to Gauss's law, the flux of **D** emerging from the flat cylinder is equal to the charge enclosed:

$$(D_{n1} - D_{n2})S = \sigma_f S. \tag{7-1}$$

The only flux of **D** is through the end faces, since the area of the cylindrical surface is arbitrarily small. Thus

$$D_{n1} - D_{n2} = \sigma_f. \tag{7-2}$$

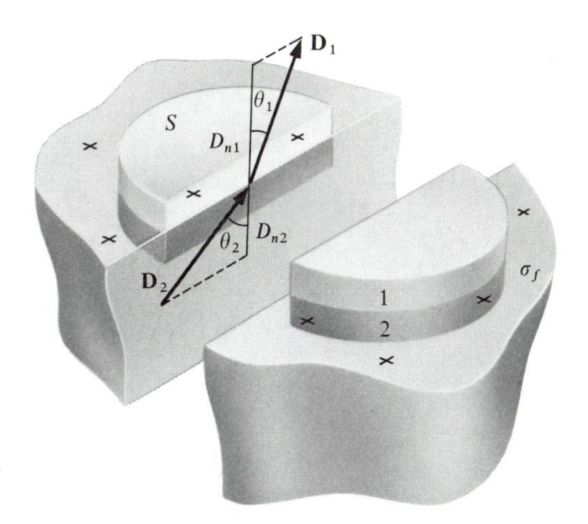

Figure 7-1 Gaussian cylinder on the interface between two media 1 and 2. The difference $D_{n1} - D_{n2}$ between the *normal* components of **D** is equal to the surface density of free charge σ_f.

At the boundary between two dielectric media the free surface charge density σ_f is generally zero and then D_n is continuous across the boundary. On the other hand, if the boundary is between a conductor and a dielectric, and if the electric field is constant, $\mathbf{D} = 0$ in the conductor and $D_n = \sigma_f$ in the dielectric, σ_f being the free charge density on the surface of the conductor.

7.1.3 THE TANGENTIAL COMPONENT OF E

This boundary condition follows from the fact that the line integral of $\mathbf{E} \cdot \mathbf{dl}$ around any closed path is zero for electrostatic fields. Consider the path shown in Fig. 7-2, with two sides parallel to the boundary, of length L, and arbitrarily close to it. The other two sides are perpendicular to the boundary. In calculating this line integral, only the first two sides of the path are important, since the lengths of the other two approach zero. If the path is small enough, E_t does not vary significantly over it and

$$E_{t1}L - E_{t2}L = 0, \tag{7-3}$$

or

$$E_{t1} = E_{t2}. \tag{7-4}$$

The tangential component of $\mathbf{E}$ is therefore continuous across the boundary.

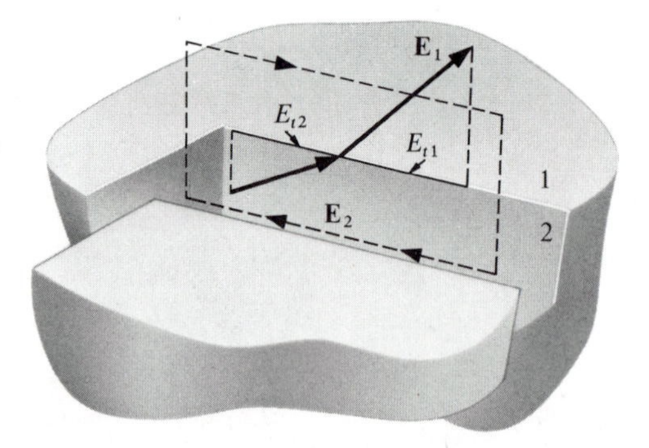

Figure 7-2 Closed path of integration crossing the interface between two media 1 and 2. Whatever be the surface charge density σ_f the *tangential* components of $\mathbf{E}$ on either side of the interface are equal: $E_{t1} = E_{t2}$.

If the boundary lies between a dielectric and a conductor, then $\mathbf{E} = 0$ in the conductor and $E_t = 0$ in both media. With static charges, $\mathbf{E}$ is therefore normal to the surface of a conductor.

7.1.4 BENDING OF LINES OF FORCE

It follows from the boundary conditions that the **D** and **E** vectors change direction at the boundary between two dielectrics. In Fig. 7-3, using Eq. 7-2 with $\sigma_f = 0$,

$$D_1 \cos \theta_1 = D_2 \cos \theta_2, \tag{7-5}$$

or

$$\epsilon_{r1}\epsilon_0 E_1 \cos \theta_1 = \epsilon_{r2}\epsilon_0 E_2 \cos \theta_2 \tag{7-6}$$

and, from Eq. 7-4,

$$E_1 \sin \theta_1 = E_2 \sin \theta_2. \tag{7-7}$$

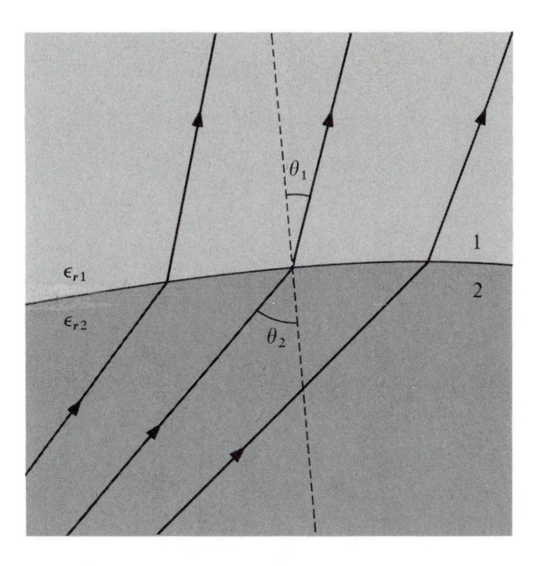

Figure 7-3 Lines of **D** or of **E** crossing the interface between two media 1 and 2. The lines change direction in such a way that $\epsilon_{r1} \tan \theta_2 = \epsilon_{r2} \tan \theta_1$.

Then, from the last two equations,

$$\frac{\tan\theta_1}{\tan\theta_2} = \frac{\epsilon_{r1}}{\epsilon_{r2}}. \tag{7-8}$$

The larger angle from the normal is in the medium with the larger relative permittivity.

7.1.5 *EXAMPLES: POINT CHARGE NEAR A DIELECTRIC, SPHERE OF DIELECTRIC IN AN ELECTRIC FIELD*

Figure 7-4 shows two examples of the bending of lines of force at the boundary between two dielectrics. The lines of **D** are broken but continuous, since lines of **D**

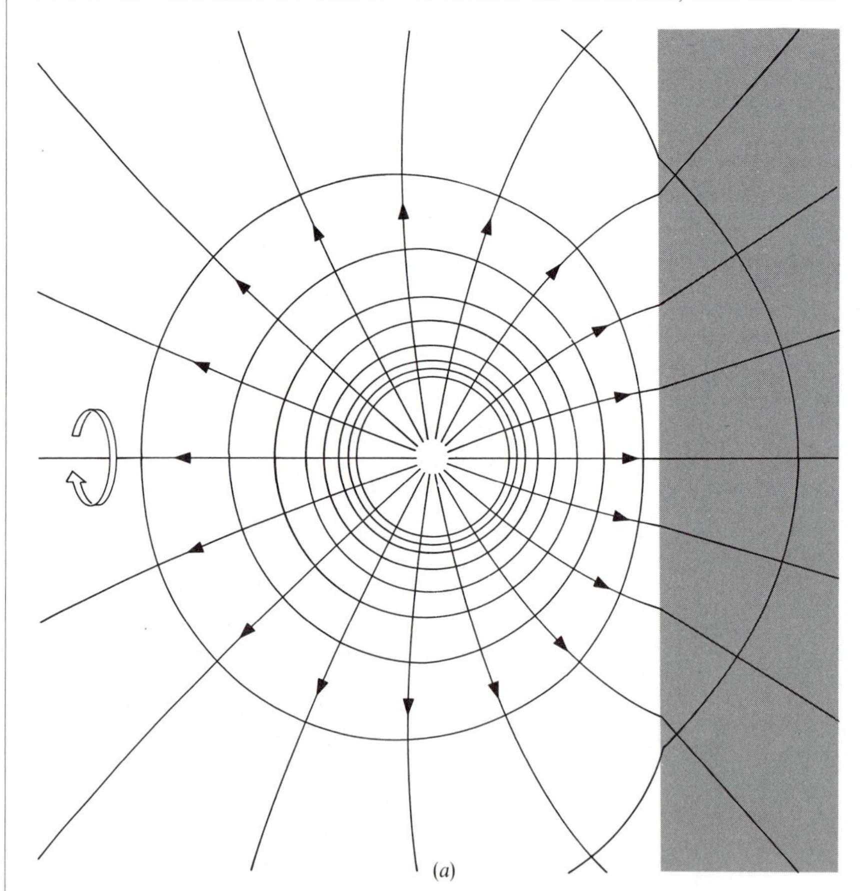

Figure 7-4 (*a*) Lines of **D** and equipotentials for a point charge in air near a dielectric. They are not shown near the charge because they get too close together.

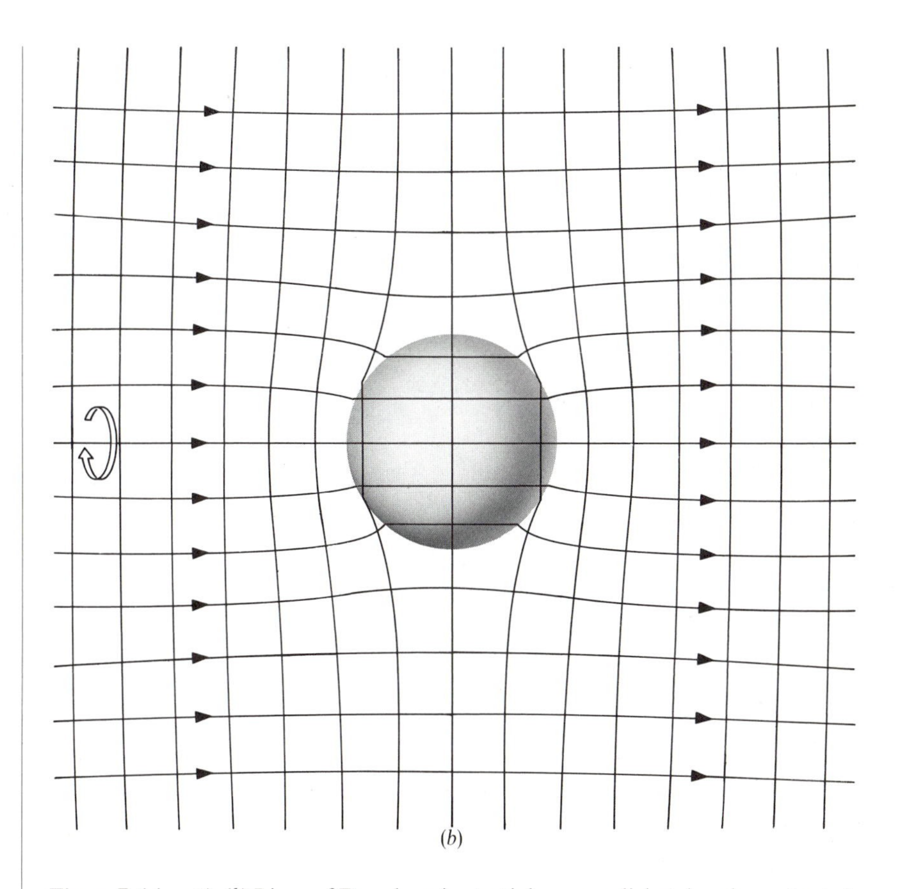

Figure 7-4 (*cont.*) (*b*) Lines of **D** and equipotentials near a dielectric sphere situated in air in a uniform electric field. In both cases, the lines of **D** are indicated by arrows, and the equipotential surfaces are generated by rotating the figures around the horizontal axis.

terminate only on free charges. Some lines of **E** either originate or terminate at the interface, according as to whether σ_b is positive or negative.

In Fig. 7-4b, the lines of **D** crowd into the sphere, showing that **D** is larger inside than outside. However, the equipotentials spread out inside, and **E** is weaker inside than outside. The density of the lines of **E** is lower inside than outside, since some of the lines of **E** coming from outside terminate at the surface.

Note that, in the case of Fig. 7-4b, the field inside the sphere is uniform. Note also that the field outside is hardly disturbed at distances larger than one radius from the surface.

7.2 POTENTIAL ENERGY OF A CHARGE DISTRIBUTION, ENERGY DENSITY

To calculate the potential energy associated with an electric field in dielectric material, we refer to the parallel-plate capacitor as in Sec. 4.3.

We now have a dielectric-filled capacitor as in Fig. 6-5. If we add charges dQ until the top plate carries a charge Q and is at a potential V, we must expend an energy

$$\frac{1}{2}QV = \frac{1}{2}\frac{Q^2}{C} = \frac{1}{2}Q^2\frac{s}{\epsilon_r\epsilon_0 A}, \tag{7-9}$$

$$= \frac{1}{2}\left(\frac{Q}{A}\right)\left(\frac{1}{\epsilon_r\epsilon_0}\frac{Q}{A}\right)sA, \tag{7-10}$$

$$= \frac{1}{2}DEsA, \tag{7-11}$$

where D is the electric displacement (Sec. 6.5), E is the electric field intensity, and sA is the volume of the dielectric.

The *energy density* is now $DE/2$ or, if the dielectric is linear, as is usually the case, $\epsilon_r\epsilon_0 E^2/2$. The potential energy is then given by the integral of $\epsilon_r\epsilon_0 E^2/2$ evaluated over the volume occupied by the field,

$$\frac{\epsilon_r\epsilon_0}{2}\int_\tau E^2\,d\tau.$$

7.3 FORCES ON CONDUCTORS IN THE PRESENCE OF DIELECTRICS

The forces between conductors in the presence of dielectrics can be calculated by the method of virtual work that we used in Sec. 4.5.

When the conductors are immersed in a *liquid* dielectric, the forces are always found to be *smaller* than those in air by the factor ϵ_r if the *charges* are the same in both cases. They are *larger* than in air by the factor ϵ_r if the *electric fields* (and hence the *voltages*) are the same.

The case of solid dielectrics will be illustrated in Probs. 7-9 and 7-10.

7.3.1 EXAMPLE: PARALLEL-PLATE CAPACITOR IMMERSED IN A LIQUID DIELECTRIC

We found in Sec. 4.5.2 that the force between the plates of an air-insulated parallel-plate capacitor is $\epsilon_0 E^2 A/2$, or $(\sigma^2/2\epsilon_0)A$, where σ is the surface charge density and A is the area of one plate. We can perform a similar calculation for a parallel-plate capacitor immersed in a liquid dielectric. The force per unit area is equal to the energy density (Sec. 4.5). Then the mechanical force holding the plates apart is

$$F_m = \epsilon_r \frac{\epsilon_0}{2} E^2 A. \tag{7-12}$$

For a given electric field strength E, or for a given voltage difference between the plates, the force is *larger* than in air.

Also,

$$F_m = \frac{D^2}{2\epsilon_r \epsilon_0} A = \frac{\sigma^2}{2\epsilon_r \epsilon_0} A, \tag{7-13}$$

$$= \frac{1}{\epsilon_r} \frac{Q^2}{2\epsilon_0 A}, \tag{7-14}$$

and, *for given charges* $+Q$ *and* $-Q$ *on the plates*, the force is *smaller* than in air.

7.4 ELECTRIC FORCES ON DIELECTRICS

When a piece of dielectric is placed in an electric field, its dipoles are subjected to electric forces and torques.

If the field is uniform, and if the molecules have a permanent dipole moment $\mathbf{p}$, each dipole experiences a torque,

$$\mathbf{T} = \mathbf{p} \times \mathbf{E}, \tag{7-15}$$

that tends to align it with the field, but the net force is zero.

If the field is non-uniform, then the forces on the two charges of a dipole are unequal, and there is also a net force.

Let us consider a dielectric whose molecules do *not* have a permanent dipole moment. We can find a general expression for the electric force per unit volume in a non-uniform field by considering the dielectric-filled cylindrical capacitor of Fig. 7-5.

The electric field is radial, and its magnitude varies as $1/r$. The net electric force on a dipole is inward, because the inward force on $-Q$ is

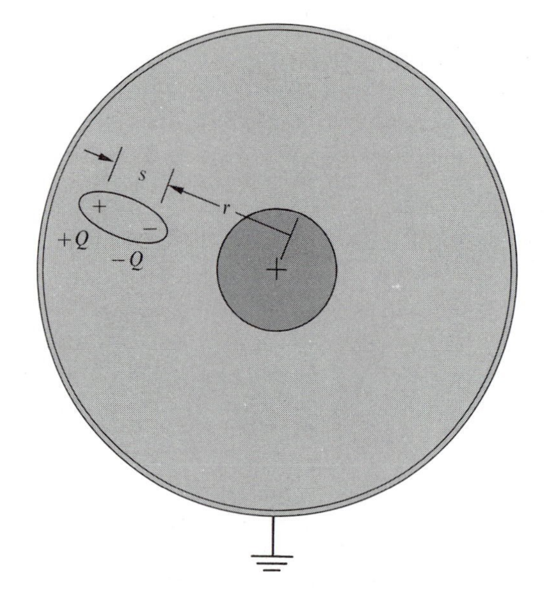

Figure 7-5 Dielectric-filled cylindrical capacitor.

larger than the outward force on $+Q$. This net inward force is

$$f = QE_r - QE_{r+dr} = Q\frac{dE}{dr}s, \tag{7-16}$$

$$= p\frac{dE}{dr}, \tag{7-17}$$

where $p = Qs$ is the dipole moment.

Therefore, if there are N dipoles per unit volume, the electric force per unit volume is

$$F_1 = Np\frac{dE}{dr} = P\frac{dE}{dr}, \tag{7-18}$$

$$= (\epsilon_r - 1)\epsilon_0 E\frac{dE}{dr} = (\epsilon_r - 1)\epsilon_0\frac{d}{dr}(E^2/2), \tag{7-19}$$

$$= \left(1 - \frac{1}{\epsilon_r}\right)\frac{d}{dr}(\epsilon_r\epsilon_0 E^2/2). \tag{7-20}$$

More generally, the electric force per unit volume is proportional to the gradient of the energy density:

$$\mathbf{F}_1 = \left(1 - \frac{1}{\epsilon_r}\right) \nabla(\epsilon_r \epsilon_0 E^2/2). \tag{7-21}$$

The dielectric tends to move to where the field is strongest. In the above expression, E is the electric field intensity inside the dielectric, of course.

7.4.1 *EXAMPLE: CURVE TRACER*

A curve tracer is an electro-mechanical device that draws curves; the device is under the control of a computer. The curves are drawn on a sheet of paper that must stay rigidly fixed to the platen. One way of holding the paper is shown in Fig. 7-6. Parallel wires are embedded in the plastic platen about two millimeters apart and charged to plus and minus 300 volts. This gives a highly inhomogeneous field at the surface, with lines of force converging toward the charged wires. Thus ∇E^2 inside the paper has a vertical component directed into the platen.

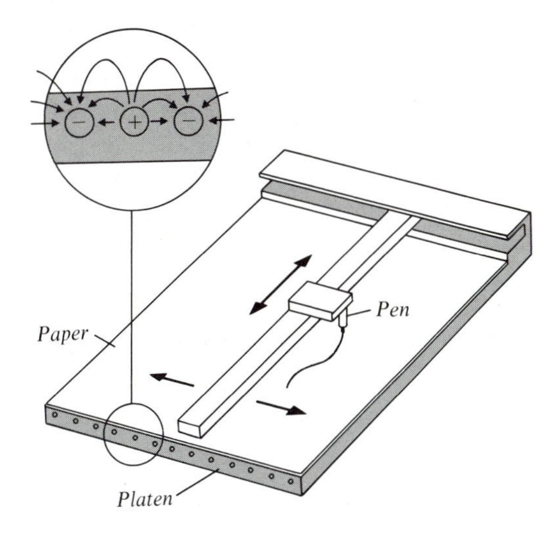

Figure 7-6 Curve tracer. The electric field produced by the charged wires embedded in the platen attracts the electrically neutral sheet of paper toward the platen.

7.5 THE POLARIZATION CURRENT DENSITY $\partial \mathbf{P}/\partial t$

When a dielectric is placed in an electric field that is a function of time, the motion of the bound charges gives a *polarization current*. We can find the polarization current density $\mathbf{J}_b$ as follows.

For any volume τ bounded by a surface S, the rate at which bound charge flows out through S must be equal to the rate of decrease of the bound charge within S (Sec. 5.1.1):

$$\int_S \mathbf{J}_b \cdot \mathbf{da} = -\frac{\partial}{\partial t}\int \rho_b \, d\tau. \tag{7-22}$$

Using the divergence theorem on the left and substituting the value of ρ_b on the right from Eq. 6-2,

$$\int_\tau \nabla \cdot \mathbf{J}_b \, d\tau = \frac{\partial}{\partial t}\int_\tau \nabla \cdot \mathbf{P} \, d\tau = \int_\tau \nabla \cdot \frac{\partial \mathbf{P}}{\partial t} \, d\tau. \tag{7-23}$$

We have put the $\partial/\partial t$ under the integral sign because it is immaterial whether the derivative with respect to the time is calculated first or last. Since τ is any volume, the integrands must be equal and the polarization current density is

$$\mathbf{J}_b = \frac{\partial \mathbf{P}}{\partial t}. \tag{7-24}$$

We could have added a constant of integration independent of the coordinates, but such a constant would be of no interest. We have used the partial derivative with respect to t because $\mathbf{P}$ can be a function both of the time and of the space coordinates x, y, z.

7.6 THE DISPLACEMENT CURRENT DENSITY $\partial \mathbf{D}/\partial t$

The *displacement current density* is

$$\frac{\partial \mathbf{D}}{\partial t} = \frac{\partial}{\partial t}(\epsilon_0 \mathbf{E} + \mathbf{P}) = \frac{\partial}{\partial t}\epsilon_0 \mathbf{E} + \frac{\partial \mathbf{P}}{\partial t}, \tag{7-25}$$

where $\partial \mathbf{P}/\partial t$ is the polarization current density of the previous section.

Note the first term: it can exist even in a vacuum.

7.6.1 EXAMPLE: DIELECTRIC-INSULATED PARALLEL-PLATE CAPACITOR

Let us calculate the displacement current in a dielectric-insulated parallel-plate capacitor.

Consider a current I flowing through a capacitor as in Fig. 7-7. We assume that positive charge flows from left to right, that $Q = 0$ at $t = 0$, and that edge effects are negligible.

On the left-hand side of the capacitor, Q increases and $dQ/dt = I$. On the right, a charge of equal magnitude but opposite sign builds up and, again, $dQ/dt = I$.

Now $Q = A\sigma$, $D = \sigma = Q/A$, where A is the area of one plate, and σ is the surface charge density.

The displacement current in the capacitor is

$$A \frac{dD}{dt} = \frac{dQ}{dt} = I. \tag{7-26}$$

The displacement current $A(dD/dt)$ is composed of two parts, first

$$A \frac{d}{dt}(\epsilon_0 E) = A \frac{d}{dt}\left(\frac{D}{\epsilon_r}\right) = A \frac{d}{dt}\left(\frac{\sigma}{\epsilon_r}\right) = \frac{1}{\epsilon_r}\frac{dQ}{dt}, \tag{7-27}$$

and the polarization current

$$A \frac{dP}{dt} = A \frac{d}{dt}(D - \epsilon_0 E) = A\left(1 - \frac{1}{\epsilon_r}\right)\frac{d\sigma}{dt} = \left(1 - \frac{1}{\epsilon_r}\right)\frac{dQ}{dt}. \tag{7-28}$$

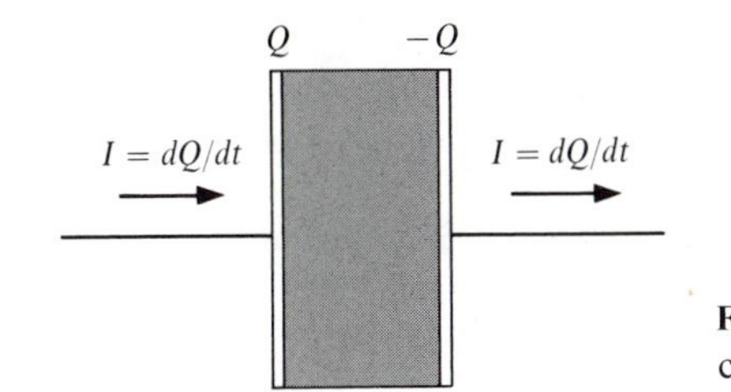

Figure 7-7 Current flowing through a capacitor.

7.7 FREQUENCY AND TEMPERATURE DEPENDENCE, ANISOTROPY

There are three basic polarization processes. (a) In *induced* or *electronic* polarization, the center of negative charge in a molecule is displaced, relative to the center of positive charge, when an external field is applied. (b) In *orientational* polarization, molecules with a permanent dipole moment tend to be aligned by an external field, the magnitude of the susceptibility being inversely proportional to the temperature. (c) Finally, *ionic* polarization occurs in ionic crystals: ions of one sign may move, with respect to ions of the other sign, when an external field is applied.

For a given magnitude of **E**, both the magnitude and the phase of **p** are functions of the frequency, first because of the various polarization processes that come into play as the frequency changes, and also because of the existence of resonances. Thus, since ϵ_r is a function of the frequency, ϵ_r is strictly definable only for a pure sinusoidal wave. See Table 6-1.

In many substances the relative permittivity decreases by a large factor as the temperature is lowered through the freezing point.

Anisotropy is another departure from the ideal dielectric behavior. Crystalline solids commonly have different dielectric properties in different crystal directions because the charges that constitute the atoms of the crystal are able to move more easily in some directions than in others. The polarization **P** is then parallel to **E** only when **E** is along certain preferred directions.

7.7.1 *EXAMPLES: WATER, SODIUM CHLORIDE, NITROBENZENE, COMPOUNDS OF TITANIUM*

Water has a relative permittivity of 81 in an electrostatic field, and of about 1.8 at optical frequencies. The large static value is attributable to the orientation of the permanent dipole moments, but the rotational inertia of the molecules is much too large for any significant response at optical frequencies.

Similarly, the relative permittivity of *sodium chloride* is 5.6 in an electrostatic field and 2.3 at optical frequencies. The larger static value is attributed to ionic motion, which again is impossible at high frequencies.

In *nitrobenzene*, ϵ_r falls from about 35 to about 3 in changing from the liquid to the solid state at 279 kelvins. In the solid state, the permanent dipoles of the nitrobenzene molecules are fixed rigidly in the crystal lattice and cannot rotate under the influence of an external field.

Ceramic capacitors utilize various *compounds of titanium* as dielectrics. These compounds are used because of their large relative permittivities ranging up to 10,000. However, ϵ_r varies with the temperature, with the applied voltage, and with the operating frequency.

7.8 *SUMMARY*

At the boundary between two media, both V and the tangential component of **E** are continuous, but the difference between the normal components of **D** is equal to the surface density of free charge σ_f.

The potential energy stored in an electric field can be calculated from the energy density $\epsilon_r \epsilon_0 E^2/2$. Hence the force per unit area exerted on a conductor submerged in a *liquid* dielectric is $\epsilon_r \epsilon_0 E^2/2$.

The force per unit volume on a dielectric is

$$\mathbf{F}_1 = \left(1 - \frac{1}{\epsilon_r}\right) \nabla(\epsilon_r \epsilon_0 E^2/2). \tag{7-21}$$

The *displacement current density* at a point is

$$\frac{\partial \mathbf{D}}{\partial t} = \frac{\partial}{\partial t} \epsilon_0 \mathbf{E} + \frac{\partial \mathbf{P}}{\partial t}, \tag{7-25}$$

where $\partial \mathbf{P}/\partial t$ is the *polarization current density*.

Because of the nature of polarization processes, ϵ_r is a function of frequency. For most commonly used dielectrics, however, ϵ_r changes very little, even when the frequency changes by many orders of magnitude (Table 6-1). In anisotropic media, **P** is parallel to **E** only when **E** is along certain preferred directions.

PROBLEMS

7-1E *CONTINUITY CONDITIONS AT AN INTERFACE*

Discuss the continuity conditions at the surface of the dielectric cylinder of Prob. 6-7.

7-2E *CONTINUITY CONDITIONS AT AN INTERFACE*

Discuss the continuity conditions at the surface of the dielectric sphere of Prob. 6-9.

7-3E *ENERGY STORAGE IN CAPACITORS*

A one-microfarad capacitor is charged to a potential of one kilovolt.

How high could you lift a one-kilogram mass with the stored energy, if you could achieve 100% efficiency?

7-4E *ENERGY STORAGE IN CAPACITORS*

The maximum stored energy per cubic meter in a capacitor is $\epsilon_r \epsilon_0 a^2/2$, where a is the dielectric strength of the dielectric. The dielectric strength is the maximum electric field intensity before rupture. See Prob. 6-5.

It is suggested that a small vehicle could be propelled by an electric motor fed by charged capacitors. Comment on this suggestion, assuming that the only problem is one of energy storage.

A good dielectric to use would be Mylar, which has a dielectric strength of 1.5×10^8 volts per meter and a relative permittivity of 3.2.

Note that the use of a dielectric increases the capacitance by ϵ_r and also permits the use of a higher voltage. The dielectric strength of air is only 3×10^6 volts per meter.

7-5 *E, D, σ, AND W_1 FOR THE THREE CAPACITORS OF FIG. 7-8*

Figure 7-8 shows three capacitors. In all three the electrodes have an area S and are separated by a distance s. In A, the dielectric is air. In B, the dielectric has a relative permittivity ϵ_r. In C, we have the same dielectric, plus a thin film of air.

a) Find E, D, the surface charge density σ on the electrodes, and the energy density W_1 for A and B.

Express your results for B in terms of those found for A.

Solution: For capacitor A,

$$E_A = V/s, \tag{1}$$

$$D_A = \epsilon_0 E_A = \epsilon_0 V/s, \tag{2}$$

$$\sigma_A = D_A = \epsilon_0 V/s, \tag{3}$$

$$W_{1A} = \frac{1}{2}\epsilon_0 E_A^2 = \frac{1}{2}\epsilon_0 V^2/s^2. \tag{4}$$

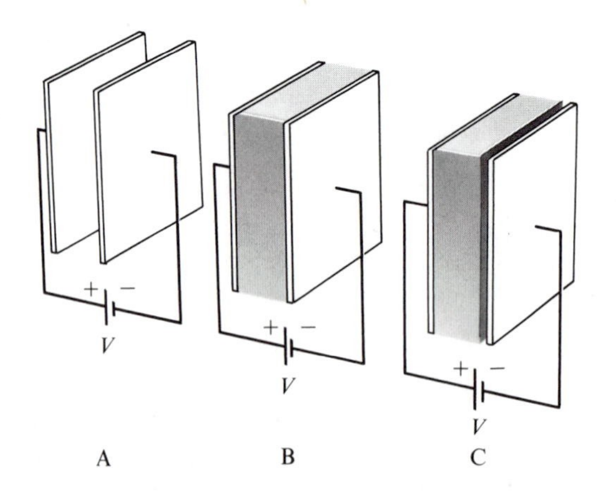

Figure 7-8 Three capacitors: A, without dielectric; B, with dielectric; C, with dielectric and an air film. See Prob. 7-5.

For capacitor B,

$$E_B = V/s = E_A, \tag{5}$$

$$D_B = \epsilon_r \epsilon_0 E_B = \epsilon_r \epsilon_0 V/s = \epsilon_r D_A, \tag{6}$$

$$\sigma_B = D_B = \epsilon_r \epsilon_0 V/s = \epsilon_r \sigma_A, \tag{7}$$

$$W_{1B} = \frac{1}{2} \epsilon_r \epsilon_0 E_B^2 = \frac{1}{2} \epsilon_r \epsilon_0 V^2/s^2 = \epsilon_r W_{1A}. \tag{8}$$

b) Find E, D, W_1 for the dielectric in C, and σ on the left-hand electrode. Express your results in terms of the quantities found for capacitor A.

Solution: Since the air film in capacitor C is thin, we may set

$$E_{Cd} = V/s = E_A. \tag{9}$$

Then D, σ on the left-hand electrode, and W_1 are the same as for capacitor B:

$$D_{Cd} = \epsilon_r D_A, \tag{10}$$

$$\sigma_{Cd} = \epsilon_r \sigma_A, \tag{11}$$

$$W_{1Cd} = \epsilon_r W_{1A}. \tag{12}$$

c) Find E, D, W_1 for the air film of capacitor C, and σ on the right-hand electrode. Express again your results in terms of the corresponding quantities found for capacitor A.

Solution: The electric displacement D in the air film is the same as in the dielectric, since there are no free charges between the plates. Then

$$D_{Ca} = \epsilon_r D_A, \tag{13}$$

$$\sigma_{Ca} = D_{Ca} = \epsilon_r D_A = \epsilon_r \sigma_A, \tag{14}$$

$$E_{Ca} = D_{Ca}/\epsilon_0 = \epsilon_r D_A/\epsilon_0 = \epsilon_r E_A, \tag{15}$$

$$W_{1Ca} = \frac{1}{2} \epsilon_0 E_{Ca}^2 = \frac{1}{2} \epsilon_0 \epsilon_r^2 E_A^2 = \epsilon_r^2 W_{1A}. \tag{16}$$

7-6 *BOUND SURFACE CHARGE DENSITY*

Find the bound surface charge densities on the dielectric of capacitor B of Prob. 7-5 in terms of V/s.

7-7E *EXAMPLE OF A LARGE ELECTRIC FORCE*

One author claims that he can attain fields of 4×10^7 volts per meter over a 2.5 millimeter gap in purified nitrobenzene ($\epsilon_r = 35$), and that the resulting electric force per unit area on the electrodes is then more than two atmospheres.

Is the force really that large?

One atmosphere equals about 10^5 pascals.

7-8 *PERPETUAL-MOTION MACHINE*

In Prob. 7-5 we showed that, for the capacitor C, the energy density in the dielectric and in the air film are, respectively,

$$\frac{\epsilon_r \epsilon_0}{2} \frac{V^2}{s^2} \quad \text{and} \quad \frac{\epsilon_r^2 \epsilon_0}{2} \frac{V^2}{s^2}.$$

Now this is disturbing because it appears to mean that the net force on the capacitor is not zero. The forces on the capacitor plates are

$$\frac{\epsilon_r \epsilon_0}{2} \frac{V^2}{s^2} S \quad \text{and} \quad \frac{\epsilon_r^2 \epsilon_0}{2} \frac{V^2}{s^2} S.$$

Is there a force on the dielectric? Well, the **E** inside is uniform and $\nabla E^2 = 0$, which means that the force density is zero. So it seems that the net force on the capacitor is

$$(\epsilon_r - 1) \frac{\epsilon_r \epsilon_0}{2} \frac{V^2}{s^2} S.$$

With Mylar ($\epsilon_r = 3.2$), $s = 0.1$ millimeter, $S = 1$ meter^2, and $V = 1$ kilovolt, the force is about 3×10^3 newtons! One could therefore propel a large vehicle indefinitely with a set of capacitors like this, fed by a small battery supplying a voltage V at zero current! Where have we erred?

Hint: (i) The values given above for the forces on the electrodes are correct. (ii) *Inside* the dielectric, there is no force. (iii) The forces on the faces of the dielectric are not zero, however. Remember that Coulomb's law applies to any net accumulation of charge, irrespective of the presence of matter.

7-9 *SELF-CLAMPING CAPACITOR*

A certain capacitor is formed of two aluminum plates of area A, separated by a sheet of dielectric of thickness t, and connected to a source of voltage V.

For various reasons, it is impossible to apply a force of more than a few tens of newtons to press the two plates together mechanically. Since neither the plates nor the dielectric are perfectly flat, the plates are spaced by a distance t, plus a good fraction of a millimeter. This is highly objectionable because the capacitance must be as large as possible.

You are asked to investigate whether the electrostatic force of attraction on the plates might not be sufficiently large to reduce the air film thickness to a negligible value.

Set $A = 4.38 \times 10^{-2}$ square meter, $t = 0.762$ millimeter, $\epsilon_r = 3.0$, $V =$ 60 kilovolts.

See also the next problem.

7-10 *ELECTROSTATIC CLAMPS*

Electrostatic clamps are sometimes used for holding work pieces while they are being machined. They utilize an insulated conducting plate charged to several thousand volts and covered with a thin insulating sheet. The work piece is placed on the sheet and grounded.

One particular type operates at 3000 volts and is advertised as having a holding power of 2×10^5 pascals.

The insulator is Mylar ($\epsilon_r = 3.2$).

a) Suppose first that there is a film of air on either side of the Mylar.

Use the results of Prob. 7-5 to show that the Mylar film is 15 micrometers thick.

b) Now calculate the electric field intensity in the air film. You should find over 6×10^8 volts per meter, which is over 200 times the dielectric strength of air (Prob. 7-4). Sparking could occur in the air film, and the dielectric could deteriorate. Mylar operates satisfactorily under these conditions, however.

c) Now assume that the air film is replaced by a film of transformer oil with a high dielectric strength and a permittivity that is not much different from that of the Mylar.

What is the thickness of the Mylar now?

7-11 *CALCULATING AN ELECTRIC FORCE BY THE METHOD OF VIRTUAL WORK*

Figure 7-9 shows a pair of parallel conducting plates immersed in a dielectric. The lower plate is fixed in position and grounded; the upper one can slide horizontally and is maintained at a fixed voltage V.

Use the method of virtual work (Sec. 4.5) to calculate the horizontal force on the upper plate when $V = 1000$ volts, $s = 1$ millimeter, $l = 100$ millimeters, $\epsilon_r = 3.0$.

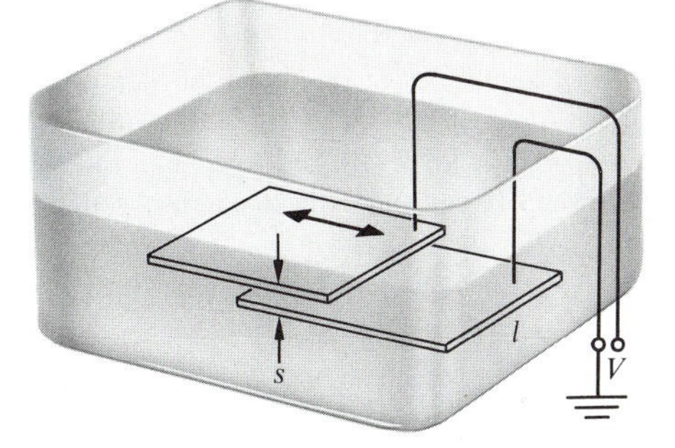

Figure 7-9 Pair of conducting plates immersed in a dielectric. The lower one is fixed, while the upper one can slide horizontally. It is possible to find the horizontal force on the upper plate by the method of virtual work. See Prob. 7-11.

7-12 *ELECTRIC FORCE*

Show that ∇E^2 is in the direction of ∇E.

7-13 *CALCULATING AN ELECTRIC FORCE BY THE METHOD OF VIRTUAL WORK*

Use the method of virtual work (Sec. 4.5) to calculate the force on the dielectric sheet in Fig. 7-10 when $V = 1000$ volts, $s = 1$ millimeter, $l = 100$ millimeters, $\epsilon_r = 3.0$.

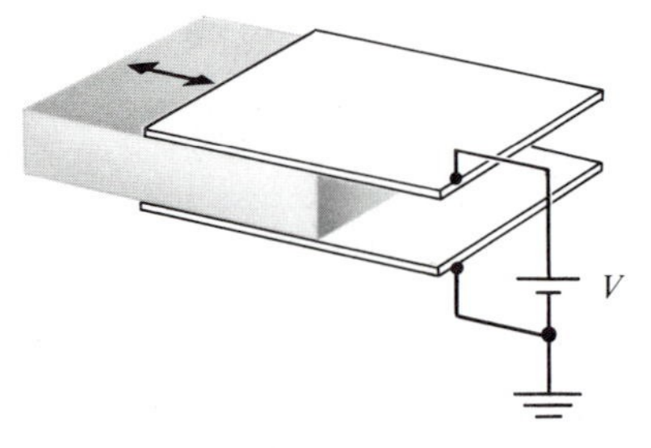

Figure 7-10 Sheet of dielectric partly inserted between a pair of conducting plates. The force on the dielectric can again be calculated from the principle of virtual work. See Prob. 7-13.

7-14 *HIGH-VOLTAGE TRANSMISSION LINES*

Losses on high-voltage transmission lines are higher than normal during foggy or rainy weather. One reason for this is that water collects on the wires and forms droplets. Since a droplet carries a charge of the same sign as that of the wire, electrostatic repulsion forces the droplet to elongate and form a sharp point where the electric field intensity can be high enough to ionize the air. The result is a *corona discharge* in the surrounding air, in which the ions are accelerated by the electric field and collide with neutral molecules, forming more ions and heating the air.

Corona discharges are objectionable because they dissipate energy, mostly by heating the air. Over long distances, they can cause heavy losses. They also cause radio and television interference.

Droplets collect on the wire in drifting by. They are also attracted by the non-uniform electric field.

You are asked to evaluate the importance of the non-uniform electric field. To do this, assume that the air is stagnant and calculate within what distance of a wire the electrostatic force will be larger than the gravitational force. Clearly, if that distance is much larger than the conductor radius, then electrostatic attraction is important.

The transmission line under consideration consists of a pair of conductors 11.7 millimeters in diameter, separated by a distance of 2 meters and operating at 100 kilovolts (± 50 kilovolts).

You can find an approximate value for the field at a distance r near one wire as follows. If λ is the charge per meter,

$$E \approx \frac{\lambda}{2\pi\epsilon_0 r}.$$

The expression on the right is the field near a single isolated wire, which is a good approximation as long as r is much less than the distance to the other wire. Now the capacitance per meter for this line is approximately $\pi\epsilon_0/5$† and

$$\lambda = (\pi\epsilon_0/5)10^5 \text{ coulombs/meter.}$$

The expression for the force per unit volume given in Sec. 7.4 involves ϵ_r and the value of E inside a water droplet. For water, at low frequencies, $\epsilon_r = 81$ (Table 6-1). To estimate the E inside a droplet, one may use the formula for a dielectric sphere in a uniform field.‡ This is not a bad approximation, because even a raindrop is small, compared to the conductor radius, so that E does not change much over one drop diameter. Then

$$E_{\text{water}} \approx \left(\frac{3}{\epsilon_r + 2}\right)\frac{\lambda}{2\pi\epsilon_0 r}.$$

Solution: Let us first calculate E inside a water droplet situated at a distance r from the center of one of the wires:

$$E_{\text{water}} = \left(\frac{3}{81 + 2}\right)\frac{(\pi\epsilon_0/5)10^5}{2\pi\epsilon_0 r} = 361.4/r \text{ volts/meter.} \tag{1}$$

The electrostatic force per cubic meter on a droplet is

$$F' = \left(1 - \frac{1}{81}\right)\frac{d}{dr}\left[\frac{1}{2} \times 81 \times 8.85 \times 10^{-12} \times \left(\frac{361.4}{r}\right)^2\right], \tag{2}$$

$$= 4.625 \times 10^{-5}\frac{d}{dr}\left(\frac{1}{r^2}\right) = -9.250 \times 10^{-5}\frac{1}{r^3} \text{ newtons/meter}^3. \tag{3}$$

The negative sign indicates that the force is attractive.

Now the gravitational force per cubic meter is 1000 g newtons, so

$$\frac{F'}{1000g} = \frac{9.250 \times 10^{-5}}{9800r^3}. \tag{4}$$

This ratio is equal to unity for $r = 2.113$ millimeters, which is smaller than the radius of the wire.

The electrostatic force of attraction on a droplet is therefore negligible.

---

† *Electromagnetic Fields and Waves*, Eq. B-55.

‡ *Electromagnetic Fields and Waves*, Eq. 4-174.

7-15 *ELECTRIC FORCE ON A DIELECTRIC*

a) Calculate the electric force per cubic meter on the dielectric of a coaxial cable whose inner conductor has a radius of 1 millimeter and whose outer conductor has an inner radius of 5 millimeters. The dielectric has a relative permittivity of 2.5. The outer conductor is grounded, and the inner conductor is maintained at 25 kilovolts.

b) Show that the electric force near the inner conductor is about 300 times larger than the gravitational force if the dielectric has the density of water, namely 10^3 kilograms per cubic meter.

7-16 *DISPLACEMENT AND POLARIZATION CURRENTS*

A capacitor with plates of area A, separated by a dielectric of thickness s and relative permittivity ϵ_r, is connected to a voltage source V through a resistance R.

Calculate the displacement and the polarization currents as functions of the time.

7-17 *DIRECT ENERGY CONVERSION*

It is possible to convert thermal energy into electrical energy with the circuit of Fig. 7-11.

With X connected to Z, the battery B charges the barium titanate capacitor C. Then X is disconnected from Z and the capacitor is heated. The relative permittivity of the titanate decreases and the voltage on C increases. Then X is connected to Y, and C recharges the battery B, while supplying energy to R. The capacitor is then cooled and the cycle repeated.

Such devices have been built to feed electronic circuits on board satellites. The capacitors are in thermal contact with panels on the outer surface of the satellite that are heated periodically by the sun as the satellite rotates about its own axis in space.

Let V_B be the potential supplied by the battery; C_1 the capacitance at the temperature T_1; C_2 and V_2 the capacitance and voltage at T_2; W_{th} the thermal energy required to heat the capacitor; and W_e the electric energy fed to R.

Calculate the efficiency W_e/W_{th} when $V_B = 700$ volts, $V_2 = 3500$ volts, $\epsilon_{r1} = 8000$, $\epsilon_{r2} = 1600$, area of one capacitor electrode 1 square meter, thickness of dielectric 0.2 millimeter, specific heat of barium titanate 2.9×10^6 joules per cubic meter per degree, and $T_2 - T_1 = 30°C$.

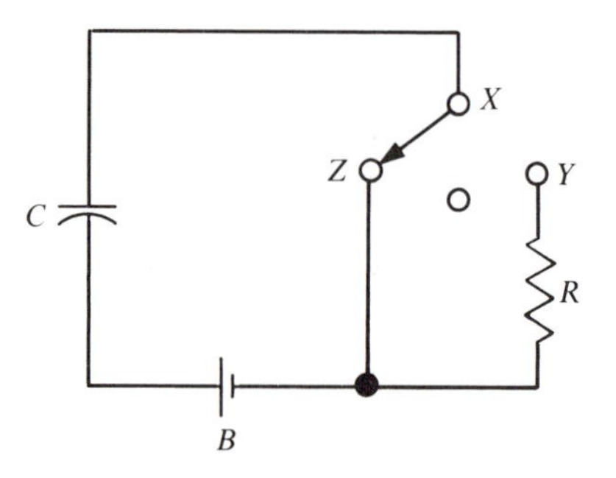

Figure 7-11 Circuit for transforming thermal energy into electric energy on board satellites. See Prob. 7-17.

CHAPTER 8

MAGNETIC FIELDS: I

The Magnetic Induction **B** *and the Vector Potential* **A**

The next eight chapters will deal with the magnetic fields of electric currents and of magnetized matter.

We shall start here by studying the magnetic induction **B** and the vector potential **A**. These two quantities correspond, respectively, to the electric field intensity **E** and to the potential V.

For the moment, we consider only constant conduction currents and non-magnetic materials.

8.1 *THE MAGNETIC INDUCTION* B

Figure 8-1 shows a circuit carrying a current I. We define the *magnetic induction* at the point P as follows:

$$\mathbf{B} = \frac{\mu_0}{4\pi} I \oint \frac{\mathbf{dl} \times \mathbf{r}_1}{r^2}, \tag{8-1}$$

where the integration is carried out over the closed circuit. As usual, the unit vector $\mathbf{r}_1$ points *from* the source *to* the point of observation: it points *from* the element $\mathbf{dl}$ *to* the point P. Magnetic inductions are expressed in *teslas*.

The constant μ_0 is *defined* as follows:

$$\mu_0 \equiv 4\pi \times 10^{-7} \text{ tesla meter/ampere}, \tag{8-2}$$

and is called the *permeability of free space*.

If the current I is distributed in space with a current density $\mathbf{J}$ amperes per square meter, then I becomes $J\,da$ and must be put under the integral

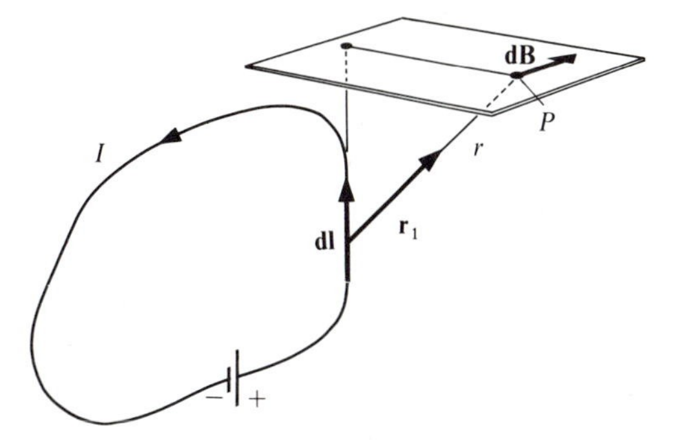

Figure 8-1 The magnetic induction $\mathbf{dB} = (\mu_0/4\pi)I\,\mathbf{dl} \times \mathbf{r}_1/r^2$ produced by an element $I\,\mathbf{dl}$ of the current I in a circuit.

sign. Then $J\,da\,\mathbf{dl}$ can be written as $\mathbf{J}\,d\tau'$, where $d\tau'$ is an element of volume, and

$$\mathbf{B} = \frac{\mu_0}{4\pi} \int_{\tau'} \frac{\mathbf{J} \times \mathbf{r}_1}{r^2}\,d\tau', \tag{8-3}$$

as in Fig. 8-2. The integration is carried out over the volume τ' occupied by the currents.

We assume that $\mathbf{J}$ is not a function of time and that there are no magnetic materials in the field.

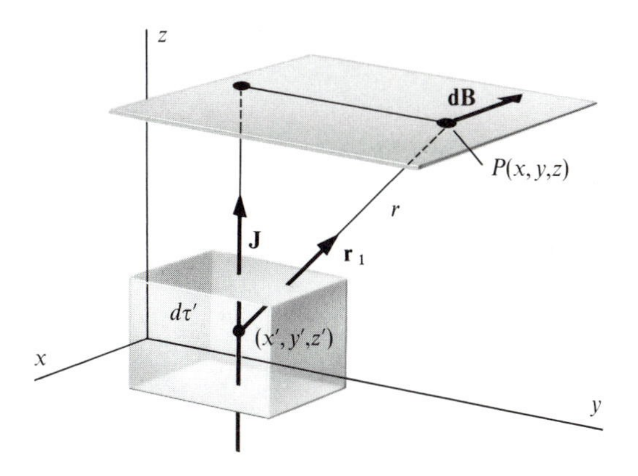

Figure 8-2 The magnetic induction $\mathbf{dB}$ due to an element $\mathbf{J}\,d\tau'$ of a volume distribution of current.

As in electrostatics, we can describe a magnetic field by drawing *lines of* **B** that are everywhere tangent to **B**.

Similarly, it is convenient to use the concept of flux, the *flux of the magnetic induction* **B** through a surface S, defined as the normal component of **B** integrated over S:

$$\Phi = \int_S \mathbf{B} \cdot \mathbf{da}. \tag{8-4}$$

The flux Φ is expressed in *webers*. Thus the tesla is one weber per square meter.

8.1.1 *EXAMPLE: LONG STRAIGHT WIRE CARRYING A CURRENT*

An element **dl** of a long straight wire carrying a current I, as in Fig. 8-3, produces a magnetic induction

$$\mathbf{dB} = \frac{\mu_0 I}{4\pi} \frac{dl \sin \theta}{r^2} \boldsymbol{\varphi}_1, \tag{8-5}$$

where $\boldsymbol{\varphi}_1$ is the unit vector pointing in the azimuthal direction. The positive directions for $\boldsymbol{\varphi}_1$ and I are related by the right-hand screw rule.

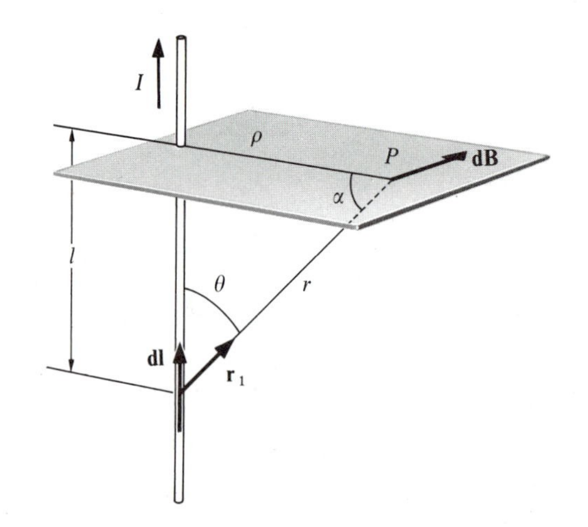

Figure 8-3 The magnetic induction **dB** produced by an element I **dl** of the current I in a long straight wire.

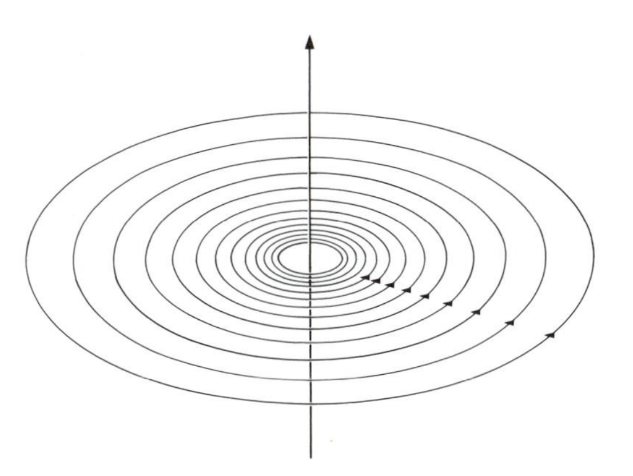

Figure 8-4 Lines of **B** in a plane perpendicular to a long straight wire carrying a current I. The density of the lines is inversely proportional to the distance to the wire. Lines close to the wire are not shown.

Expressing dl, $\sin\theta$, and r^2 in terms of α and ρ,

$$\mathbf{B} = \frac{\mu_0 I}{4\pi\rho}\int_{-\pi/2}^{+\pi/2}\cos\alpha\,d\alpha\,\boldsymbol{\varphi}_1 = \frac{\mu_0 I}{2\pi\rho}\boldsymbol{\varphi}_1. \tag{8-6}$$

The magnitude of **B** thus falls off inversely as the first power of the distance from an infinitely long wire. The lines of **B** are concentric circles lying in a plane perpendicular to the wire, as in Fig. 8-4.

8.1.2 *EXAMPLE: CIRCULAR LOOP, MAGNETIC DIPOLE MOMENT* m

A circular loop of radius a carries a current I, as in Fig. 8-5.

An element $I\,\mathbf{dl}$ of current produces a $\mathbf{dB}$ having a component dB_z on the axis, as indicated in the figure. By symmetry, the total **B** is along the axis, and

$$dB_z = \frac{\mu_0 I\,dl}{4\pi\,r^2}\cos\theta, \tag{8-7}$$

$$B_z = \frac{\mu_0 I}{4\pi}\frac{2\pi a}{r^2}\cos\theta = \frac{\mu_0 I a^2}{2(a^2 + z^2)^{3/2}}. \tag{8-8}$$

Therefore, on the axis, the magnetic induction is maximum at the center of the ring and drops off as z^3 for $z^2 \gg a^2$.

Figure 8-6 shows lines of **B** in a plane containing the axis of the loop.

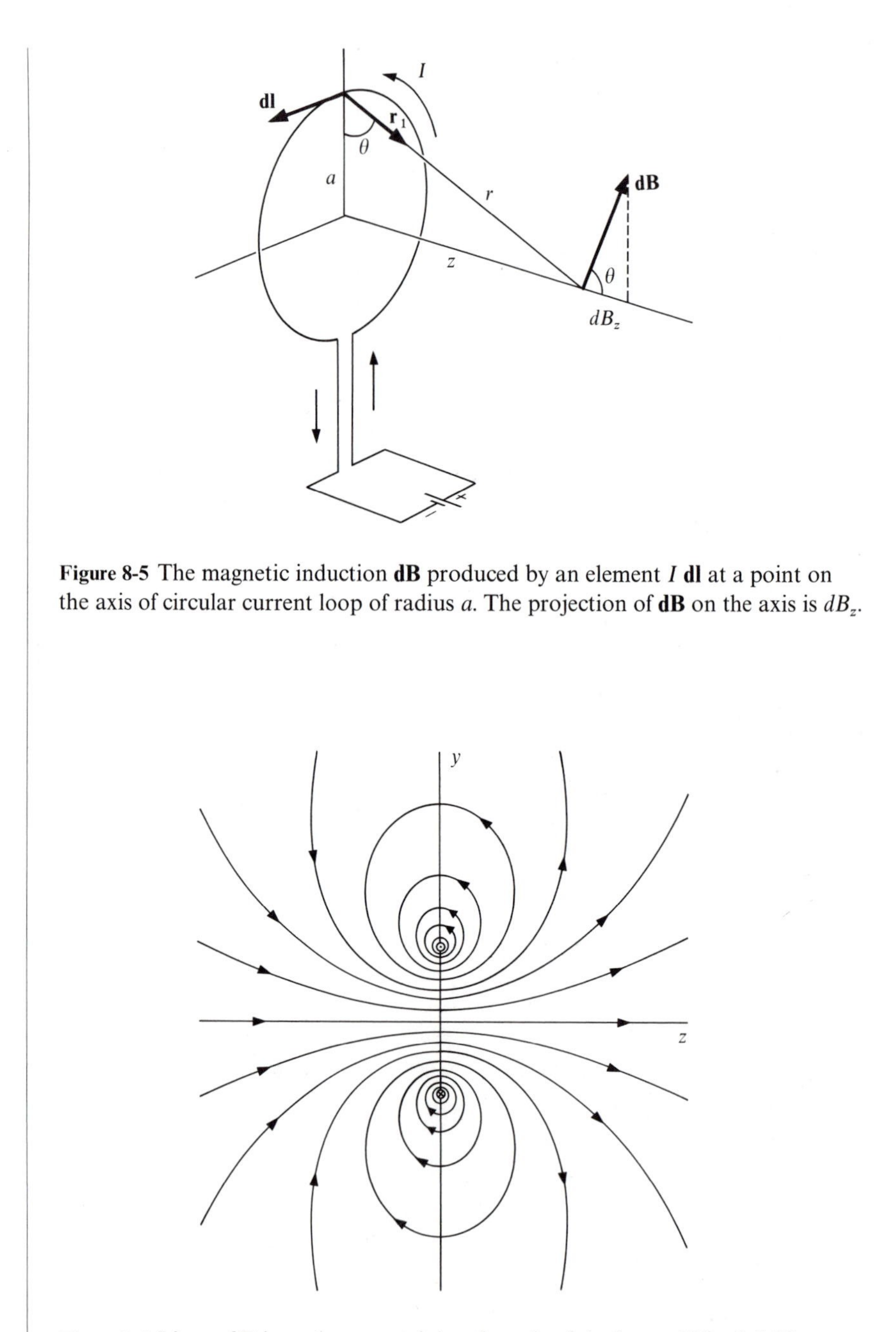

Figure 8-5 The magnetic induction **dB** produced by an element I **dl** at a point on the axis of circular current loop of radius a. The projection of **dB** on the axis is dB_z.

Figure 8-6 Lines of **B** in a plane containing the axis of the loop of Fig. 8-5. The direction of the current in the loop is shown by the dot and the cross.

Far from the loop, the field is the same as that of an electric dipole (Sec. 2.5.1), except that the factor $1/4\pi\epsilon_0$ is replaced by $\mu_0/4\pi$, and that the electric dipole moment **p** is replaced by the magnetic dipole moment **m**. The *magnetic dipole moment* **m** is a vector whose magnitude is $\pi a^2 I$ and that is normal to the plane of the loop, in the direction given by the right-hand screw rule with respect to the current I.

8.2 *THE DIVERGENCE OF* B

Consider a current element $I\,\mathbf{dl}$ and a volume $d\tau$ containing a point P as in Fig. 8-7. The current element produces at P a magnetic induction

$$\mathbf{dB} = \frac{\mu_0}{4\pi}\frac{I\,\mathbf{dl} \times \mathbf{r}_1}{r^2}. \tag{8-9}$$

As we saw in Sec. 8.1.1, the lines of **B** due to the element $I\,\mathbf{dl}$ are circles situated in planes perpendicular to a line through **dl** and centered on it as

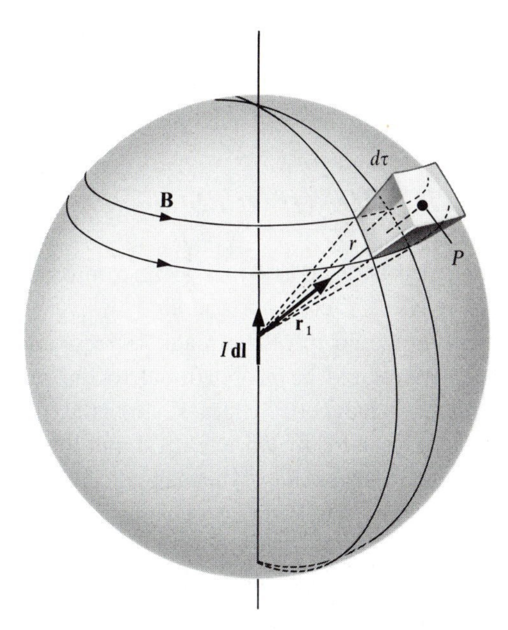

Figure 8-7 The current element I**dl** produces a magnetic induction **B**. The net outward flux of **B** through the surface of the element of volume $d\tau$ is zero.

in Fig. 8-7. From the figure, it is obvious that the net outward flux of the **B** due to $I\,\mathbf{dl}$ through the surface of the volume $d\tau$ is zero. Now, any volume can be subdivided into volume elements of the same kind as $d\tau$. Therefore, for any volume τ bounded by a surface S,

$$\int_S \mathbf{B} \cdot \mathbf{da} = 0. \tag{8-10}$$

This equation is true of all magnetic fields. Then, using the divergence theorem (Sec. 1.10),

$$\int_\tau \boldsymbol{\nabla} \cdot \mathbf{B}\, d\tau = 0. \tag{8-11}$$

So

$$\boxed{\boldsymbol{\nabla} \cdot \mathbf{B} = 0.} \tag{8-12}$$

The derivatives contained in the operator $\boldsymbol{\nabla}$ are with respect to the field point where the magnetic induction is **B**. This is another of Maxwell's equations.

8.3 MAGNETIC MONOPOLES

The statement that $\boldsymbol{\nabla} \cdot \mathbf{B}$ is equal to zero implies that magnetic fields are due solely to electric currents and that magnetic "charges" do not exist. Otherwise, one would have the magnetic equivalent of Eq. 6-12, and the divergence of **B** would be proportional to the magnetic "charge" density.

Elementary magnetic "charges" are called *magnetic monopoles*. Their existence was postulated by Dirac in 1931, but they have never been observed to date. The theoretical value of the magnetic monopole is $2h/e$, or $8.271\,17 \times 10^{-15}$ weber, where h is Planck's constant ($6.625\,6 \times 10^{-34}$ joule second), and e is the charge of the electron, $1.602\,1 \times 10^{-19}$ coulomb.

The surface integral of $\mathbf{B} \cdot \mathbf{da}$ over a closed surface is equal to the enclosed magnetic charge.

Problems 8-9 and 12-6 concern two methods that have been used for detecting monopoles in bulk matter. See also Prob. 19-5.

8.4 THE VECTOR POTENTIAL A

Since $\nabla \cdot \mathbf{B} = 0$, it is reasonable to assume that there exists a vector $\mathbf{A}$ such that

$$\mathbf{B} = \nabla \times \mathbf{A}, \tag{8-13}$$

because the divergence of a curl is identically equal to zero (see Prob. 1-29). The derivatives in the del operator are evaluated at the field point.

We can find an integral for $\mathbf{A}$, starting from Eq. 8-3,

$$\mathbf{B} = \frac{\mu_0}{4\pi} \int_{\tau'} \mathbf{J} \times \frac{\mathbf{r}_1}{r^2}\, d\tau'. \tag{8-14}$$

From Probs. 1-13 and 1-27,

$$\mathbf{J} \times \frac{\mathbf{r}_1}{r^2} = \nabla\left(\frac{1}{r}\right) \times \mathbf{J} = \nabla \times \frac{\mathbf{J}}{r} - \frac{1}{r} \nabla \times \mathbf{J}. \tag{8-15}$$

Now, the last term on the right is zero, because $\mathbf{J}$ is a function of x', y', z', while ∇ contains derivatives with respect to x, y, z. Then

$$\mathbf{B} = \frac{\mu_0}{4\pi} \int_{\tau'} \nabla \times \frac{\mathbf{J}}{r}\, d\tau' = \nabla \times \left[\frac{\mu_0}{4\pi} \int_{\tau'} \frac{\mathbf{J}}{r}\, d\tau'\right], \tag{8-16}$$

and

$$\mathbf{A} = \frac{\mu_0}{4\pi} \int_{\tau'} \frac{\mathbf{J}}{r}\, d\tau'. \tag{8-17}$$

If the current is limited to a conducting wire,

$$\mathbf{A} = \frac{\mu_0 I}{4\pi} \oint \frac{\mathbf{dl}}{r}. \tag{8-18}$$

Note that we can add to this integral any vector whose curl is zero without affecting the value of $\mathbf{B}$ in any way.

Note also that $\mathbf{B}$ depends on the space derivatives of $\mathbf{A}$, and not on $\mathbf{A}$ itself. The value of $\mathbf{B}$ at a given point can thus be calculated from $\mathbf{A}$, only if $\mathbf{A}$ is known in the *region* around the point considered.

8.4.1 EXAMPLE: LONG STRAIGHT WIRE

In Sec. 8.1.1 we calculated **B** for a long straight wire carrying a current I, starting from the definition of **B**. We can also calculate **B** starting from

$$\mathbf{A} = \frac{\mu_0 I}{4\pi} \int_{-\infty}^{\infty} \frac{\mathbf{dl}}{r}. \tag{8-19}$$

One can see immediately that **A** is parallel to the wire, since the **dl**'s are all along the wire.

We first calculate A for a current of finite length $2L$. Referring to Fig. 8-8,

$$A = 2\frac{\mu_0 I}{4\pi} \int_0^L \frac{dl}{(\rho^2 + l^2)^{1/2}}, \tag{8-20}$$

$$= \frac{\mu_0 I}{2\pi} \ln\left[l + (\rho^2 + l^2)^{1/2}\right]_0^L, \tag{8-21}$$

$$= \frac{\mu_0 I}{2\pi}\left(\ln\left\{L\left[1 + \left(1 + \frac{\rho^2}{L^2}\right)^{1/2}\right]\right\} - \ln\rho\right) \tag{8-22}$$

$$\approx \frac{\mu_0 I}{2\pi} \ln\frac{2L}{\rho} \qquad (\rho^2 \ll L^2). \tag{8-23}$$

For $L \to \infty$, A tends to infinity logarithmically. We can still try to calculate B, however, because a function can be infinite and still have finite derivatives. For example, if $y = x$, $y \to \infty$ as $x \to \infty$, but dy/dx remains equal to 1.

To calculate B for $\rho^2 \ll L^2$, we use the fundamental definition of the curl given in Eq. 1-77, using the rectangular path shown in Fig. 8-8 for the line integral:

$$B = \lim \frac{1}{\Delta\rho\Delta z} \oint \mathbf{A} \cdot \mathbf{dl}, \tag{8-24}$$

as $\Delta\rho\Delta z$ tends to zero. Since **A** is parallel to the wire,

$$B = \lim \frac{1}{\Delta\rho\Delta z}\left[A(\rho) - A(\rho + \Delta\rho)\right]\Delta z, \tag{8-25}$$

$$= \frac{\mu_0 I}{2\pi} \lim \frac{1}{\Delta\rho}\left(\ln\frac{2L}{\rho} - \ln\frac{2L}{\rho + \Delta\rho}\right), \tag{8-26}$$

$$= \frac{\mu_0 I}{2\pi} \lim \frac{1}{\Delta\rho}\left(\ln\frac{\rho + \Delta\rho}{\rho}\right), \tag{8-27}$$

$$= \frac{\mu_0 I}{2\pi} \lim \frac{1}{\Delta\rho}\left[\ln\left(1 + \frac{\Delta\rho}{\rho}\right)\right]. \tag{8-28}$$

Now $\ln(1 + x) \approx x$ when x is small and, finally, B is equal to $\mu_0 I/2\pi\rho$, as previously.

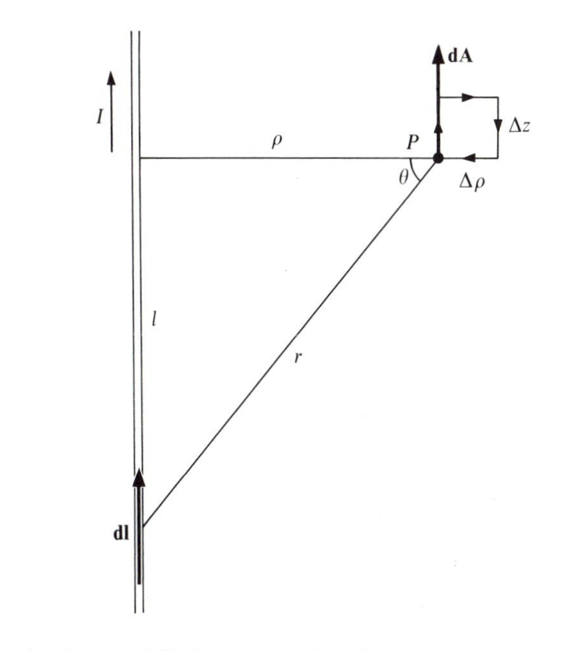

Figure 8-8 An element I **dl** of a current I in a long straight wire produces an element of vector potential **dA** at the point P. The small rectangle is the path of integration used for calculating **B**.

8.5 THE VECTOR POTENTIAL A *AND THE SCALAR POTENTIAL V*

The vector potential **A** of magnetic fields is in many respects similar to the scalar potential (or, simply, the potential) V of electric fields (Sec. 2.5).

Let us return briefly to Chapters 2 and 3.

a) With electric fields, the important quantity, in practice, is the intensity **E**: it is **E** that gives the stored energy and the forces on charged bodies and on polarized dielectrics. It is also **E** that causes breakdown in dielectrics.

b) The quantity V is related to **E** through $\mathbf{E} = -\nabla V$. So one can add to V any quantity that is independent of the coordinates without affecting **E** in any way. Also, the value of **E** at a point can be calculated only if V is known in the *region* around that point.

c) Finally, the integral for V(Eq. 2-18) is simpler to evaluate than that for **E** (Eq. 2-6). So, in general, it is easier to find **E** by first calculating V, and then ∇V, than to calculate **E** directly. However, if the geometry of the charge distribution is particularly simple, one can use Gauss's law (Sec. 3.2). Then **E** is easy to calculate directly and there is no need to go through V.

Now let us see how **A** compares with V.

a) The quantity that corresponds to **E** is the magnetic induction **B**: it is **B** that gives the stored energy and the magnetic forces (Chapter 13).

b) The vector potential **A** is related to **B** only through $\mathbf{B} = \nabla \times \mathbf{A}$. One can add to **A** any quantity whose curl is zero without changing **B**, and one can deduce from **A** the value of **B** at a point only if one knows the value of **A** in the region around that point.

c) The integral for **A** (Eq. 8-18) is also simpler to calculate than that for **B** (Eq. 8-1). However, with simple current distributions, it is much easier to use Ampère's circuital law (Sec. 9.1) to find **B** than to calculate **A** first.

So there is a great similarity between V and **A**. Later on, in Sec. 11.5, we shall find an equation that relates **E** to both V and **A**, in time-dependent fields.

8.6 *THE LINE INTEGRAL OF THE VECTOR POTENTIAL A OVER A CLOSED CURVE*

Using Stokes's theorem (Sec. 1.13),

$$\oint_C \mathbf{A} \cdot \mathbf{dl} = \int_S (\nabla \times \mathbf{A}) \cdot \mathbf{da}, \tag{8-29}$$

where S is any surface bounded by the closed curve C. But, from Eq. 8-13, the curl of **A** is **B**. Then

$$\oint_C \mathbf{A} \cdot \mathbf{dl} = \int_S \mathbf{B} \cdot \mathbf{da} = \Phi; \tag{8-30}$$

the line integral of $\mathbf{A} \cdot \mathbf{dl}$ over a closed curve C is equal to the magnetic flux through C.

8.7 *SUMMARY*

The *magnetic induction* **B** due to a circuit C carrying a current I is defined as follows:

$$\mathbf{B} = \frac{\mu_0}{4\pi} I \oint_C \frac{\mathbf{dl} \times \mathbf{r}_1}{r^2}. \tag{8-1}$$

Magnetic inductions are measured in *teslas*, and μ_0 is defined to be *exactly* $4\pi \times 10^{-7}$ tesla meter per ampere.

The *magnetic flux* through a surface S is

$$\Phi = \int_S \mathbf{B} \cdot \mathbf{da}. \tag{8-4}$$

Magnetic flux is measured in *webers*, and a tesla is one weber per square meter.

The divergence of **B** is zero:

$$\boxed{\nabla \cdot \mathbf{B} = 0,} \tag{8-12}$$

since the magnetic flux through a closed surface is always identically equal to zero, if magnetic monopoles do not exist. This is one of Maxwell's equations.

It follows from this that

$$\mathbf{B} = \nabla \times \mathbf{A}, \tag{8-13}$$

where **A** is the *vector potential*

$$\mathbf{A} = \frac{\mu_0 I}{4\pi} \oint \frac{\mathbf{dl}}{r}. \tag{8-18}$$

The line integral of $\mathbf{A} \cdot \mathbf{dl}$ over any closed curve C is equal to the magnetic flux through any surface S bounded by C:

$$\oint_C \mathbf{A} \cdot \mathbf{dl} = \Phi. \tag{8-30}$$

PROBLEMS

8-1E *MAGNETIC FIELD ON THE AXIS OF A CIRCULAR LOOP*

Plot a curve of B as a function of z on the axis of a circular loop of 100 turns having a mean radius of 100 millimeters and carrying a current of one ampere.

8-2 *SQUARE CURRENT LOOP*

Compute the magnetic induction B at the center of a square current loop of side a carrying a current I.

8-3 *MAGNETIC FIELD OF A CHARGED ROTATING DISK*

An insulating disk of radius R carries a uniform free surface charge density σ on one face. It rotates about its axis at an angular velocity ω.

a) What is the magnitude of the electric field intensity E, close to the disk?

b) Show that the surface current density α at the radius r is $\omega r \sigma$.

c) Show that, at the center of the charged surface,

$$B = \frac{1}{2} \mu_0 \omega R \sigma.$$

d) Calculate E and B for $R = 0.1$ meter, $\sigma = 10^{-6}$ coulomb per square meter, and $\omega = 1000$ radians per second.

8-4 *SUNSPOTS*

See Prob. 8-3.

The Zeeman effect† observed in the spectra of sunspots reveals the existence of magnetic inductions as large as 0.4 tesla.

Let us assume that the magnetic field is due to a disk of electrons 10^7 meters in radius rotating at an angular velocity of 3×10^{-2} radian per second. The thickness of the disk is small compared to its radius.

a) Show that the density of electrons required to achieve a B of 0.4 tesla is about 10^{19} per square meter.

b) Show that the current is about 3×10^{12} amperes.

c) In view of the enormous size of the Coulomb forces, such charge densities are clearly impossible. Then how could such currents exist?

8-5E *HELMHOLTZ COILS*

One often wishes to obtain a uniform magnetic field over an appreciable volume. In such cases one uses a pair of Helmholtz coils as in Fig. 8-9.

Show that the field on the axis of symmetry, at the midpoint between the two coils is

$$(0.8)^{3/2} \mu_0 NI/a,$$

where N is the number of turns in each coil.

If you have the courage to calculate the field as a function of z, and its derivatives, you will find that the first, second, and third derivatives are all zero at $z = 0$!

Figure 8-10 shows B_z as a function of z for values of r up to $0.16a$.

Roughly speaking, B_z is uniform, within 10%, inside a sphere of radius $0.1a$.

---

† When a gas is subjected to a strong magnetic field, its spectral lines are split into several components. The splitting is a measure of the magnetic induction.

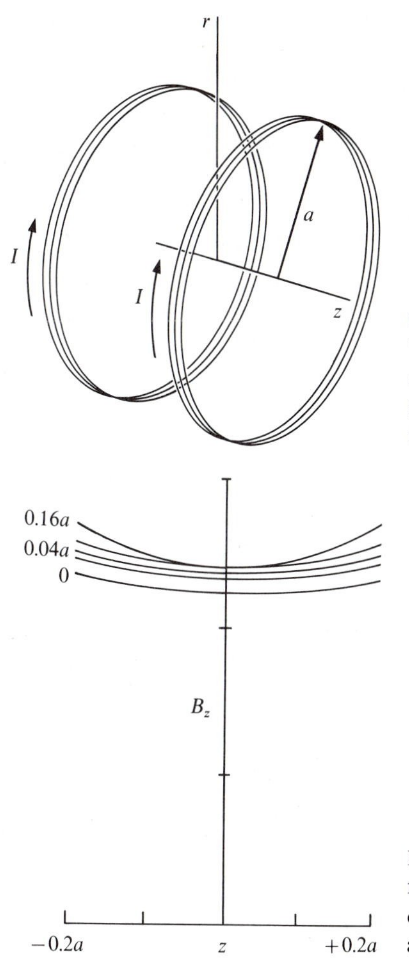

Figure 8-9 Pair of Helmholtz coils. When the spacing between the coils is equal to one radius a, the magnetic field near the center is remarkably uniform. See Prob. 8-5.

Figure 8-10 Axial component of **B** near the center of a pair of Helmholtz coils as in Fig. 8-9, as a function of z and for various values of r.

8-6 *HELMHOLTZ COILS*

You are asked to design a pair of Helmholtz coils (Prob. 8-5) that will cancel the earth's magnetic field, within 10 percent, over a spherical volume having a radius of 100 millimeters.

The magnetic induction of the earth in the laboratory is 5×10^{-5} tesla and forms an angle of 70 degrees with the horizontal. The laboratory is situated in the Northern Hemisphere. See the footnote to Prob. 8-11.

a) How must the coils be oriented, and in what direction must the current flow?
b) Specify the coil diameter and spacing.
c) What is the total number of ampere-turns required?
d) The laboratory has a good supply of No. 18 enameled copper wire on hand.

This wire has a cross-section of 0.823 square millimeter and a resistance of 21.7 ohms per kilometer. An adjustable power supply is also available. It can supply from 0 to 50 volts at a maximum of two amperes.

How many turns should each coil have?

Specify the operating current and voltage.

Will it be necessary to cool the coils? If so, what do you suggest?

8-7 *LINEAR DISPLACEMENT TRANSDUCER*

Draw a curve of B_z as a function of z on the axis of a pair of Helmholtz coils (Prob. 8-5) with the currents flowing in opposite directions and with the coils separated by a distance $2a$, instead of a, for values of z ranging from $-a$ to $+a$.

Note the linearity of the curve over most of this region.

One could build a linear displacement transducer for measuring the position of an object by fixing to it a Hall probe (Sec. 10.8.1) and having it move in the field of such a pair of opposing Helmholtz coils.

8-8E *THE SPACE DERIVATIVES OF* **B** *IN A STATIC FIELD*

Consider the field of a circular loop, as in Sec. 8.1.2.

On the right and somewhat above the axis, the lines of **B** flare out, so B_z decreases with z and $\partial B_z/\partial z$ is negative. At the same point, B_y increases with y, so $\partial B_y/\partial y$ is positive. Similarly, $\partial B_x/\partial x$ is positive.

a) Why is this, mathematically? How are these derivatives related?

b) Compare these derivative for other points on Fig. 8-6.

8-9E *MAGNETIC MONOPOLES*

It is predicted theoretically that a magnetic charge Q^* situated in a magnetic field would be subjected to a force Q^*B/μ_0.

Calculate the energy acquired by a magnetic monopole in a field of 10 teslas over a distance of 0.16 meter. Express your answer in gigaelectron-volts (10^9 electron-volts).

In one experiment for detecting magnetic monopoles, various samples, such as deep-sea sediments, were subjected to such a field, as in Fig. 8-11. None were found.

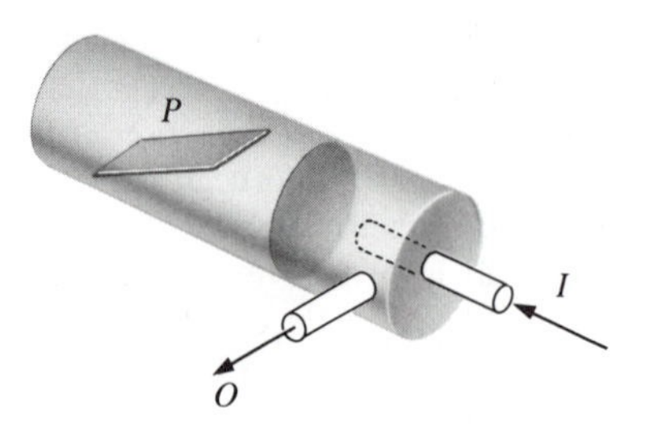

Figure 8-11 Apparatus used in an unsuccessful attempt to observe magnetic monopoles. The device is inserted inside a coil producing an axial **B** of 10 teslas. A slurry of deep-sea sediment flows into the chamber on the right at *I* and out at *O*. It was hoped that magnetic monopoles in the sample would be accelerated in the magnetic field and would leave tracks in the photographic plate *P*. See Prob. 8-9.

8-10 *MAGNETIC FIELD OF A CHARGED ROTATING SPHERE*

A conducting sphere of radius R is charged to a potential V and spun about a diameter at an angular velocity ω as in Fig. 8-12.

a) Show that the surface charge density σ is $\epsilon_0 V/R$.

b) Now show that the surface current density is

$$\alpha = \epsilon_0 \omega V \sin\theta = M \sin\theta,$$

where M is $\epsilon_0 \omega V$, and θ is the polar angle.

c) Show that the magnetic induction at the center is

$$B = (2/3)\epsilon_0 \mu_0 \omega V = (2/3)\mu_0 M.$$

It turns out that B is uniform inside the sphere.

d) What would be the value of B for a sphere 0.1 meter in radius, charged to 10 kilovolts, and spinning at 10^4 revolutions per minute?

e) Show that the dipole moment is

$$m = (4/3)\pi R^3 M.$$

f) What is the dipole moment for the above sphere?

g) What current flowing through a loop 0.1 meter in diameter would give the same dipole moment?

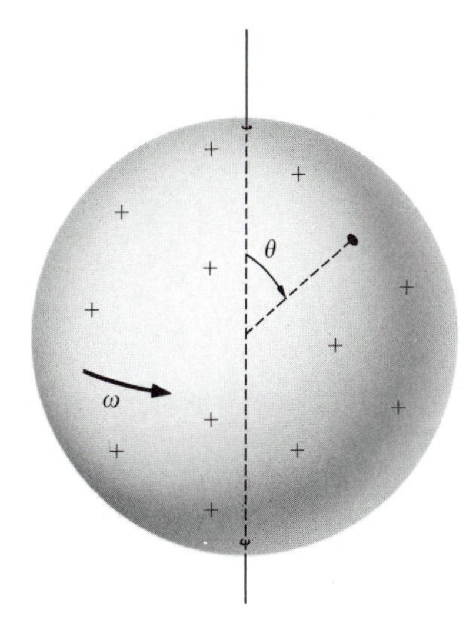

Figure 8-12 Charged sphere spinning about a diameter. See Probs. 8-10 and 8-11.

THE EARTH'S MAGNETIC FIELD

The origin of the earth's magnetic field is still largely unknown. According to one model, the earth would carry a surface charge, producing an azimuthal current because of the rotation, and hence the magnetic field, as in Prob. 8-10.

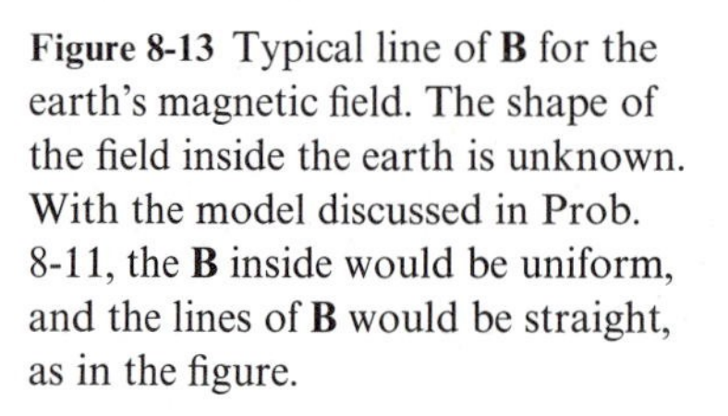

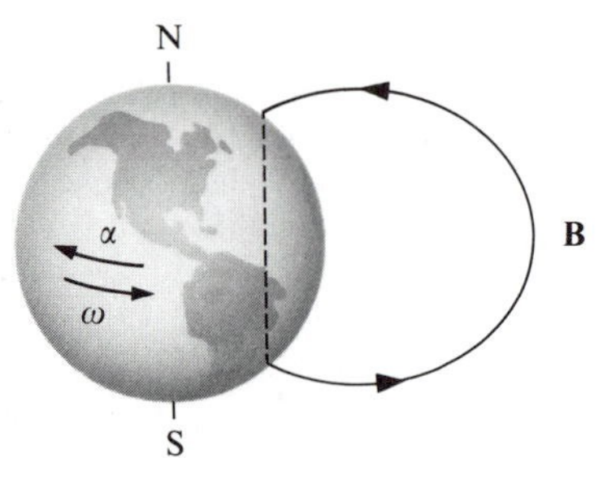

Figure 8-13 Typical line of **B** for the earth's magnetic field. The shape of the field inside the earth is unknown. With the model discussed in Prob. 8-11, the **B** inside would be uniform, and the lines of **B** would be straight, as in the figure.

This model is attractive, at first sight, because it gives a magnetic field that has the correct configuration, or nearly so, except that one has to disregard the fact that the magnetic and geographic poles do not coincide. As we shall see, this model requires an impossibly large surface charge density σ.

a) What must be the sign of σ?†

Solution: Figure 8-13 shows the earth with its North and South geographic poles, a line of **B**, and the direction of the angular velocity ω.

The current must be in the direction shown, and hence σ must be negative.

b) With this current distribution, it can be shown that the **B** inside the earth is uniform.

Use the results of Prob. 8-10 to deduce the value of M from the fact that the vertical component of **B** at the poles is 6.2×10^{-5} tesla. At the poles, the value of **B** is the same, immediately above and immediately below the surface.

Solution: Since

$$B = \frac{2}{3}\mu_0 M = 6.2 \times 10^{-5}, \tag{1}$$

$$M = \frac{1.5}{\mu_0} 6.2 \times 10^{-5} = 74 \text{ amperes/meter}. \tag{2}$$

c) Calculate the magnetic dipole moment of the earth.

The earth has a radius of 6.4×10^6 meters.

Solution: Again from Prob. 8-10,

$$m = (4/3)\pi R^3 M, \tag{3}$$

$$= (4/3)\pi(6.4 \times 10^6)^3 74 = 8.1 \times 10^{22} \text{ ampere meters}^2. \tag{4}$$

† Remember that the magnetic pole that is situated in the Northern Hemisphere is a "South" pole: the "North" pole of a magnetic needle points North.

d) Calculate the surface charge density required to give the surface current density calculated above.

If there were such a surface charge, what would be the value of the vertical electric field intensity?

There does exist a negative charge at the surface of the earth. It gives an E of about 100 volts per meter. See Prob. 4-2. Also, the maximum electric field intensity that can be sustained in air at normal temperature and pressure is 3×10^6 volts per meter.

Solution: At the equator, $\theta = 0$

$$\alpha = \sigma v = M = 74, \tag{5}$$

where v is the tangential velocity. Thus

$$\sigma = \frac{74}{(2\pi \times 6.4 \times 10^6)/(24 \times 60 \times 60)} = 0.16 \text{ coulomb/meter}^2. \tag{6}$$

This surface charge density would give an electric field intensity

$$E = \frac{\sigma}{\epsilon_0} = \frac{0.16}{8.85 \times 10^{-12}} = 1.8 \times 10^{10} \text{ volts/meter}, \tag{7}$$

from Sec. 3.2, which is absurdly large.

CHAPTER 9

MAGNETIC FIELDS: II

Ampère's Circuital Law

This chapter is devoted to Ampère's circuital law and to the curl of **B**. Ampère's law is used to calculate magnetic inductions in much the same way as Gauss's law is used to calculate electric field intensities. The value of $\nabla \times \mathbf{B}$ will follow immediately from Ampère's law by using Stokes's theorem.

We still limit ourselves to constant conduction currents and to nonmagnetic media.

9.1 *AMPERE'S CIRCUITAL LAW*

We have seen in Sec. 8.1.1 that the magnetic induction vector **B** near a long straight wire is azimuthal and that its magnitude is $\mu_0 I/2\pi\rho$, where ρ is the distance from the wire to the point considered. Thus, over a circle of radius ρ centered on the wire as in Fig. 9-1,

$$\oint \mathbf{B} \cdot d\mathbf{l} = \frac{\mu_0 I}{2\pi\rho} 2\pi\rho = \mu_0 I. \tag{9-1}$$

This is a general result and, for any closed curve C,

$$\oint_C \mathbf{B} \cdot d\mathbf{l} = \mu_0 I, \tag{9-2}$$

where I is the current enclosed by C. The positive direction for the integration is related to the positive direction for I by the right-hand screw rule as in Fig. 9-1.

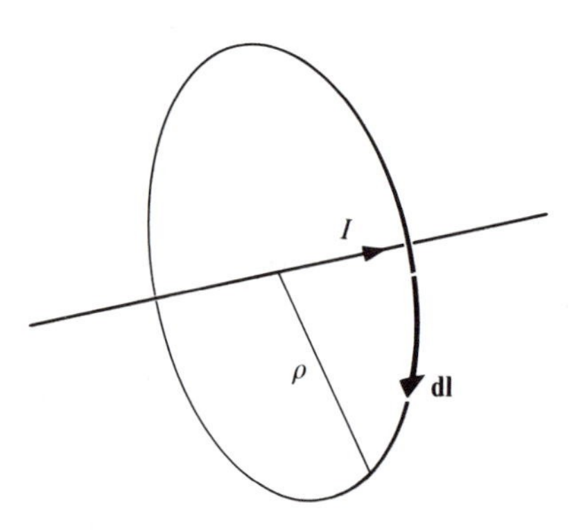

Figure 9-1 Positive direction for the integration path around a current I.

For a volume distribution of current,

$$\oint_C \mathbf{B} \cdot \mathbf{dl} = \mu_0 \int_S \mathbf{J} \cdot \mathbf{da}, \tag{9-3}$$

where $\mathbf{J}$ is the current density through any surface S bounded by the curve C as in Fig. 9-2a. This is *Ampère's circuital law.*

In many cases the same current crosses the surface bounded by the curve C several times. With a solenoid, for example, C could follow the axis and return outside the solenoid, as in Fig. 9-2b. The total current crossing the surface is then the current in each turn multiplied by the number of turns, or the number of *ampere-turns.*

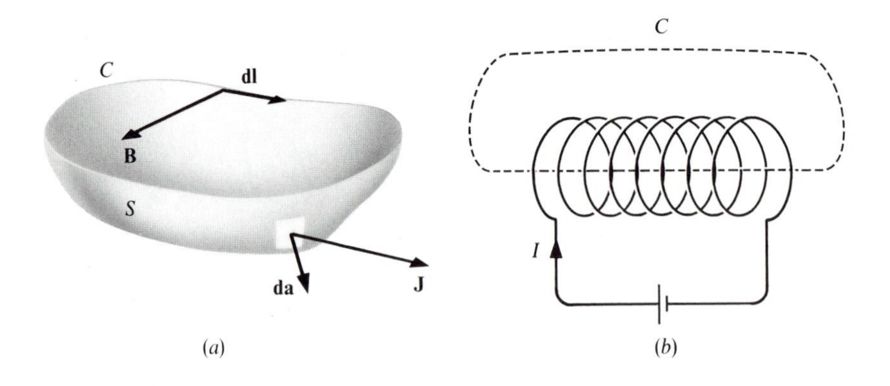

Figure 9-2 (*a*) Ampère's circuital law states that the line integral of $\mathbf{B} \cdot \mathbf{dl}$ around C is equal to μ_0 times the current through any surface bounded by C. (*b*) Path of integration C for a solenoid. In this case the line integral of $\mathbf{B} \cdot \mathbf{dl}$ is $8\mu_0 I$.

The circuital law can be used to calculate B when its magnitude is the same, all along the path of integration. This law is thus somewhat similar to Gauss's law, which is used to compute E, when E is uniform over a surface.

9.1.1 EXAMPLE: LONG CYLINDRICAL CONDUCTOR

The long cylindrical conductor of Fig. 9-3 carries a current I uniformly distributed over its cross-section with a density

$$J = I/\pi R^2. \tag{9-4}$$

Outside the conductor, **B** is azimuthal and independent of φ, so that, according to the circuital law,

$$B = \mu_0 I/2\pi\rho. \tag{9-5}$$

Inside the conductor, for a circular path of radius ρ,

$$B = \frac{\mu_0 J \pi \rho^2}{2\pi\rho} = \frac{\mu_0 J \rho}{2} = \frac{\mu_0 I \rho}{2\pi R^2}. \tag{9-6}$$

The magnetic induction B therefore increases linearly with ρ inside the conductor. Outside the conductor, B decreases as $1/\rho$. The curve of B as a function of ρ is shown in Fig. 9-4.

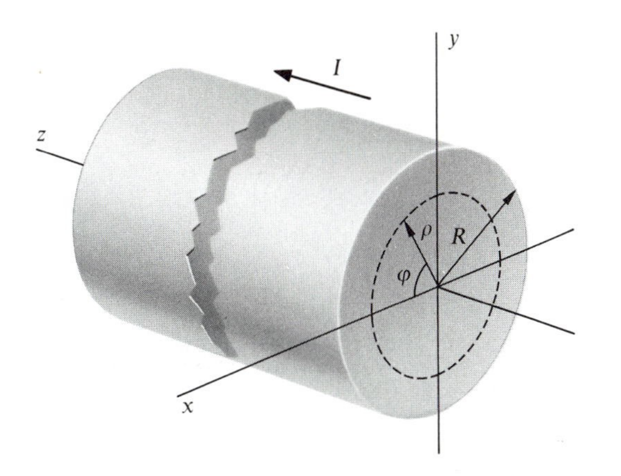

Figure 9-3 Long cylindrical conductor carrying a current I.

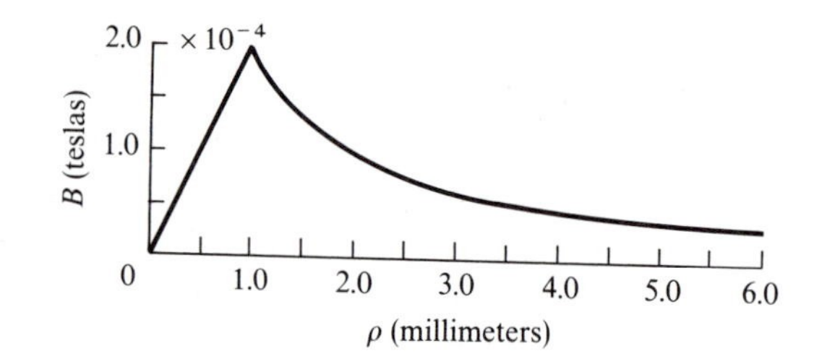

Figure 9-4 The magnetic induction B as a function of radius for a wire of 1 millimeter radius carrying a current of 1 ampere.

9.1.2 *EXAMPLE: TOROIDAL COIL*

A close-wound toroidal coil of square cross-section, as in Fig. 9-5, carries a current I.

Along path a, the line integral of **B** is zero, since there is no current linking this path. Then the azimuthal B is zero in this region. The same applies to c and to any similar path outside the toroid. Then the azimuthal B is zero everywhere outside.

Inside, along path b,

$$2\pi\rho B = \mu_0 NI, \tag{9-7}$$

where N is the total number of turns, and

$$B = \mu_0 NI/2\pi\rho. \tag{9-8}$$

There exist non-azimuthal components of the magnetic induction outside the toroid. For a path such as d in Fig. 9-5, the area bounded by the path is crossed

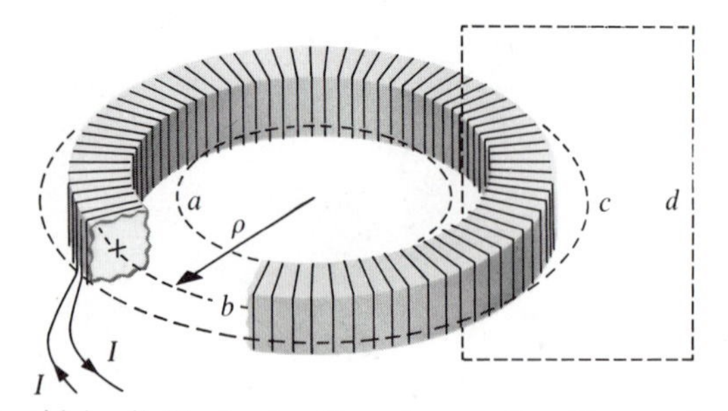

Figure 9-5 Toroidal coil. The broken lines show paths of integration.

once by the current in the toroidal winding and, at a distance large compared to the outer radius of the toroid, the magnetic induction is that of a single turn along the mean radius.

Although the magnetic induction **B** outside the toroid is essentially zero, the vector potential **A** is not. This will be evident if one remembers that **A** is a constant times the integral of $I\,\mathbf{dl}/r$, where r is the distance between the element **dl** and the point where **A** is calculated (Sec. 8.4). For example, at a point close to the winding, **A** is due mostly to the nearby turns, it is parallel to the current, and it has approximately the same value inside and outside.

The vector potential **A** can therefore exist in a region where there is no **B** field. This simply means that we can have at the same time $\mathbf{A} \neq 0$ and $\boldsymbol{\nabla} \times \mathbf{A} = 0$. For example, $\mathbf{A} = k\mathbf{i}$, where k is a constant, satisfies this condition. We are already familiar with a similar situation in electrostatics: the electric potential can have any uniform value in a region where $\mathbf{E} = -\boldsymbol{\nabla} V = 0$.

9.1.3 *EXAMPLE: LONG SOLENOID*

For the long solenoid of Fig. 9-6, we select a region remote from the ends, so that end effects will be negligible; also, we assume that the pitch of the winding is small. Let us choose cylindrical coordinates with the z-axis coinciding with the axis of symmetry of the solenoid, as in Fig. 9-7.

1. We first note that **B** has the following characteristics, both *inside and outside* the solenoid.

a) By symmetry, all three components B_ρ, B_z, B_φ are functions neither of φ nor of z.

b) Moreover, $B_\rho = 0$ for the following reason. Consider an axial cylinder of length l and radius ρ, either larger than the solenoid radius, or smaller, as in Fig. 9-6. The integral of $\mathbf{B} \cdot \mathbf{da}$ over its surface is simply $2\pi\rho l B_\rho$ since the integrals over the two end faces cancel. But, according to Eq. 8-10, the integral of $\mathbf{B} \cdot \mathbf{da}$ over any closed surface is zero. Then $B_\rho = 0$.

c) Consider a rectangular path a inside the solenoid as in Fig. 9-6. By Ampère's circuital law, the line integral of $\mathbf{B} \cdot \mathbf{dl}$ around the path is zero. Since $B_\rho = 0$, as we have just found, the line integrals on the vertical sides must cancel, which means that, inside the solenoid, B_z is also independent of ρ. The same applies to a similar path entirely outside the solenoid.

2. *Outside* the solenoid.

a) We have shown above that, outside the solenoid, B_z is independent of all three coordinates ρ, φ, z. However, **B** is zero at infinity. Therefore $B_z = 0$ outside the solenoid.

b) A path such as b is linked once by the current and, outside the solenoid,

$$B_\varphi = \mu_0 I / 2\pi\rho. \tag{9-9}$$

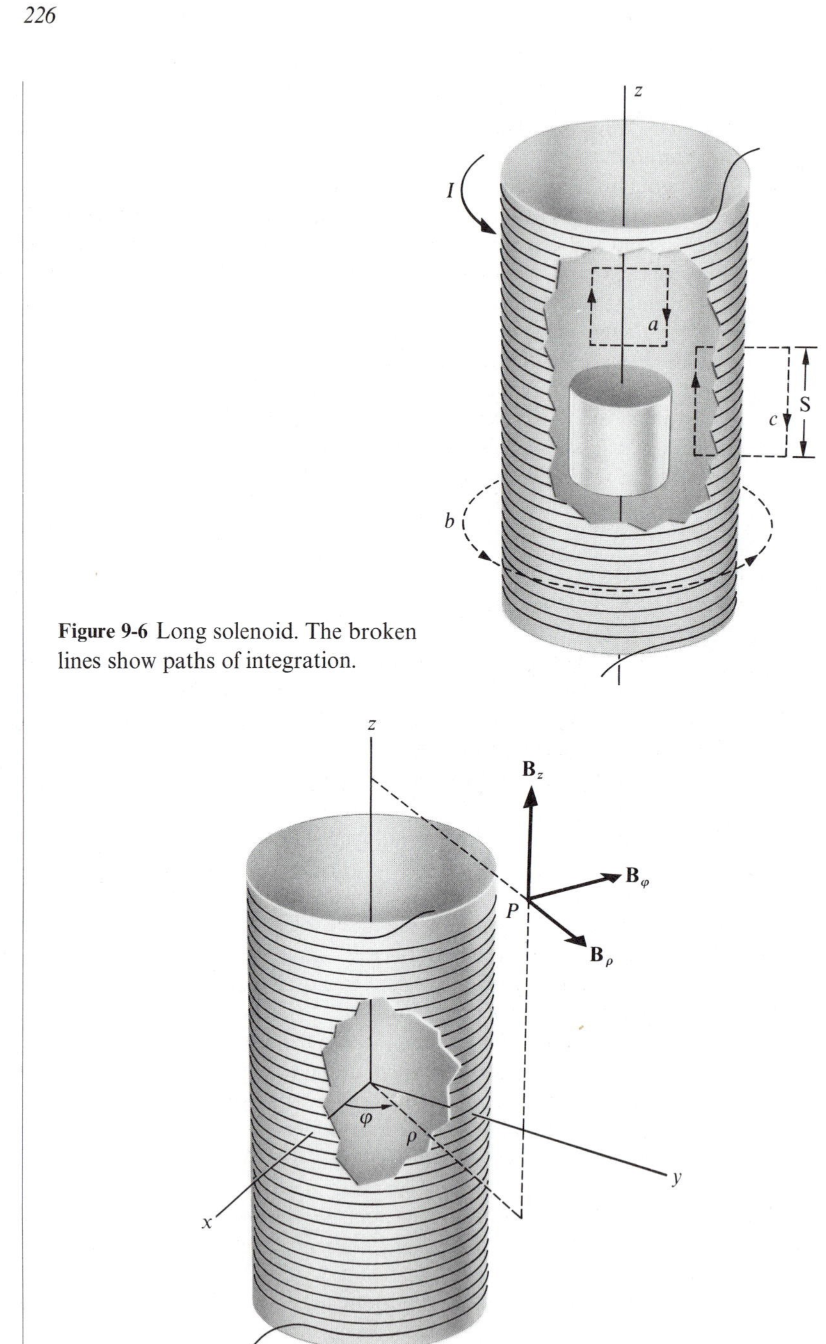

Figure 9-6 Long solenoid. The broken lines show paths of integration.

Figure 9-7 Components of **B** at P, in cylindrical coordinates.

3. *Inside* the solenoid.

a) $B_\varphi = 0$ inside because the line integral of $\mathbf{B} \cdot \mathbf{dl}$ over a circle of radius ρ, say the top edge of the small cylinder shown in Fig. 9-6, is $2\pi\rho B_\varphi$, and this must be zero according to Eq. 9-2 because there is no current enclosed by the path.

b) Considering now path c in Fig. 9-6, and remembering that $B_\rho = 0$ both inside and outside, and that $B_z = 0$ outside, we see that $B_z S = \mu_0 N'IS$, where N' is the number of turns per meter. Therefore

$$B_z = \mu_0 N'I. \tag{9-10}$$

The magnetic induction inside a long solenoid in the region remote from the ends is therefore axial, uniform, and equal to μ_0 times the number of ampere-turns per meter $N'I$. It is much larger than the azimuthal B outside, as long as $(1/N')/2\pi\rho \ll 1$, or as long as the pitch of the winding $1/N'$, divided by its circumference $2\pi\rho$, is very small.

It is obvious, from Fig. 9-10, that B_z outside a *finite* solenoid is not zero. This outside B_z can be larger or smaller than the outside B_φ, depending on the geometry of the solenoid, but both are much smaller than B_z inside.

9.1.4 *EXAMPLE: REFRACTION OF LINES OF* **B** *AT A CURRENT SHEET*

Imagine a thin conducting sheet carrying a current density of $\boldsymbol{\alpha}$ amperes per meter, as in Fig. 9-8. We can understand the refraction of lines of **B** at a current sheet by proceeding as in Sec. 7.1. Since the divergence of **B** is zero, the normal component

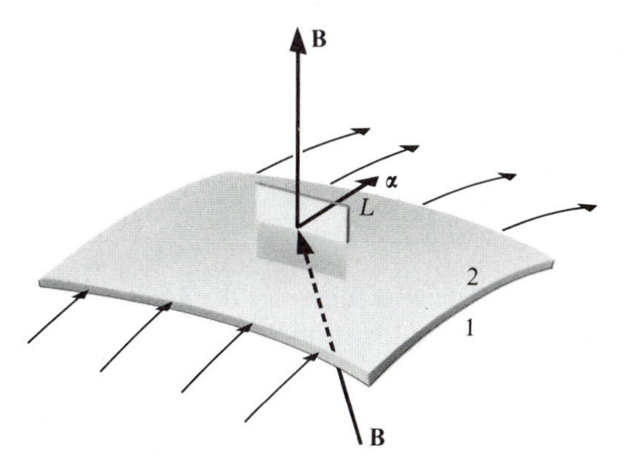

Figure 9-8 Conducting sheet carrying a current density of $\boldsymbol{\alpha}$ amperes per meter. Since $\boldsymbol{\nabla} \cdot \mathbf{B} = 0$, the normal component of **B** is the same on both sides of the sheet. According to the circuital law, however, the tangential component is not conserved and a line of **B** is deflected in the direction shown.

of **B** is conserved:

$$B_{n1} = B_{n2}. \tag{9-11}$$

Also, if we apply Ampère's circuital law to a path of length L that is perpendicular to the sheet as in the figure,

$$B_{t1}L - B_{t2}L = \mu_0 \alpha L, \tag{9-12}$$

$$B_{t2} = B_{t1} - \mu_0 \alpha. \tag{9-13}$$

The line of force is therefore rotated in the clockwise direction for an observer looking in the direction of the vector $\boldsymbol{\alpha}$.

We could have arrived at this result in another way. The magnetic induction **B** is due to the current sheet itself and to other currents flowing elsewhere. According to Ampère's circuital law, the current sheet produces, just below itself in the figure, a **B** that is directed to the left and whose magnitude is $\mu_0\alpha/2$ (Prob. 9-3). Similarly, the **B** just above the sheet is directed to the right and has the same magnitude. If we add this field to that of the other currents, we see that the tangential components of **B** must differ as above.

9.1.5 *EXAMPLE: SHORT SOLENOID*

We can calculate B on the axis of the short solenoid of Fig. 9-9 by summing the contributions of the individual turns, using Eq. 8-8. If the length of the solenoid is l and if its radius is a, the magnetic induction at the center is

$$B = \frac{\mu_0}{2} \int_{-l/2}^{+l/2} \frac{a^2 N'I \, dz}{(a^2 + z^2)^{3/2}}, \tag{9-14}$$

$$= \frac{\mu_0 a^2 N'I}{2} \left[\frac{z}{a^2(a^2 + z^2)^{1/2}} \right]_{-l/2}^{+l/2} = \mu_0 N'I \sin \theta_m, \tag{9-15}$$

as in Fig. 9-9. We have assumed that the solenoid is close wound.

For a long solenoid, $\theta_m \to \pi/2$, and $B \to \mu_0 N'I$, as in Sec. 9.1.3.

At one end, again on the axis,

$$B = \mu_0 N'I \sin \theta_e/2. \tag{9-16}$$

The magnetic induction thus decreases at both ends of the solenoid, and this is of course due to the fact that the lines of **B** flare out as in Fig. 9-10.

Upon crossing the winding, the radial component of **B** remains unchanged, but the axial component changes both its magnitude and its sign. For example,

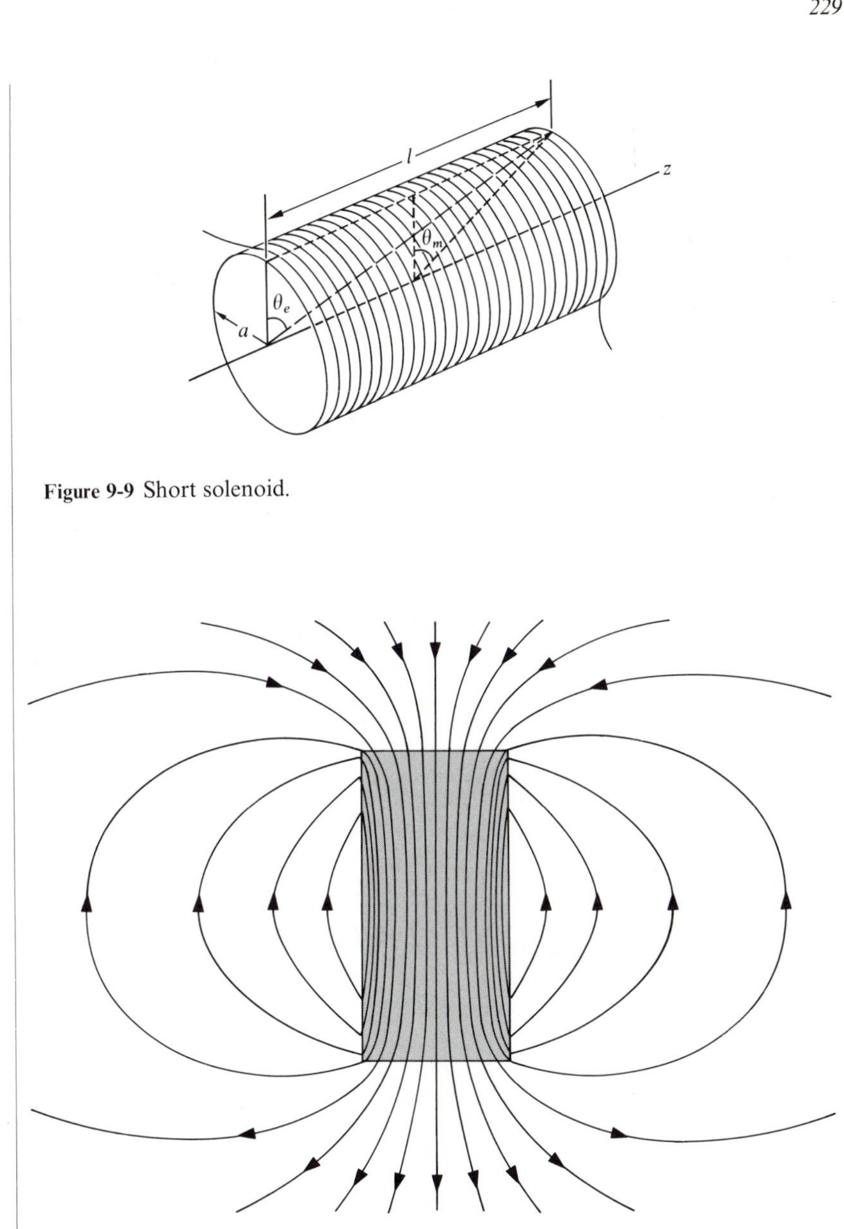

Figure 9-9 Short solenoid.

Figure 9-10 Lines of **B** for a solenoid whose length is twice its diameter.

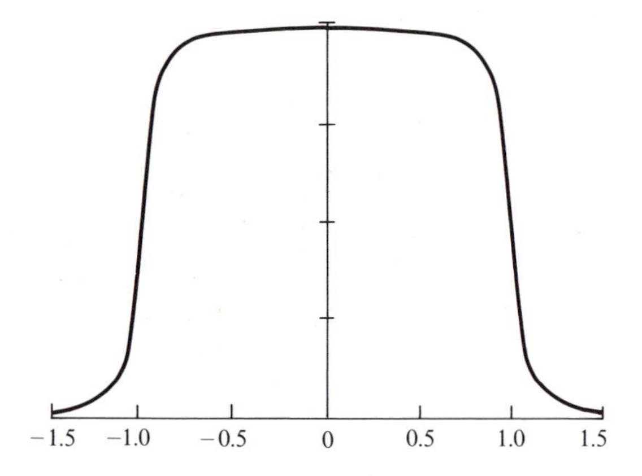

Figure 9-11 The magnetic induction B as a function of z on the axis of a solenoid with a length equal to 5 times its diameter. Note how sharply the field drops off at the ends.

the axial component in the upper left-hand side of the solenoid in Fig. 9-10 changes from, say, $-0.9\mu_0 N'I$ to $+0.1\mu_0 N'I$, since the surface current density λ is $N'I$ (Sec. 9.1.4).

Figure 9-11 shows B as a function of z for a solenoid with $l = 10a$. See Prob. 9-5.

9.2 *THE CURL OF THE MAGNETIC INDUCTION* B

Applying Stokes's theorem to Eq. 9-3, we find that

$$\int_S (\nabla \times \mathbf{B}) \cdot \mathbf{da} = \mu_0 \int_S \mathbf{J} \cdot \mathbf{da} \tag{9-17}$$

for any bounded surface S. Then

$$\nabla \times \mathbf{B} = \mu_0 \mathbf{J}. \tag{9-18}$$

This is another of Maxwell's equations, but not in its final form yet, because we are still limited to static fields, and we have not yet considered magnetic materials.

9.3 SUMMARY

Ampère's circuital law states that the line integral of $\mathbf{B} \cdot \mathbf{dl}$ over a closed curve C is equal to the current flowing through any surface S bounded by C:

$$\oint_C \mathbf{B} \cdot \mathbf{dl} = \mu_0 \int_S \mathbf{J} \cdot \mathbf{da}. \tag{9-3}$$

Upon applying Stokes's theorem to this equation, one finds that

$$\nabla \times \mathbf{B} = \mu_0 \mathbf{J}. \tag{9-18}$$

PROBLEMS

9-1E *DEFINITION OF μ_0*

Show that the permeability of free space μ_0 can be defined as follows: If an infinitely long solenoid carries a current density of one ampere per meter, then the magnetic induction in teslas inside the solenoid is numerically equal to μ_0.

9-2 *MAGNETIC FIELD OF A CURRENT-CARRYING TUBE*

A length of tubing carries a current I in the longitudinal direction.

a) What is the value of B outside?
b) How is $\mathbf{A}$ oriented outside?
c) What is the value of B inside?
d) Show that $\mathbf{A}$ is uniform inside.

9-3E *MAGNETIC FIELD CLOSE TO A CURRENT SHEET*

A conducting sheet carries a current density of α amperes per meter.

Show that, very close to the sheet, the magnetic induction B due to the current in the sheet is $\mu_0\alpha/2$ in the direction perpendicular to the current and parallel to the sheet.

9-4E *VAN DE GRAAFF HIGH-VOLTAGE GENERATOR*

In a Van de Graaff high-voltage generator, a charged insulating belt is used to transport electric charges to the high-voltage electrode.

a) Calculate the current carried by a belt 0.5 meter wide driven by a pulley 0.1 meter in diameter and rotating at 60 revolutions per second, if the electric field intensity at the surface of the belt is 2 kilovolts per millimeter.

b) Calculate the magnetic induction close to the surface of the belt, neglecting edge effects. See Prob. 9-3.

9-5 *FIELD ON THE AXIS OF A SHORT SOLENOID*

A short solenoid carries a current I and has N' turns per meter.

Show that, at any point on the axis,

$$B = \frac{1}{2}\mu_0 I N'(\cos\alpha_1 + \cos\alpha_2),$$

where α_1 and α_2 are the angles subtended at the point by a radius R at either end of the solenoid.

For example, if the coil has a length $2L$, and if the point is situated at a distance x from the center,

$$\cos\alpha_1 = \frac{L - x}{[R^2 + (L - x)^2]^{1/2}}, \qquad \cos\alpha_2 = \frac{L + x}{[R^2 + (L + x)^2]^{1/2}}.$$

9-6 *FIELD AT THE CENTER OF A COIL*

A coil has an inner radius R_1, an outer radius R_2, and a length L.

a) Show that the magnetic induction at the center, when the coil carries a current I, is

$$B = \frac{\mu_0 n I L}{2} \ln \frac{\alpha + (\alpha^2 + \beta^2)^{1/2}}{1 + (1 + \beta^2)^{1/2}},$$

where

$$\alpha = \frac{R_2}{R_1}, \qquad \beta = \frac{L}{2R_1},$$

and n is the turn density, or the number of wires per square meter.

b) Show that the length of the wire is

$$l = Vn = 2\pi n(\alpha^2 - 1)\beta R_1^3,$$

where V is the volume occupied by the wire.

9-7 *CURRENT DISTRIBUTION GIVING A UNIFORM* **B**

A long straight conductor has a circular cross-section of radius R and carries a current I. Inside the conductor, there is a cylindrical hole of radius a whose axis is parallel to the axis of the conductor and at a distance b from it.

Show that the magnetic induction inside the hole is uniform and equal to

$$\frac{\mu_0 b I}{2\pi(R^2 - a^2)}.$$

See also the next problem.

Hint: Use a uniform current distribution throughout the circle of radius R, plus a current in the opposite direction in the hole.

9-8D† *SADDLE COILS*

Figure 9-12 shows the construction of a pair of saddle coils. The windings occupy the two outer regions of a pair of overlapping circles.

a) In what general direction is the field oriented in the region between the conductors?

b) It is said that the magnetic induction is uniform in the hollow. Is this true? What is the value of B?

Assume that the length of the coil is infinite.

You can solve this one by straight integration if you wish, but there is a much easier way. See the hint for Prob. 9-7.

Such coils are used for magnetohydrodynamic generators. See Prob. 10-16.

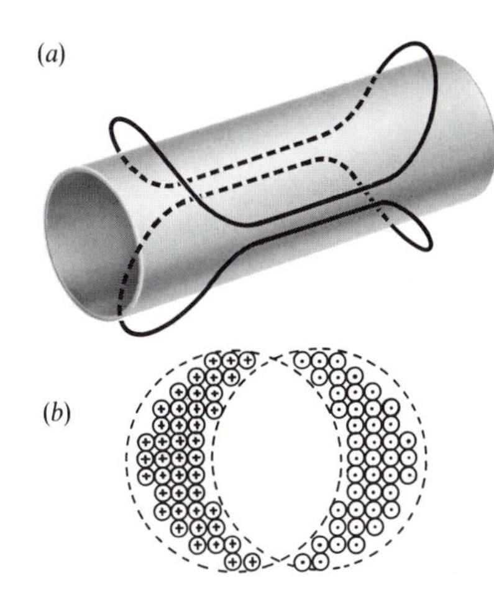

Figure 9-12 (*a*) Pair of saddle coils for producing a transverse magnetic field inside a tubular enclosure. Only one turn of each coil is shown. (*b*) Cross-section of the windings. See Prob. 9-8.

9-9D *TOROIDAL COIL*

A toroidal coil of N turns has a major radius R and a minor radius r.

a) Calculate the magnetic flux by integrating B over a cross-section.

b) At what radius does B have its mean value?

† The letter D indicates that the problem is relatively difficult.

CHAPTER 10

MAGNETIC FIELDS: III

Transformation of Electric and Magnetic Fields

In this chapter, we shall start with the long-known fact that a moving charged particle is deflected by a magnetic field. For example, the electron beam in a television tube is deflected both horizontally and vertically by the magnetic fields of two sets of coils carrying rapidly varying currents.

Now why should a magnetic field exert a force on an electrically charged particle? The explanation is that the particle "sees" an *electric* field. In other words, the electrons in the television tube "see" an *electric* field when they pass between the deflecting coils.

Let us consider a more general case. Suppose one has some configuration of electric charges and electric currents that produce an electric field intensity $\mathbf{E}_1$ and a magnetic induction $\mathbf{B}_1$, in a laboratory. Both $\mathbf{E}_1$ and $\mathbf{B}_1$ are functions of the coordinates x, y, z. Now imagine an observer moving at some constant velocity $\mathbf{v}$ with respect to the laboratory. This observer is equipped with appropriate instruments for measuring the electric field intensity and the magnetic induction. At every point, he will find a field $\mathbf{E}_2$, $\mathbf{B}_2$ that is *different* from $\mathbf{E}_1$, $\mathbf{B}_1$ (except if $\mathbf{v}$, $\mathbf{E}_1$, $\mathbf{B}_1$ are all parallel). This chapter concerns the equations that are used to find $\mathbf{E}_2$, $\mathbf{B}_2$, given $\mathbf{E}_1$, $\mathbf{B}_1$, and inversely.

Whenever one expresses $\mathbf{E}_1$, $\mathbf{B}_1$ in terms of $\mathbf{E}_2$, $\mathbf{B}_2$, or inversely, one *transforms* the electromagnetic field from one reference frame to another.

Most of the problems at the end of this chapter concern magnetic forces exerted on *particles*; *macroscopic* manifestations of magnetic forces will be dealt with in Chapters 13 and 15.

10.1 THE LORENTZ FORCE

It is observed experimentally that a charge Q moving at a velocity $\mathbf{v}$ in a region where the magnetic induction is $\mathbf{B}$, is subjected to a *magnetic force*

$$\mathbf{F} = Q\mathbf{v} \times \mathbf{B}. \tag{10-1}$$

Since this force is perpendicular to $\mathbf{v}$, $\mathbf{F} \cdot \mathbf{v} = 0$ and the power supplied to the particle is zero. The $Q\mathbf{v} \times \mathbf{B}$ force therefore changes the direction of $\mathbf{v}$ without changing its magnitude.

More generally, if there is also an electric field $\mathbf{E}$,

$$\mathbf{F} = Q(\mathbf{E} + \mathbf{v} \times \mathbf{B}). \tag{10-2}$$

This is the *Lorentz force.*

10.1.1 EXAMPLE: THE CROSSED-FIELD MASS SPECTROMETER

The crossed-field mass spectrometer is illustrated in Fig. 10-1. The positive ions of mass m and charge Q have a velocity v given by

$$\frac{1}{2}mv^2 = QV. \tag{10-3}$$

In the region of the deflecting plates, they are submitted to an upward force QE and to a downward force QvB. The net force is zero for ions having a velocity

$$v = E/B, \tag{10-4}$$

or a mass

$$m = 2QVB^2/E^2. \tag{10-5}$$

The mass spectrum is obtained by observing the collector current I as a function of E, for a constant B.

This type of mass spectrometer has a poor resolution, mostly because the ion beam is defocused in the fringing electric and magnetic fields. It is nonetheless a simple and useful device if one is concerned only with the very lightest elements.

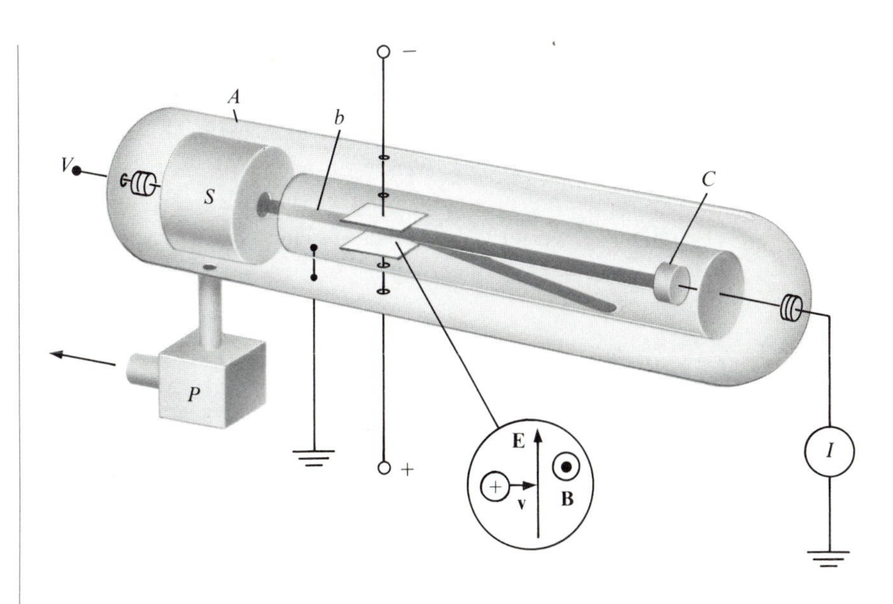

Figure 10-1 Crossed-field mass spectrometer. Positive ions produced in a source S maintained at a potential V are focused into an ion beam b. In passing through the superposed electric and magnetic fields, the beam splits vertically into its various components. Ions of the proper velocity are undeflected and are collected at C. The spectrometer is enclosed in a vacuum vessel A evacuated by a pump P.

10.2 *EQUIVALENCE OF* E *AND* v × B

The Lorentz force of Eq. 10-2 is intriguing. Why should $\mathbf{v} \times \mathbf{B}$ have the same effect as the electric field $\mathbf{E}$? Clearly, from Eq. 10-2, the particle cannot tell whether it "sees" an $\mathbf{E}$ or a $\mathbf{v} \times \mathbf{B}$ term. This is illustrated in the above example. It is also illustrated in Prob. 10-6, where two types of mass spectrometer are compared. The ion trajectory is circular in both types, but the deflecting force is $Q\mathbf{E}$ in one, and $Q\mathbf{v} \times \mathbf{B}$ in the other. Thus $\mathbf{v} \times \mathbf{B}$ is, somehow, an electric field intensity.

As we shall see in the next sections, the explanation is provided by the theory of relativity. It has to do with reference frames.

10.3 *REFERENCE FRAMES, OBSERVERS, AND RELATIVITY*

By *observer*, we mean either a human being equipped with proper instruments, or some device that can make measurements, take photographs, and so on, either automatically or under remote control.

An observer takes his measurements with respect to his *reference frame*. For example, one normally measures a magnetic induction with an instrument that is at rest with respect to the earth. This particular frame is usually called the *laboratory reference frame*.

Special relativity is concerned with the observations made by two observers, one of whom has a constant velocity with respect to the other.

10.4 THE GALILEAN TRANSFORMATION

Imagine two reference frames S_1 and S_2 as in Fig. 10-2, with S_2 moving at a velocity $v\mathbf{i}$ with respect to S_1. Now imagine some event, say a nuclear reaction, occurring at a certain instant at a certain point in space. An observer in S_1 notes the coordinates of the event as x_1, y_1, z_1, t. An observer in S_2 notes, likewise, x_2, y_2, z_2, t, and

$$x_1 = x_2 + vt, \tag{10-6}$$

$$y_1 = y_2, \tag{10-7}$$

$$z_1 = z_2. \tag{10-8}$$

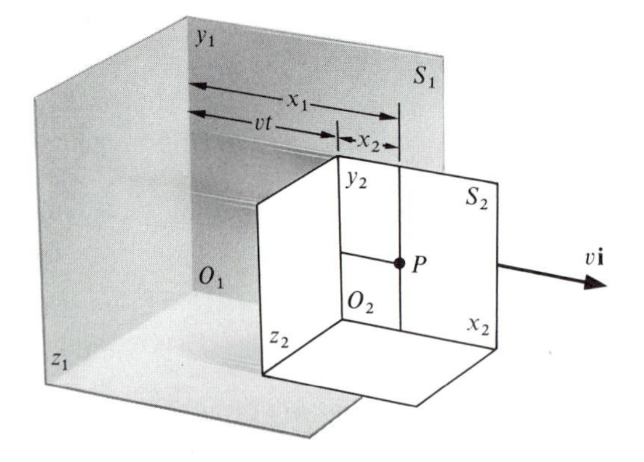

Figure 10-2 Two Cartesian coordinate systems, one moving at a velocity $v\mathbf{i}$ with respect to the other in the positive direction of the common x-axis. The two systems overlap when the origins O_1 and O_2 coincide. *We shall always refer to these two coordinate systems whenever we discuss relativistic effects.*

We have set $t = 0$ at the instant when the two reference frames overlap. These equations constitute the *Galilean transformation.*

The inverse transformation, giving the coordinates x_2, y_2, z_2 in terms of x_1, y_1, z_1, is given by interchanging the subscripts 1 and 2, and changing the sign of v.

10.5 THE LORENTZ TRANSFORMATION

Although the Galilean transformation is adequate for everyday phenomena, it is only approximate, and the correct equations are those of the *Lorentz transformation*:

$$x_1 = \gamma(x_2 + vt_2), \tag{10-9}$$

$$y_1 = y_2, \tag{10-10}$$

$$z_1 = z_2, \tag{10-11}$$

$$t_1 = \gamma\left(t_2 + \frac{v}{c^2}x_2\right), \tag{10-12}$$

where

$$\gamma = \frac{1}{[1 - (v/c)^2]^{1/2}}, \tag{10-13}$$

and c is the *velocity of light in a vacuum*, $2.997\,924\,58 \times 10^8$ meters per second.

The inverse relationship, giving the coordinates in S_1 in terms of those in S_2, is again obtained by interchanging the subscripts 1 and 2, and changing the sign of v.

Note that, in general, $t_1 \neq t_2$. In other words, the observer on S_1 does *not* agree with the observer on S_2 as to the time of occurrence of a given event.

The Lorentz transformation forms the basis of special relativity. One can deduce from it equations of transformation for the length of an object, the duration of an event, a velocity, an acceleration, a force, a mass, and so on.†

10.6 INVARIANCE OF THE VELOCITY OF LIGHT c

Experiments designed to measure the velocity of light c in a vacuum, with respect to reference frames moving at various velocities, always give the same value. Thus c is said to be *invariant*.

10.6.1 EXAMPLE: THE VELOCITY OF LIGHT IS UNAFFECTED BY THE EARTH'S ORBITAL VELOCITY

The orbital velocity of the earth around the sun is 3×10^4 meters per second. This is two orders of magnitude larger than the tangential velocity due to the rotation of the earth about its own axis. At noontime, the orbital velocity is westward while, at midnight, it is eastward. If one measures c in the east-west direction in the laboratory, first at noontime and then at midnight, one finds precisely the same value, within the experimental error.

10.7 INVARIANCE OF ELECTRIC CHARGE

Electric charge is also invariant. This is again an experimental fact.

10.7.1 EXAMPLE: THE ELECTRON CHARGE

In the *Millikan oil-drop experiment*, one measures the velocity of an electrically charged microscopic oil drop in an electric field. The velocity is of the order of millimeters per minute and is a measure of the electric charge on the drop. The charge carried by a drop is always a multiple of the electron charge, 1.602×10^{-19} coulomb.

If now one measures in the laboratory frame the charge on an electron emerging from an accelerator with a velocity approaching the velocity of light, by deflecting it in a known magnetic field, one finds again 1.602×10^{-19} coulomb.

† *Electromagnetic Fields and Waves*, Chapter 5.

10.8 TRANSFORMATION OF ELECTRIC AND MAGNETIC FIELDS

The expression for the Lorentz force in Eq. 10-2 gives us a clue as to how to transform a magnetic field. In this expression, $\mathbf{F}$, Q, $\mathbf{E}$, $\mathbf{v}$, $\mathbf{B}$ are all measured in the same reference frame S_1, which is normally the laboratory frame:

$$\mathbf{F}_1 = Q(\mathbf{E}_1 + \mathbf{v} \times \mathbf{B}_1). \tag{10-14}$$

Now call S_2 the reference frame of the particle, moving at the velocity $\mathbf{v}$ at a certain instant.

What is the field in S_2? Let us assume for the moment that $v^2 \ll c^2$, where c is the velocity of light. In that case, the force is the same in both reference frames. In S_2, the particle velocity is zero and there can be no magnetic force. Then the force in S_2 must be $Q\mathbf{E}_2$, where

$$\mathbf{E}_2 = \mathbf{E}_1 + \mathbf{v} \times \mathbf{B}_1. \tag{10-15}$$

So the *magnetic* field $\mathbf{B}_1$ in frame 1 becomes an *electric* field $\mathbf{v} \times \mathbf{B}_1$ in frame 2.

The above equation is valid only for $v^2 \ll c^2$. When v approaches c, it is shown in relativity theory that the forces in the two reference frames are different, and the correct value of $\mathbf{E}_2$ is as follows:†

$$\mathbf{E}_{2\perp} = \gamma(\mathbf{E}_{1\perp} + \mathbf{v} \times \mathbf{B}_1), \tag{10-16}$$

$$\mathbf{E}_{2\|} = \mathbf{E}_{1\|}, \tag{10-17}$$

where the subscripts refer to the components that are either perpendicular or parallel to the velocity $v\mathbf{i}$ of frame S_2 with respect to S_1, as in Fig. 10-3; γ is as defined in Eq. 10-13.

The magnetic field in S_2 is given by

$$\mathbf{B}_{2\perp} = \gamma\left(\mathbf{B}_{1\perp} - \frac{1}{c^2}\mathbf{v} \times \mathbf{E}_1\right), \tag{10-18}$$

$$\mathbf{B}_{2\|} = \mathbf{B}_{1\|}. \tag{10-19}$$

Equations 10-16 to 10-19 are the *equations of transformation for* **E** *and* **B**.

---

† *Electromagnetic Fields and Waves*, pages 262 and 288.

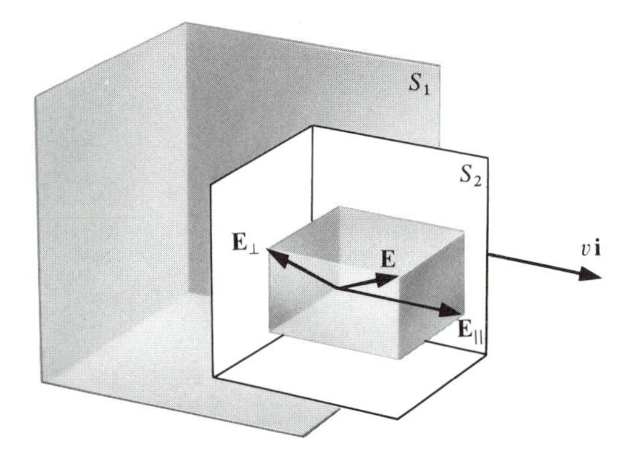

Figure 10-3 The components $\mathbf{E}_{||}$ and $\mathbf{E}_\perp$ of $\mathbf{E}$.

Thus, given a field $\mathbf{E}_1$, $\mathbf{B}_1$ in frame S_1, one can calculate the field $\mathbf{E}_2$, $\mathbf{B}_2$ in frame S_2. The inverse transformation is obtained, as usual, by interchanging the subscripts 1 and 2, and changing the sign of v.

When $v^2 \ll c^2$, $\gamma \approx 1$ and

$$\mathbf{E}_2 = \mathbf{E}_1 + \mathbf{v} \times \mathbf{B}_1, \tag{10-20}$$

$$\mathbf{B}_2 = \mathbf{B}_1 - \frac{1}{c^2} \mathbf{v} \times \mathbf{E}_1. \tag{10-21}$$

Note that the expression for the Lorentz force given in Eq. 10-2 remains true, even if $v \approx c$.

10.8.1 *EXAMPLE: THE HALL EFFECT*

Semiconductors contain either one or both of two types of mobile charges, namely conduction electrons and holes (Sec. 5.1). When a current flows through a bar of semiconductor in the presence of a transverse magnetic field $\mathbf{B}$, as in Fig. 10-4, the mobile charges drift, not only in the direction of the applied electric field $\mathbf{E}$ but also in a direction perpendicular to both the applied electric and magnetic fields, because of the $Q\mathbf{v} \times \mathbf{B}$ force. This gives rise to a voltage difference V between the upper and lower electrodes. The drift is similar to the motion of the ions in the mass-spectrometer of Sec. 10.1.1.

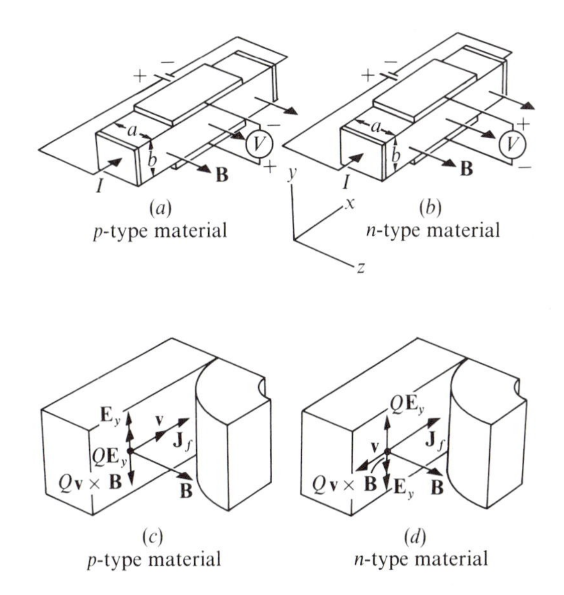

Figure 10-4 Hall effect in semiconductors. In *p*-type materials conduction is due to the drift of positive charges (holes) and the Hall voltage is as shown in (*a*). In *n*-type materials conduction is due to the conduction electrons and the Hall voltage has the opposite polarity as in (*b*). Ordinary conductors such as copper behave as in (*b*). Figure (*c*) shows the two opposing transverse forces $Q\mathbf{E}_y$ and $Q\mathbf{v} \times \mathbf{B}$ on a positive charge drifting along the axis of the bar at a velocity $\mathbf{v}$. Figure (*d*) shows how these forces are reversed in *n*-type material.

If the voltmeter V draws a negligible current, the plates charge up until their field E_y is sufficient to stop the transverse drift. This transverse electric field is called the *Hall field*.

We shall first calculate the magnitude and direction of the Hall field E_y by using the Lorentz force. Then, as an exercise, we shall find the field $\mathbf{E}_2$, $\mathbf{B}_2$ in the reference frame of the charge carriers, to arrive again at E_y. We assume that the material is *n*-type as in Fig. 10-4b and 10-4d: the charge carriers are negative.

a) The field E_y must be such that Eq. 10-4 is satisfied. Then its magnitude is vB. Since $\mathbf{v} \times \mathbf{B}$ is in the positive direction of the y-axis (Fig. 10-4d), E_y must point in the negative direction and

$$E_y = -vB, \tag{10-22}$$

as in Figs. 10-4b and 10-4d.

b) In the laboratory frame S_1, the electric field intensity is $E_x\mathbf{i} + E_y\mathbf{j}$, and the magnetic induction is $B\mathbf{k}$. We have omitted the subscripts 1 for simplicity.

To find the field in the reference frame S_2 of the moving charges, we use Eqs. 10-20 and 10-21. For n-type material, the velocity is $-v\mathbf{i}$, where v is a positive quantity. Then

$$\mathbf{E}_2 = E_x\mathbf{i} + E_y\mathbf{j} - v\mathbf{i} \times B\mathbf{k}, \quad (10\text{-}23)$$

$$= E_x\mathbf{i} + (E_y + vB)\mathbf{j}, \quad (10\text{-}24)$$

$$\mathbf{B}_2 = B\mathbf{k} + (1/c^2)v\mathbf{i} \times (E_x\mathbf{i} + E_y\mathbf{j}), \quad (10\text{-}25)$$

$$= [B + (vE_y/c^2)]\mathbf{k}. \quad (10\text{-}26)$$

The force in frame 2 is $Q\mathbf{E}_2$, and since it has no y component, $E_y = -vB$ as previously.

If there are n electrons per cubic meter, each carrying a charge e,

$$I = abJ = ab(nev), \quad (10\text{-}27)$$

$$V = vBb = BI/nea. \quad (10\text{-}28)$$

Note that charge carriers of either sign are swept *down* by the magnetic field.

The Hall effect is commonly used for measuring magnetic inductions and for various other purposes, some of which are described in Probs. 8-7, 10-15, and 17-12.

10.9 SUMMARY

A charge Q moving at a velocity $\mathbf{v}$ in superposed electric and magnetic fields $\mathbf{E}$ and $\mathbf{B}$ is submitted to the *Lorentz force*

$$\mathbf{F} = Q(\mathbf{E} + \mathbf{v} \times \mathbf{B}), \quad (10\text{-}2)$$

where $\mathbf{F}$, $\mathbf{E}$, $\mathbf{v}$, and $\mathbf{B}$ are all measured with respect to the same reference frame S_1.

To find the field perceived by the charge Q in its own reference frame S_2 moving at the velocity $v\mathbf{i}$ with respect to S_1, one must use special relativity.

which is based on the *Lorentz transformation:*

$$x_1 = \gamma(x_2 + vt_2), \tag{10-9}$$

$$y_1 = y_2, \tag{10-10}$$

$$z_1 = z_2, \tag{10-11}$$

$$t_1 = \gamma\left(t_2 + \frac{v}{c^2}x_2\right), \tag{11-12}$$

with

$$\gamma = \frac{1}{[1 - (v/c)^2]^{1/2}}, \tag{10-13}$$

and c equal to the velocity of light in a vacuum, 2.997 924 58 $\times$ 10^8 meters per second.

It is possible to deduce from this set of equations other transformation equations for a mass, a velocity, etc. A given electric charge Q and the velocity of light c always have the same value, whatever the velocity of the reference frame with respect to which they are measured.

For a given field $\mathbf{E}_1$, $\mathbf{B}_1$ in frame S_1, the field in S_2, moving at a velocity $\mathbf{v}$ with respect to S_1, is given by

$$\mathbf{E}_{2\perp} = \gamma(\mathbf{E}_{1\perp} + \mathbf{v} \times \mathbf{B}_1), \tag{10-16}$$

$$\mathbf{E}_{2\parallel} = \mathbf{E}_{1\parallel}, \tag{10-17}$$

$$\mathbf{B}_{2\perp} = \gamma\left(\mathbf{B}_{1\perp} - \frac{1}{c^2}\mathbf{v} \times \mathbf{E}_1\right), \tag{10-18}$$

$$\mathbf{B}_{2\parallel} = \mathbf{B}_{1\parallel}. \tag{10-19}$$

If $v^2 \ll c^2$, then $\gamma = 1$ and

$$\mathbf{E}_2 = \mathbf{E}_1 + \mathbf{v} \times \mathbf{B}_1, \tag{10-20}$$

$$\mathbf{B}_2 = \mathbf{B}_1 - \frac{1}{c^2}\mathbf{v} \times \mathbf{E}_1. \tag{10-21}$$

PROBLEMS

10-1E *THE CYCLOTRON FREQUENCY*

A particle of mass m, charge Q, and velocity v, describes a circle of radius R in a plane perpendicular to the direction of a uniform magnetic field $\mathbf{B}$.

a) Show that

$$BQv = mv^2/R.$$

b) Show that the angular velocity

$$\omega = BQ/m.$$

The frequency $BQ/2\pi m$ is called the *cyclotron frequency* because it is the frequency at which an ion circulates in a cyclotron.

Note how ω is independent of the velocity v of the particle. This is not strictly true, however, because m is itself a function of the velocity:

$$m = \frac{m_0}{[1 - (v^2/c^2)]^{1/2}},$$

where c is the velocity of light in a vacuum and m_0 is the *rest mass*. In practice, ω may be considered to be independent of v, for $v^2 \ll c^2$. The mass m is 10 percent larger than m_0 when v is about $0.4c$.

The above three equations are valid at all velocities.

c) Calculate the cyclotron frequency for a deuteron in a field of one tesla, for $v^2 \ll c^2$.

10-2E *MOTION OF A CHARGED PARTICLE IN A UNIFORM* B

A charged particle moves in a region where $\mathbf{E} = 0$ and $\mathbf{B}$ is uniform.

Show that the particle describes either a circle or a helix.

10-3E *MAGNETIC MIRRORS*

Figure 10-5a shows a cross-section of a solenoid having a uniform turn density, except near the ends where extra turns are added to obtain a higher magnetic induction than near the center.

Show qualitatively that, if the axial velocity is not too large, a charged particle that spirals around the axis in a vacuum inside the solenoid will be reflected back when it reaches the higher magnetic field. The regions of higher magnetic field are called *magnetic mirrors*.

Magnetic mirrors are used for confining high-temperature plasmas. (A *plasma* is a highly ionized gas. Its net charge density is approximately zero.)

In the Van Allen radiation belts, charged particles are trapped in the earth's magnetic field, oscillating north and south along the magnetic field lines. They are reflected near the North and South magnetic poles, where the lines crowd together, forming magnetic mirrors, as in Fig. 10-5b.

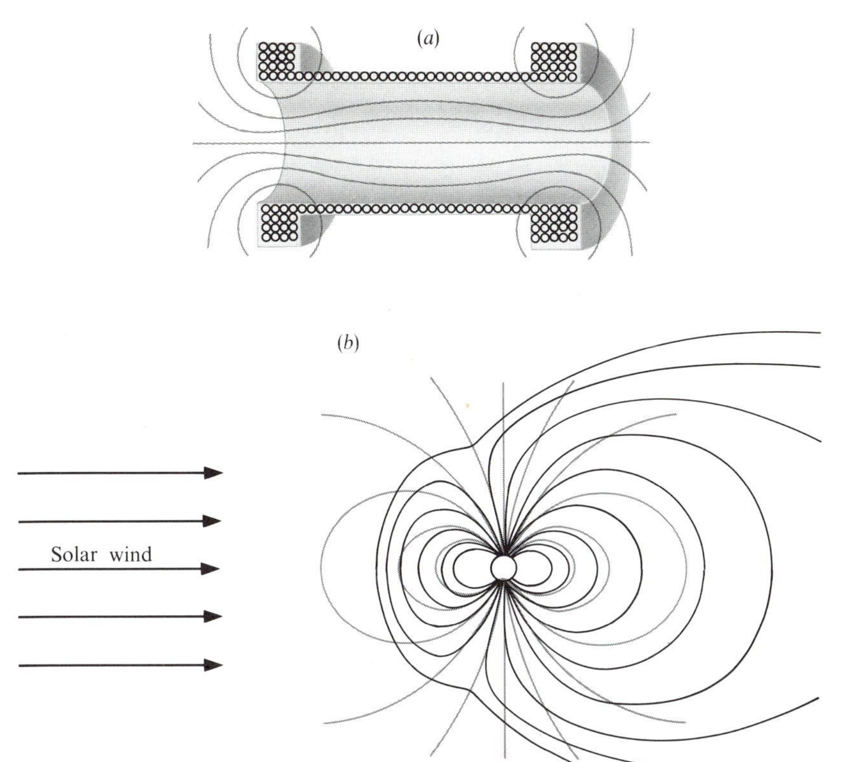

Figure 10-5 (*a*) Cross-section of a solenoid with magnetic mirrors at the ends. (*b*) Gray lines: magnetic field of the earth. The earth acts as a magnetic dipole. Black lines: magnetic field near the earth, as observed by means of satellites. The solar wind† is composed of protons and electrons evaporated from the surface of the sun. Since the interplanetary pressure is low, there are few collisions and the conductivity is high. In moving through the magnetic field of the earth, the charge particles are subjected to a $\mathbf{v} \times \mathbf{B}$ electric field, and currents flow according to Lenz's law. The black lines show the net field.

10-4E HIGH-ENERGY ELECTRONS IN THE CRAB NEBULA

Some astrophysicists believe that there exists, within the Crab Nebula, electrons with energies of about 2×10^{14} electron-volts spiralling in a magnetic field of 2×10^{-8} tesla. See Prob. 10-1.

a) Calculate the energy W of such an electron in joules.

† *Electromagnetic Fields and Waves*, p. 502.

b) Calculate its mass from the relation

$$W = mc^2,$$

where c is the velocity of light in a vacuum, 3×10^8 meters per second.

How does this mass compare with that of a low-velocity ($v^2 \ll c^2$) electron, which is 9.1×10^{-31} kilogram?

c) Calculate the radius of its orbit, setting $v = c$ and neglecting the velocity component parallel to **B**.

d) Calculate the number of days required to complete one turn.

10-5E *MAGNETIC FOCUSING*

There exist many devices that utilize fine beams of charged particles. The cathode-ray tube that is used in television receivers and in oscilloscopes is the best-known example. The electron microscope is another example. In these devices the particle beam is focused and deflected in much the same way as a light beam in an optical instrument.

Beams of charged particles can be focused and deflected by properly shaped *electric* fields. See, for example, Probs. 2-11 and 2-12. The electron beam in a TV tube is focused with electric fields and deflected with magnetic fields. Beams can be deflected along a circular path in a *magnetic* field as in Prob. 10-1. Let us see how they can be focused by a magnetic field.

Figure 10-6 shows an electron gun situated inside a long solenoid. The electrons that emerge from the hole in the anode have a small transverse velocity component and, if there is no current in the solenoid, they spread out as in the figure. Let us see what happens when we turn on the magnetic field.

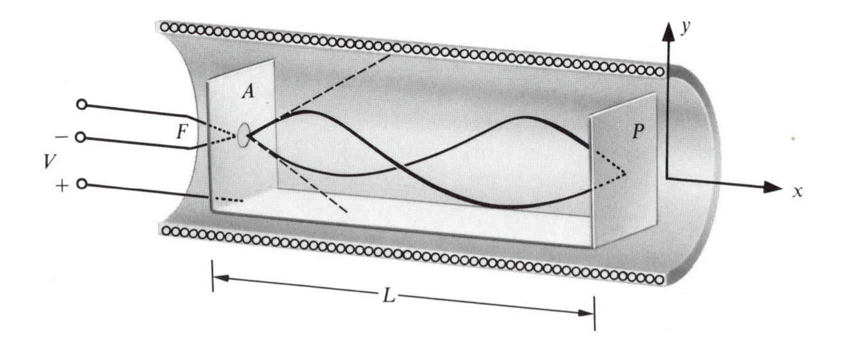

Figure 10-6 Focusing of an electron beam in a uniform magnetic field: F is a heated filament, A is the anode, and the electron beam is focused at P. The accelerating voltage is V. A filament supply of a few volts is connected between the top two terminals. When the solenoid is not energized, the beam diverges as shown by the dashed lines. See Prob. 10-5.

Let the velocity components at the hole be v_x and v_y, with $v_x^2 \gg v_y^2$. Then

$$eV = \frac{1}{2}m(v_x^2 + v_y^2) \approx \frac{1}{2}mv_x^2.$$

If $v_y = 0$, the electron continues parallel to the axis. If $v_y \neq 0$, it follows a helical path at an angular velocity ω as in Prob. 10-1. After one complete turn, it has returned to the axis. So, if B is adjusted correctly, all the electrons will converge at the same point P, as in the figure, and form an image of the hole in the anode.

a) Show that this will occur if

$$B = 2^{3/2}\pi \frac{(mV/e)^{1/2}}{L}.$$

b) Calculate the number of ampere turns per meter IN' in the solenoid, for an accelerating voltage V of 10 kilovolts and a distance L of 0.5 meter.

In actual practice, magnetic focusing is achieved with short coils, and not with solenoids.

10-6E *DEMPSTER MASS-SPECTROMETER*

Figure 2-10 shows an electrostatic velocity analyser. It was shown in Prob. 2-11 that a particle of charge Q, mass m, and velocity v passing through the first slit with the proper orientation will reach the detector if

$$v = QER/m.$$

Figure 10-7 shows a *Dempster mass-spectrometer*. Here again, the ion describes a circle, except that the centripetal force is now $Q\mathbf{v} \times \mathbf{B}$, instead of $Q\mathbf{E}$. The magnetic field is provided by an electromagnet.

a) Show that

$$m = QR^2B^2/2V.$$

b) In one particular experiment, a mass-spectrometer of this latter type was used for the hydrogen ions H_1^+, H_2^+, H_3^+, with $R = 60.0$ millimeters and $V = 1000$ volts.

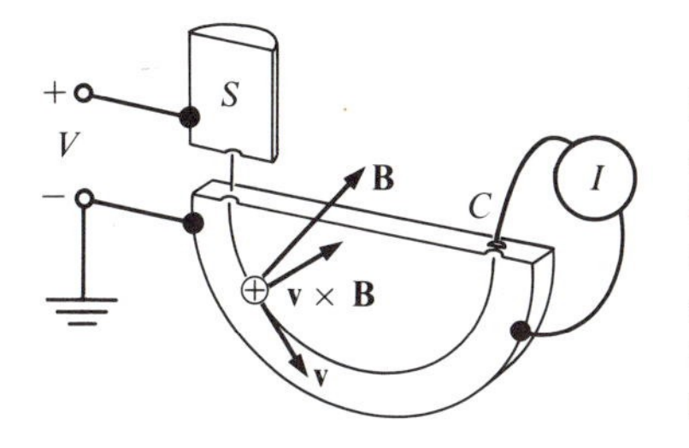

Figure 10-7 Cross-section of a mass spectrometer of the Dempster type. Ions produced at the sourse S are collected at C. The ions are accelerated through a difference of potential V and describe a semicircle of radius R. See Prob. 10-6.

The H_1^+ ion is a proton, H_2^+ is composed of two protons and one electron, and H_3^+ is composed of three protons and two electrons.

Find the values of B for these three ions.

c) Draw a curve of B as a function of m under the above conditions, from hydrogen to uranium.

Would you use this type of spectrometer for heavy ions? Why?

10-7 *MASS SPECTROMETER*

Figure 10-8 shows a mass spectrometer that separates ions, both according to their velocies and according to their charge-to-mass ratios.

Show that an ion of charge Q, mass m, and velocity v is collected at the point

$$x = \frac{2}{B}\left(\frac{m}{Q}\right)v, \qquad y = \frac{\pi^2 E}{2B^2}\left(\frac{m}{Q}\right).$$

Ions with a given Q/m ratio are all collected at the same y, and they are spread out along x according to their velocities.

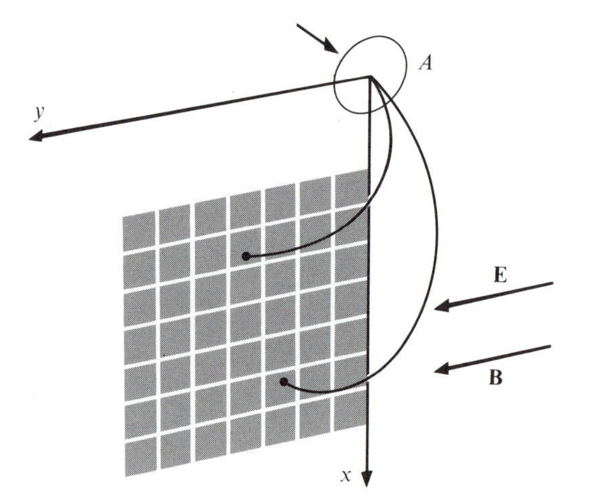

Figure 10-8 Mass spectrometer. Ions injected at A follow helical paths in the **E** and **B** fields that are both parallel to the y-axis. Each square is a separate collector. See Prob. 10-7.

10-8E *HIGH-TEMPERATURE PLASMAS*

One method of injecting and trapping ions in a high-temperature plasma is illustrated in Fig. 10-9. Molecular ions of deuterium, D_2^+ (two deuterons plus one electron, the deuteron being composed of one proton and one neutron), are injected into a magnetic field and are dissociated into pairs of D^+ ions (deuterons) and electrons

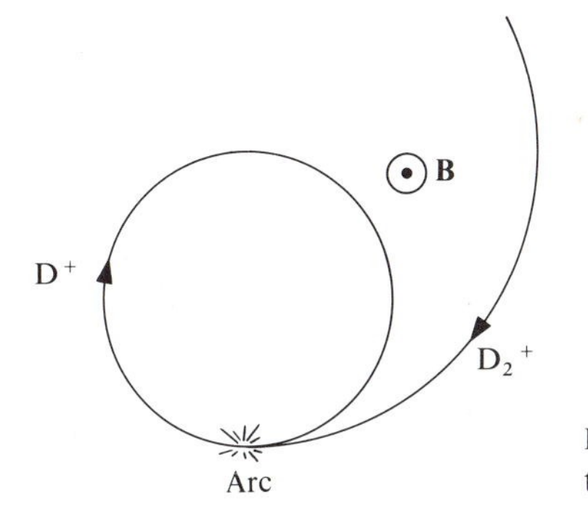

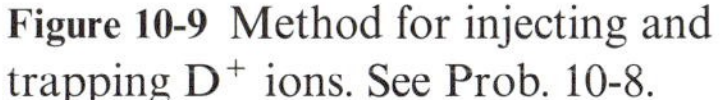
Figure 10-9 Method for injecting and trapping D^+ ions. See Prob. 10-8.

in a high-intensity arc. The radius of the trajectory is reduced and the ions are trapped. See Prob. 10-1.

a) Calculate the radius of curvature R for D_2^+ ions having a kinetic energy of 600 kiloelectron-volts in a B of 1.00 tesla.

b) Calculate R for the D^+ ions produced in the arc. The D^+ ions have one half the kinetic energy of the D_2^+ ions.

10-9 *HIGH-TEMPERATURE PLASMAS*

The type of discharge shown in Fig. 10-10 has been used to produce high-temperature plasmas. The discharge has the shape of a cylindrical shell and is situated in the space between two conducting coaxial cylinders that act as return paths for the current.

a) Is there a magnetic field outside the outer cylinder? Inside the inner cylinder? Why?

b) Draw a figure showing lines of **B**.

c) Consider the upper part of the discharge. Suppose a positive ion, moving toward the right, has an upward component of velocity. How is its trajectory affected by the magnetic field after it has emerged from the discharge?

d) What happens to a positive ion that tends to leave the upper part of the discharge by moving downward?

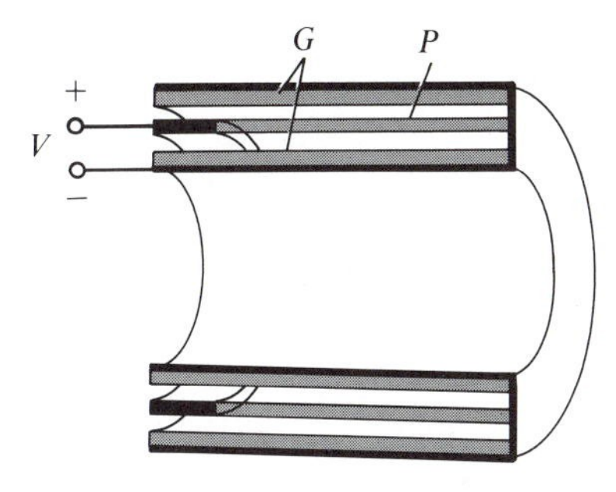

Figure 10-10 Section through a device that has been utilized for producing a high-temperature plasma P. The copper enclosure, represented by heavy lines, is separated from the plasma by a pair of glass cylinders G. See Prob. 10-9.

e) Electrons move toward the left. What happens to them when they leave the discharge, by moving either upward or downward?

10-10E ION-BEAM DIVERGENCE

Let us calculate the forces acting on a particle in a beam of positively charged particles having a velocity v and carrying charges Q. We assume that there are no externally applied electric or magnetic fields. We shall consider positive particles, but it will be a simple matter to apply our results to negative ones.

First, there is an outward **E**, as in Fig. 10-11. This is simply a case of electrostatic repulsion. Also, the beam current produces an azimuthal **B**, and $\mathbf{v} \times \mathbf{B}$ points inward, as in the figure.

Thus, there is an outward electric force $Q\mathbf{E}$ and an inward magnetic force $Q\mathbf{v} \times \mathbf{B}$. Does the beam converge or diverge? Experimentally, electron and ion beams always diverge, when left to themselves, but the divergence is slight when the particle velocity v approaches the velocity of light c.

We consider a particle situated at the edge of a beam of radius R. The current is I and the particle velocity is v.

Calculate

a) the charge λ per meter of beam,

b) the outward electric force QE,

c) the inward magnetic force QvB,

d) the net force.

You should find that the net force points outward and is proportional to $1 - \epsilon_0\mu_0 v^2$, or to $1 - (v/c)^2$ (see Eq. 20-24). Then the net force tends to zero as $v \to c$.

If the particles are negative, **E** is inward instead of outward, but $Q\mathbf{E}$ is again outward. Also, **B** points in the opposite direction and $Q\mathbf{v} \times \mathbf{B}$ points again inward.

In practice, a vacuum is never perfect. Let us say the ions are positive. If their energy is of the order of tens of electron-volts or more, they ionize the residual gas, forming low-energy positive ions and low-energy electrons. These positive ions drift

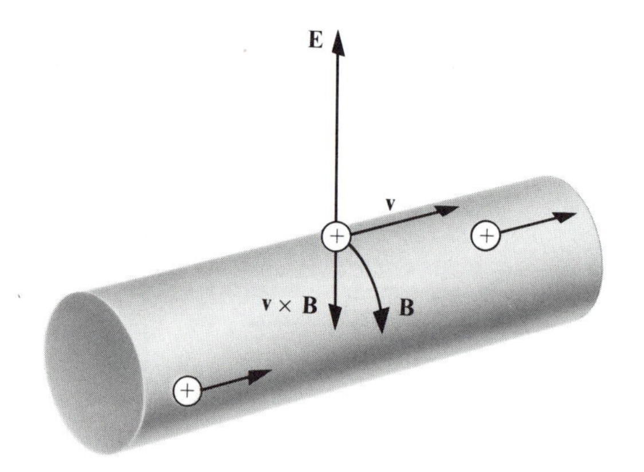

Figure 10-11 Portion of a positive-ion beam and its electric and magnetic fields. See Prob. 10-10.

away from the positive beam. The low-energy electrons, however, remain trapped in the beam and neutralize part of its space charge, thereby reducing E. The magnetic force then tends to pinch the beam. This phenomenon is called the *pinch effect*. It is also called *gas focusing*. With gas focusing, some of particles in the beam are scattered away, in colliding with the gas molecules.

10-11 *ION THRUSTER*

Figure 10-12 shows an ion thruster that utilizes the magnetic force. Other types of thruster are described in Probs. 2-14, 2-15, and 5-3. An arc A ionizes the gas, which enters from the left, and the ions are blown into the crossed electric and magnetic fields. If the current flowing between the electrodes C and D is I, then the thrust F is BIs, which is the force exerted on the gas ions. We assume that $\mathbf{B}$ and s are uniform throughout the thrust chamber, and we neglect the fringing fields.

If m' is the mass of gas flowing per unit time, and if v is the exhaust velocity with respect to the vehicle, then F is $m'v$, and the *kinetic power* communicated to the gas is

$$P_G = \frac{1}{2} m'v^2 = \frac{1}{2} Fv = \frac{1}{2} BIsv.$$

If the efficiency is defined as

$$\eta = \frac{P_G}{P_G + P_D},$$

where P_D is the power dissipated as heat between C and D, show that

$$\eta = \frac{1}{1 + \dfrac{2m'}{\sigma B^2 \tau}} = \frac{1}{1 + \dfrac{2J}{\sigma B v}} = \frac{1}{1 + \dfrac{2E}{Bv}},$$

where σ is the electrical conductivity, τ is the volume of the plasma in the crossed fields, J is the current density, and E is the transverse electric field produced by the electrodes C and D.

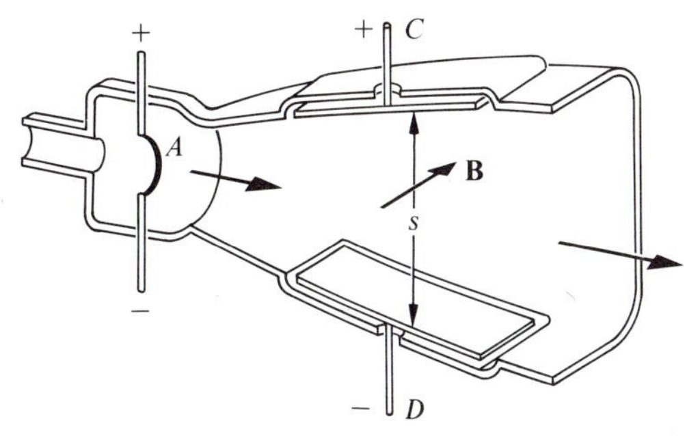

Figure 10-12 Ion thruster. See Prob. 10-11.

10-12E *GAMMA*

For what value of v is the value of γ one percent larger than unity?

10-13E *REFERENCE FRAMES*

Consider two reference frames, 1 and 2, as in Fig. 10-2, where frame 2 has a velocity $v = c/2$ with respect to 1.

An event is found to occur at $x_1 = y_1 = z_1 = 1$ meter, $t_1 = 1$ second.

Find the coordinates x_2, y_2, z_2, t_2.

10-14 *REFERENCE FRAMES*

See the preceding problem.

If the event occurs at $x_2 = y_2 = z_2 = 1$ meter, $t_2 = 1$ second.

Find x_1, y_1, z_1, t_1.

10-15 *HALL EFFECT*

Let us investigate more closely the Hall effect described in Sec. 10.8.1. We assume again that the current is carried by electrons of charge $-e$. Their effective mass is m^*. The *effective mass* takes into account the periodic forces exerted on the electrons, as they travel through the crystal lattice. As a rule the effective mass is *smaller* than the mass of an isolated electron.

The electrons are subjected to a force

$$\mathbf{F} = -e(\mathbf{E} + \mathbf{v} \times \mathbf{B}),$$

where $\mathbf{E}$ has two components, the applied field E_x and the Hall field E_y. The average drift velocity is

$$\mathbf{v} = \mathscr{M} \frac{\mathbf{F}}{e},$$

where $\mathscr{M}$ is their mobility. The mobility of a particle in a medium is its average drift velocity, divided by the electric field applied to it. The law $\mathbf{F} = m^*\mathbf{a}$ applies only between collisions with the crystal lattice.

a) Show that

$$v_x = -\mathscr{M}(E_x + v_y B),$$

$$v_y = -\mathscr{M}(E_y - v_x B),$$

$$v_z = 0.$$

b) If n is the number of conduction electrons per cubic meter, the current density is

$$J = -nev.$$

Show that

$$J_x = ne\mathcal{M}\frac{E_x - \mathcal{M}E_yB}{1 + \mathcal{M}^2B^2},$$

$$J_y = ne\mathcal{M}\frac{E_y + \mathcal{M}E_xB}{1 + \mathcal{M}^2B^2}.$$

If $J_y = 0$, as in Sec. 10.8.1,

$$E_y = -\mathcal{M}E_xB,$$

or

$$V_y = \frac{b}{a}\mathcal{M}V_xB.$$

Note that the Hall voltage V_y is proportional to the *product* of the applied voltage V_x and B. This fact makes the Hall effect useful for multiplying one variable by another. See Prob. 17-12.

When connected in this way, the Hall element has four terminals and is called a *Hall generator*, or a *Hall probe*.

As a rule, n-type materials are used because of their relatively high mobility. However, the Hall effect exists in *all* conducting bodies.

One important advantage of Hall elements is that they can be made small. For certain applications they are incorporated into integrated circuits.

c) Calculate V_y for $b = 1$ millimeter, $a = 5$ millimeters, $\mathcal{M} = 7$ meters squared per volt second (indium antimonide), $V_x = 1$ volt, $B = 10^{-4}$ tesla.

d) Show that, if $E_y = 0$,

$$\frac{\Delta R}{R_0} = \mathcal{M}^2B^2,$$

where R_0 is the resistance of the element in the x-direction when $B = 0$, and ΔR is the increase in resistance upon application of the magnetic field.

The Hall field E_y is made equal to zero by making c small, say a few micrometers, and plating conducting strips parallel to the y axis, as in Fig. 10-13. The element then has only two terminals and is called a *magnetoresistor*.

Magnetoresistors are used for measuring magnetic inductions. See Probs. 15-3 and 15-5.

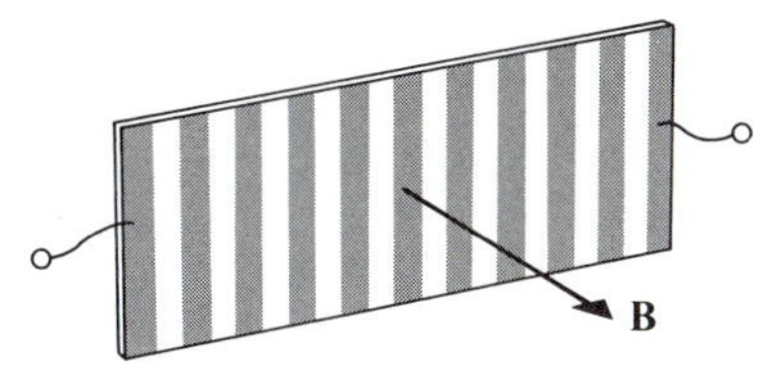

Figure 10-13 Magnetoresistor. See Prob. 10-15.

10-16

MAGNETOHYDRODYNAMIC GENERATOR

Figure 10-14 shows schematically the principle of operation of a *magnetohydrodynamic*, or *MHD generator*. The function of an MHD generator is to transform the kinetic energy of a hot gas directly into electric energy.

A very hot gas is injected on the left at a high velocity v. The gas is made conducting by injecting a salt such as K_2CO_3 that ionizes readily at high temperature, forming positive ions and free electrons. Conductivities of the order of 100 siemens per meter are thus achieved, at temperatures of about 3000 kelvins. (The conductivity of copper is 5.8×10^7 siemens per meter). Of course $v^2 \ll c^2$.

The ions and the electrons are deflected in the magnetic field. With the B shown, the positive ions are deflected downward and the electrons upward. The resulting current flows through a load resistance R, giving a voltage difference V, and hence an electric field E between the plates. One such MHD generator is planned to have a power output in the 500-megawatt range. It would use a B of several teslas over a region 20 meters long and 3 meters in diameter.

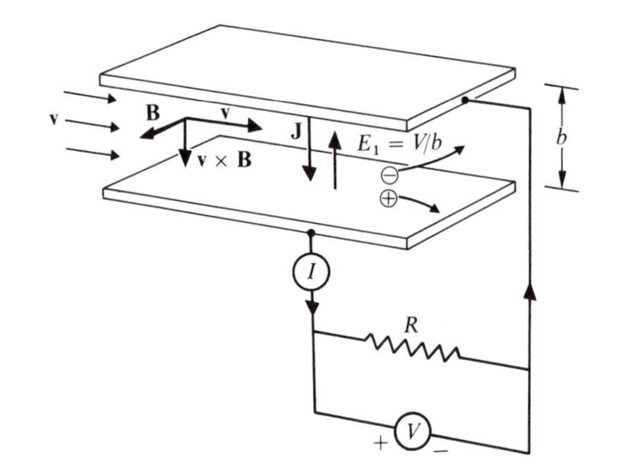

Figure 10-14 Schematic diagram of a magnetohydrodynamic generator. The kinetic energy of a very hot gas, injected on the left at a velocity **v**, is transformed directly into electric energy. The magnetic field **B** is that of a pair of coils, outside the chamber. See Prob. 9-8. The moving ions are deflected either up or down, depending on their sign. See Prob. 10-16.

a) Let us suppose that **E**, **B**, and the velocity **v** are uniform inside the chamber. These are rather crude assumptions. In particular, **v** is not uniform because (a) the ions and electrons have a vertical component of velocity, and (b) the vertical velocities give $Q\mathbf{v} \times \mathbf{B}$ forces that point backward, for both types of particle; these braking forces slow down the horizontal drift of the charged particles. It is through this deceleration that part of the initial kinetic energy of the gas is transformed into electric energy.

The electrodes each have a surface area A and are separated by a distance b. Show that

$$I = \frac{V'}{R + R_i},$$

where R_i is the internal resistance of the generator, and V' is the value of V when R is infinite. Thévenin's theorem (Sec. 5.12) therefore applies here.

Note that, because $v^2 \ll c^2$, the current density is the same in the reference frame of the laboratory as it is in the reference frame of the moving gas.

Solution: We have two reference frames, that of the laboratory and that of the moving gas.

In the reference frame of the gas,

$$\mathbf{E}_2 = \mathbf{E} + \mathbf{v} \times \mathbf{B}. \tag{1}$$

We omit the subscript 1 for the quantities measured in the reference frame of the laboratory, for simplicity. Thus

$$\mathbf{J}_2 = \sigma \mathbf{E}_2 = \sigma(\mathbf{E} + \mathbf{v} \times \mathbf{B}), \tag{2}$$

$$J = J_2 = \sigma(vB - E) = \sigma\left(vB - \frac{V}{b}\right), \tag{3}$$

$$I = JA = \sigma A\left(vB - \frac{IR}{b}\right), \tag{4}$$

$$= \frac{\sigma A v B}{1 + \dfrac{\sigma A R}{b}} = \frac{vBb}{\dfrac{b}{\sigma A} + R} = \frac{V'}{R + R_i}, \tag{5}$$

and

$$V' = vBb, \qquad R_i = b/\sigma A. \tag{6}$$

Note that V' is $|\mathbf{v} \times \mathbf{B}|b$, and that R_i is the resistance of a conductor of conductivity σ, length b, and cross-section A.

b) Find an expression for the efficiency

$$\mathscr{E} = \frac{\text{power dissipated in } R}{\text{total power dissipated}}, \tag{7}$$

as a function of I, that does not involve the resistance R.

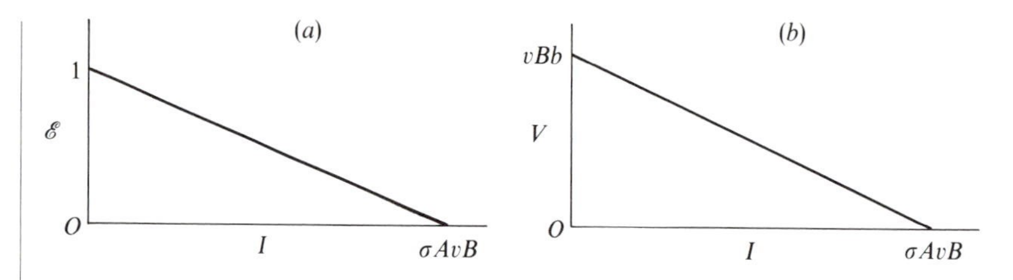

Figure 10-15 (*a*) Efficiency as a function of load current for the magnetohydrodynamic generator of Fig. 10-14. (*b*) Output voltage as a function of load current.

What is the efficiency at $I = 0$?
What is the value of the current when the efficiency is zero?
Sketch a curve of $\mathscr{E}$ as a function of I.

Solution:

$$\mathscr{E} = \frac{I^2 R}{I^2 R + I^2 R_i} = \frac{R}{R + R_i}, \tag{8}$$

$$= \frac{R}{R + (b/\sigma A)} = \frac{\sigma A R}{\sigma A R + b}. \tag{9}$$

Since

$$I = \frac{\sigma A v B b}{\sigma A R + b}, \tag{10}$$

$$\mathscr{E} = \frac{\sigma A R}{\sigma A v B b} I = \frac{R I}{v B b}, \tag{11}$$

$$= \frac{V' - R_i I}{v B b} = \frac{v B b - (b/\sigma A) I}{v B b} = 1 - \frac{I}{\sigma A v B}. \tag{12}$$

The efficiency is equal to unity when $I = 0$.

The efficiency is zero when $I = \sigma A v B$. In that case $R = 0$, from Eq. 4, the output voltage V is zero, and $I = V'/R_i$, from Eq. 5.

Figure 10-15a shows $\mathscr{E}$ as a function of I.

c) Find an expression for V as a function of I that does not involve R.
What is the value of V when the current is zero?
What is the value of the current when the voltage is zero?
Sketch a curve of V as a function of I.

Solution:

$$V = IR = vBb - \frac{b}{\sigma A} I = vBb\left(1 - \frac{I}{\sigma A v B}\right). \tag{13}$$

When the current is zero ($R \to \infty$), the output voltage is vBb. The ions then flow through the generator horizontally, undeflected. See Sec. 10.1.1.

When the voltage is zero, $R = 0$ and I is equal to σAvB as above.

Figure 10-15b shows V as a function of I.

10-17E *ELECTROMAGNETIC FLOWMETERS*

Electromagnetic flowmeters operate as follows. A fluid, which must be at least slightly conducting (blood, for example), flows in a nonconducting tube between the poles of a magnet. The ions in the fluid are then subjected to magnetic forces $Q\mathbf{v} \times \mathbf{B}$ in the direction perpendicular to the velocity $\mathbf{v}$ of the fluid and to the magnetic induction $\mathbf{B}$. Electrodes placed on either side of the tube and in contact with the fluid thus acquire charges of opposite signs; the resulting voltage difference is a measure of $\mathbf{v}$.

Note that, in the electromagnetic flowmeter, ions of both signs have the same velocity $\mathbf{v}$ and the polarity of the electrodes is always the same for a given direction of flow and for a given direction of $\mathbf{B}$. Compare with the Hall effect discussed in Sec. 10.8.1.

Faraday attempted to measure the velocity of the Thames river in this way in 1832. The magnetic field was of course that of the earth.

Consider an idealized case where the tube has a rectangular cross-section ab, with side b parallel to $\mathbf{B}$, and where the velocity is the same throughout the cross-section.

What is the voltage between the electrodes?

In fact, the velocity is maximum on the axis and zero at the inner surface of the tube. Moreover, the non-uniform $\mathbf{v} \times \mathbf{B}$ field causes currents to flow within the fluid. It turns out, curiously enough, that, for a tube of circular cross-section and radius a, the voltage between diametrically opposite electrodes is $2avB$, where v is the average velocity, which is just what one would expect if the velocity were uniform.

CHAPTER 11

MAGNETIC FIELDS: IV

The Faraday Induction Law

In this chapter we shall consider the line integral of $\mathbf{E} \cdot d\mathbf{l}$ around a closed circuit. This will lead us to still another of Maxwell's equations.

In Chapter 2 we saw that, in electrostatic fields, the line integral of $\mathbf{E} \cdot d\mathbf{l}$ around a closed circuit C is zero. This is *not* a general rule. If the circuit C, or part of it, moves in a magnetic field, then one must take into account the $\mathbf{v} \times \mathbf{B}$ term of Chapter 10. This will give us the Faraday induction law, according to which the above integral is equal to minus the time derivative of the enclosed magnetic flux.

What if, in a given reference frame, we have a changing magnetic field? Then electric field intensity is equal, not to $-\nabla V$ as in Chapter 2, but rather to $-\nabla V - \partial \mathbf{A}/\partial t$, where $\mathbf{A}$ is the vector potential of Sec. 8.4, and Faraday's induction law again applies.

11.1 THE INTEGRAL OF $\mathbf{E} \cdot d\mathbf{l}$

We have seen in Sec. 2.5 that an electro*static* field is conservative, or that

$$\oint \mathbf{E} \cdot d\mathbf{l} = 0. \tag{11-1}$$

Thus the work performed by an electro*static* field is zero when a charge moves around a closed path.

However, there are many cases where one has a closed, conducting circuit, without sources, with part, or all of the circuit moving in a magnetic field. The conduction electrons in the moving conductors then feel an electric field $\mathbf{v} \times \mathbf{B}$, as in Sec. 10.1. In such cases, the above integral, evaluated over the circuit, is not zero.

The $\mathbf{v} \times \mathbf{B}$ field in a conductor moving in a magnetic field is called the *induced electric field intensity* and the integral of $\mathbf{E} \cdot \mathbf{dl}$ around a closed circuit is called the *induced electromotance.*

11.1.1 EXAMPLE: THE EXPANDING LOOP

The expanding loop of Fig. 11-1 has one side that can slide to the right at a velocity $\mathbf{v}$ in a region of uniform $\mathbf{B}$. Both $\mathbf{v}$ and $\mathbf{B}$ are measured in the laboratory reference frame S_1. In the moving wire, there is an electric field $\mathbf{E}_2 = \mathbf{v} \times \mathbf{B}$. The line integral of $\mathbf{E} \cdot \mathbf{dl}$ around the circuit is thus evaluated partly in S_1 and partly in S_2. In the counter-clockwise direction,

$$\oint \mathbf{E} \cdot \mathbf{dl} = \int_{abcd} \mathbf{E}_1 \cdot \mathbf{dl} + \int_{da} \mathbf{E}_2 \cdot \mathbf{dl}, \tag{11-2}$$

$$= 0 + \int_{da} (\mathbf{v} \times \mathbf{B}) \cdot \mathbf{dl}, \tag{11-3}$$

$$= vwB. \tag{11-4}$$

This is the induced electromotance. It acts in the counter-clockwise direction in the figure. If, at a certain instant, the loop has a resistance R, then the current is vwB/R.

We have assumed that R is large so as to make the magnetic field of the current flowing around the loop negligible compared to B.

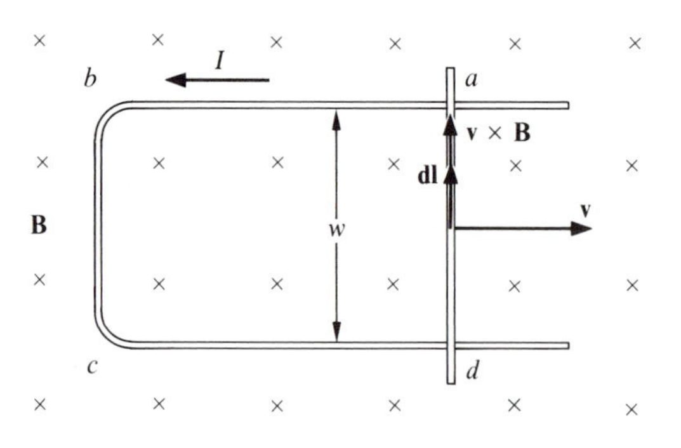

Figure 11-1 A conducting wire ad slides at a velocity v on a pair of conducting rails in a region of uniform magnetic induction $\mathbf{B}$. The magnetic force on the electrons in the wire produces a current I in the circuit.

11.2 THE FARADAY INDUCTION LAW

The right-hand side of Eq. 11-4 is the area swept by the wire per unit time, multiplied by B, or $d\Phi/dt$, where Φ is the magnetic flux linking the circuit. Now it is the custom to choose the positive direction for Φ relative to the positive direction for the line integral, according to the right-hand screw rule. Then, since the integral has been evaluated in the counter-clockwise direction while Φ points into the paper,

$$\oint \mathbf{E} \cdot \mathbf{dl} = -\frac{d\Phi}{dt}. \tag{11-5}$$

This is the *Faraday induction law*: the electromotance induced in a circuit is equal to minus the rate of change of the magnetic flux linking the circuit.

The path of integration may be chosen at will and need not lie in conducting material.

If there are no sources in a circuit, the current is equal to the induced electromotance divided by the resistance of the circuit, exactly as if the electromotance were replaced by a battery of the same voltage and polarity. The induced electromotance is expressed in volts, and it adds algebraically to the voltages of the other sources that may be present in the circuit.

If one has a rigid loop and a time-dependent **B**, Eq. 11-5 still applies.

If one has a rigid *coil* and a time-dependent **B**, then one must add the fluxes through each turn. The sum of these fluxes is called the *flux linkage*. We shall use the symbol Λ for flux linkage, keeping Φ for the surface integral of $\mathbf{B} \cdot \mathbf{da}$ over a simple loop. For example, if a coil has N turns with all the turns close together as in Fig. 11-2, Λ is N times the Φ through one of the turns:

$$\Lambda = N\Phi. \tag{11-6}$$

Thus, for circuits other than simple loops, Faraday's induction law becomes

$$\oint \mathbf{E} \cdot \mathbf{dl} = -\frac{d\Lambda}{dt}. \tag{11-7}$$

Flux linkage is expressed in weber-turns.

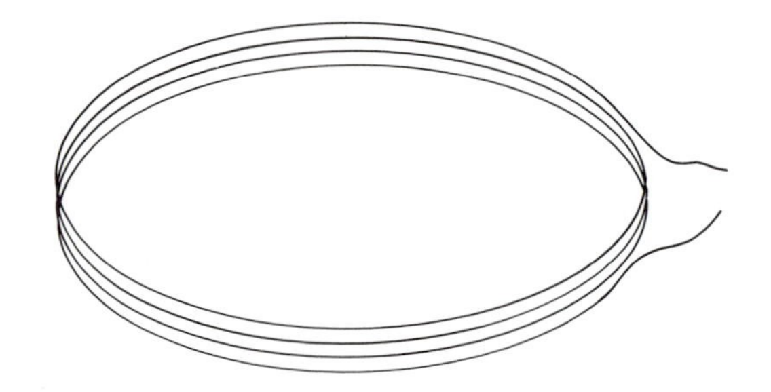

Figure 11-2 Coil with several turns, all close together.

Equation 11-7 should be used with care. It is *not* as general as it is commonly said to be. It is correct (a) if the circuit is rigid and the magnetic induction **B** is time-dependent, or (b) if all or part of the circuit moves *in such a way as to produce an induced electric field intensity* $\mathbf{v} \times \mathbf{B}$. See Prob. 11-4. As we shall see in Sec. 11.5, the **E** in the first case is $-\partial \mathbf{A}/\partial t$.

11.2.1 EXAMPLE: LOOP ROTATING IN A MAGNETIC FIELD

Figure 11-3 shows a loop rotating in a constant and uniform magnetic field at an angular velocity ω. We shall calculate the induced electromotance using the $\mathbf{v} \times \mathbf{B}$ field in the wire, and then verify the Faraday induction law. This will illustrate the principle of operation of an *electric generator*.

The induced electromotance is

$$\mathcal{V} = \oint (\mathbf{v} \times \mathbf{B}) \cdot \mathbf{dl}, \tag{11-8}$$

where the integral is evaluated around the loop.

Along the top and bottom sides, $\mathbf{v} \times \mathbf{B}$ is perpendicular to $\mathbf{dl}$ and the term under the integral is zero. Setting $\theta = \omega t$, and remembering that $v = \omega(a/2)$, the electromotance induced in the vertical sides is

$$\mathcal{V} = 2\omega(a/2)(\sin \omega t)bB = \omega abB \sin \omega t. \tag{11-9}$$

This electromotance is counterclockwise in the figure. It is positive because it is in the positive direction around the loop, with respect to the direction of **B**, according to the right-hand screw rule.

Now

$$\Phi = abB \cos \theta = abB \cos \omega t \tag{11-10}$$

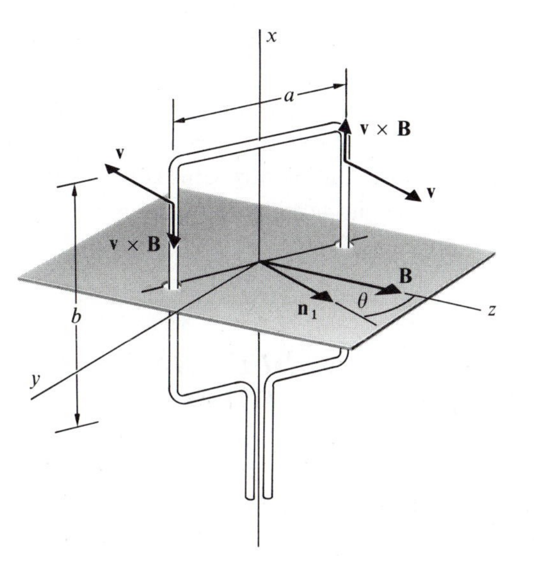

Figure 11-3 Loop rotating in a constant and uniform magnetic field **B**. The vector $\mathbf{n}_1$ is normal to the loop.

and

$$\mathcal{V} = -\frac{d\Phi}{dt}, \tag{11-11}$$

as predicted by the Faraday induction law.

The electromotance is zero when the plane of the loop is perpendicular to **B**, since **v** along the vertical sides is then parallel to **B**, and $\mathbf{v} \times \mathbf{B}$ is zero all around the loop.

Note that the voltage $\mathcal{V}$ of Eq. 11-9 is a sinusoidal function of t. Such voltages are said to be *alternating*. Chapters 16 to 18 are devoted to alternating voltages and currents.

11.3 LENZ'S LAW

If the flux linkage Λ increases, $d\Lambda/dt$ is positive, and the electromotance is negative, that is, the induced electromotance is in the negative direction. On the other hand, if Λ decreases, $d\Lambda/dt$ is negative, and the electromotance is in the positive direction.

The direction of the induced current is always such that it produces a magnetic field that opposes, to a greater or lesser extent, the *change* in flux, depending on the resistance in the circuit. Thus, if Λ increases, the induced current produces an opposing flux. If Λ decreases, the induced current produces an aiding flux. This is Lenz's law.

The currents induced by changing magnetic fields in conductors other than wires are called *eddy currents.*

11.4 THE FARADAY INDUCTION LAW IN DIFFERENTIAL FORM

The Faraday law stated in Eqs. 11-5 and 11-7 gives the electromotance induced in a complete circuit when the flux linkage is a function of the time. It applies to both fixed and deformable paths in both constant and time-dependent magnetic fields, with the restriction stated at the end of Sec. 11.2.

We shall now find an important equation that concerns the value of the $\mathbf{E}$ induced at a given point by a time-dependent $\mathbf{B}$.

If, in a given reference frame, we have a time-dependent $\mathbf{B}$, then we can state the Faraday induction law in differential form as follows. Using Stokes's theorem, Eq. 11-5 becomes

$$\int_S (\nabla \times \mathbf{E}) \cdot \mathbf{da} = -\frac{d\Phi}{dt} = -\frac{d}{dt}\int_S \mathbf{B} \cdot \mathbf{da}, \tag{11-12}$$

where S is any surface bounded by the integration path. If the path is fixed in space, we may interchange the order of differentiation and integration on the right-hand side and

$$\int_S (\nabla \times \mathbf{E}) \cdot \mathbf{da} = -\int_S \frac{\partial \mathbf{B}}{\partial t} \cdot \mathbf{da}. \tag{11-13}$$

We have used the partial derivative of $\mathbf{B}$ because we now require the rate of change of $\mathbf{B}$ with time at a fixed point.

Since the above equation is valid for arbitrary surfaces, the integrands must be equal at every point, and

$$\boxed{\nabla \times \mathbf{E} = -\frac{\partial \mathbf{B}}{\partial t}.} \tag{11-14}$$

This is a general law for stationary media. We have here another of the four Maxwell equations. We have already found two others, namely Eqs. 6-12 and 8-12:

$$\boxed{\nabla \cdot \mathbf{E} = \frac{\rho_f + \rho_b}{\epsilon_0},} \qquad \boxed{\nabla \cdot \mathbf{B} = 0.} \tag{11-15}$$

Equation 11-14 is a differential equation that relates the space derivatives of **E** at a particular point to the time rate of change of **B** at the same point. Both **E** and **B** are measured in the *same* reference frame. The equation does *not* give the value of **E**, unless it can be integrated.

11.5 THE ELECTRIC FIELD INTENSITY E *IN TERMS OF V AND* A

Equation 11-14 will give us an important expression for **E** in terms of the potentials V and **A**. Since

$$\mathbf{B} = \nabla \times \mathbf{A}, \tag{11-16}$$

$$\nabla \times \mathbf{E} = -\frac{\partial}{\partial t}(\nabla \times \mathbf{A}) = -\nabla \times \frac{\partial \mathbf{A}}{\partial t}, \tag{11-17}$$

or

$$\nabla \times \left(\mathbf{E} + \frac{\partial \mathbf{A}}{\partial t}\right) = 0. \tag{11-18}$$

The term between parentheses must be equal to a quantity whose curl is zero, namely a gradient. Then we can set

$$\mathbf{E} = -\nabla V - \frac{\partial \mathbf{A}}{\partial t}. \tag{11-19}$$

For steady currents, **A** is a constant and this equation reduces to Eq. 2-13. The quantity V is therefore the electric potential of Sec. 2.5, which is also known as the *scalar potential.*

Equation 11-19 is a general expression for **E**. It states that an electric field intensity can arise both from accumulations of charge, through the $-\nabla V$ term, and from changing magnetic fields, through the $-\partial \mathbf{A}/\partial t$ term. All three quantities, **E**, **A**, and V, are measured in the *same* reference frame.

Thus, if we have a field $\mathbf{E}_1$, $\mathbf{B}_1$ in reference frame 1, the electric field in frame 2, moving at a constant velocity $\mathbf{v}$ with respect to 1, is given by Eqs. 10-16 and 10-17, or by Eq. 10-20. However, in any given reference frame, **E** is given by Eq. 11-19 and **B** is equal to the curl of **A** as in Eq. 8-13.

11.5.1 *EXAMPLE: THE ELECTROMOTANCE INDUCED IN A LOOP BY A PAIR OF LONG PARALLEL WIRES CARRYING A VARIABLE CURRENT*

A pair of parallel wires, as in Fig. 11-4, carries equal currents I in opposite directions, and I increases at the rate dI/dt. We shall first calculate the induced electromotance from Eq. 11-5 and then from Eq. 11-19.

a) From Sec. 9.1.1, the current I in wire a produces a magnetic induction

$$B_a = \mu_0 I/2\pi\rho_a, \tag{11-20}$$

and a similar relation exists for wire b. The flux through the loop in the direction shown in Fig. 11-4 is thus

$$\Phi = \frac{\mu_0 I}{2\pi}\left(\int_{r_a}^{r_a+w} \frac{h\,d\rho_a}{\rho_a} - \int_{r_b}^{r_b+w} \frac{h\,d\rho_b}{\rho_b}\right), \tag{11-21}$$

$$= \frac{\mu_0 h I}{2\pi} \ln\left[\frac{r_b(r_a + w)}{r_a(r_b + w)}\right]. \tag{11-22}$$

From Eq. 11-5, the electromotance induced in the clockwise direction is $-\partial\Phi/\partial t$:

$$\oint \mathbf{E}\cdot d\mathbf{l} = -\frac{\mu_0 h}{2\pi}\frac{dI}{dt} \ln\left[\frac{r_b(r_a + w)}{r_a(r_b + w)}\right]. \tag{11-23}$$

Thus I' flows in the counterclockwise direction. This is in agreement with Lenz's law: the current I' produces a magnetic field that opposes the increase in Φ.

b) Let us now use Eq. 11-19 to calculate this same electromotance from the time derivative of the vector potential **A**, V being equal to zero in this case. From Sec.

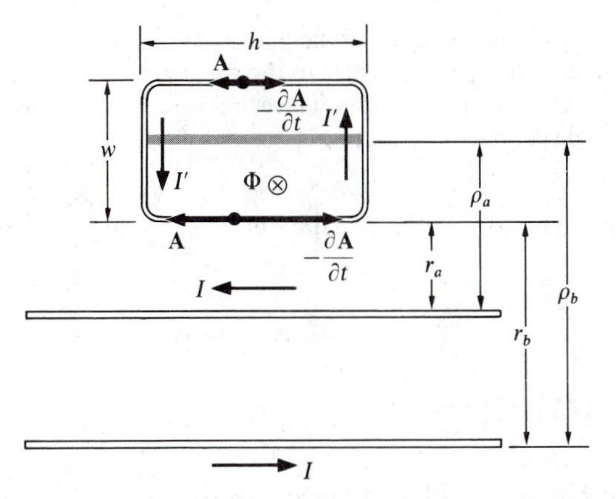

Figure 11-4 Pair of parallel wires carrying equal currents I in opposite directions in the plane of a closed rectangular loop of wire. When I increases, the induced electromotance gives rise to a current I' in the direction shown. The vector potential **A** and the induced electric field $-\partial \mathbf{A}/\partial t$ are shown on the horizontal sides of the loop. The induced current I' flows in the counterclockwise direction because $-\partial \mathbf{A}/\partial t$ is larger on the lower wire than on the upper wire.

8.4.1, **A** is parallel to the wires and, if we choose the leftward direction as positive,

$$A_L = \frac{\mu_0 I}{2\pi} \ln \frac{r_b}{r_a}, \tag{11-24}$$

$$A_U = \frac{\mu_0 I}{2\pi} \ln \left(\frac{r_b + w}{r_a + w} \right) \tag{11-25}$$

along the lower and upper sides of the loop, respectively. Thus

$$E_L = -\frac{\mu_0}{2\pi} \frac{dI}{dt} \ln \frac{r_b}{r_a}, \tag{11-26}$$

$$E_U = -\frac{\mu_0}{2\pi} \frac{dI}{dt} \ln \left(\frac{r_b + w}{r_a + w} \right), \tag{11-27}$$

and, in the clockwise direction,

$$\oint \mathbf{E} \cdot \mathbf{dl} = -\frac{\mu_0 h}{2\pi} \frac{dI}{dt} \ln \left[\frac{r_b(r_a + w)}{r_a(r_b + w)} \right] \tag{11-28}$$

as previously.

We have disregarded the field $-\partial \mathbf{A}/\partial t$ along the left-hand and right-hand sides, because it is perpendicular to the wires.

To find the electromotance induced in the loop by a changing current in a single conductor, we set $r_b \to \infty$, and then

$$\oint \mathbf{E} \cdot d\mathbf{l} = \frac{\mu_0 h}{2\pi} \frac{dI}{dt} \ln \left(\frac{r_a}{r_a + w} \right). \tag{11-29}$$

11.6 SUMMARY

The *Faraday induction law* can be stated as follows: For a circuit C linked by magnetic flux Φ,

$$\oint_C \mathbf{E} \cdot d\mathbf{l} = -\frac{d\Phi}{dt}. \tag{11-5}$$

The integral on the left is called the *induced electromotance.* The positive directions chosen for Φ and for the integration around C are related according to the right-hand screw rule. This law is true if either $\mathbf{B}$ changes, giving a $-\partial \mathbf{A}/\partial t$ field, or if the conductor moves, giving a $\mathbf{v} \times \mathbf{B}$ field. If the integral is evaluated over a coil, the flux Φ must be replaced by the flux linkage Λ.

Lenz's law states that the induced current tends to oppose the *change* in flux.

In differential form, the Faraday induction law is

$$\boxed{\nabla \times \mathbf{E} = -\frac{\partial \mathbf{B}}{\partial t}.} \tag{11-14}$$

This is another of Maxwell's equations. Both $\mathbf{E}$ and $\mathbf{B}$ are measured in the same reference frame.

The *electric field intensity* $\mathbf{E}$ can be due to accumulations of charge, or to changing magnetic fields, or both. In general,

$$\mathbf{E} = -\nabla V - \frac{\partial \mathbf{A}}{\partial t}, \tag{11-19}$$

where $\mathbf{E}$, V, and $\mathbf{A}$ are all measured in the same reference frame.

PROBLEMS

11-1E *BOAT TESTING TANK*

A carriage runs on rails on either side of a long tank of water equipped for testing boat models. The rails are 3.0 meters apart and the carriage has a maximum speed of 20 meters per second.

a) Calculate the maximum voltage between the rails if the vertical component of the earth's magnetic field is 2.0×10^{-5} tesla.

b) What would be the voltage if the tank were situated at the magnetic equator?

11-2 *EXPANDING LOOP*

A conducting bar slides at a constant velocity v along conducting rails separated by a distance s in a region of uniform magnetic induction **B** perpendicular to the plane of the rails. A resistance R is connected between the rails. The resistance of the rest of the circuit is negligible.

a) Calculate the current I flowing in the circuit.

b) How much power is required to move the bar?

c) How does this power compare with the power loss in the resistance R?

11-3E *INDUCED CURRENTS*

A bar magnet is pulled through a conducting ring at a constant velocity as in Fig. 11-5.

Sketch curves of (a) the magnetic flux Φ, (b) the current I, (c) the power dissipated in the ring, as functions of the time.

Use the positive directions for Φ and I shown in the figure.

This phenomenon has been used to measure the velocities of projectiles. The projectile, with a tiny permanent magnet inserted into its nose, is made to pass through two coils in succession, separated by a distance of about 100 millimeters. The time delay between the pulses is a measure of the velocity. The method has been used up to velocities of 5 kilometers per second.

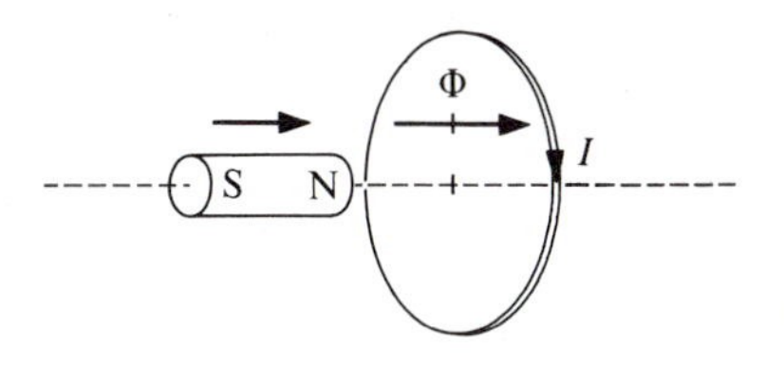

Figure 11-5 See Prob. 11-3.

11-4 *INDUCED CURRENTS*

Figure 11-6a shows a conducting disk rotating in front of a bar magnet.

a) In what direction does the current flow in the wire, clockwise or counterclockwise? Why?

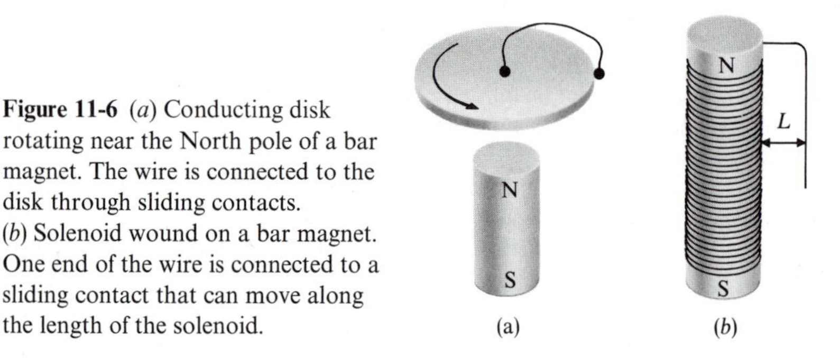

Figure 11-6 (*a*) Conducting disk rotating near the North pole of a bar magnet. The wire is connected to the disk through sliding contacts. (*b*) Solenoid wound on a bar magnet. One end of the wire is connected to a sliding contact that can move along the length of the solenoid.

b) What happens if both the polarity of the magnet and the direction of rotation are reversed?

c) Show that there is no current generated with the set-up of Fig. 11-6b, despite the fact that the magnetic flux linking the circuit is not constant.

11-5E *INDUCED ELECTROMOTANCE*

A loop of wire is situated in a time-dependent magnetic field with

$$B = 1.00 \times 10^{-2} \cos (2\pi \times 60)t$$

perpendicular to the plane of the loop.

Calculate the induced electromotance in a 100-turn square loop 100 millimeters on the side.

11-6 *ELECTROMAGNETIC PROSPECTION*

In electromagnetic prospection, a coil carrying an alternating current causes induced currents to flow in conducting ore bodies, and the alternating magnetic field of these induced currents is detected by means of a second coil placed some distance away from the first one, as in Fig. 11-7.

Let us consider a simple example of induced currents.

A thin conducting disk of thickness h, radius R, and conductivity σ is placed in a uniform alternating magnetic field

$$B = B_0 \cos \omega t$$

parallel to the axis of the disk as in Fig. 11-8.

a) Find the induced current density J as a function of the radius. Assume that the conductivity σ is small enough to render the magnetic field of the induced current negligible, compared to the applied B.

Choose the positive directions for **B** and for **J** in the directions shown in the figure.

b) Sketch curves of B and J as functions of the time. Explain why B and J are so related.

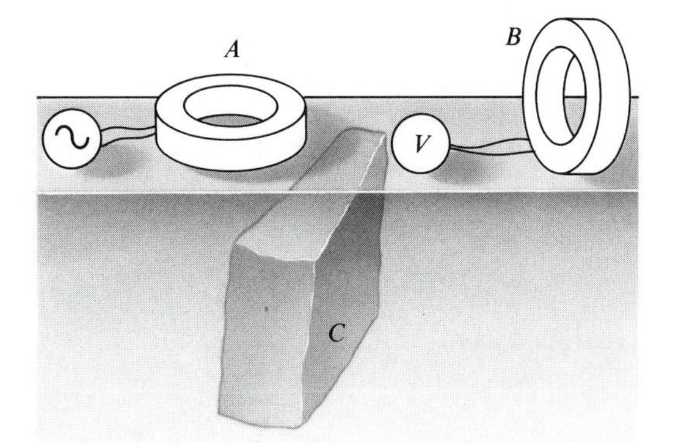

Figure 11-7 Electromagnetic prospection. Coil A, connected to a source of alternating current, produces an alternating magnetic field. Coil B is held perpendicular to A and is sensitive only to the magnetic field due to the currents induced by A in the ore body C. The figure is not drawn to scale; the coil diameters are of the order of one meter or less.

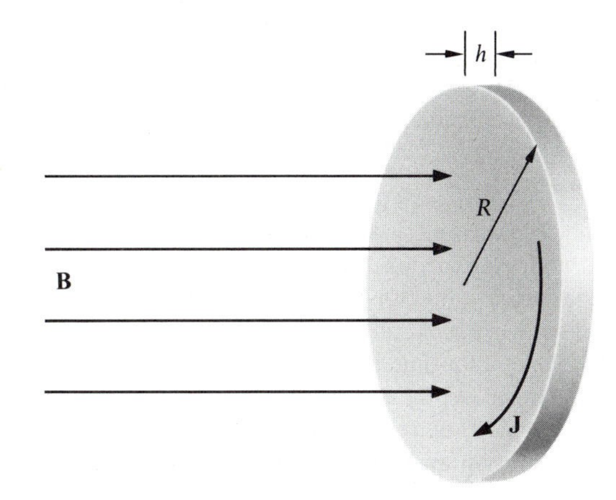

Figure 11-8 See Prob. 11-6.

11-7E *INDUCTION HEATING*

It is common practice to heat conductors by subjecting them to an alternating magnetic field. The changing flux induces *eddy currents* that heat the conductor by the Joule effect (Sec. 5.2.1). The method is called *induction heating*. It is used extensively for melting metals and for hardening and forging steel.

The powers used range from watts to megawatts, and the frequencies from 60 hertz to several hundred kilohertz.

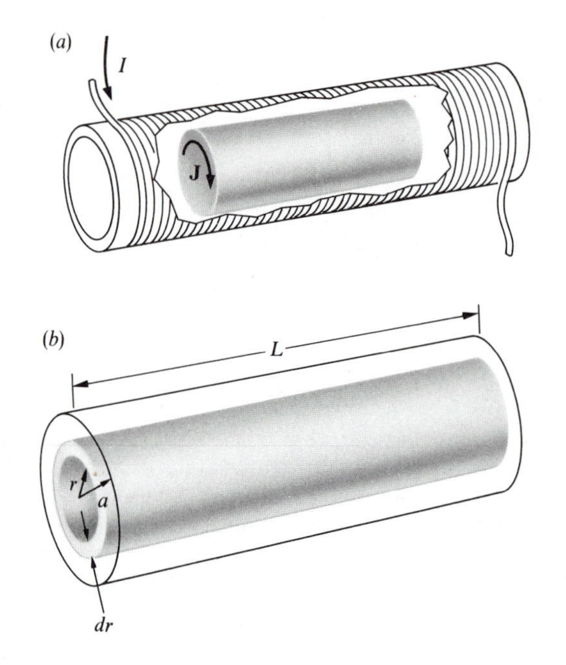

Figure 11-9 (*a*) Heating a conducting cylinder by induction. (*b*) Ring inside the cylinder of (*a*). See Prob. 11-7.

Induction heating has the advantage of convenience and of not contaminating the metal with combustion gases. It even permits heating a conductor enclosed in a vacuum enclosure. Induction heating has another major advantage. At high frequencies, currents flow near the surface of a conductor. This is the *skin effect*. See Probs. 16-17 and 18-14. So, by choosing the frequency correctly, one can apply a brief heat treatment down to a known depth. This is particularly important because there are numerous purposes for which one requires steel parts with a hard skin and a soft core: the hard skin resists abrasion and the soft core reduces breakage. Plowshares are heat-treated in this way.

Induction furnaces are used for melting metals. They consist of large crucibles with capacities ranging up to 30 tons, thermally insulated and surrounded by current-carrying coils. Operation is usually started with part of the load already molten.

Let us consider Fig. 11-9a where a rod of radius a, length L, and conductivity σ is placed inside a solenoid having N' turns per meter and carrying an alternating current

$$I = I_0 \cos \omega t.$$

As usual, we neglect end effects.

We also assume that the frequency is low enough to avoid the complications due to the skin effect. This assumption is well satisfied at 60 hertz with a rod of graphite

($\sigma = 1.0 \times 10^5$ siemens per meter) having a radius of 60 millimeters. In other words, the magnetic induction of the induced currents is negligible. We also assume that the conductor is non-magnetic.

This is admittedly a highly simplified illustration of induction heating. Moreover, the power dissipated in the graphite is only a few watts, which is absurdly small for such a large piece. This should nonetheless be a useful exercise on induced currents.

Consider a ring of radius r, thickness dr, and length L inside the conductor, as in Fig. 11-9b.

a) Show that the electromotance induced in the ring is

$$\mu_0 \pi r^2 \omega N' I_0 \sin \omega t.$$

b) Show that, for a current flowing in the azimuthal direction, the ring has a resistance

$$R = 2\pi r / \sigma L\, dr.$$

c) Show that the average power dissipated in the ring is

$$\frac{1}{4}(\mu_0 \omega N' I_0)^2 \pi \sigma L r^3\, dr.$$

d) Show that the total average power dissipated in the cylinder is

$$\frac{1}{16}(\mu_0 \omega N' I_0)^2 \pi \sigma L a^4.$$

e) Calculate the power dissipated in the graphite rod, setting $L = 1$ meter, $I_0 = 20$ amperes, $N' = 5000$ turns per meter.

11-8D *INDUCED ELECTROMOTANCE*

A magnetic field is described by

$$\mathbf{B} = B_0 \sin \frac{2\pi y}{\lambda} (\sin \omega t)\mathbf{i}.$$

In this field a square loop of side $\lambda/4$ lies in the yz-plane with its sides parallel to the y- and z-axes. The loop moves at a constant velocity $v\mathbf{j}$.

Calculate the electromotance induced in the loop as a function of the time if the trailing edge of the loop is at $y = 0$ at $t = 0$.

Set $\omega = 2\pi v/\lambda$.

11-9 *BETATRON*

Figure 11-10 shows the principle of operation of the betatron. The betatron produces electrons having energies of the order of millions of electron-volts. Electrons are held in a circular orbit in a vacuum chamber by a magnetic field **B**. The electrons

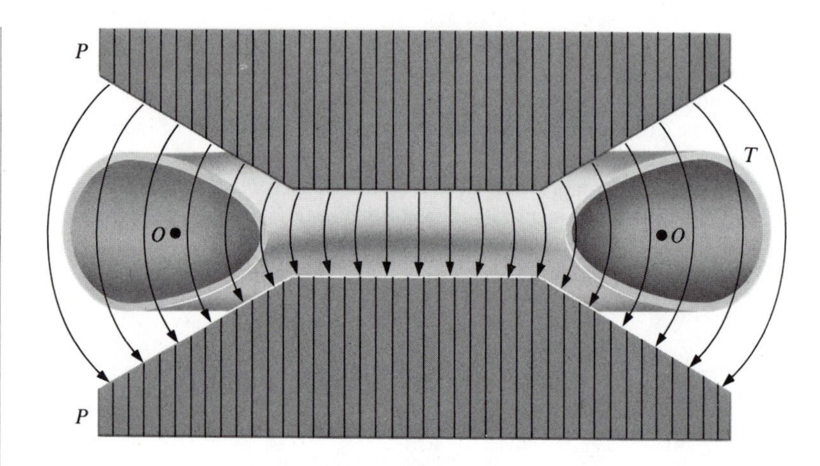

Figure 11-10 Cross-section through the central region of a betatron. The electrons describe a circular orbit in a plane perpendicular to the paper at O—O, in the field of an electromagnet and inside an evacuated ceramic torus T. Only the pole pieces P of the electromagnet are shown. The core is laminated. The pole pieces are shaped so that the average B inside the orbit is equal to twice that at the orbit. The electromagnet is operated on alternating current, and the electrons are accelerated only during that part of the cycle when both B and dB/dt have the proper signs. See Prob. 11-9 and Fig. 11-11.

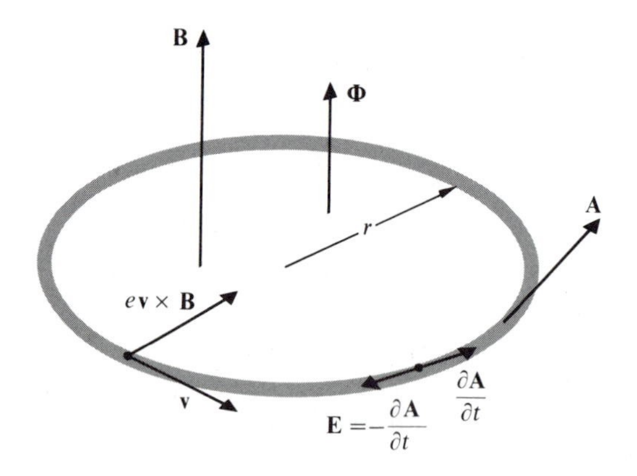

Figure 11-11 Circular electron orbit in the betatron. It is assumed that the magnetic flux is in the direction shown and that it increases. Here, $e = -1.60 \times 10^{-19}$ coulomb.

are accelerated by increasing the magnetic flux linking the orbit. Betatrons serve as sources of x-rays. There are very few that are still in use today.

Show that the average magnetic induction over the plane of the orbit must be twice the magnetic induction at the orbit, if the orbit radius is to remain fixed when both **B** and the electron energy increase.

Solution: The various vectors we are concerned with are shown in Fig. 11-11. In the reference frame of the laboratory, an electron is subjected to the Lorentz force

$$e(\mathbf{E} + \mathbf{v} \times \mathbf{B}).$$

The electric field **E** is due to the time-dependent magnetic field and is $-\partial \mathbf{A}/\partial t$. We do not know the value of **A**, but we do know, from Faraday's induction law, that

$$E = -\frac{1}{2\pi r}\frac{d\Phi}{dt}. \tag{1}$$

We assume that the magnetic flux Φ increases. This electric field E provides the tangential acceleration that increases the electron energy. The centripetal force $e\mathbf{v} \times \mathbf{B}$ keeps the electron on its circular orbit.

So, in the azimuthal direction,

$$\frac{d}{dt}(mv) = eE = -\frac{e}{2\pi r}\frac{d\Phi}{dt}, \tag{2}$$

while, in the radial direction,

$$mv^2/r = -Bev. \tag{3}$$

In these two equations, all the quantities are positive except e.

Equating the two values of mv obtained from Eqs. 2 and 3,

$$mv = -\frac{e}{2\pi r}\Phi = -Ber, \tag{4}$$

$$\Phi = 2\pi r^2 B, \tag{5}$$

where B is the magnetic induction at the radius r of the electron orbit. Now, by definition,

$$\Phi = \pi r^2 \overline{B}, \tag{6}$$

where $\overline{B}$ is the average magnetic induction within the circle of radius r. Then, comparing Eqs. 5 and 6, $\overline{B}$ is equal to $2B$.

11-10 THE TOLMAN AND BARNETT EFFECTS

If a conductor is given an acceleration $\mathbf{a}$, the conduction electrons of mass m and charge $-e$ are subjected to inertia forces $-m\mathbf{a}$.

Show that, if $\mathbf{E}'$ is the equivalent total electric field intensity and if $\mathbf{B}$ is the magnetic induction in the conductor,

$$\nabla \times \mathbf{E}' = -\frac{\partial \mathbf{B}}{\partial t} + \frac{m}{e} \nabla \times \mathbf{a}.$$

This effect was predicted by Maxwell, and was observed for the first time in 1916 by Tolman and collaborators.

The inverse effect, namely the acceleration of a body carrying a variable current, was also predicted by Maxwell, and was first observed by Barnett and others in 1930.

11-11E ELECTRIC CONDUITS

The U.S. National Electrical Code rules that both conductors of a circuit operating on alternating current, if enclosed in a metallic conduit, must be run in the same conduit.

Let us suppose that we have a single conductor carrying an alternating current and enclosed within a conducting tube.

Show that both $\mathbf{A}$ and $\partial \mathbf{A}/\partial t$ are longitudinal in the tube.

So, with a single wire, a longitudinal current is induced in the tube. This causes a needless power loss, and may even cause sparking at faulty joints. With two conductors carrying equal and opposite currents, both $\mathbf{A}$ and $\partial \mathbf{A}/\partial t$ are essentially zero in the conduit.

11-12E THE POTENTIALS V AND $\mathbf{A}$

We have seen that

$$\mathbf{E} = -\nabla V - \partial \mathbf{A}/\partial t, \qquad \mathbf{B} = \nabla \times \mathbf{A}.$$

Show that neither $\mathbf{E}$ nor $\mathbf{B}$ are affected if the potentials V and $\mathbf{A}$ are replaced by

$$V + \partial G/\partial t \qquad \text{and} \qquad \mathbf{A} - \nabla G,$$

where G is any function of x, y, z, t whose second derivatives exist and are continuous.

CHAPTER 12

MAGNETIC FIELDS: V

Mutual Inductance M and Self-Inductance L

In this chapter we shall see how one utilizes the Faraday induction law to calculate induced voltages and currents in electric circuits. In Chapter 5 we ascribed to each branch of a circuit a certain resistance R. If the current flowing in a branch produces an appreciable magnetic field, then the branch also possesses a *self-inductance* L. If the magnetic field of one branch produces a flux linkage in another branch, then there exists a *mutual inductance* M between the two branches. Thus, the magnetic properties of the branches are characterized by the two quantities L and M.

We shall return to mutual inductance later on, in Chapter 18, which deals with power transfer and transformers.

12.1 MUTUAL INDUCTANCE M

To calculate the electromotance induced in one circuit when the current changes in another circuit, it is convenient to express the flux linkage in the first one in terms of the current in the second and of a geometrical factor involving both circuits.

Consider the two circuits of Fig. 12-1. The current I_a in circuit a produces in b a flux linkage Λ_{ab} that is proportional to I_a:

$$\Lambda_{ab} = M_{ab} I_a. \tag{12-1}$$

Similarly, if there is a current I_b through circuit b, the magnetic flux of b linking a is

$$\Lambda_{ba} = M_{ba} I_b. \tag{12-2}$$

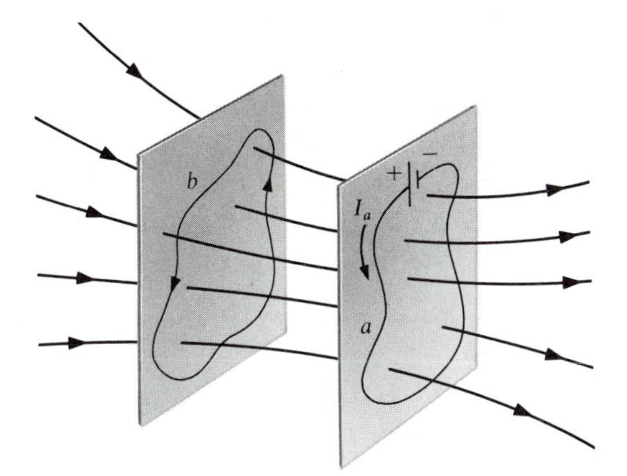

Figure 12-1 Two circuits a and b. The flux Φ_{ab} shown linking b and originating in a is positive. This is because its direction is related by the right-hand screw rule to the direction chosen to be positive around b.

Since both circuits can be of any shape, one naturally does not expect to find a general relationship between M_{ab} and M_{ba}. On the contrary, it can be shown that $M_{ab} = M_{ba}$.† The factor of proportionality $M = M_{ab} = M_{ba}$ is called the *mutual inductance* between the two circuits.

Mutual inductance depends solely on the geometry of the two closed circuits and on the position and orientation of one with respect to the other. When multiplied by the current in one circuit, M gives the flux linkage in the other.

A similar situation exists in relation with the capacitance between two conductors. The capacitance C depends solely on the geometry of the two conductors and on the position and orientation of one with respect to the other. If one of the conductors is grounded, then the charge induced on it is equal to CV, where V is the voltage applied to the other.

Since the mutual inductance is the flux linkage in one circuit per unit of current in the other, inductance is measured in weber-turns per ampere, or in *henrys*.

The mutual inductance between two circuits is one henry when a current of one ampere in one of the circuits produces a flux linkage of one weber-turn in the other.

† *Electromagnetic Fields and Waves*, p. 343.

The sign of M is chosen as follows. The mutual inductance between two circuits a and b is positive if a current in the positive direction in a produces in b a flux that is in the same direction as one due to a positive current in b. This is illustrated in Fig. 12-1. The positive directions for the currents are chosen arbitrarily.

A pair of coils designed so as to possess a mutual inductance is called a *mutual inductor*. Mutual inductors are also called *transformers*.

12.1.1 *EXAMPLE: MUTUAL INDUCTANCE BETWEEN TWO COAXIAL SOLENOIDS*

Let us calculate the mutual inductance between two coaxial solenoids as in Fig. 12-2. We assume that both windings are long, compared to their common diameter $2R$, that they have the same number of turns per meter N', and that they are wound in the same direction, as in the figure. We set $l_a \geq l_b$.

Let us assume a current I_a in coil a. Then the flux of coil a linking each turn of coil b is

$$\Phi_{ab} = \mu_0 \pi R^2 N' I_a, \tag{12-3}$$

and the mutual inductance is

$$M = \Lambda_{ab}/I_a = N_b \Phi_{ab}/I_a = \mu_0 \pi R^2 N' N_b, \tag{12-4}$$

$$= \mu_0 \pi R^2 N_a N_b / l_a. \tag{12-5}$$

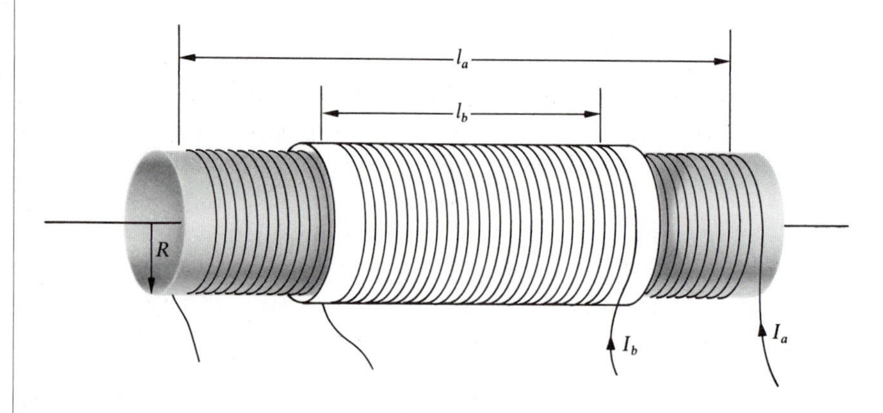

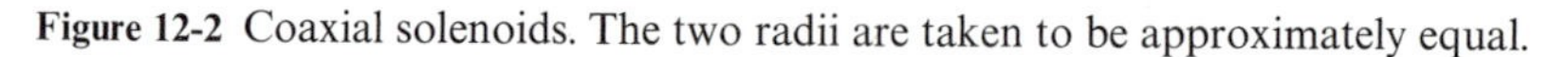
Figure 12-2 Coaxial solenoids. The two radii are taken to be approximately equal.

We can also calculate the mutual inductance by assuming a current I_b in coil b. Then

$$\Phi_{ba} = \mu_0 \pi R^2 N' I_b. \tag{12-6}$$

This flux links only $l_b N' = N_b$ turns of coil a, since B falls rapidly to zero beyond the end of a long solenoid, as we saw in Sec. 9.1.5. Then the mutual inductance is

$$M = \Lambda_{ba}/I_b = N_b \Phi_{ba}/I_b = \mu_0 \pi R^2 N' N_b = \mu_0 \pi R^2 N_a N_b / l_a, \tag{12-7}$$

as previously.

12.2 INDUCED ELECTROMOTANCE IN TERMS OF MUTUAL INDUCTANCE

From Sec. 11.2, the electromotance induced in circuit b by a change in I_a in Fig. 12-1 is

$$\oint_b \mathbf{E} \cdot d\mathbf{l} = -\frac{d\Lambda_{ab}}{dt} = -M \frac{dI_a}{dt}. \tag{12-8}$$

In this case, $\mathbf{E}$ is $-\partial \mathbf{A}/\partial t$, where $\mathbf{A}$ is the value of the vector potential of the current I_a at a point on circuit b (Sec. 11.5).

Similarly, the electromotance induced in circuit a by a change in I_b is

$$\oint_a \mathbf{E} \cdot d\mathbf{l} = -\frac{d\Lambda_{ba}}{dt} = -M \frac{dI_b}{dt}. \tag{12-9}$$

These equations are convenient for computing the induced electromotance, since they involve only the mutual inductance and dI/dt, both of which can be measured.

We now have a second definition of the henry: the mutual inductance between two circuits is one henry if a current changing at the rate of one ampere per second in one circuit induces an electromotance of one volt in the other.

12.2.1 EXAMPLE: THE VECTOR POTENTIAL OF THE INNER SOLENOID

It is paradoxical that a varying current in the inner solenoid should induce an electromotance in the outer one, since we have shown (Sec. 9.1.3) that the magnetic induction outside a long solenoid is zero. The explanation is that the induced electric field intensity at any given point is equal to the negative time derivative of the vector potential at that point (Eq. 11-19) and that the vector potential **A** does not vanish outside an infinite solenoid, despite the fact that $\mathbf{B} = \nabla \times \mathbf{A}$ does. See also Sec. 9.1.2.

We can actually calculate the vector potential just outside the inner solenoid from the mutual inductance. If we consider the direction of the current I_a in the inner solenoid as positive, then the electromotance induced in the outer one is $-M\, dI_a/dt$, or

$$-\mu_0 \pi R^2 N' N_b \frac{dI_a}{dt},$$

and the induced electric field intensity $-\partial A/\partial t$ is $2\pi R N_b$ times smaller:

$$-\frac{\partial A}{\partial t} = -\frac{\mu_0 R N'}{2} \frac{dI_a}{dt}. \qquad (12\text{-}10)$$

Integrating, we have the vector potential at any point close to the solenoid:

$$A = \frac{\mu_0 R N'}{2} I_a \qquad (12\text{-}11)$$

in the azimuthal direction.

12.3 SELF-INDUCTANCE L

A single circuit carrying a current I is of course linked by its own flux, as in Fig. 12-3. The flux linkage Λ is proportional to the current:

$$\Lambda = LI, \qquad (12\text{-}12)$$

where L is called the *self-inductance*, or simply the *inductance* of the circuit. Self-inductance, like mutual inductance, depends solely on the geometry and is measured in henrys. It is always positive.

A circuit that is designed so as to have a self-inductance is called an *inductor*.

An inductor has a self-inductance of one henry if a current of one ampere produces a flux linkage of one weber-turn.

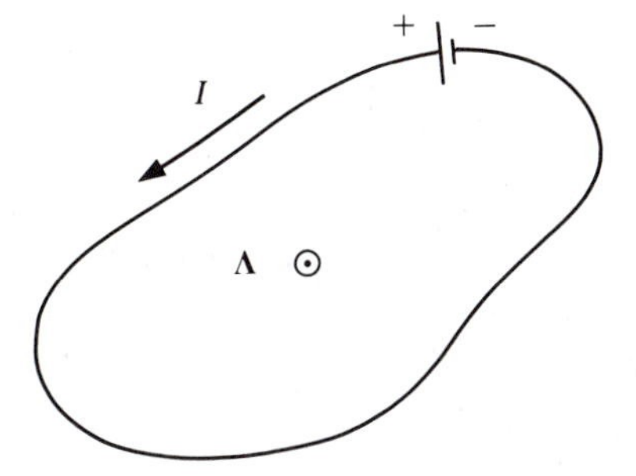

Figure 12-3 Isolated circuit carrying a current I and its flux linkage Λ.

A change in the current flowing through a circuit produces within the circuit itself an induced electromotance

$$\oint \mathbf{E} \cdot \mathbf{dl} = -\frac{d\Lambda}{dt} = -L\frac{dI}{dt}. \tag{12-13}$$

The induced electromotance tends to oppose the *change* in current, according to Lenz's law, and adds to whatever other voltages are present. If a varying current flows through an inductance L, the voltage across it is $L\,dI/dt$, as in Fig. 12-4.

An inductor therefore has a self-inductance of one henry if the current flowing through it changes at the rate of one ampere per second when the voltage difference between its terminals is one volt.

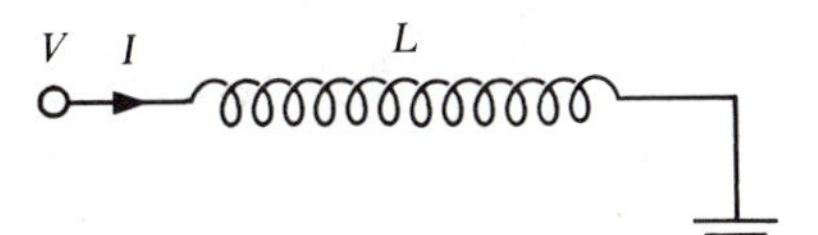

Figure 12-4 Idealized inductor with zero resistance. The voltage V is $L\,dI/dt$.

12.3.1 *EXAMPLE: LONG SOLENOID*

It was shown in Sec. 9.1.3 that the magnetic induction inside a long solenoid, neglecting end effects, is uniform, and that

$$B = \mu_0 N'I, \tag{12-14}$$

where N' is the number of turns per meter. Thus

$$\Phi = \frac{\mu_0 NI}{l}\pi R^2, \tag{12-15}$$

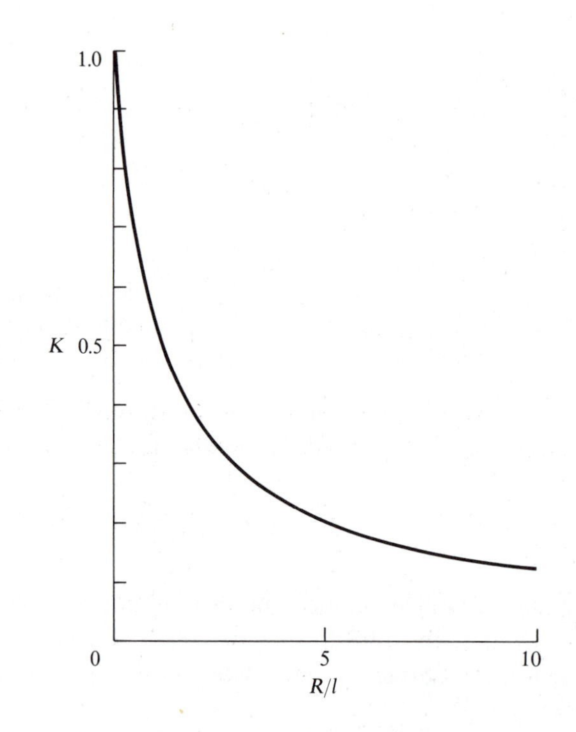

Figure 12-5 Factor K for calculating the inductance of a short solenoid.

where N is the total number of turns, l is the length of the solenoid, and R is its radius. Then

$$L = \frac{\Lambda}{I} = \frac{N\Phi}{I} = \frac{\mu_0 N^2}{l} \pi R^2. \qquad (12\text{-}16)$$

The self-inductance of a long solenoid is thus proportional to the *square* of the number of turns and to its cross-section, and inversely proportional to its length.

The inductance of a *short solenoid* is smaller by a factor K, which is a function of R/l, as in Fig. 12-5.

12.3.2 *EXAMPLE: TOROIDAL COIL*

A toroidal coil of N turns is wound on a form of non-magnetic material having a square cross-section, as in Fig. 12-6.

According to the circuital law, or from Eq. 9-8, the magnetic induction in the azimuthal direction, at a radius ρ inside the toroid, is

$$B = \mu_0 NI/2\pi\rho. \qquad (12\text{-}17)$$

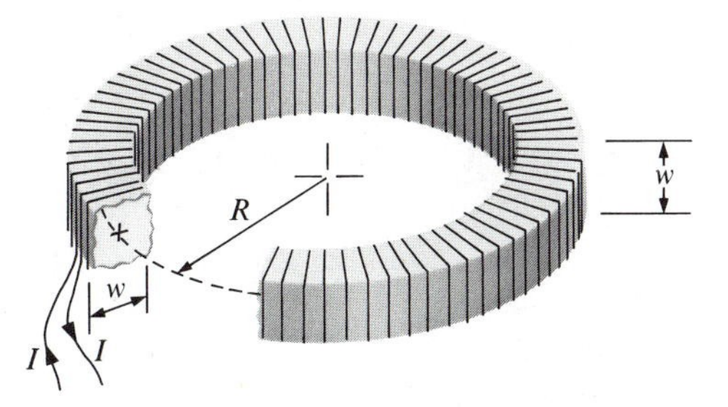

Figure 12-6 Toroidal coil of square cross-section and mean radius R.

Thus the flux linkage is

$$\Lambda = \frac{\mu_0 N^2 I}{2\pi} \int_{R-w/2}^{R+w/2} \frac{w\,d\rho}{\rho}, \tag{12-18}$$

$$= \frac{\mu_0 N^2 I}{2\pi} w \ln\left[\frac{2R + w}{2R - w}\right], \tag{12-19}$$

$$L = \frac{\mu_0 N^2 w}{2\pi} \ln\left[\frac{2R + w}{2R - w}\right]. \tag{12-20}$$

The self-inductance of a toroidal coil is proportional to the square of the number of turns, like that of a long solenoid.

12.4 COEFFICIENT OF COUPLING k

In Sec. 12.1.1 we found that the mutual inductance between the two solenoids of Fig. 12-2 is

$$M = \mu_0 \pi R^2 N_a N_b / l_a. \tag{12-21}$$

Now, from Eq. 12-16,

$$L_a = \mu_0 \pi R^2 N_a^2 / l_a, \tag{12-22}$$

$$L_b = \mu_0 \pi R^2 N_b^2 / l_b, \tag{12-23}$$

and

$$M = \left(\frac{l_b}{l_a}\right)^{1/2} (L_a L_b)^{1/2}. \tag{12-24}$$

The mutual inductance is thus proportional to the geometrical mean of the self-inductances.

More generally,

$$M = k(L_a L_b)^{1/2}, \tag{12-25}$$

where k is the *coefficient of coupling*, which can be either positive or negative. Moreover,

$$|k| \leq 1. \tag{12-26}$$

For example, in the present case, $k = (l_b/l_a)^{1/2} \leq 1$, by hypothesis.

When $|k| \approx 1$, the two circuits are said to be *tightly coupled*; they are *loosely coupled* when $|k| \ll 1$.

12.5 INDUCTORS CONNECTED IN SERIES

12.5.1 ZERO MUTUAL INDUCTANCE

Consider two inductors connected in series as in Fig. 12-7a or 12-7b. Their mutual inductance is approximately zero. Then

$$V = L_1 \frac{dI}{dt} + L_2 \frac{dI}{dt} = (L_1 + L_2) \frac{dI}{dt}, \tag{12-27}$$

and the effective inductance is simply the sum of the inductances. The same rule applies to any number of inductors connected in series, as long as the mutual inductances are all zero.

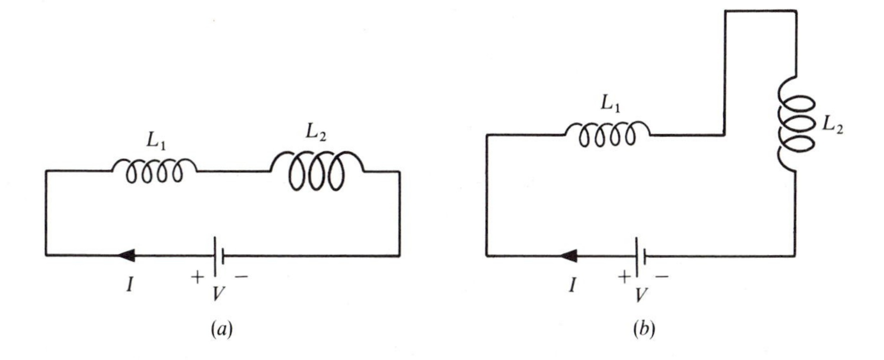

Figure 12-7 (*a*) Two coils connected in series. They are assumed to be far from each other, and their mutual inductance is approximately zero. (*b*) Two coils connected in series, one perpendicular to the other. The flux linkage and the mutual inductance are again approximately zero.

12.5.2 *NON-ZERO MUTUAL INDUCTANCE*

If the coupling coefficient between the two inductors connected in series is not zero, as in Fig. 12-8, then the voltage on coil 1 is L_1 times dI/dt, plus M times dI/dt in coil 2, and the voltage on coil 2 is given by a similar expression. Thus

$$V = (L_1 + M)\frac{dI}{dt} + (L_2 + M)\frac{dI}{dt}, \tag{12-28}$$

$$= (L_1 + L_2 + 2M)\frac{dI}{dt}. \tag{12-29}$$

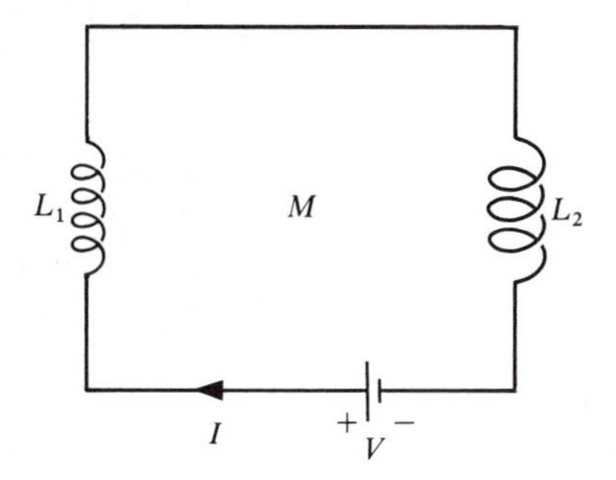

Figure 12-8 Two coils connected in series with a non-zero mutual inductance M.

The effective inductance is thus

$$L = L_1 + L_2 + 2M. \tag{12-30}$$

Remember that M can be either positive or negative (Sec. 12.1).

12.5.3 EXAMPLE: COAXIAL SOLENOIDS

Figure 12-9a shows the coaxial solenoids of Fig. 12-2 connected in series, with the left-hand wires connected together. Let us calculate the inductance between a and b.

Suppose the current flows from terminal A to terminal B. Then the flux of a points left, while that of b points right. So the flux of b partly cancels that of a, and

$$L = L_a + L_b - 2M, \tag{12-31}$$

where L_a, L_b, M are given in Eqs. 12-22 to 12-24 and M is a positive quantity, here.

If the solenoids are connected as in Fig. 12-9b,

$$L = L_a + L_b + 2M. \tag{12-32}$$

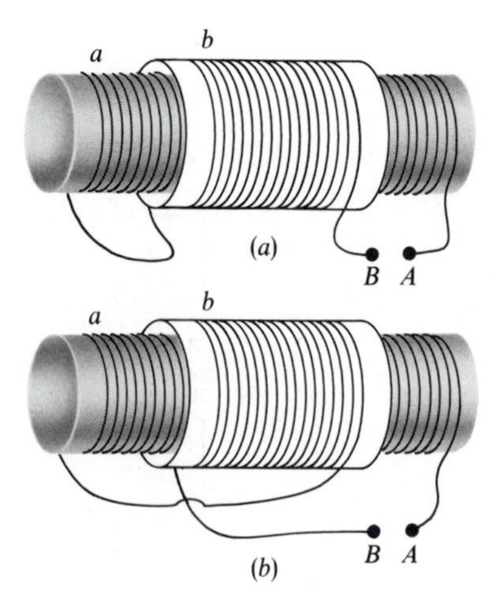

Figure 12-9 The coaxial solenoids of Fig. 12-2, connected in series, (a) with M negative, and (b) with M positive.

12.6 INDUCTORS CONNECTED IN PARALLEL

12.6.1 ZERO MUTUAL INDUCTANCE

We have just seen that two inductors connected in series are treated like resistors in series (Sec. 5.4) when their mutual inductance is zero. One might guess that inductors in parallel are calculated like resistors in parallel (Sec. 5.5) when M is zero. As we shall see, this is correct.

We now have two inductors connected in parallel as in Fig. 12-10, with zero M. Then

$$V = L_1 \frac{dI_1}{dt} = L_2 \frac{dI_2}{dt}. \tag{12-33}$$

The total current I is $I_1 + I_2$. If L is the self-inductance that is equivalent to L_1 and L_2 in parallel,

$$V = L \frac{dI}{dt}, \tag{12-34}$$

$$= L\left(\frac{dI_1}{dt} + \frac{dI_2}{dt}\right), \tag{12-35}$$

$$= L\left(\frac{1}{L_1} + \frac{1}{L_2}\right) V, \tag{12-36}$$

$$= L\left(\frac{L_1 + L_2}{L_1 L_2}\right) V, \tag{12-37}$$

and

$$L = \frac{L_1 L_2}{L_1 + L_2}, \tag{12-38}$$

as we guessed at the beginning.

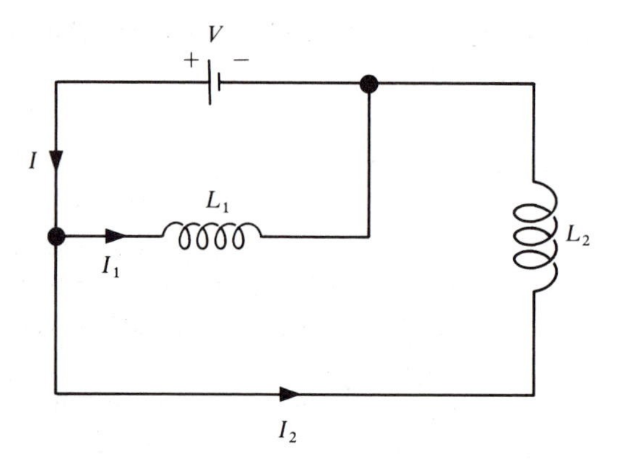

Figure 12-10 Two coils connected in parallel with approximately zero mutual inductance.

12.6.2 NON-ZERO MUTUAL INDUCTANCE

Let us now consider Fig. 12-11, which shows two coupled inductors, connected in parallel. If L is the equivalent inductance,

$$V = L \frac{dI}{dt}, \tag{12-39}$$

$$= L \left(\frac{dI_1}{dt} + \frac{dI_2}{dt} \right). \tag{12-40}$$

Also, for the left-hand coil,

$$V = L_1 \frac{dI_1}{dt} + M \frac{dI_2}{dt} \tag{12-41}$$

and, for the right-hand coil,

$$V = L_2 \frac{dI_2}{dt} + M \frac{dI_1}{dt}. \tag{12-42}$$

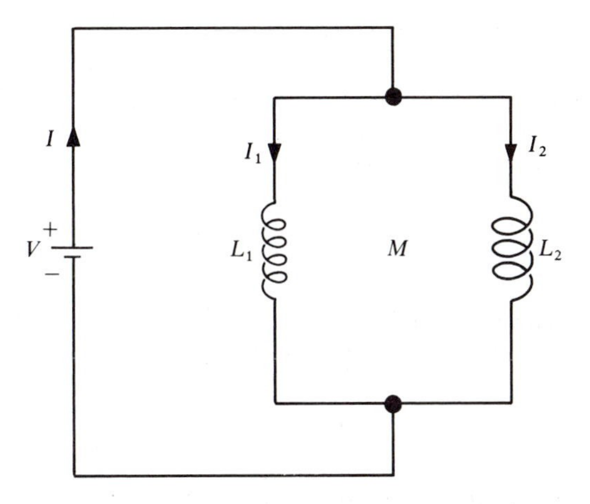

Figure 12-11 Two coils connected in parallel with a non-zero mutual inductance M. Here, M is positive: a positive (downward) current in L_1 produces a downward B in L_2, while a positive (downward) current in L_2 also produces a downward B in L_2.

From the last two equations,

$$(L_1 - M)\frac{dI_1}{dt} = (L_2 - M)\frac{dI_2}{dt}. \tag{12-43}$$

Then, from Eqs. 12-40 and 12-41,

$$L\left(1 + \frac{L_1 - M}{L_2 - M}\right)\frac{dI_1}{dt} = \left(L_1 + M\frac{L_1 - M}{L_2 - M}\right)\frac{dI_1}{dt}, \tag{12-44}$$

and

$$L = \frac{L_1(L_2 - M) + M(L_1 - M)}{(L_2 - M) + (L_1 - M)}, \tag{12-45}$$

$$= \frac{L_1 L_2 - M^2}{L_1 + L_2 - 2M}. \tag{12-46}$$

When $M = 0$, we revert to the previous case.

12.6.3 EXAMPLE: COAXIAL SOLENOIDS

Figure 12-12 shows again the coaxial solenoids of Fig. 12-2, this time connected in parallel. With the connections shown in Fig. 12-12a, M is positive and the inductance between terminals A and B is

$$L = \frac{L_a L_b - M^2}{L_a + L_b - 2M} \tag{12-47}$$

with the L's and the M on the right as in Eqs. 12-22 to 12-24.

If the connections to *one* of the solenoids are interchanged as in Fig. 12-12b,

$$L = \frac{L_a L_b - M^2}{L_a + L_b + 2M}. \tag{12-48}$$

Here again, M is the positive quantity calculated as in Eq. 12-24.

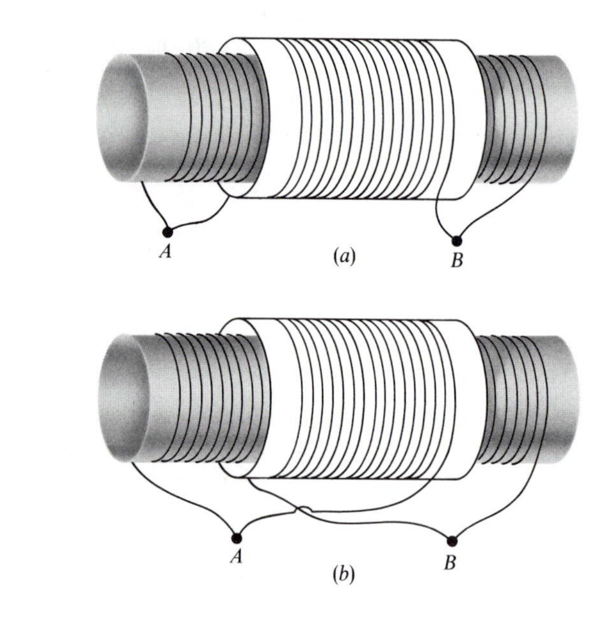

Figure 12-12 The pair of coaxial solenoids of Fig. 12-2, connected in parallel, (*a*) with M positive, (*b*) with M negative.

12.7 TRANSIENTS IN RL CIRCUITS

When a circuit comprising resistances and inductances is disturbed in some way, whether by changing the applied voltages or currents, or by modifying the circuit, the currents take some time to adjust themselves to their new steady-state values. We discussed a similar phenomenon in relation with RC circuits in Sec. 5.14.

12.7.1 EXAMPLE: INDUCTOR L CONNECTED TO A VOLTAGE SOURCE V_0

Figure 12-13 shows an inductor L connected to a voltage source V_0. The inductor and its connecting wires are made of a single continuous wire of conductivity σ and cross-section a.

a) *The current is constant.*

If the wire is in air ($\epsilon_r \approx 1$), its surface charge density is $\epsilon_0 E_n$, where E_n is the normal component of $\mathbf{E}$ just outside the wire. This follows from Gauss's law (Sec. 3.2) because $E_n = 0$ inside the wire, as we shall show below. Similarly, if the wire is submerged in a dielectric, the surface charge density is $\epsilon_r \epsilon_0 E_n$.

At any given point on the surface of the wire, the field E_n is proportional to V_0. It depends on the configuration of the circuit, as well as on the position and configuration of the neighboring bodies.

Inside the wire, the volume charge density is zero, as we saw in Sec. 5.1.2. Also,

$$\mathbf{E} = -\nabla V = \mathbf{J}/\sigma. \tag{12-49}$$

Thus E and $|\nabla V|$, like J, are the same everywhere inside the wire. The lines of force inside the wire are parallel and, if the length of the wire is l,

$$l|\nabla V| = l\frac{J}{\sigma} = l\frac{I}{a\sigma} = IR = V_0. \tag{12-50}$$

Thus $|\nabla V|$ inside the wire is simply V_0/l.

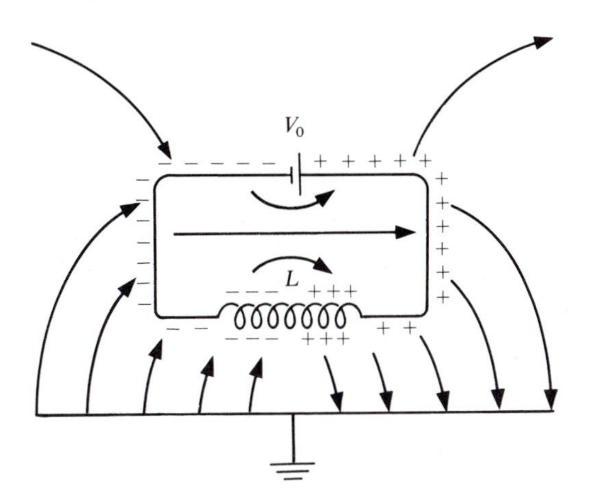

Figure 12-13 Inductor L connected to a voltage source.

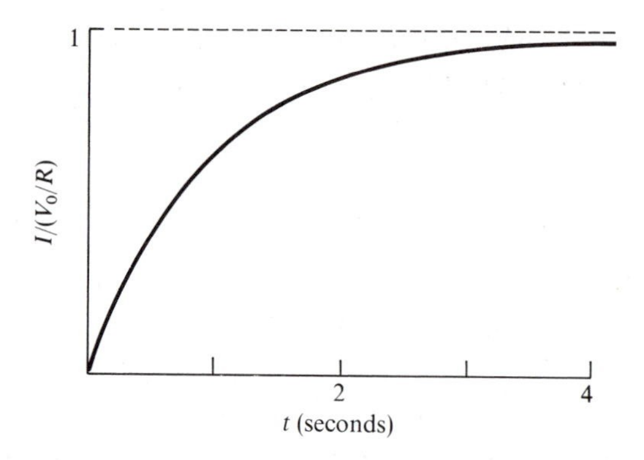

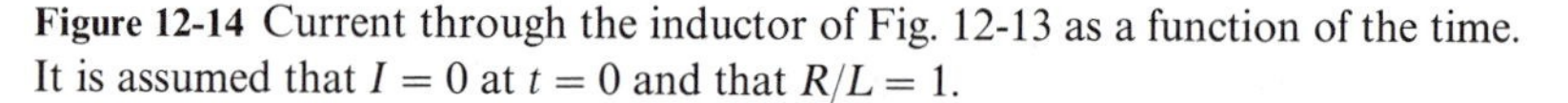

Figure 12-14 Current through the inductor of Fig. 12-13 as a function of the time. It is assumed that $I = 0$ at $t = 0$ and that $R/L = 1$.

b) *The current builds up from zero to its steady state value* V_0/R.

We assume that the time the field takes to propagate from the source to the inductor is negligible. We also neglect the stray capacitance between the turns of the inductor and between the wire and ground.

Since the wire has both a resistance R and an inductance L, Kirchoff's voltage law (Sec. 5.7) gives us that

$$L\frac{dI}{dt} + RI = V_0. \tag{12-51}$$

We saw in Sec 12.3 that the voltage across a pure inductance is $L\,dI/dt$. From Sec. 5.13, the solution of this equation is

$$I = \frac{V_0}{R} + Ke^{-(R/L)t}, \tag{12-52}$$

where K is a constant of integration. By hypothesis, $I = 0$ at $t = 0$, and thus $K = -V_0/R$. Then

$$I = \frac{V_0}{R}(1 - e^{-(R/L)t}). \tag{12-53}$$

The current I increases with time as in Fig. 12-14. Note the analogy with Fig. 5-23.

The current density J is again the same throughout the wire since, by hypothesis, there is zero capacitance, and hence no accumulation of charge at the surface of the wire. Thus

$$E = \frac{J}{\sigma} = \frac{I}{a\sigma} = \frac{R}{l}I, \tag{12-54}$$

and E is uniform throughout the wire. It builds up gradually, like I. Also,

$$RI = El. \tag{12-55}$$

From Eq. 11-19,

$$RI = \oint \left(-\nabla V - \frac{\partial \mathbf{A}}{dt} \right) \cdot \mathbf{dl}, \tag{12-56}$$

$$= -\oint \nabla V \cdot \mathbf{dl} - \frac{d}{dt} \oint \mathbf{A} \cdot \mathbf{dl}, \tag{12-57}$$

where the integration is evaluated along the wire, in the direction of **E**, clockwise in Fig. 12-13.

Now, from Eq. 8-54, the line integral of $\mathbf{A} \cdot \mathbf{dl}$ around a simple circuit is equal to the enclosed flux. More generally, the integral is equal to the flux linkage:

$$\oint \mathbf{A} \cdot \mathbf{dl} = \Lambda \tag{12-58}$$

and

$$RI = -\oint \nabla V \cdot \mathbf{dl} - \frac{d\Lambda}{dt}, \tag{12-59}$$

$$= -\oint \nabla V \cdot \mathbf{dl} - \frac{L\,dI}{dt}, \tag{12-60}$$

from Eq. 12-12. Comparing with Eq. 12-51, we find that

$$V_0 = -\oint \nabla V \cdot \mathbf{dl}. \tag{12-61}$$

This integral is therefore a constant. The negative sign comes from the fact that the integral is the voltage of the left-hand terminal with respect to the right-hand one, or $-V_0$.

Inside the wires that lead from the source to the coil, the magnetic field is weak and $|\nabla V|$ is approximately equal to the E of Eq. 12-54, with I a function of the time as in Eq. 12-53.

Inside the coiled part of the wire, **A** is not negligible. The $-\nabla V$ field points in the same direction as the current, while $-\partial \mathbf{A}/\partial t$ points in the opposite direction. At $t = 0$, when the source is connected, $\mathbf{A} = 0$ but $|\partial \mathbf{A}/\partial t|$ is large, $I = 0$, and the two terms on the right in Eqs. 12-56, 12-57, 12-59, 12-60 cancel. At the beginning, most of the voltage drop appears across the coil while, under steady-state conditions, the potential drop is uniformly distributed along the wire.

12.7.2 EXAMPLE: HORIZONTAL BEAM DEFLECTION IN THE CATHODE-RAY TUBE OF A TELEVISION RECEIVER

The electron beam sweeps horizontal lines across the screen, one below the other, until the whole screen has been covered. The process then repeats itself. The image is obtained by modulating the beam intensity.

The beam is deflected by the magnetic fields of two pairs of coils, one for the horizontal motion, and one for the vertical. The beam sweeps horizontally at a constant velocity, and returns at a much higher velocity.

To sweep horizontally at a constant velocity, the current through the pair of horizontal deflection coils must increase linearly with time. This is achieved by applying a constant voltage V to the coils, for then

$$L\frac{dI}{dt} \approx V, \qquad I \approx \frac{V}{L}t. \tag{12-62}$$

12.8 SUMMARY

The *mutual inductance* M between two circuits is equal to the magnetic flux linking one circuit per unit current flowing in the other:

$$\Lambda_{ab} = M_{ab}I_a. \tag{12-1}$$

The mutual inductance depends solely on the geometry and on the relative positions and orientations of the two circuits. It has the same value, whichever circuit produces the magnetic flux linking the other. Mutual inductance is measured in *henrys*.

Mutual inductance can be either positive or negative. The mutual inductance between two circuits a and b is positive if a current in the positive direction in a produces in b a flux that is in the same direction as that produced by a positive current in b.

The electromotance induced in circuit b by a change in current in circuit a is

$$\oint_b \mathbf{E} \cdot \mathbf{dl} = -M\frac{dI_a}{dt}. \tag{12-8}$$

The *self-inductance* L of a circuit is its flux linkage per unit current.

Thus

$$\oint \mathbf{E} \cdot \mathbf{dl} = -L\frac{dI}{dt}. \tag{12-13}$$

The voltage across an inductance L carrying a current I is $L\, dI/dt$.

The *coefficient of coupling* k between two circuits is defined by

$$M = k(L_a L_b)^{1/2}, \tag{12-25}$$

and can have values ranging from -1 to $+1$. It is zero if none of the flux of one circuit links the other.

The effective inductance of two *inductors connected in series* is

$$L = L_1 + L_2 + 2M, \tag{12-30}$$

where M is the mutual inductance, which can be either positive or negative.

The effective inductance of two *inductors connected in parallel is*

$$L = \frac{L_1 L_2 - M^2}{L_1 + L_2 - 2M}. \tag{12-46}$$

When a voltage is applied to an inductive circuit, the currents take some time to adjust to their steady-state values.

PROBLEMS

12-1E Starting from Eq. 12-5, show that μ_0 is expressed in henrys per meter.

12-2E *MUTUAL INDUCTANCE*

A long straight wire is situated on the axis of a toroidal coil of N turns as in Fig. 12-15.

Show that

$$M = \frac{\mu_0 b N^2}{2\pi} \ln\left(1 + \frac{b}{a}\right).$$

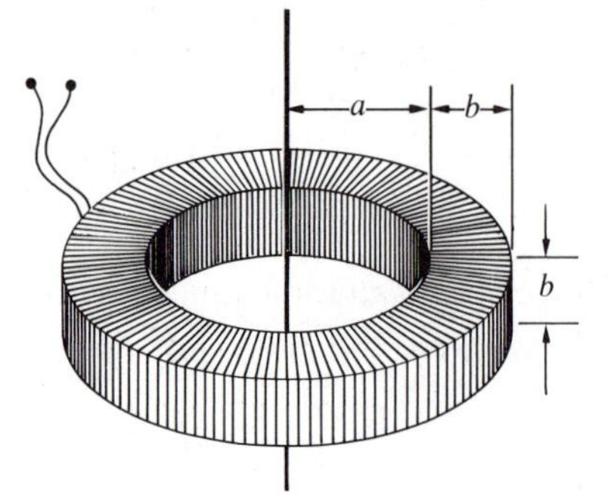

Figure 12-15 See Prob. 12-2.

12-3E MUTUAL INDUCTANCE

a) Show that the mutual inductance between two coaxial coils, separated by a distance z as in Fig. 12-16, one of radius a and N_a turns, and the other of radius $b \ll a$ and N_b turns is

$$\frac{\pi\mu_0 N_a N_b a^2 b^2}{2(a^2 + z^2)^{3/2}}.$$

We shall use this result in Prob. 14-7.

You can calculate this in one of two ways. You can either assume a current in coil a and calculate the flux through coil b, or assume a current in b and calculate the flux through a. From Sec. 12.1, M is the same, either way. Now, if the current is in a, you know how to calculate the field at b, from Sec. 8.1.2, since $b \ll a$. If the current is in b, then the calculation is vastly more difficult because you have to calculate **B** away from the axis. So you should calculate the magnetic flux through coil b due to a current through coil a.

b) How does the mutual inductance vary when the small coil is rotated around the vertical axis?

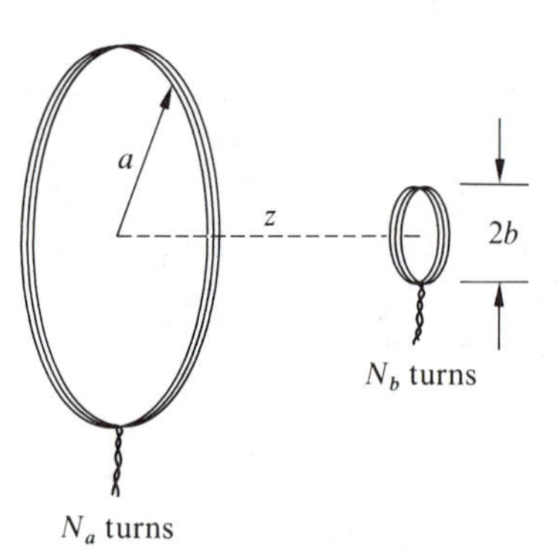

Figure 12-16 See Prob. 12-3.

c) What if one rotated the large coil around a vertical axis? Would the mutual inductance vary in the same way?

12-4E **A** *OUTSIDE A SOLENOID*

Show that the vector potential at a distance r from the axis *outside* an infinite solenoid of radius R and N' turns per meter carrying a current I is

$$A = \mu_0 N' R^2 I/2r.$$

The vector potential is azimuthal.

12-5E **A** *INSIDE A SOLENOID*

Show that the vector potential at a distance r from the axis, *inside* an infinite solenoid of radius R and N' turns per meter carrying a current I is

$$A = \mu_0 N' I r/2.$$

Here also, the vector potential is azimuthal.

12-6 *MAGNETIC MONOPOLES*

Magnetic monopoles are the magnetic equivalent of electric charges. See Sec. 8.3.

Figure 12-17 shows schematically one device that has been used in the search for magnetic monopoles in bulk matter. At the beginning of the experiment, SW is closed and there is zero current through the N-turn superconducting coil C. A sample of matter SA, with a mass of the order of one kilogram, is made to follow a closed path P traversing C.

Imagine that the sample contains one monopole of charge e^* weber with radial lines of **B**. As it approaches a given turn of C, the flux through that turn is $e^*/2$ weber

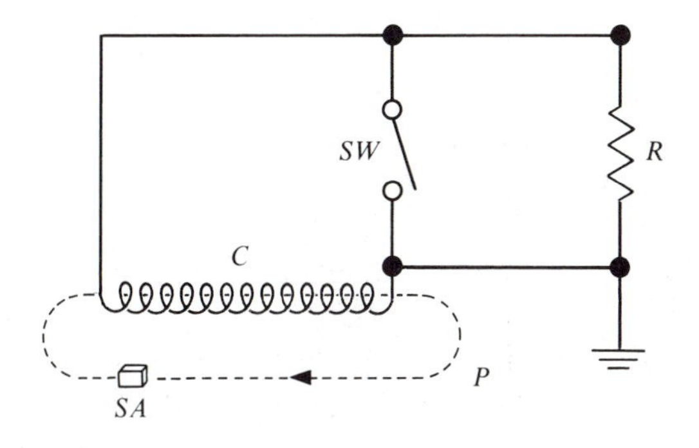

Figure 12-17 Magnetic monopole detector. The sample SA is placed on a conveyor and follows path P with the switch SW closed. After hundreds of passes, the switch is opened and the voltage pulse across R is observed on an oscilloscope. The coil C is superconducting. See Prob. 12-6.

to the right, and, a moment later, it is $e^*/2$ weber to the left. Thus, on emerging at the right-hand end of C, the flux linkage through C due to the monopole has increased by Ne^*. However, the resistance of the coil is zero and the induced current keeps the flux linkage zero.

Calculate the current per monopole after 100 passes through a coil of 1200 turns with a self-inductance of 75 millihenrys.

The current is detected by opening the switch SW and observing the voltage pulse on R.

All the measurements to date have given null results.

12-7 ROGOWSKI COIL

Figure 12-18 shows a Rogowski coil surrounding a current-carrying wire. The coil has N turns, and $r \gg r'$. Rogowski coils are used for measuring large fluctuating currents, particularly large current pulses.

a) If the value of R is chosen large enough, V is unaffected by that part of the circuit that is to the right of the vertical dotted line. Show that

$$V = \mu_0 \frac{Nr'^2}{2r} \frac{dI}{dt}. \tag{1}$$

Show that, if C is now chosen large enough to make $V' \ll V$, then

$$V' = \mu_0 \frac{Nr'^2}{2r} \frac{I}{RC}. \tag{2}$$

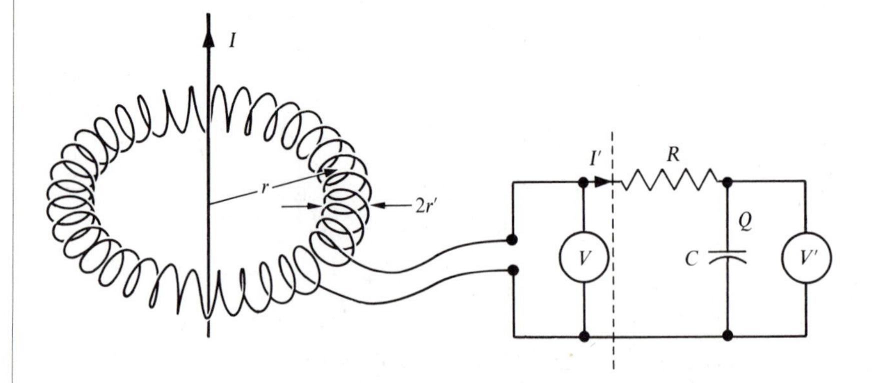

Figure 12-18 Rogowski coil for measuring a heavy current pulse in the axial wire. With the RC circuit shown, the voltage V' is proportional to I. See Prob. 12-7.

See Prob. 5-30. The product RC must be large compared to $1/f$, where f is the repetition frequency, dI/dt being of the order of If. The current I need not be sinusoidal.

Solution: Under those conditions, V is the induced electromotance:

$$V = N\frac{d\Phi}{dt} = N\frac{d}{dt}\left(\pi r'^2 \frac{\mu_0 I}{2\pi r}\right), \tag{3}$$

$$= \frac{\mu_0 N r'^2}{2r}\frac{dI}{dt}. \tag{4}$$

Thus the mutual inductance is $\mu_0 N r'^2/2r$.

Also,

$$V = I'R + V' \approx I'R = \frac{dQ}{dt}R, \tag{5}$$

$$\approx RC\frac{dV'}{dt}. \tag{6}$$

Comparing now Eqs. 4 and 6,

$$V' \approx \mu_0 \frac{Nr'^2}{2r}\frac{I}{RC}. \tag{7}$$

We have set V' equal to zero when I is zero. For example, the current I could be in the form of short pulses and, between pulses, both I and V' would be zero.

b) Show that, if the capacitor is shorted and if R is made small so that V is decreased by a factor of, say, 20 or more, then $I' \approx I/N$.

The self-inductance of the coil is

$$L \approx \mu_0 N^2 r'^2/2r. \tag{8}$$

Solution: The induced electromotance is equal to the sum of the voltage drops across the inductance L of the coil and the resistance R:

$$\frac{\mu_0 N r'^2}{2r}\frac{dI}{dt} = L\frac{dI'}{dt} + RI'. \tag{9}$$

If R were very large, then RI' would be nearly equal to the induced electromotance. On the other hand, if R is small, we may set

$$L\frac{dI'}{dt} \approx \frac{\mu_0 N r'^2}{2r}\frac{dI}{dt}. \tag{10}$$

Substituting now the value of L,

$$\frac{\mu_0 N^2 r'^2}{2r}\frac{dI'}{dt} \approx \frac{\mu_0 N r'^2}{2r}\frac{dI}{dt}, \tag{11}$$

$$I' \approx I/N. \tag{12}$$

12-8E *INDUCED CURRENTS*

An azimuthal current is induced in a conducting tube of length l, average radius a, and thickness b, with $b \ll a$.

a) Show that its resistance and inductance are given by

$$R = 2\pi a/\sigma b l, \qquad L = \mu_0 \pi a^2/l.$$

b) Calculate L, R, L/R for a copper tube ($\sigma = 5.8 \times 10^7$ siemens per meter), one meter long, 10 millimeters in diameter, and with a wall thickness of 1 millimeter.

12-9 *COAXIAL LINES*

Coaxial lines, as in Fig. 12-19, are widely used for the interconnection of electronic equipment and for long-distance telephony. They can be used with either direct or alternating currents, up to very high frequencies, where the wavelength c/f is of the same order of magnitude as the diameter of the line. At these frequencies, of the order of 10^{10} hertz, the field inside the line becomes much more complicated than that shown in the figure and quite unmanageable.

There is zero electric field outside the line. Since the outer conductor carries the same current as the inner one, there is also zero magnetic field, from Ampere's circuital law (Sec. 9.1).

A coaxial line acts as a cylindrical capacitor. We found its capacitance per meter in Prob. 6-5:

$$C' = \frac{2\pi\epsilon_r\epsilon_0}{\ln(R_2/R_1)},$$

where R_1 and R_2 are the radii of the inner and outer conductors, respectively.

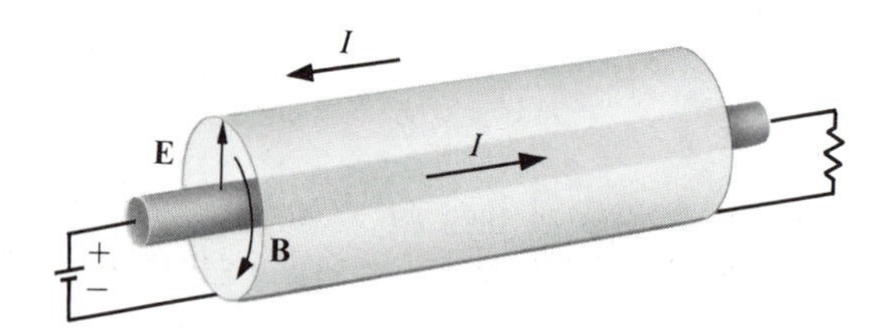

Figure 12-19 Coaxial line. See Prob. 12-9.

A coaxial line is also inductive, since there is an azimuthal magnetic field in the annular space between the two conductors.

Show that the self-inductance per meter is

$$L' = \frac{\mu_0}{2\pi} \ln \frac{R_2}{R_1}.$$

12-10 *COAXIAL LINES*

It is known that high-frequency currents do not penetrate into a conductor as do low-frequency currents. This is called the skin effect. See Prob. 11-7.

Would you expect the self-inductance of a coaxial line to increase or to decrease with increasing frequency?

12-11 *LONG SOLENOID WITH CENTER TAP*

Consider a long solenoid with connections at both ends, A and C, and with a center tap B. The two halves are in series, and thus

$$L_{AC} = L_{AB} + L_{BC} + 2M,$$

where M is their mutual inductance.

Now, if one uses Sec. 12.3.1 to calculate the three L's, one finds that $M = 0$. This is absurd because the coefficient of coupling is surely not zero.

Can you explain this strange result?

12-12 *TRANSIENT IN RLC CIRCUIT*

The switch in the circuit of Fig. 12-20 is initially open, and then closed at $t = 0$.

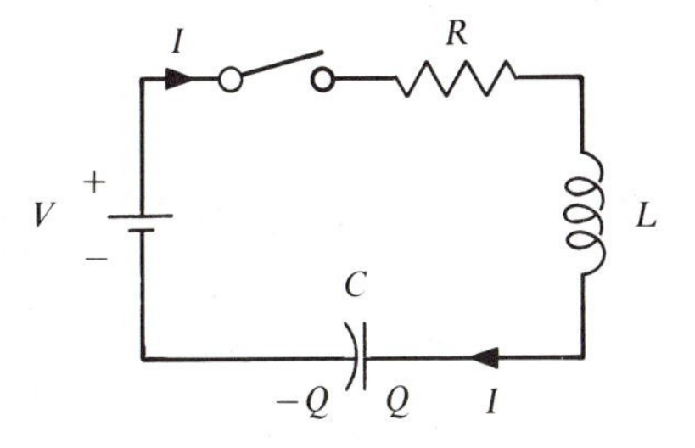

Figure 12-20 See Prob. 12-12.

Draw a curve of Q as a function of t for $V = 100$ volts, $L = 1$ henry, $C = 1$ microfarad, $R = 8000$ ohms, with t varying from 0 to 20 milliseconds.

Solution: From Kirchoff's voltage law,

$$RI + L\frac{dI}{dt} + \frac{Q}{C} = V, \tag{1}$$

or, since $I = dQ/dt$,

$$L\frac{d^2Q}{dt^2} + R\frac{dQ}{dt} + \frac{Q}{C} = V. \tag{2}$$

The particular solution of this equation is

$$Q = CV = 10^{-4}, \tag{3}$$

and the complementary function is the solution of

$$L\frac{d^2Q}{dt^2} + R\frac{dQ}{dt} + \frac{Q}{C} = 0. \tag{4}$$

Let us set

$$Q = Ae^{nt}. \tag{5}$$

Then, substituting in Eq. 4 and dividing by Q,

$$\left(Ln^2 + Rn + \frac{1}{C}\right) = 0, \tag{6}$$

$$n = -\frac{R}{2L} \pm \left[\frac{R^2}{4L^2} - \frac{1}{LC}\right]^{1/2}. \tag{7}$$

Note that there are two values of n, say n_1 and n_2. Then

$$Q = Ae^{n_1 t} + Be^{n_2 t}, \tag{8}$$

where A and B are unknown constants of integration.

Here $R/2L = 4000$, the bracket of Eq. 7 is 1.5×10^7, and the general solution is

$$Q = e^{-4000t}(Ae^{3873t} + Be^{-3873t}) + 10^{-4}, \tag{9}$$

$$= Ae^{-127t} + Be^{-7873t} + 10^{-4}. \tag{10}$$

Since $Q = 0$ at $t = 0$, $A + B = -10^{-4}$. Also,

$$I = \frac{dQ}{dt} = -127Ae^{-127t} - 7873Be^{-7873t} \tag{11}$$

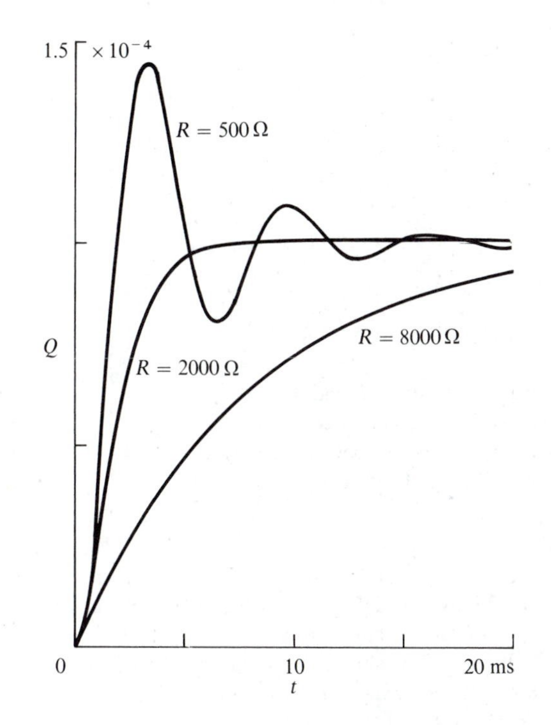

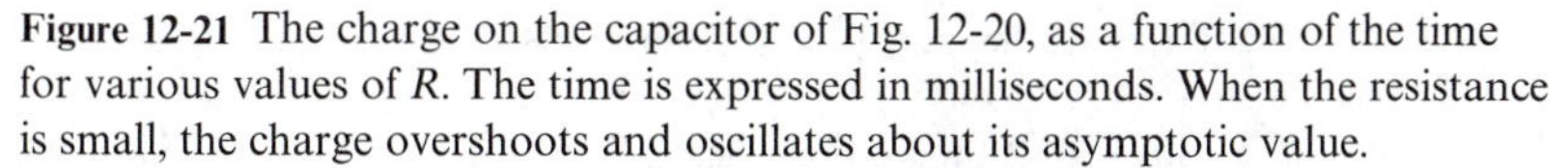

Figure 12-21 The charge on the capacitor of Fig. 12-20, as a function of the time for various values of R. The time is expressed in milliseconds. When the resistance is small, the charge overshoots and oscillates about its asymptotic value.

and $I = 0$ at $t = 0$, so that

$$127A + 7873B = 0, \tag{12}$$

$$A = -1.016 \times 10^{-4}, \qquad B = 1.640 \times 10^{-6}, \tag{13}$$

$$Q = -1.016 \times 10^{-4} e^{-127t} + 1.640 \times 10^{-6} e^{-7873t} + 10^{-4}. \tag{14}$$

As a check, we can show that $L\,dI/dt = 100$ at $t = 0$. At that instant, $I = 0$ and the voltage drop on R is zero.

Figure 12-21 shows Q as a function of t for $R = 500$, 2000, and 8000 ohms. With $R = 2000$ ohms, the bracket is zero. The right-hand side of Eq. 8 then becomes $(A + Bt) \exp nt$. With $R = 500$ ohms, the bracket is negative and we again have another type of solution.

12-13 *VOLTAGE SURGES ON INDUCTORS*

If the current through a large inductor is interrupted suddenly, dI/dt is large and transient voltages appear across the switch and across the coil. The voltages can

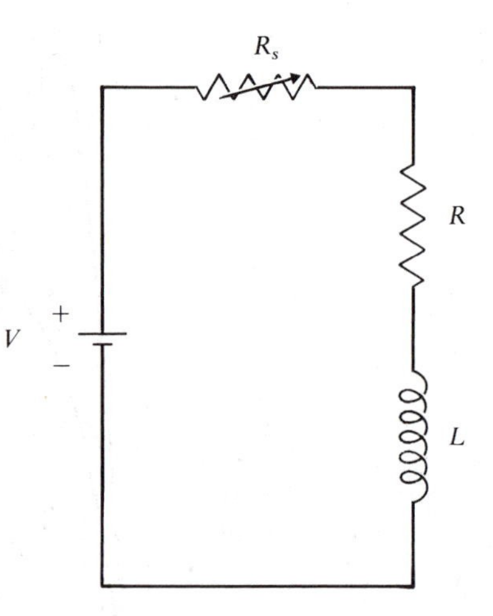

Figure 12-22 Inductor L and its resistance R fed by a source V through a switch represented here by a resistor R_s. See Prob. 12-13.

be large enough to damage both. *Care must therefore be taken to reduce the current slowly*. This is particularly important when using large electromagnets. No problem arises, however, on connecting the source to the inductor. See Sec. 12.7.1.

Figure 12-22 shows the inductor L and its resistance R fed by a source V. The resistance R_s plays the role of the switch. At first, R_s is zero and $I = V/R$. Then R_s is quickly increased to some value much larger than R.

Set $L = 10$ henrys, $R = 10$ ohms, $V = 100$ volts, $dI/dt = -10^3$ amperes per second.

a) Show that the voltages across the switch and across the inductor both rise to about 10 kilovolts.

Of course the insulation of the switch and the insulation between turns on the inductor break down much before the voltage can rise to such a high value.

b) It is possible to suppress the transient by using a diode. A *diode* is a two-terminal device that has either a very low or a very high resistance, depending on the polarity of the applied voltage, as in Fig. 12-23. In other words, a diode will pass a current in only one direction. Of course, if the inverse voltage is sufficiently high, current will pass in the forbidden direction and the diode will be damaged.

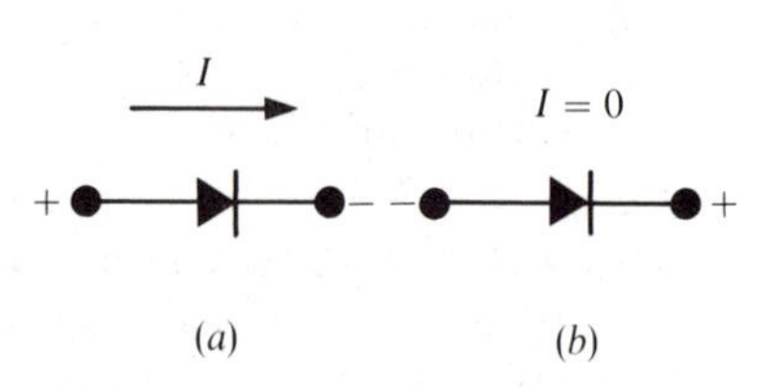

Figure 12-23 The symbol for a diode is a triangle and a bar. In (*a*), the diode has a low resistance and can usually be considered as a short circuit. In (*b*), the resistance is high and is usually considered to be infinite. The curve of I as a function of the applied voltage is non-linear, even for forward voltages.

A given type of diode can be used up to a rated maximum current, and up to a rated maximum inverse voltage.

You are given a diode that is rated at 20 amperes and 200 volts peak inverse voltage. How should it be used?

What is the current through the diode as a function of the time if the switch is opened at $t = 0$?

What happens to the energy stored in the magnetic field of the inductor after the switch is opened?

12-14D *TRANSIENT IN RLC CIRCUIT*

One wishes to charge a capacitor as quickly as possible. The source supplies an open-circuit voltage V_s and has an internal resistance R (Sec. 5.12), as in Fig. 12-24a.

It is suggested that the rate of charge could be increased by adding an inductor with $L = R^2C/4$ in series with the capacitor as in Fig. 12-24b.

Draw curves of the voltage V across the capacitor as a function of time for V_s = 100 volts, R = 100 ohms, C = 1 microfarad, with and without the inductor.

Is the suggestion valid?

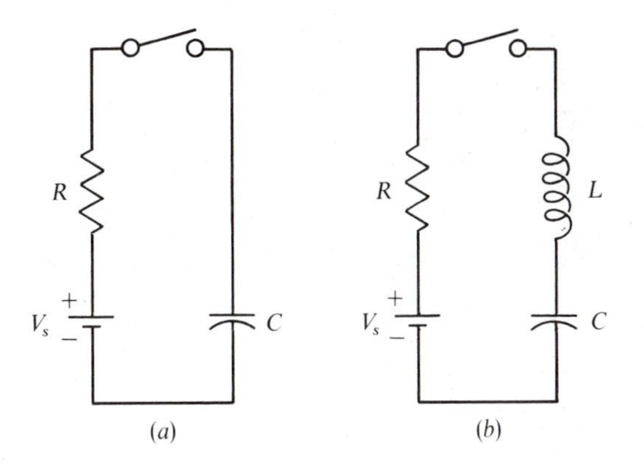

Figure 12-24 Capacitor C charged by a source V_s having an internal resistance R. The inductor L in (b) has been added in the hope that the capacitor will charge more rapidly. See Prob. 12-14.

12-15D *TRANSIENT IN RLC CIRCUIT*

In the circuit of Fig. 12-25 the switch is initially open. The capacitor is charged to the voltage V and no current is drawn from the source.

Set V = 100 volts, L = 1 henry, R = 10 ohms, C = 1 microfarad.

a) The switch is closed. Show that

$$I = 10(1 - e^{-10t}).$$

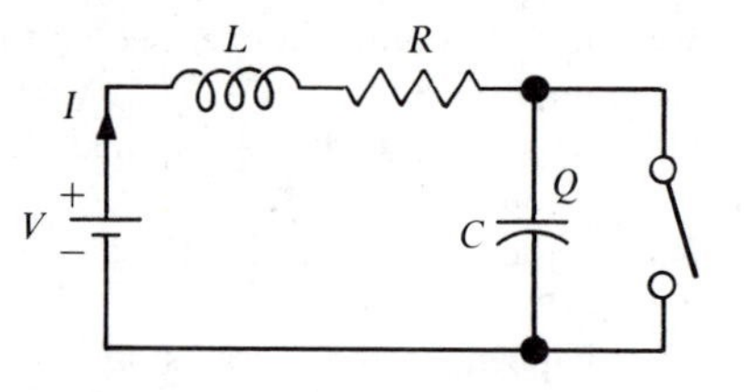

Figure 12-25 See Prob. 12-15.

b) After a time $t \gg L/R$, the switch is opened. Show that, from that time on,

$$Q \approx -10^{-4}e^{-5t}\cos 10^3 t + 10^{-4},$$

$$I \approx 0.1e^{-5t}\sin 10^3 t.$$

CHAPTER 13

MAGNETIC FIELDS: VI

Magnetic Forces

In Chapters 4 and 7, we studied the energy stored in an electric field, and the resulting electric forces. Then, in Chapter 10, we studied the magnetic forces on individual charged particles. Now we must study magnetic energy and macroscopic magnetic forces. Magnetic forces are, in practice, so much larger than electric forces that they lend themselves to many more applications. For example, electric motors almost invariably use magnetic forces.†

We continue to assume that the materials in the field are non-magnetic. We shall deal with the magnetic forces exerted by electromagnets in Chapter 15.

13.1 FORCE ON A WIRE CARRYING A CURRENT IN A MAGNETIC FIELD

Consider a wire of cross-sectional area a, carrying a current I as in Fig. 13-1. There are n conduction electrons per unit volume, each one carrying a charge $-e$ at a velocity $\mathbf{v}$.

If the wire is situated in a magnetic field $\mathbf{B}$, due to currents flowing elsewhere, the magnetic force on one electron is $-e\mathbf{v} \times \mathbf{B}$ (Sec. 10.1) and,

---

† Although electrostatic motors are exceedingly rare, they have been studied extensively. See, for example, A. D. Moore, Editor, *Electrostatics and Its Applications*, John Wiley, New York, 1973.

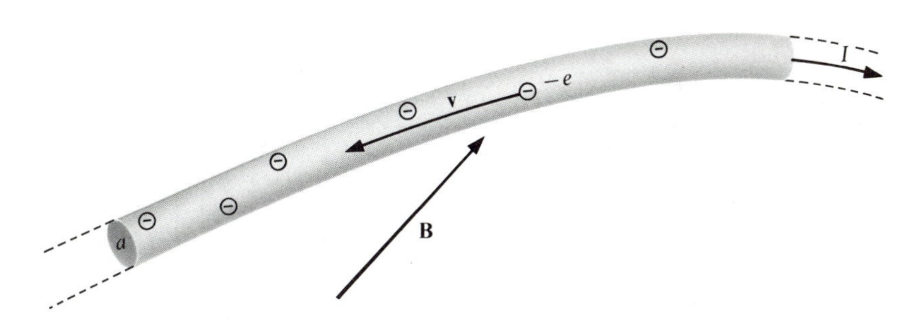

Figure 13-1 Wire of cross-section a carrying a current I. The charges $-e$ move at a velocity **v**. The magnetic field **B** is due to currents flowing elsewhere.

for a length dl containing $na\,dl$ conduction electrons,

$$\mathbf{dF} = na\,dl(-e\mathbf{v} \times \mathbf{B}). \tag{13-1}$$

Now the charge passing through a given cross-section per second is the charge on the carriers contained in a length v of the wire and

$$\mathbf{I} = -nae\mathbf{v}. \tag{13-2}$$

Then

$$\mathbf{dF} = \mathbf{I}\,dl \times \mathbf{B} = I\,\mathbf{dl} \times \mathbf{B}, \tag{13-3}$$

where **dl** is in the direction of the current.

For a straight length l of wire carrying a current I in a uniform magnetic field **B**,

$$\mathbf{F} = I\mathbf{l} \times \mathbf{B}. \tag{13-4}$$

Figure 13-2 shows the lines of **B** near a current-carrying wire, perpendicular to a uniform magnetic field. The force on a length l is then simply IlB, in the direction shown.

As for electric fields, it is useful to imagine that *lines of* **B** really exist, are under tension, and repel each other laterally. This simple model gives the correct direction for magnetic forces, as can be seen from Fig. 13-2.

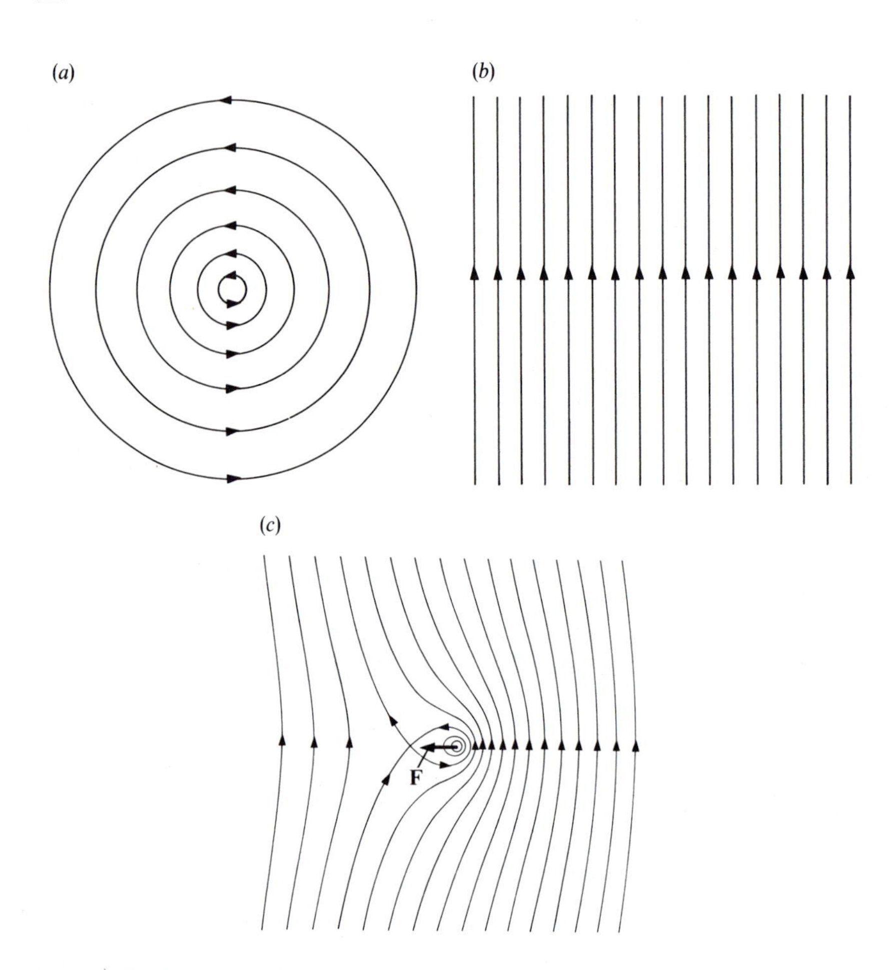

Figure 13-2 Magnetic field near a current-carrying wire situated in a uniform magnetic field. (*a*) Lines of **B** for a current perpendicular to the paper. (*b*) Lines of **B** for a uniform field parallel to the paper. (*c*) Superposition of fields in (*a*) and (*b*) and the resulting magnetic force **F**. The wire carries a current of 10 amperes, the uniform field has a **B** of 2×10^{-4} tesla, and the point where the lines of **B** are broken is at 10 millimeters from the center of the wire.

13.1.1 EXAMPLE: THE HODOSCOPE

The *hodoscope* is a device that simulates the trajectory of a charged particle in a magnetic field. The principle involved is simple: if the charged particle, of mass m, charge Q, and velocity v, is replaced by a light wire fixed at the two ends of the trajectory and carrying a current I, the wire will follow the trajectory if

$$mv/Q = T/I, \tag{13-5}$$

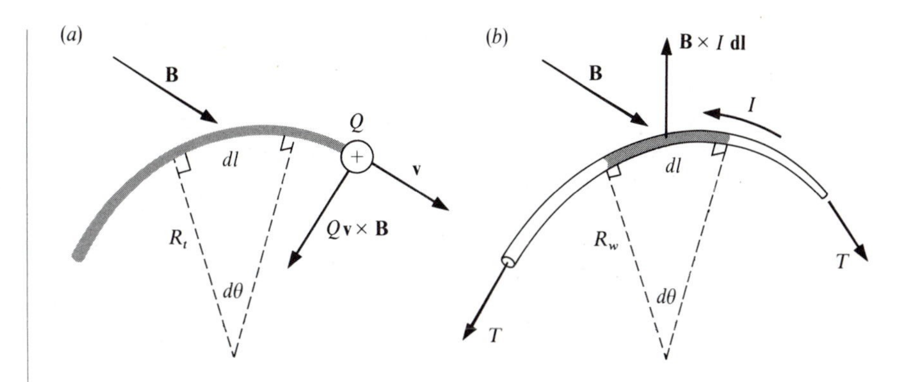

Figure 13-3 (*a*) Charge Q moving at a velocity $\mathbf{v}$ in a magnetic field $\mathbf{B}$. The radius of curvature of the trajectory is R_t. (*b*) Light wire carrying a current I in the opposite direction in the same magnetic field. The tension in the wire is T, and its radius of curvature is R_w. It is shown that the wire has the same radius of curvature as the trajectory if mv/Q is equal to T/I. The angle $d\theta$ is infinitesimally small.

where T is the tension in the wire. This statement is by no means obvious, but we shall demonstrate its validity.

The advantage of the so-called *floating wire* lies in the fact that it is much easier to experiment with a wire than with an ion beam.

Let us consider a region where the ion beam is perpendicular to **B** as in Fig. 13-3a. Then the charge Q is subjected to a force QvB. Thus

$$QvB = mv^2/R_t, \tag{13-6}$$

where R_t is the radius of curvature of the trajectory, and

$$R_t = mv/BQ. \tag{13-7}$$

In Fig. 13-3b we have replaced the charged particles by a light wire carrying a current I flowing in the *opposite* direction. If the element dl is in equilibrium, the outward force $BI\,dl$ is compensated by the inward component of the tension forces T:

$$BI\,dl = 2T\sin(d\theta/2) = 2T\,d\theta/2 = T\,dl/R_w. \tag{13-8}$$

The radius of curvature of the wire is thus

$$R_w = T/BI, \tag{13-9}$$

and the two radii of curvature will be the same if

$$mv/Q = T/I. \tag{13-10}$$

For example, if we have a proton ($m = 1.7 \times 10^{-27}$ kilogram) beam with a kinetic energy of 10^6 electron-volts ($10^6 \times 1.6 \times 10^{-19}$ joule), then T/I is about 0.15 and, if $I = 1$ ampere, T is 0.15 newton.

Note that the particle will be deflected downward if the magnetic force is downward, but the wire will curve downward if the force is *upward*. The magnetic forces must therefore be directed in opposite directions.

If the magnetic field is not uniform, and if the beam is not perpendicular to $\mathbf{B}$, then the wire does not always follow the trajectory. For example, magnetic fields are often used both to deflect and focus an ion beam. In such cases the focusing forces on the beam can become defocusing forces on the wire, which is then deflected away from the trajectory.

13.2 *MAGNETIC PRESSURE*

In Chapter 10 we considered the magnetic forces acting on individual *particles*, and we have just seen how one can calculate the force on a current-carrying *wire*. If, now, we have a current *sheet*, it is appropriate to think in terms of magnetic *pressure*.

Let us imagine a flat current sheet carrying $\boldsymbol{\alpha}$ amperes per meter and situated in a uniform tangential magnetic field $\mathbf{B}/2$ due to currents flowing elsewhere, as in Fig. 13-4a, with $\boldsymbol{\alpha}$ normal to $\mathbf{B}$.

We adjust $\boldsymbol{\alpha}$ until it produces an aiding field $\mathbf{B}/2$ on one side and an opposing field $\mathbf{B}/2$ on the other side, as in Fig. 13-4b. Then, from Prob. 9-3, $\alpha = B/\mu_0$ and the force per unit area, or the pressure, is α times the field $B/2$ due to the currents flowing elsewhere, or $B^2/2\mu_0$.

So, whenever one has a current flowing through a conducting sheet, with a magnetic induction B on one side and zero magnetic induction on the other side, the sheet is subjected to a pressure $B^2/2\mu_0$. This pressure is exerted in the same direction as if the magnetic field were replaced by a compressed gas.

We have arrived at this result for a flat current sheet, but the same applies to any current sheet when $\mathbf{B} = 0$ on only one side.

Now we saw in Sec. 13.1 that, qualitatively, one can explain magnetic forces by assuming that the lines of $\mathbf{B}$ (a) repell each other laterally, and (b) are under tension. The magnetic pressure we have here is clearly ascribable to the lateral repulsion, since the lines are roughly parallel to the current sheet.

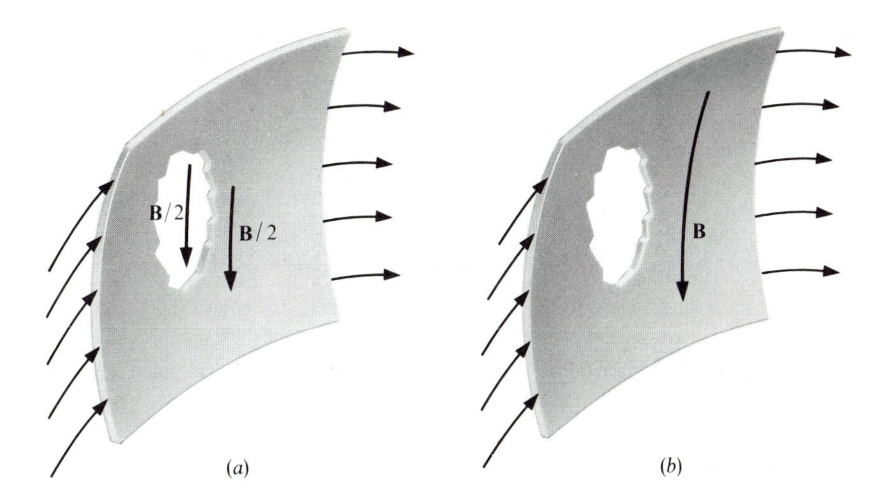

Figure 13-4 (*a*) A current sheet carries a current density of α amperes per meter of its width and is situated in a uniform magnetic field $\mathbf{B}/2$ due to currents flowing elsewhere. (*b*) The total magnetic field is that of (*a*), plus that of the current sheet. By adjusting α so that it just cancels the magnetic field on the left-hand side of the sheet, we have a total field $\mathbf{B}$ on the right.

Since the lines are under tension, there should also be a force of attraction in the direction of $\mathbf{B}$. This is correct, as we shall see later on in Sec. 15.3. In fact, along the lines of $\mathbf{B}$, the force per unit area is again $B^2/2\mu_0$.

13.3 MAGNETIC ENERGY DENSITY

In Sec. 4.5 we found that, in an electrostatic field, the force per unit area on a conductor and the energy density in the field are both equal to $\epsilon_0 E^2/2$. Now we have just found that the magnetic pressure is $B^2/2\mu_0$. One might therefore expect the energy density in a magnetic field to be equal to $B^2/2\mu_0$. This turns out to be correct.†

Figure 13-5 shows the magnetic pressure, or the magnetic energy density, $B^2/2\mu_0$, as a function of B.

† *Electromagnetic Fields and Waves*, p. 367.

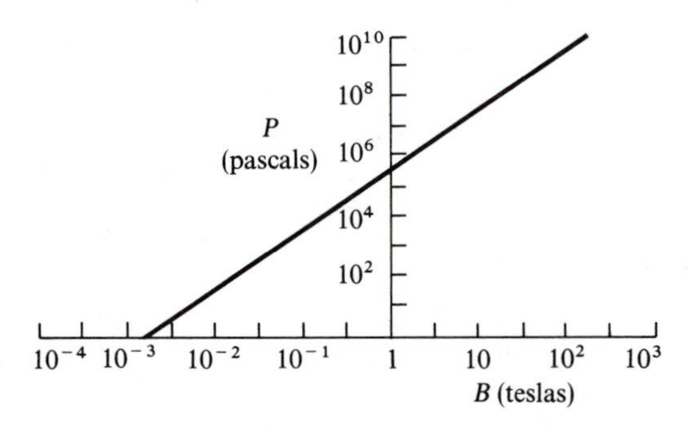

Figure 13-5 Magnetic pressure $B^2/2\mu_0$ as a function of B. The magnetic pressure is equal to the magnetic energy density. Atmospheric pressure at sea level is about 10^5 pascals.

13.4 *MAGNETIC ENERGY*

Since the magnetic energy density is $B^2/2\mu_0$, the energy stored in a magnetic field is

$$W_m = \frac{1}{2\mu_0} \int_\infty B^2 \, d\tau, \tag{13-11}$$

where the integral is evaluated over all space.

13.4.1 *EXAMPLE: THE LONG SOLENOID*

The magnetic energy stored in the field of a long solenoid of N turns, radius R, and length l, carrying a current I is

$$W_m = \frac{1}{2\mu_0} \left(\frac{\mu_0 N I}{l} \right)^2 (\pi R^2 l), \tag{13-12}$$

$$= \frac{\mu_0 N^2 \pi R^2}{2l} I^2. \tag{13-13}$$

The magnetic pressure is

$$p_m = \frac{1}{2\mu_0} \left(\frac{\mu_0 N I}{l} \right)^2 = \frac{\mu_0 N^2}{2l^2} I^2. \tag{13-14}$$

13.5 MAGNETIC ENERGY IN TERMS OF THE CURRENT I AND OF THE INDUCTANCE L

If one compares Eq. 13-13 for the magnetic energy stored in the field of a solenoid with Eq. 12-16 for the inductance of a solenoid, one finds that

$$W_m = \frac{1}{2}\left(\frac{\mu_0 N^2 \pi R^2}{l}\right) I^2 = \frac{1}{2} L I^2. \tag{13-15}$$

This is a general result: the magnetic energy stored in the field of a circuit of inductance L carrying a current I is $\frac{1}{2}LI^2$.

You will recall from Sec. 4.3 that the energy stored in the electric field of a capacitor of capacitance C charged to a voltage V is $\frac{1}{2}CV^2$.

If one has two circuits with inductances L_a and L_b, and having a mutual inductance M, then the magnetic energy is

$$W_m = \frac{1}{2} L_a I_a^2 + \frac{1}{2} L_b I_b^2 + M I_a I_b. \tag{13-16}$$

This is shown in Prob. 13-16.

13.5.1 EXAMPLE: COAXIAL SOLENOIDS

We can easily verify the above equation for coaxial solenoids.

We first evaluate the right-hand side. From Eqs. 12-16 and 12-7,

$$\frac{1}{2} L_a I_a^2 + \frac{1}{2} L_b I_b^2 + M I_a I_b,$$

$$= \frac{1}{2}\frac{\mu_0 N_a^2 \pi R^2}{l_a} I_a^2 + \frac{1}{2}\frac{\mu_0 N_b^2 \pi R^2}{l_b} I_b^2 + \frac{\mu_0 N_a N_b \pi R^2}{l_a} I_a I_b, \tag{13-17}$$

$$= \frac{\mu_0 \pi R^2}{2}\left(\frac{N_a^2 I_a^2}{l_a} + \frac{N_b^2 I_b^2}{l_b} + \frac{2 N_a N_b I_a I_b}{l_a}\right). \tag{13-18}$$

To calculate W_m, we use the fact that the energy density is $B^2/2\mu_0$. Over the length l_b we have a total B of

$$\mu_0\left(\frac{N_a I_a}{l_a} + \frac{N_b I_b}{l_b}\right),$$

giving a magnetic energy

$$\frac{1}{2\mu_0}\mu_0^2\left(\frac{N_aI_a}{l_a}+\frac{N_bI_b}{l_b}\right)^2\pi R^2 l_b.$$

Over the length $l_a - l_b$, we have only the field of coil a and a magnetic energy

$$\frac{\mu_0}{2}\left(\frac{N_aI_a}{l_a}\right)^2\pi R^2(l_a - l_b).$$

Then the magnetic energy stored in the field of the coaxial solenoids is

$$W_m = \frac{\mu_0\pi R^2}{2}\left(\frac{N_a^2I_a^2l_b}{l_a^2}+\frac{N_b^2I_b^2}{l_b}+2\frac{N_aN_bI_aI_b}{l_a}+\frac{N_a^2I_a^2}{l_a}-\frac{N_a^2I_a^2l_b}{l_a^2}\right), \quad (13\text{-}19)$$

which reduces to the expression on the right-hand side of Eq. 13-18.

13.6 MAGNETIC FORCE BETWEEN TWO ELECTRIC CURRENTS

It is well known that circuits carrying electric currents exert forces on each other. In Fig. 13-6, the force exerted *by* circuit a *on* circuit b is obtained by integrating Eq. 13-3 over circuit b:

$$\mathbf{F}_{ab} = I_b\oint_b d\mathbf{l}_b \times \mathbf{B}_a, \quad (13\text{-}20)$$

where **B** is defined as in Eq. 8-1.

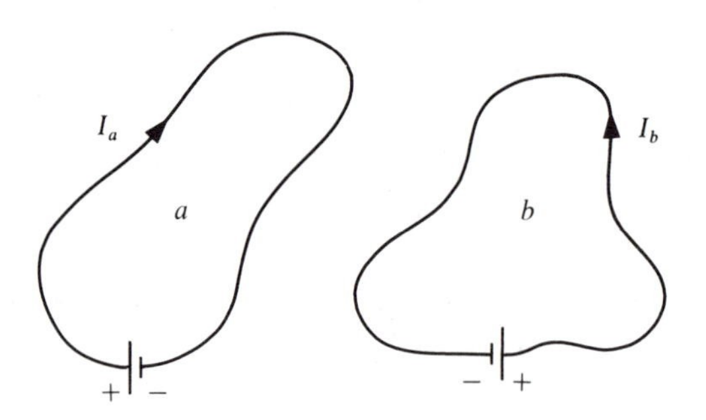

Figure 13-6 Pair of circuits carrying currents I_a and I_b.

13.6.1 EXAMPLE: THE FORCE BETWEEN TWO INFINITELY LONG PARALLEL WIRES

The force between two infinitely long parallel wires carrying currents I_a and I_b, separated by a distance ρ, as in Fig. 13-7, is calculated as follows. The force acting on an element $I_b\,\mathbf{dl}_b$ is

$$\mathbf{dF} = I_b(\mathbf{dl}_b \times \mathbf{B}_a), \tag{13-21}$$

$$dF = I_b\, dl_b \mu_0 I_a / 2\pi\rho, \tag{13-22}$$

and the force per unit length is

$$\frac{dF}{dl_b} = \mu_0 I_a I_b / 2\pi\rho. \tag{13-23}$$

The force is attractive if the currents are in the same direction; and it is repulsive if they are in opposite directions.

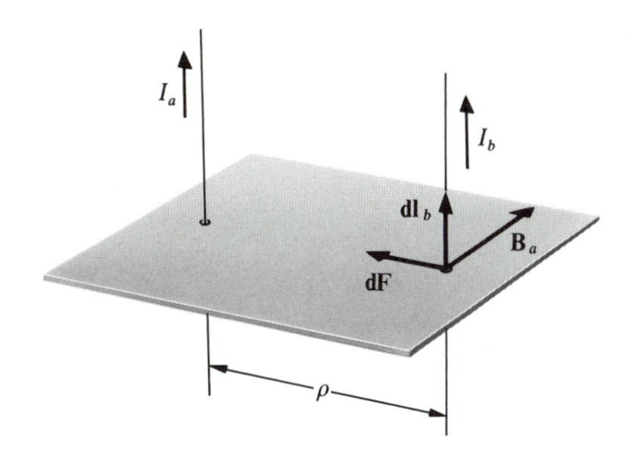

Figure 13-7 Two long parallel wires carrying currents in the same direction. The element of force **dF** acting on element $\mathbf{dl}_b$ is in the direction shown.

13.7 MAGNETIC FORCES WITHIN AN ISOLATED CIRCUIT

In an isolated circuit the current flows in its own magnetic field and is therefore subjected to a force. This is illustrated in Fig. 13-8: the interaction of

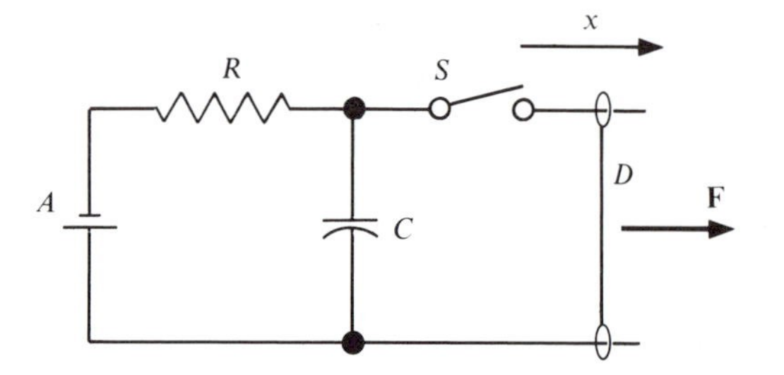

Figure 13-8 Schematic diagram of a rail gun. The battery A charges, through a resistance R, the capacitor C, which can be made to discharge through the line by closing switch S. The role of the capacitor is to store electric charge and to supply a very large current to the loop for a very short time. If side D is allowed to move, it moves to the right under the action of the magnetic force **F**.

the current I with its own magnetic induction **B** produces a force that tends to extend the circuit to the right.

Magnetic forces within an isolated circuit are usually calculated by postulating a small displacement and using the principle of conservation of energy. See Prob. 13-25.

13.8 SUMMARY

The force on an element of wire of length **dl** carrying a current I in a region where the magnetic induction is **B**, is

$$\mathbf{dF} = I\,\mathbf{dl} \times \mathbf{B}. \qquad (13\text{-}3)$$

If a conducting sheet situated in a magnetic field carries a current that cancels the magnetic field on one side, then it is subjected to a *magnetic pressure* $B^2/2\mu_0$, where B is the magnetic induction on the other side. This expression also gives the *energy density* in the magnetic field.

The *magnetic energy* stored in a magnetic field is obtained by integrating the energy density:

$$W_m = \frac{1}{2\mu_0}\int_\infty B^2\,d\tau. \qquad (13\text{-}11)$$

If the field is due to an inductance L, then

$$W_m = (1/2)LI^2 \tag{13-15}$$

or, if we have two inductances L_a and L_b with a mutual inductance M, then

$$W_m = (1/2)L_a I_a^2 + (1/2)L_b I_b^2 + MI_a I_b. \tag{13-16}$$

The *magnetic force* exerted by a circuit a on a circuit b is

$$\mathbf{F}_{ab} = I_b \oint_b \mathbf{dl}_b \times \mathbf{B}_a. \tag{13-20}$$

Magnetic forces also exist within an isolated circuit. They are calculated by postulating a small displacement and using the principle of the conservation of energy.

PROBLEMS

13-1E *MAGNETIC FORCE*

Calculate the magnetic force on an arc 50 millimeters long carrying a current of 400 amperes in a direction perpendicular to a uniform $\mathbf{B}$ of 5×10^{-2} tesla.

High-current *circuit breakers* often comprise coils that generate a magnetic field to blow out the arc that forms when the contacts open.

13-2E *MAGNETIC FORCE*

a) Find the current density necessary to float a copper wire in the earth's magnetic field at the equator.

Assume a field of 10^{-4} tesla. The density of copper is 8.9×10^3 kilograms per cubic meter.

b) Will the wire become hot?

The conductivity of copper is 5.80×10^7 siemens per meter.

c) In what direction must the current flow? See Prob. 8-11.

d) What would happen if the experiment were performed at one of the magnetic poles?

13-3E *MAGNETIC FORCE*

Calculate the force due to the earth's magnetic field on a horizontal wire 100 meters long carrying a current of 50 amperes due north.

Set $B = 5 \times 10^{-5}$ tesla, pointing downward at an angle of 70 degrees with the horizontal.

See Prob. 8-11.

13-4E *MAGNETIC FORCE*

Show that the total force on a closed circuit carrying a current I in a uniform magnetic field is zero.

13-5E *ELECTROMAGNETIC PUMPS*

The conduction current density in a liquid metal is

$$\mathbf{J}_f = \sigma(\mathbf{E} + \mathbf{v} \times \mathbf{B}),$$

where σ is the conductivity, **E** is the electric field intensity, **v** is the velocity of the fluid, and **B** is the magnetic induction, all quantities measured in the frame of reference of the laboratory.

Show that the magnetic force per unit volume of fluid is

$$\sigma(\mathbf{E} + \mathbf{v} \times \mathbf{B}) \times \mathbf{B}.$$

For example, in crossed electric and magnetic fields, a conducting fluid is pushed in the direction of $\mathbf{E} \times \mathbf{B}$. The field $\mathbf{v} \times \mathbf{B}$ then opposes **E**. This is the principle of operation of *electromagnetic pumps*.

13-6E *HOMOPOLAR GENERATOR AND HOMOPOLAR MOTOR*

Figure 13-9 shows a *homopolar generator*. It consists in a conducting disk rotating in an axial magnetic field.

In one particular case, $B = 1$ tesla, $R = 0.5$ meter, and the angular velocity is 3000 revolutions per minute.

Calculate the output voltage.

Homopolar generators are inherently high-current, low-voltage devices. Homopolar generators using superconducting coils and with power outputs of several megawatts are used for the purification of metals by electrolysis.

If a voltage is applied to the output terminals of a homopolar generator, it becomes a *homopolar motor*. Homopolar motors can provide large torques. They are suitable, in particular, for ship propulsion.

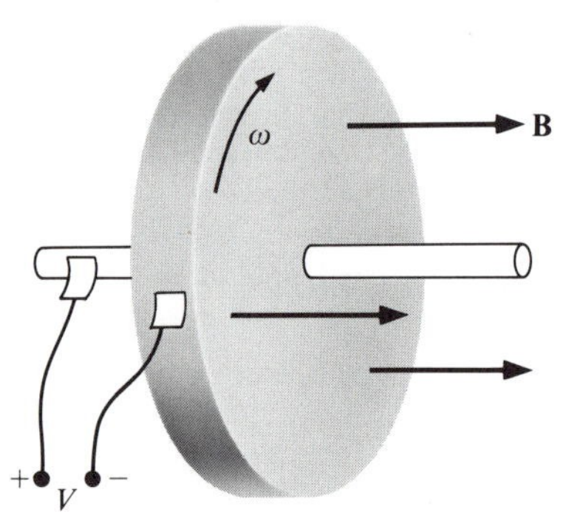

Figure 13-9 Homopolar generator. See Prob. 13-6.

13-7 *HOMOPOLAR MOTOR*

See Prob. 13-6. Now consider the device illustrated in Fig. 13-10. Will the wheel turn when the switch is closed? If so, in which direction and why?[†]

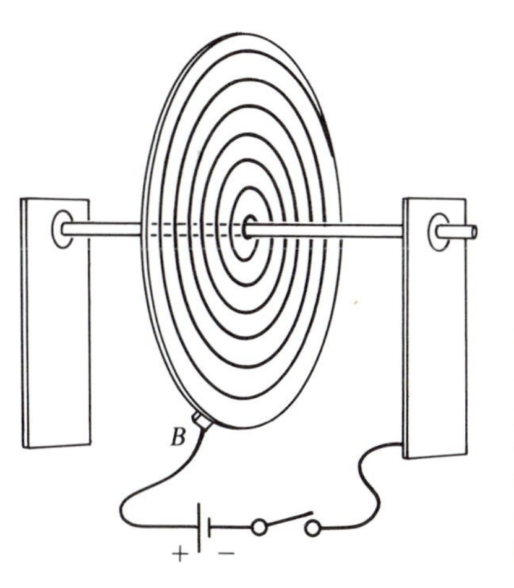

Figure 13-10 Spiral coil. A voltage is applied between the axis and the periphery. Contact to the periphery is made with a carbon brush like that used on motors. Does the disk turn? See Prob. 13-7. The disk could be made with a printed-circuit board.

13-8 *MAGNETIC PRESSURE*

So as to clarify the concept of magnetic pressure, let us consider, first a single solenoid, and then a pair of coaxial solenoids.

a) Figure 13-11a shows a longitudinal section through a solenoid. The small element identified by vertical bars (1) produces its own magnetic field, both outside and inside the solenoid, and (2) is situated in the field produced by the rest of the winding.

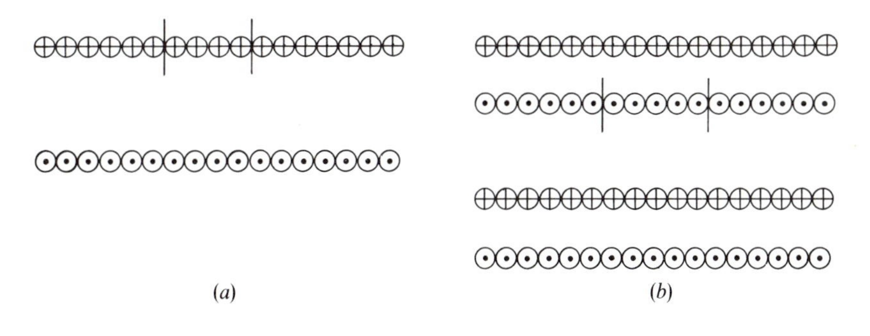

Figure 13-11 (*a*) Longitudinal section through a long solenoid. In Prob. 13-8, we consider the fields close to the small element identified by vertical bars. (*b*) Section through a pair of coaxial solenoids.

[†] See the *American Journal of Physics*, Volume 38, 1970, p. 1273.

Show the first field by means of solid arrows, and the second one by means of dashed arrows.

We are of course thinking here of the fields infinitely close to an infinitely thin winding.

Now can you explain why the magnetic pressure is $B^2/2\mu_0$?

b) Figure 13-11b shows a pair of coaxial solenoids carrying equal currents in opposite directions. In this example we have three fields.

Show, on either side of the element, (a) solid arrows for the field of the element, (b) dashed arrows for the field of the rest of the inner solenoid, (c) wavy arrows for the field of the outer solenoid.

What is the magnetic pressure on the inner solenoid?

13-9E *MAGNETIC PRESSURE*

a) Show that the magnetic pressure is about $4B^2$ atmospheres. One atmosphere is about 10^5 pascals.

b) Draw a log-log plot of the electric force per unit area $\frac{1}{2}\epsilon_0 E^2$, in pascals, as a function of E. The maximum electric field intensity that can be maintained in air at normal temperature and pressure is 3×10^6 volts per meter.

c) Discuss the similarities and differences between the magnetic pressure and the electric force per unit area.

13-10E *MAGNETIC PRESSURE*

It is possible to attain very high pressures by discharging a large capacitor through a hollow wire.

a) Show that, for a thin tube of radius R carrying a current I, the inward magnetic pressure is

$$p_m = \mu_0 I^2/8\pi^2 R^2.$$

b) Calculate the pressure for a current of 30 kiloamperes in a tube one millimeter in diameter.

One atmosphere is approximately 10^5 pascals.

13-11E *MAGNETIC PRESSURE*

Magnetic fields are used for performing various mechanical tasks that require a high power level for a very short time.

For example, if a light aluminum tube is inserted axially in a solenoid, and if the solenoid, is suddenly connected to a charged capacitor, the induced electromotance $d\Phi/dt$ produces a large current in the tube, which collapses under the magnetic pressure. The tube can act as a shutter to turn off a beam of light or of soft X-rays.

Let us calculate the pressure exerted on a conducting tube placed inside a long solenoid. If the current I in the solenoid is increased gradually from zero to some arbitrarily large value, the current induced in the tube is small and the magnetic pressure is negligible. Let us assume that dI/dt is so large that the induced current in the tube maintains zero magnetic field inside it. Then there is a magnetic field only in the annular region between the solenoid and the conducting tube.

a) Calculate the pressure on the tube for $B = 1$ tesla.

b) What would the pressure be if the conducting tube were parallel to the axis, but some distance away?

13-12E *ENERGY STORAGE*

Compare the energies per unit volume in (a) a magnetic field of 1.0 tesla and (b) an electrostatic field of 10^6 volts per meter.

13-13E *MAGNETIC PRESSURE*

Imagine that the current in a long solenoid is maintained constant, while the magnetic pressure increases the radius from R to $R + dR$. Then the magnetic energy *in*creases by

$$d\left(\frac{\mu_0 N^2 \pi R^2}{2l} I^2\right) = \frac{1}{l} \pi \mu_0 I^2 N^2 R \, dR.$$

a) Show that the mechanical work done by the magnetic pressure p_m is $2\pi R l p_m \, dR$.

b) Show that the *in*crease in magnetic energy is equal to the mechanical work done.

c) During the expansion, the magnetic induction inside the solenoid remains constant because B depends only on the number of turns per meter and on the current. So, as the radius increases, the flux linkage also increases.

Show that the extra energy supplied by the source during the expansion, $I(N \, d\Phi)$, is just twice the mechanical work done.

We conclude that one half of the energy supplied by the source serves to perform mechanical work, and the other half serves to increase the magnetic energy. See Prob. 4-13.

13-14 *FLUX COMPRESSION*

Flux compression is one method of obtaining a magnetic field with a very large B. For example, imagine a light conducting tube situated inside a solenoid producing a steady B_0. The annular space between the tube and the solenoid is filled with an explosive, and the solenoid is reinforced externally. If the tube is now imploded, it will be crushed, an azimuthal current will flow, and the internal magnetic pressure $B^2/2\mu_0$ will build up until it is close to the external gas pressure.

a) Show that, if the radius of the tube shrinks very rapidly,

$$B \approx B_0(R_0/R)^2,$$

where B is the magnetic induction when the radius has been reduced to R.

b) Show that the surface current density in the tube must be about 10^9 amperes per meter to achieve a field of 10^3 teslas.

c) If the initial B is 10 teslas, and if the tube has initially a diameter of 100 millimeters, what should be the value of B when the tube is compressed to a diameter of about 10 millimeters?

d) Calculate the resulting increase in magnetic energy. Assume that the cylinder is 200 millimeters long and that the current in the solenoid is constant. Neglect end effects.

Flux compression can also be achieved by means of a conducting piston shot axially into a solenoid.

If the radius of the solenoid is R_0, and if the radius of the piston is R, the magnetic induction in the annular region between the piston and the solenoid, becomes

$$B \approx \frac{B_0}{1 - (R/R_0)^2}.$$

13-15E *PULSED MAGNETIC FIELDS*

Extremely high magnetic fields can be obtained by discharging a capacitor through a low-inductance coil. The capacitor leads must of course have a low inductance. Such fast capacitors cost approximately two dollars per joule of stored-energy capacity.

a) Estimate the cost of a capacitor that could store an energy equal to that of a 100-tesla magnetic field occupying a volume of one liter.

b) Estimate the cost of the electricity required to charge the capacitors.

c) Calculate the magnetic pressure in atmospheres, at 100 teslas. One atmosphere is about 10^5 pascals.

13-16E *MAGNETIC ENERGY*

It was stated in Sec. 13.5 that, if we have two inductances L_a and L_b, carrying currents I_a and I_b,

$$W_m = \frac{1}{2} L_a I_a^2 + \frac{1}{2} L_b I_b^2 + M I_a I_b.$$

You can prove this quite easily.

Let coil a produce flux linkages

$$\Lambda_{aa} \text{ in coil } a \qquad \text{and} \qquad \Lambda_{ab} \text{ in coil } b.$$

Similarly, let coil b produce flux linkages

$$\Lambda_{ba} \text{ in coil } a \qquad \text{and} \qquad \Lambda_{bb} \text{ in coil } b.$$

a) Find W_m in terms of the currents and of the flux linkages.

b) Verify the above equation.

13-17E *ENERGY STORAGE*

a) Show that the magnetic energy stored in an isolated circuit carrying a current I and with a flux linkage Λ is

$$W_m = (1/2) I \Lambda.$$

b) Calculate W_m for a long solenoid using this formula.

13-18 *ENERGY STORAGE*

a) A constant-voltage source set at a voltage V is connected to an inductance L of negligible resistance at $t = 0$.

Find, as functions of the time, (i) the current I, (ii) the energy supplied by the source, and (iii) the energy stored in the magnetic field.

b) A constant-current source set at a current I is connected to a capacitor C at $t = 0$.

Find, as functions of the time, (i) the voltage V on the capacitor, (ii) the energy supplied by the source, and (iii) the energy stored in the electric field.

13-19 *ENERGY STORAGE*

Electrical utilities must be able to meet peak power demands. It is therefore desirable to store energy during periods of low demand and feed it back into the grid when needed. One way of storing energy is to pump water into an elevated reservoir.

It has been suggested that large amounts of energy could also be stored in the magnetic fields of superconducting coils.

a) Compare the energy densities in (i) an electric field of 10^8 volts per meter in Mylar, for which $\epsilon_r = 3.2$, and (ii) a magnetic field of 8 teslas. These fields are about as high as present technology will permit. The magnetic energy density is larger by two orders of magnitude.

b) It is suggested that 10 gigawatt-hours be stored in a solenoid with a length-to-diameter ratio of 20. The solenoid would be made in four parts, joined end to end to form a quasi-toroidal coil located in a tunnel excavated in solid rock. Disregard the toroidal shape in your calculation. The magnetic induction would be 8 teslas. (i) Calculate the total length and the diameter. (ii) Calculate the magnetic pressure in atmospheres. (iii) Calculate the surface area that would require cryogenic thermal insulation.

13-20 *SUPERCONDUCTING POWER TRANSMISSION LINE*

A superconducting power transmission line has been proposed that would have the following characteristics. It would carry 10^{11} watts at 200 kilovolts DC over a distance of 10^3 kilometers. The conductors would have a circular cross-section of 5 square centimeters and would be held 5 centimeters apart, center to center.

a) Calculate the magnetic force per meter.

It so happens that the force of attraction is the same as if the conductors were replaced by fine wires, 5 centimeters apart.

b) Calculate $B^2/2\mu_0$ mid-way between the two conductors.

c) Calculate the stored energy and its cost at 0.5 cent per kilowatt-hour.

The self-inductance of such a line is 0.66 microhenry per meter.

13-21 *ELECTRIC MOTORS AND MOVING-COIL METERS*

Electric motors, as well as moving-coil voltmeters and ammeters, utilize the torque exerted on a current-carrying coil situated in a magnetic field.

Consider an N-turn rectangular coil situated in a uniform magnetic field as in Fig. 13-12.

a) Calculate the forces on sides 1, 2, 3, 4.

b) Show that the torque

$$T = NIBab \sin \theta.$$

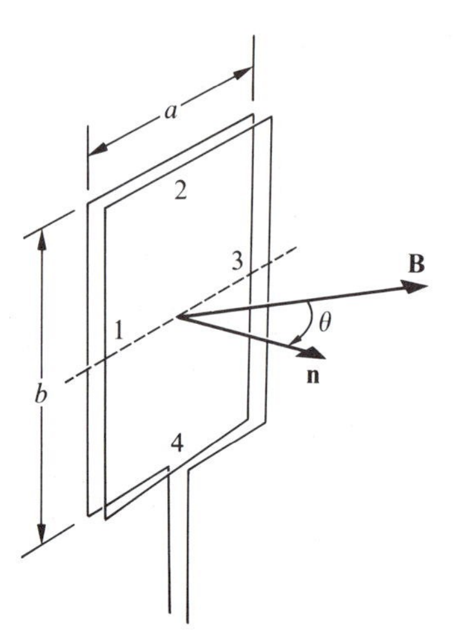

Figure 13-12 Rectangular coil whose normal **n** forms an angle θ with a uniform magnetic field. See Prob. 13-21.

In an electric motor, we have essentially a number of such coils, offset by, say, 15 degrees, around the axis, and connected to a set of contacts called a *commutator*, fixed to the axis. Connection to the coils is made by *carbon brushes* that ride on the commutator. The coils and their commutator form the *armature* of the motor. The magnetic field is supplied by the *stator coils*. The magnetic flux is carried by laminated iron, both in the armature and in the stator *yoke*.

Moving coil voltmeters and ammeters use a uniform radial magnetic field as in Fig. 13-13. The torque is then simply $NIBab$ and is independent of θ. A light spiral spring exerts a restoring torque that is proportional to θ, giving a deflection θ proportional to I.

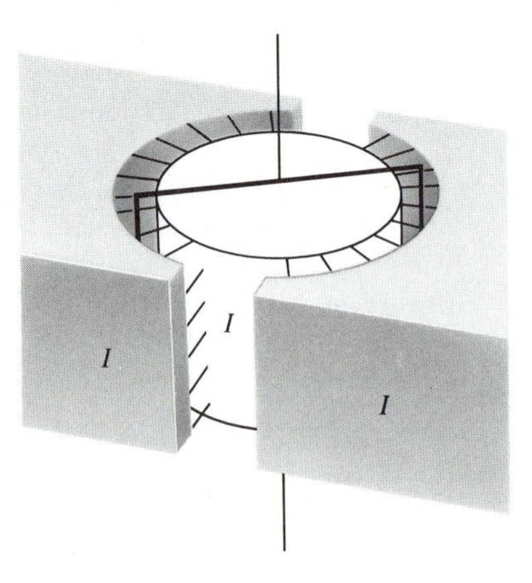

Figure 13-13 Moving coil and magnetic field in a voltmeter or ammeter. The magnetic field is supplied by a permanent magnet. The parts marked I are made of iron.

13-22 *MAGNETIC TORQUE*

Consider a square single-turn coil of side a, carrying a current I.

a) Show that it tends to orient itself in a magnetic field in such a way that the total magnetic flux linking the coil is *maximum*.

b) Show that the torque exerted on the coil is $\mathbf{m} \times \mathbf{B}$, where $\mathbf{m}$ is the magnetic moment of the coil (Sec. 8.1.2), and $\mathbf{B}$ is the magnetic induction when the current in the coil is zero.

This result is independent of the shape of the coil.

13-23 *ATTITUDE CONTROL FOR SATELLITES*

See the previous problem.

Many satellites require attitude control to keep them properly oriented. For example, communication satellites must keep their antenna systems on target. Attitude control requires a method for exerting appropriate torques as they are required.

Attitude control can be achieved to a certain extent by means of coils whose magnetic fields interact with that of the earth.

a) Show that the torque exerted by such a coil is

$$NIBA \sin \theta,$$

where N is the number of turns, I is the current, B is the magnetic induction due to the earth, A is the area of the coil, and θ is the angle between the earth's magnetic field and the normal to the coil.

b) Calculate the number of ampere-turns required for a coil wound around the outside surface of a satellite whose diameter is 1.14 meters. The torque required at $\theta = 5$ degrees is 10^{-3} newton-meter, and the magnetic induction at an altitude of 700 kilometers over the equator, where the orbit of the satellite will be situated, is 4.0×10^{-5} tesla.

13-24 *MECHANICAL FORCES ON AN ISOLATED CIRCUIT*

Show that, if the geometry of an isolated active circuit is altered, the energy supplied by the source is equal to twice the increase in magnetic energy. Thus the mechanical work performed is equal to the increase in magnetic energy. Assume that the current is kept constant.

It follows that, on this assumption, the force on an element of an active circuit is equal to the rate of change of magnetic energy.

See Probs. 13-13 and 13-25.

13-25 *RAIL GUN, OR PLASMA GUN*

Calculate the force on the movable link D in the circuit of Fig. 13-8.

Instead of being a metallic rod, the link can also be an electric arc. The device then accelerates blobs of plasma and is called a *plasma gun*. Statellite thrusters can be made in this way. See Probs. 2-14, 2-15, 5-3, and 10-11.

Solution: We cannot solve this problem by integrating the magnetic force $\mathbf{J} \times \mathbf{B}$ over the volume of the link, since neither $\mathbf{J}$ nor $\mathbf{B}$ are known, or even easily calculated. Instead, we shall find the force by investigating the magnetic and mechanical energies involved.

If we set $x = 0$ at the initial position of the link, then the inductance L in the circuit is $L_0 + L'x$, where $L'x$ is the inductance associated with the magnetic flux linking the rails, L' being the self-inductance per meter.

During the discharge of the capacitor C we can neglect the battery and the resistance R, since the current flowing in that part of the circuit is negligible. Let the resistance on the right-hand side of the circuit be only that of the arc, R'.

At any instant, the power supplied by the capacitor serves to (a) increase the magnetic energy $\frac{1}{2}LI^2$, (b) increase the kinetic energy, and (c) dissipate energy in the resistance R'. Thus

$$VI = \frac{d}{dt}\left(\frac{1}{2}LI^2\right) + Fv + I^2R', \tag{1}$$

where F is the driving force and v is the velocity of the link. Note that both L and I are functions of the time.

Now what is the value of V? The voltage V (a) increases the flux linkage in the inductance and (b) gives a voltage drop IR' on the resistor R':

$$V = \frac{d}{dt}(LI) + IR', \tag{2}$$

where LI is the flux linkage.

Combining Eqs. 1 and 2,

$$\left(L\frac{dI}{dt} + I\frac{dL}{dt} + IR'\right)I = LI\frac{dI}{dt} + \frac{1}{2}I^2\frac{dL}{dt} + Fv + I^2R', \tag{3}$$

$$\frac{1}{2}I^2\frac{dL}{dt} = Fv, \tag{4}$$

$$Fv = \frac{1}{2}I^2\frac{dL}{dx}\frac{dx}{dt} = \frac{1}{2}I^2\frac{dL}{dx}v, \tag{5}$$

$$F = \frac{1}{2}L'I^2. \tag{6}$$

CHAPTER 14

MAGNETIC FIELDS: VII

Magnetic Materials

Thus far we have studied only those magnetic fields that are attributable to the motion of free charges. Now, on the atomic scale, all bodies contain spinning electrons that move around in orbits, and these electrons also produce magnetic fields.

Our purpose in this chapter is to express the magnetic fields of these atomic currents in macroscopic terms.

Magnetic materials are similar to dielectrics in that individual charges or systems of charges can possess magnetic moments (Sec. 8.1.2), and these moments, when properly oriented, produce a resultant magnetic moment in a macroscopic body. Such a body is then said to be *magnetized.*

In most atoms the magnetic moments associated with the orbital and spinning motions of the electrons cancel. If the cancellation is not complete, the material is said to be *paramagnetic*. When such a substance is placed in a magnetic field, its atoms are subjected to a torque that tends to align them with the field, but thermal agitation tends to destroy the alignment. This phenomenon is analogous to the alignment of polar molecules in dielectrics.

In *diamagnetic* materials, the elementary moments are not permanent but are induced according to the Faraday induction law (Sec. 11.2). All materials are diamagnetic, but orientational magnetization may predominate.

Magnetic devices use *ferromagnetic* materials, such as iron, in which the magnetization can be orders of magnitude larger than that of either para- or diamagnetic substances. This large magnetization results from electron spin and is associated with group phenomena in which all the elementary moments in a small region, known as a *domain*, are aligned. The magnetization of one domain may be oriented at random with respect to that of a neighboring domain.

There is one important difference between dielectric and magnetic materials. In most dielectrics **D** is proportional to **E** and the medium is said to be *linear*. Ferromagnetic materials are not only highly *non*-linear, but their behavior also depends on their previous history. The calculation of the fields associated with magnetic materials is therefore largely empirical. This will be the subject of the next chapter.

14.1 THE MAGNETIZATION M

The *magnetization* **M** is the magnetic moment per unit volume at a given point. If **m** is the average magnetic dipole moment per atom, and if N is the number of atoms per unit volume,

$$\mathbf{M} = N\mathbf{m}, \tag{14-1}$$

if the individual dipole moments in the element of volume considered are all aligned in the same direction.

The magnetization **M** is measured in amperes per meter, and it corresponds to the polarization **P** in dielectrics (Sec. 6.1).

14.2 THE EQUIVALENT SURFACE CURRENT DENSITY $\boldsymbol{\alpha}_e$

Figure 14-1 shows a cylinder of material divided into square cells of cross-sectional area a^2, where $a \to 0$. Imagine that each cell carries a clockwise surface current of M amperes per meter, as in the figure. Then each one-meter length of cell has a magnetic moment of a^2M ampere meter squared (Sec. 8.1.2) and the magnetic moment per cubic meter is **M** amperes per meter in the direction shown in the figure. Now the current in one cell is canceled by the currents in the adjoining cells, except at the periphery of the material. Thus, if the material has a magnetization M, the equivalent current density α_e at the surface is M.

More generally, the *equivalent, or Amperian, surface current density* is

$$\boldsymbol{\alpha}_e = \mathbf{M} \times \mathbf{n}_1, \tag{14-2}$$

where $\mathbf{n}_1$ is a unit vector normal to the surface and pointing *outward.*

These equivalent currents do not dissipate energy because they do not involve electron drift and scattering processes like those associated with conduction currents.

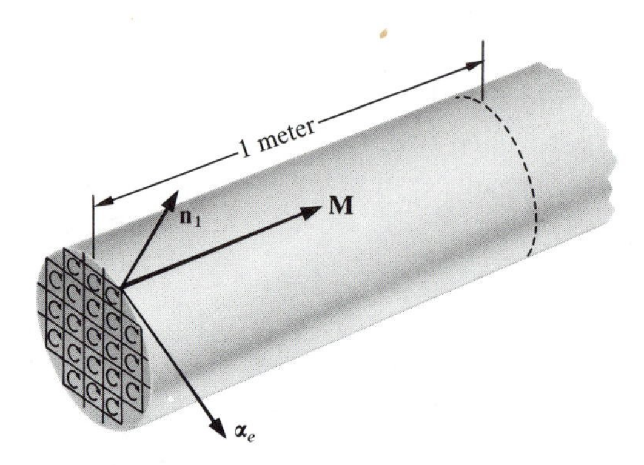

Figure 14-1 Ampère's model for the equivalent current in a cylinder of magnetized material.

14.3 *THE EQUIVALENT VOLUME CURRENT DENSITY* $\mathbf{J}_e$

Let us now suppose that the magnetization **M** is a function of the x-coordinate as in Fig. 14-2, with

$$M = px, \tag{14-3}$$

where p is a positive constant. Since M is expressed in amperes per meter, p is expressed in amperes per square meter.

At the surface of the material, the equivalent current density $\boldsymbol{\alpha}_e$ is again given by Eq. 14-2, with **M** equal to the local value of the magnetization Thus $\boldsymbol{\alpha}_e$ is a function of x, as in the figure.

Inside the material, currents parallel to the x-axis cancel.

Currents parallel to the y-axis do not cancel. Along any of the vertical planes inside, the left-hand side of a given cell carries an upward current of

$$p\left(x - \frac{a}{2}\right) \text{amperes,}$$

while the right-hand side carries a downward current of

$$p\left(x + \frac{a}{2}\right) \text{amperes.}$$

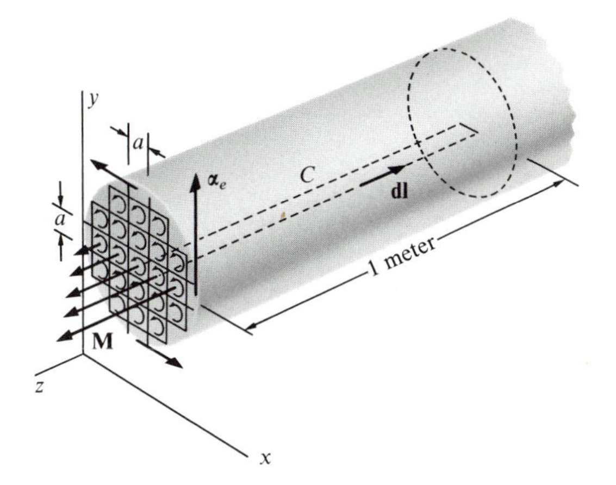

Figure 14-2 Magnetized cylinder in which **M** is a function of the x-coordinate: M increases linearly with x as in Eq. 14-3.

Along any vertical plane, we therefore have a net downward current of pa amperes.

Note that this downward current is independent of x. This is simply because we have assumed that M is directly proportional to x.

So, along every vertical division between cells, we have a net downward current of pa amperes. The net downward current density is thus p, since we have pa amperes for every interval a. In other words, we have an equivalent current density

$$\mathbf{J}_e = -p\mathbf{j}. \tag{14-4}$$

We can now show that $\mathbf{J}_e$ is equal to $\nabla \times \mathbf{M}$ in the following way. Along the curve C of Fig. 14-2, $\mathbf{M} \cdot \mathbf{dl}$ is zero over the left-hand end, because $\mathbf{M}$ is perpendicular to $\mathbf{dl}$. The same applies to the right-hand end. Thus

$$\oint_C \mathbf{M} \cdot \mathbf{dl} = -p\left(x + \frac{a}{2}\right) + p\left(x - \frac{a}{2}\right), \tag{14-5}$$

$$= -pa, \tag{14-6}$$

where the line integral is evaluated in the direction shown in the figure.

Since the area of the loop is $a \times 1$, the y-component of the curl of $\mathbf{M}$ is $-p$, from Sec. 1.12, and

$$\mathbf{J}_e = (\nabla \times \mathbf{M})_y \mathbf{j}. \tag{14-7}$$

More generally, the equivalent volume current density is

$$\mathbf{J}_e = \nabla \times \mathbf{M}. \tag{14-8}$$

14.4 *CALCULATION OF MAGNETIC FIELDS ORIGINATING IN MAGNETIZED MATERIAL*

Let us first recall the reasoning we followed in Sec. 6.3 to find the electric field due to polarized dielectric material. We found that the charge displacement that occurs on polarization results in the accumulation of net densities of bound charge, σ_b on the surfaces and ρ_b inside. We then argued that *any* net charge density gives rise to an electric field, exactly as if it were in a vacuum. As a consequence, $\mathbf{E}$ can be calculated at any point in space from σ_b and ρ_b.

We now have an entirely analogous situation in the magnetic fields originating in magnetized materials. The magnetization gives equivalent currents, with densities $\boldsymbol{\alpha}_e$ on the surfaces and $\mathbf{J}_e$ inside. These currents produce magnetic fields, both inside and outside the material, exactly as if they were situated in a vacuum.

In principle, one can calculate $\mathbf{B}$ at any point in space from the equivalent current densities $\boldsymbol{\alpha}_e$ and $\mathbf{J}_e$. In practice, the vector $\mathbf{M}$ is an unknown function of the coordinates inside the material, and one can do little more than guess its value. We shall return to this subject in the next chapter.

14.4.1 *EXAMPLE: UNIFORMLY MAGNETIZED BAR MAGNET*

The uniformly magnetized bar magnet of Fig. 14-1 is a good example to use at this point.

We have shown in Sec. 14.2 that the cylindrical surface carries an equivalent current density α_e that is equal to M. Inside the magnet, $\nabla \times \mathbf{M} = 0$, and the volume current density $\mathbf{J}_e$ is zero. Over the end faces, $\mathbf{M}$ is parallel to the normal unit vector $\mathbf{n}_1$, and $\boldsymbol{\alpha}_e$ is zero.

The **B** field of a cylinder uniformly magnetized in the direction of its axis of symmetry is therefore identical to that of a solenoid of the same dimensions with N' turns per meter and carrying a current I such that $IN' = M$, as in Fig. 9-10.

In a real bar magnet the magnetic moments of the individual atoms tend to align themselves with the **B** field, so that the magnetization **M** is weaker near the ends. The end faces also carry Amperian currents, since $\mathbf{M} \times \mathbf{n}$ is zero only on the axis. The net result is that there are two "poles," one at each end of the magnet, from which lines of **B** radiate in all directions outside the magnet. The poles are most conspicuous if the bar magnet is long and thin.

14.5 THE MAGNETIC FIELD INTENSITY H

We found in Sec. 9.2 that, for a steady current density of free charge $\mathbf{J}_f$ and for non-magnetic materials,

$$\nabla \times \mathbf{B} = \mu_0 \mathbf{J}_f. \tag{14-9}$$

Now we have just seen that magnetized material can be replaced by its equivalent currents for calculating **B**. Consequently, if we have magnetized material as well as steady currents,

$$\nabla \times \mathbf{B} = \mu_0(\mathbf{J}_f + \mathbf{J}_e), \tag{14-10}$$

where $\mathbf{J}_e$ is the equivalent volume current density. This equation is of course valid only at points where the derivatives of **B** with respect to the coordinates x, y, z exist. It is therefore not applicable at the surface of magnetic materials and we disregard $\boldsymbol{\alpha}_e$.

Substituting $\nabla \times \mathbf{M}$ for $\mathbf{J}_e$,

$$\nabla \times \mathbf{B} = \mu_0(\mathbf{J}_f + \nabla \times \mathbf{M}), \tag{14-11}$$

$$\nabla \times \left(\frac{\mathbf{B}}{\mu_0} - \mathbf{M}\right) = \mathbf{J}_f. \tag{14-12}$$

The vector between parentheses, whose curl is equal to the free current density at the point, is the *magnetic field intensity*

$$\mathbf{H} = \frac{\mathbf{B}}{\mu_0} - \mathbf{M}, \tag{14-13}$$

and is expressed in amperes per meter. Note that **H** and **M** are expressed in the same units.

Thus

$$\mathbf{B} = \mu_0(\mathbf{H} + \mathbf{M}). \tag{14-14}$$

This equation is to be compared with Eq. 6-15, which applies to dielectrics:

$$\mathbf{E} = \frac{1}{\epsilon_0}(\mathbf{D} - \mathbf{P}). \tag{14-15}$$

14.6 AMPÈRE'S CIRCUITAL LAW

Rewriting now Eq. 14-12, we have that, for steady currents, either inside or outside magnetic material,

$$\nabla \times \mathbf{H} = \mathbf{J}_f. \tag{14-16}$$

Integrating over a surface S,

$$\int_S (\nabla \times \mathbf{H}) \cdot \mathbf{da} = \int_S \mathbf{J}_f \cdot \mathbf{da} \tag{14-17}$$

or, using Stoke's theorem on the left-hand side,

$$\oint_C \mathbf{H} \cdot \mathbf{dl} = I_f, \tag{14-18}$$

where C is the curve bounding the surface S, and I_f is the current of free charges linking the curve C. Note that I_f does *not* include the equivalent currents. The term on the left is called the *magnetomotance*. A magnetomotance is expressed in amperes, or in ampere-turns.

This is a more general form of *Ampère's circuital law* (Sec. 9.1), in that it can be used to calculate $\mathbf{H}$ even in the presence of magnetic materials. It is rigorously valid, however, only for steady currents; we shall deal with variable currents in Chapter 19.

14.6.1 EXAMPLE: STRAIGHT WIRE CARRYING A CURRENT I AND EMBEDDED IN MAGNETIC MATERIAL

Figure 14-3 shows a straight wire carrying a current I and embedded in magnetic material. The magnetomotance around the path shown circling the current I is equal to I, and $H = I/2\pi r$, exactly as if the wire were situated in a vacuum.

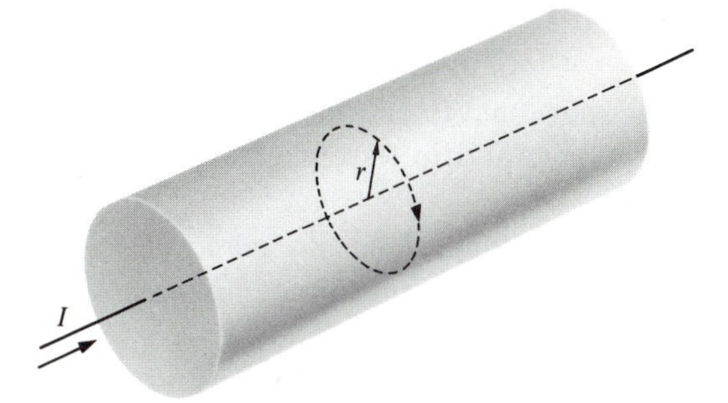

Figure 14-3 Straight wire carrying a current I and embedded in magnetic material.

If the material is isotropic, and if it has the shape of a circular cylinder with I along its axis, **M** is azimuthal, like **H**.

14.7 *MAGNETIC SUSCEPTIBILITY* χ_m, *PERMEABILITY* μ, *AND RELATIVE PERMEABILITY* μ_r

As for dielectrics (Sec. 6.6), it is convenient to define a *magnetic susceptibility* χ_m such that

$$\mathbf{M} = \chi_m \mathbf{H}. \tag{14-19}$$

Now, since

$$\mathbf{B} = \mu_0(\mathbf{H} + \mathbf{M}), \tag{14-20}$$

then

$$\mathbf{B} = \mu_0(1 + \chi_m)\mathbf{H} = \mu_0\mu_r\mathbf{H} = \mu\mathbf{H}, \tag{14-21}$$

where

$$\mu_r = 1 + \chi_m, \qquad \mu = \mu_0\mu_r. \tag{14-22}$$

The quantity μ is the permeability and μ_r is the *relative permeability*. Both χ_m and μ_r are dimensionless quantities.

Equation 14-20 is general, but Eqs. 14-21 and 14-22 are based on the assumption that the material is both isotropic and linear, in other words, that **M** is proportional to **H** and in the same direction. This assumption is unfortunately *not* valid in ferromagnetic materials, as we shall see in the next section.

The magnetic susceptibility of *paramagnetic substances* is smaller than unity by several orders of magnitude and is proportional to the inverse of the absolute temperature.

In *diamagnetic materials*, the magnetization is in the direction *opposite* to the external field; the relative permeability is *less* than unity and is independent of the temperature. If orientational magnetization predominates, the resultant relative permeability is greater than unity.

14.8 THE MAGNETIZATION CURVE

One can measure B for various values of H with a *Rowland ring* whose minor radius is much smaller than its major radius, as in Fig. 14-4. The function of winding a, which has N_a turns and carries a current I_a, is to produce a known magnetic field intensity

$$H = N_a I_a / 2\pi r \tag{14-23}$$

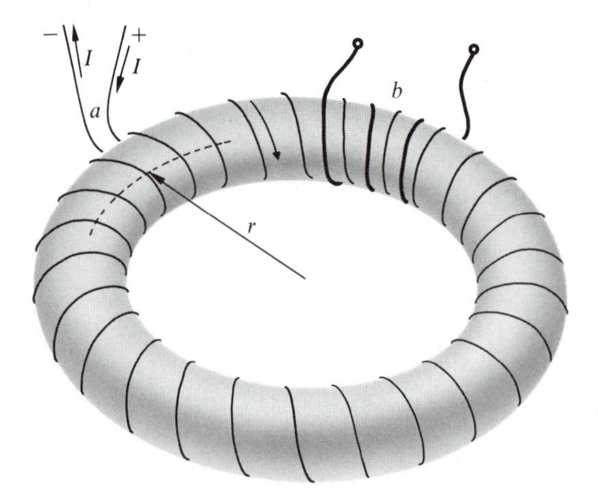

Figure 14-4 Rowland ring for the determination of B as a function of H in a ferromagnetic substance.

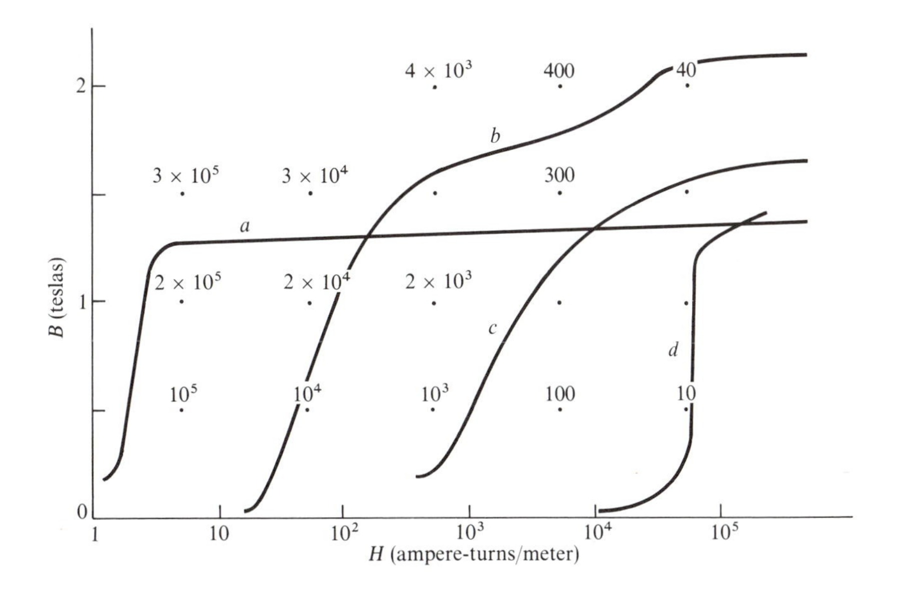

Figure 14-5 Magnetization curves for various magnetic materials: *a*, Permalloy; *b*, annealed pure iron; *c*, ductile cast iron; *d*, Alnico 5. The numbers shown are the relative permeabilities $B/\mu_0 H$.

in the sample. The magnetic induction is

$$B = \Phi/S, \tag{14-24}$$

where S is the cross-sectional area of the core, and Φ is the magnetic flux. One can measure changes in Φ, and hence changes in B, by changing I and integrating the electromotance induced in winding b as in Prob. 14-16. Both B and H are therefore easily measurable.

If we start with an unmagnetized sample and increase the current in coil a, the magnetic induction also increases, but irregularly, as in Fig. 14-5. All ferromagnetic substances have such S-shaped *magnetization curves.* Note the large variations in μ_r for a given material. Whenever one uses μ_r in relation with ferromagnetic materials, one refers to the ratio $B/\mu_0 H$, for specific values of B or of H.

A characteristic feature of the magnetization curve is the *saturation* induction beyond which M increases no further. Maximum alignment of the domains is then achieved, after which $dB = \mu_0\, dH$. The saturation induction lies in the range from 1 to 2 teslas, depending on the material.

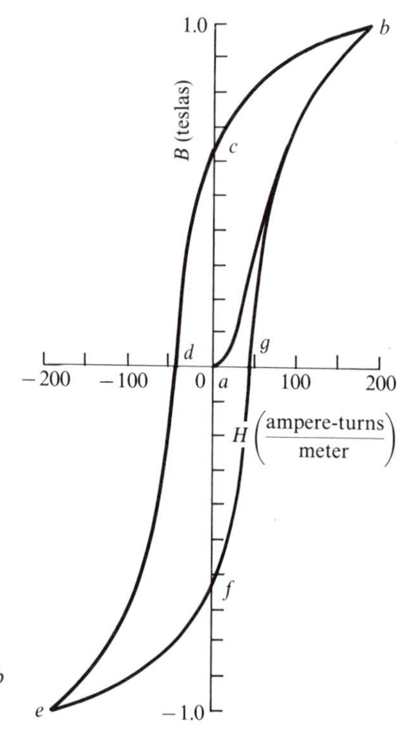

Figure 14-6 Magnetization curve *ab* and hysteresis loop *bcdefgb*.

14.9 HYSTERESIS

Consider now Fig. 14-6. Curve *ab* is the magnetization curve. Once point *b* has been reached, if the current in winding *a* of Fig. 14-4 is reduced to zero, *B* decreases along *bc*. The magnitude of the magnetic induction at *c* is the *remanence* or the *retentivity* for the particular sample of material. If the current is then reversed in direction and increased, *B* reaches a point *d* where it is reduced to zero. The magnitude of *H* at this point is known as the *coercive force*. On further increasing the current in the same direction a point *e*, symmetrical to point *b*, is reached. If the current is now reduced, reversed, and increased, the point *b* is again reached. The closed curve *bcdefgb* is known as a *hysteresis loop*. If, at any point, the current is varied in a smaller cycle, a small hysteresis loop is described.

Energy is required to describe a hysteresis cycle. This can be shown by considering Fig. 14-7. When the current in winding *a* of Fig. 14-4 is increasing, the electromotance induced in the winding opposes the increase

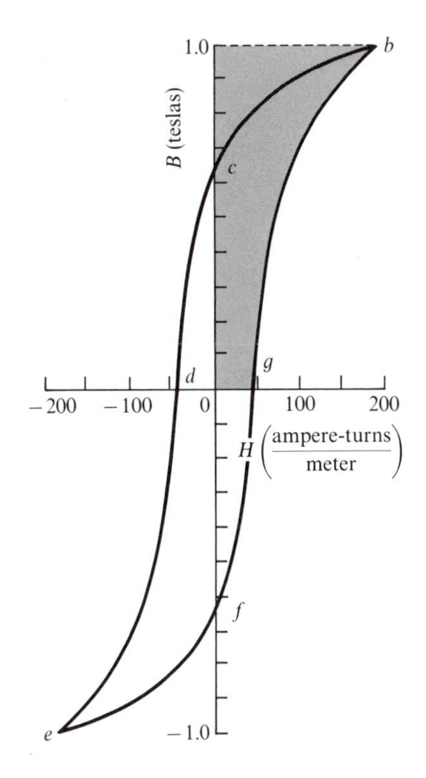

Figure 14-7 The shaded area gives the energy per unit volume required to go from g to b on the hysteresis loop. The energy per unit volume required to describe a complete hysteresis loop is equal to the area enclosed by the loop.

in current, according to Lenz's law (Sec. 11.3), and the extra power spent by the source is

$$\frac{dW}{dt} = I\left(N\frac{d\Phi}{dt}\right) = INS\frac{dB}{dt}, \tag{14-25}$$

where S is the cross-sectional area of the ring, N is the number of turns in winding a, and B is the average magnetic induction in the core. Also,

$$\frac{dW}{dt} = \left(\frac{IN}{l}\right)Sl\frac{dB}{dt} = H\tau\frac{dB}{dt}, \tag{14-26}$$

where l is the mean circumference of the ring, $\tau = Sl$ is its volume, and $IN/l = H$, as in Eq. 14-23. Thus

$$W_1 = \tau\int_g^b H\,dB \tag{14-27}$$

is the energy supplied by the source in going from the point g to the point b in Fig. 14-7. This integral corresponds to the shaded area in the figure and is equal to the energy supplied per unit volume of the magnetic core.

When the current is in the same direction but is decreasing, the polarity of the induced electromotance is reversed, according to Lenz's law, with the result that the energy

$$W_2 = \tau \int_b^c H\, dB \tag{14-28}$$

is returned to the source.

Finally, the energy supplied by the source during one cycle is

$$W = \tau \oint H\, dB, \tag{14-29}$$

where the integral is evaluated around the hysteresis loop. The area of the hysteresis loop in tesla-ampere turns per meter or in weber-ampere turns per cubic meter is therefore the number of joules dissipated per cubic meter and per cycle in the core.

Hysteresis losses can be minimized by selecting a material with a narrow hysteresis loop.

14.9.1 *EXAMPLE: TRANSFORMER IRON*

For the transformer iron of Fig. 14-7, the area enclosed by the hysteresis loop is approximately 150 weber-ampere turns per cubic meter or 150 joules per cubic meter and per cycle, or 1.1 watts per kilogram at 60 hertz. This iron is suitable for use in a power transformer, but other alloys are available with lower hysteresis losses.

14.10 *BOUNDARY CONDITIONS*

Let us examine the continuity conditions that **B** and **H** must obey at the interface between two media. We shall proceed as in Sec. 7.1.

Figure 14-8a shows a short cylindrical volume whose top and bottom faces are parallel and infinitely close to the interface. Since there is zero net flux through the cylindrical surface (Sec. 8.2), the flux through the top face must equal that through the bottom, and

$$B_{n1} = B_{n2}. \tag{14-30}$$

The normal component of **B** is therefore continuous across the interface.

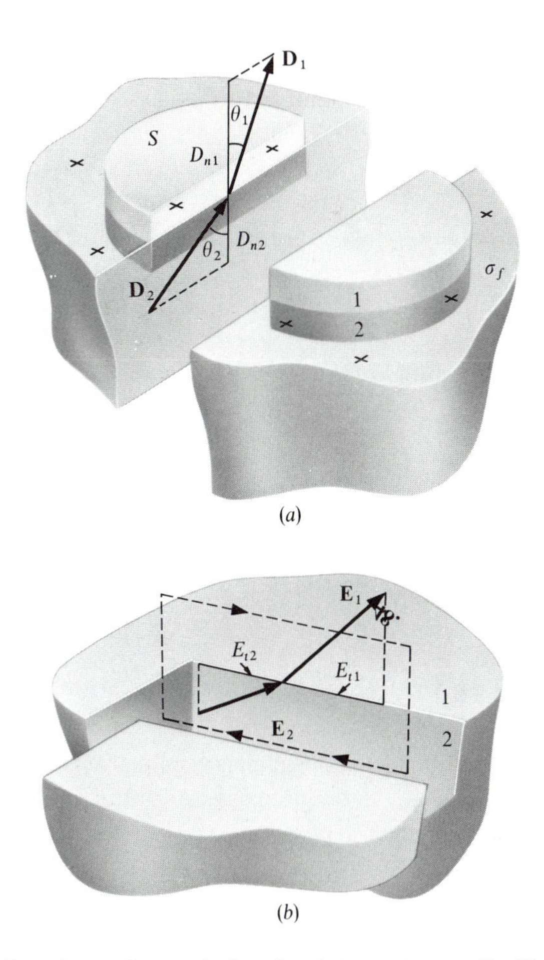

Figure 14-8 (*a*) Gaussian surface at the interface between two media. (*b*) Closed path crossing the interface.

Consider now Fig. 14-8b. The closed path has two sides parallel to the interface and close to it. From the circuital law (Sec. 14.6),

$$\oint \mathbf{H} \cdot \mathbf{dl} = I, \tag{14-31}$$

where I is the conduction current linking the path.

If the two sides parallel to the interface are infinitely close to it, I is equal to zero,

$$H_{t1} = H_{t2}, \tag{14-32}$$

and the tangential component of $\mathbf{H}$ is continuous across an interface.

If we can set **B** equal to μ**H** in both media, the relative permeabilities being those that correspond to the actual values of **H**, then

$$\mu_{r1}\mu_0 H_1 \cos\theta_1 = \mu_{r2}\mu_0 H_2 \cos\theta_2, \tag{14-33}$$

from the continuity of the normal component of **B**, and

$$H_1 \sin\theta_1 = H_2 \sin\theta_2, \tag{14-34}$$

from the continuity of the tangential component of **H**. Then

$$\frac{\tan\theta_1}{\tan\theta_2} = \frac{\mu_{r1}}{\mu_{r2}}. \tag{14-35}$$

The lines of **B**, or of **H**, are farther away from the normal in the medium having the larger permeability.

14.11 *SUMMARY*

The *magnetization* **M** is the magnetic moment per unit volume.

One can calculate the magnetic field of a piece of magnetic material by assuming an *equivalent surface current density*

$$\boldsymbol{\alpha}_e = \mathbf{M} \times \mathbf{n}_1, \tag{14-2}$$

where $\mathbf{n}_1$ is a unit vector normal to the surface and pointing outward, and an *equivalent volume current density*

$$\mathbf{J}_e = \nabla \times \mathbf{M}. \tag{14-8}$$

The *magnetic field intensity* is

$$\mathbf{H} = \frac{\mathbf{B}}{\mu_0} - \mathbf{M}, \tag{14-13}$$

and

$$\nabla \times \mathbf{H} = \mathbf{J}_f. \tag{14-16}$$

This is *Ampère's circuital law* in terms of **H**. In integral form,

$$\oint_C \mathbf{H} \cdot \mathbf{dl} = I_f, \tag{14-18}$$

where I_f is the current of free charges linking the curve C.

If a magnetic material is linear and isotropic, then

$$\mathbf{B} = \mu_0(1 + \chi_m)\mathbf{H} = \mu_0\mu_r\mathbf{H} = \mu\mathbf{H}, \tag{14-21}$$

where χ_m is the *magnetic susceptibility*, μ_r is the *relative permeability*, and μ is the *permeability*.

Ferromagnetic materials are highly non-linear, and often anisotropic. The relation between **B** and **H** depends on the previous history of the field, and is represented by the *hysteresis loop*. The area of a hysteresis loop gives the energy dissipated in the material per cubic meter and per cycle.

At the boundary between two materials, the tangential component of **H** and the normal component of **B** are continuous.

PROBLEMS

14-1E *MAGNETIC FIELD OF THE EARTH*

A sphere of radius R is uniformly magnetized in the direction parallel to a diameter. The external field of such a sphere is closely similar to that of the earth. See Prob. 8-11.

Show that the equivalent surface current density is the same as if the sphere carried a uniform surface charge density σ and rotated at an angular velocity ω such that

$$\sigma\omega R = M.$$

14-2E *EQUIVALENT CURRENTS*

An iron torus whose major radius is much larger than its minor radius is magnetized in the azimuthal direction with M uniform.

What can you say about $\boldsymbol{\alpha}_e$?

14-3E *EQUIVALENT CURRENTS*

A long tube is uniformly magnetized in the direction parallel to its axis. What is the value of **B** inside the tube?

14-4 *DIELECTRICS AND MAGNETIC MATERIALS COMPARED*

a) Imagine a parallel-plate capacitor whose plates are charged and insulated. How are **E** and **D** affected by the introduction of a dielectric between the plates?

b) Show a polar molecule of the dielectric oriented in the field.

c) Is its energy maximum or minimum?

d) Now imagine a long solenoid carrying a fixed current.

How are **B** and **H** affected by the introduction of a cylinder of ferromagnetic material inside the solenoid?

e) Show a small current loop representing the magnetic dipole moment of an atom, and show the direction of the current on the loop.

f) Is its energy maximum or minimum?

14-5 *MAGNETIC TORQUE*

Show that the torque exerted on a permanent magnet of dipole moment **m** situated in a magnetic field **B** is $\mathbf{m} \times \mathbf{B}$.

See Prob. 13-22.

14-6E *MEASUREMENT OF M*

The following method has been used to measure the magnetization M of a small sphere of material, induced by an applied uniform magnetic field $\mathbf{B}_0$ as in Fig. 14-9.

In the figure, HG is a Hall generator (Sec. 10.8.1) with its current flowing parallel to $\mathbf{B}_0$. Since Hall generators are sensitive only to magnetic fields perpendicular to their current flow, this one measures only the dipole field originating in the small sphere.

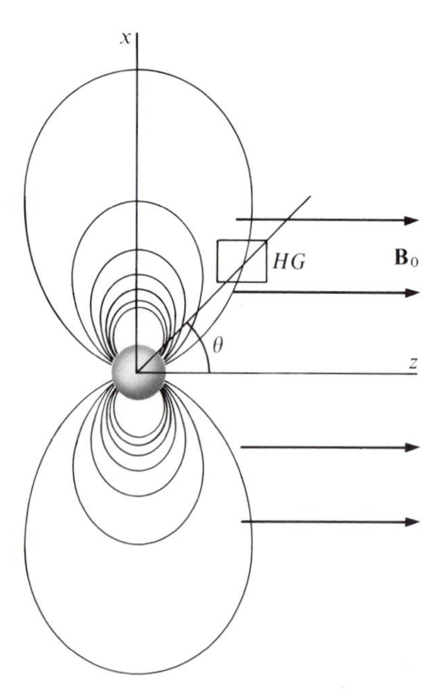

Figure 14-9 Set-up for measuring the magnetization M induced in a small spherical sample of material by a uniform field $\mathbf{B}_0$. The Hall generator HG is oriented so as to be sensitive to the x-component of the field originating in the sphere, and not to $\mathbf{B}_0$. See Prob. 14-6.

If the magnetization is M, the magnetic moment m of the sphere is $(4/3)\pi R^3 M$, and it turns out that the x-component of the dipole field of the sample at HG is

$$\frac{3\mu_0 m}{4\pi} \frac{\sin\theta\cos\theta}{r^3}.$$

Thus, at a given angle θ, this field decreases as $1/r^3$.

At what angle θ will the measured field be largest?

14-7 *MICROMETEORITE DETECTOR*

Figure 14-10 shows an instrument that has been devised to detect ferromagnetic micrometeorites falling to the ground. The instrument detects particles as small as 10 micrometers in diameter and measures their magnetic moment, which is a measure of their size and composition.

A particle of radius b is magnetized in the field of the solenoid and acquires a magnetic moment m. It is then equivalent to a small coil having a magnetic moment m.

The mutual inductance between the particle and the coil C of mean radius a shown in the figure is then the M of Prob. 12-3, with $\pi b^2 N_b I_b = m$.

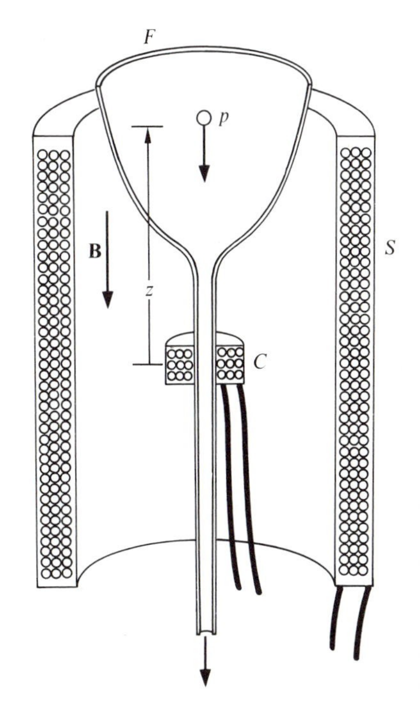

Figure 14-10 Section through a micrometeorite detector. Air is sucked through a funnel F, and a particle p is magnetized in the field $\mathbf{B}$ of a solenoid S. In passing through the tube, the particle induces a voltage in coil C. See Prob. 14-7.

a) Show that the voltage induced in the coil is

$$\frac{3\mu_0 N_a a^2}{2(a^2 + z^2)^{5/2}} zmv,$$

where v is the velocity of the particle.

b) Draw a curve of $z(a^2 + z^2)^{-5/2}$ as a function of z for a = 10 millimeters. The coordinate z should vary from -50 to $+50$ millimeters.

Since the velocity is constant, z is proportional to the time t.

14-8E *MECHANICAL DISPLACEMENT TRANSDUCER*

One often wishes to obtain a voltage that is proportional to the displacement of an object from a known reference position.

If there are no magnetic materials nearby, one can fix to the object a small permanent magnet and measure its field with a Hall generator, as in Fig. 14-11. The Hall element is sensitive only to the x-component of the magnetic field and, from Prob. 14-6,

$$B_x = \frac{3\mu_0 m}{4\pi} \frac{xz}{(x^2 + z^2)^{5/2}}.$$

a) Draw a curve of $xz/(x^2 + z^2)^{5/2}$ as a function of z for $x = 100$ millimeters and for $z = -50$ to $+50$ millimeters.

b) Over what range does B_x deviate by less than 5% from the tangent to the curve at $z = 0$?

The linear region is longer for larger values of x, but B_x decreases rapidly with x.

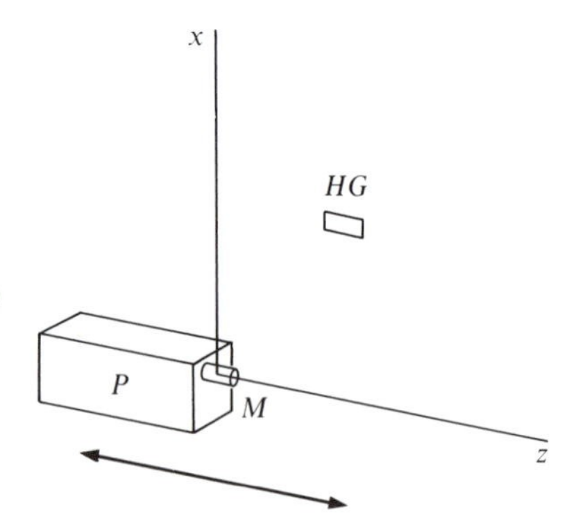

Figure 14-11 Hall generator HG used as a mechanical displacement transducer. The moving part P slides along the z-axis. The Hall generator is oriented so as to monitor the x-component of the magnetic field of the small permanent magnet M fixed to P. See Prob. 14-8.

14-9 *MAGNETIZED DISK*

A thin disk of iron of radius a and thickness t is magnetized in the direction parallel to its axis.

Calculate **H** and **B** on the axis, both inside and outside the iron.

14-10E *TOROIDAL COIL WITH MAGNETIC CORE*

Compare the fields inside two similar toroidal coils, both carrying a current I, one with a non-magnetic core and the other with a magnetic core.

How are the equivalent currents on the magnetic core oriented with respect to those in the coil?

14-11 *EQUIVALENT CURRENTS*

A long wire of radius a carries a current I and is situated on the axis of a long hollow iron cylinder of inner radius b and outer radius c.

a) Compute the flux of B inside a section of the cylinder l meters long.

b) Find the equivalent current density on the inner and outer iron surfaces, and find the direction of these equivalent currents relative to the current in the wire.

c) Find B at distances $r > c$ from the wire.

How would this value be affected if the iron cylinder were removed?

14-12E *THE DIVERGENCE OF* **H**

We have seen in Sec. 8.2 that $\nabla \cdot \mathbf{B}$ is zero. This equation is valid even in non-linear, non-homogeneous, and non-isotropic media.

Set $\mathbf{B} = \mu_r \mu_0 \mathbf{M}$. Under what conditions is $\nabla \cdot \mathbf{H} \neq 0$?

14-13E *THE MAGNETIZATION CURVE*

Use Fig. 14-5 to find the relative permeability of annealed pure iron at a magnetic induction of one tesla.

14-14 *ROWLAND RING*

A ring of ductile cast-iron (Fig. 14-4) has a major radius of 200 millimeters and a minor radius of 10 millimeters. A 500-turn toroidal coil is wound over it.

a) What is the value of B inside the ring when the current through the coil is 2.4 amperes? Use Fig. 14-5.

b) A 10-turn coil is wound over the first one. Calculate the voltage induced in it if the current in the large coil suddenly increases by a small amount at the rate of 10 amperes per second. Assume that μ_r remains constant.

14-15E *THE WEBER AMPERE-TURN*

Show that one weber ampere-turn is one joule.

14-16 *ROWLAND RING*

Figure 14-12 shows how one can measure B as a function of H to obtain the magnetization and hysteresis curves of a ring of magnetic material.

Winding a is fed by an adjustable power supply. As in Sec. 14.8,

$$H = N_a I_a / 2\pi r.$$

The voltage across winding b is integrated with the circuit of Prob. 5-32.

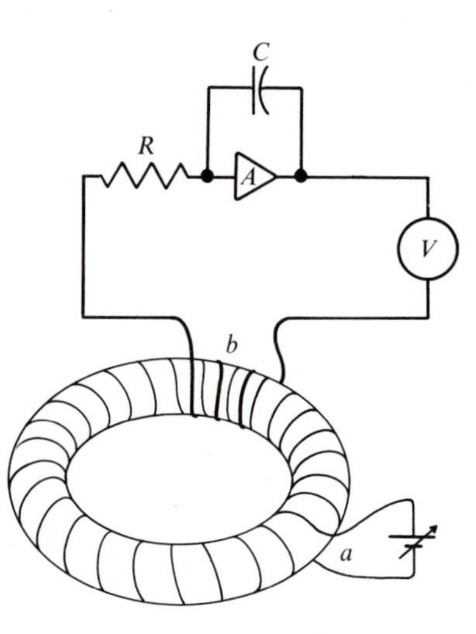

Figure 14-12 Set-up used for measuring B as a function of H in a ring-shaped sample. See Prob. 14-16.

The cross-section of the ring is S and winding b has N_b turns.
Show that

$$B = RCV/N_bS$$

if, at the beginning of the experiment, $B = V = 0$. The integrating circuit draws essentially zero current.

Solution: The voltage across winding b is

$$N_b \frac{d\Phi}{dt} = N_bS \frac{dB}{dt}, \tag{1}$$

since there is zero current in b, and hence no voltage drop caused by the resistance and inductance of the winding.

Then, from Prob. 5-32, disregarding the negative sign,

$$V = \frac{1}{RC} \int_0^t N_bS \frac{dB}{dt}\, dt = N_bSB/RC, \tag{2}$$

and

$$B = RCV/N_bS.$$

14-17 *TRANSFORMER HUM*

Why does a transformer hum?

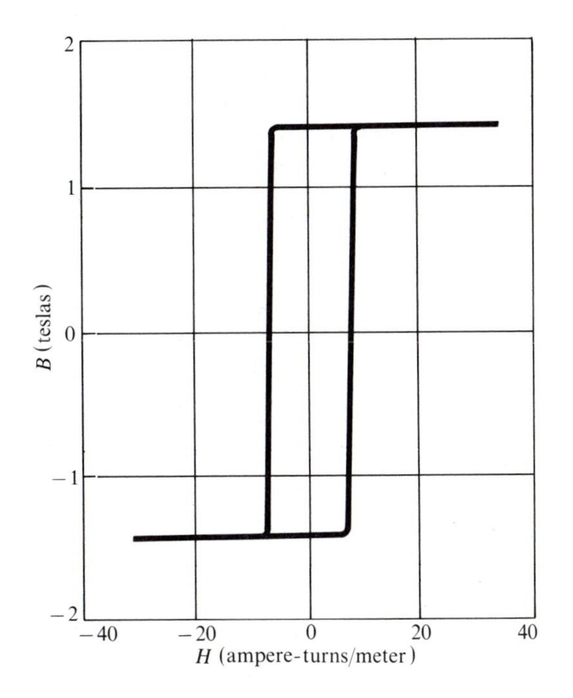

Figure 14-13 Hysteresis loop for the alloy Deltamax. See Prob. 14-18.

14-18E *POWER LOSS DUE TO HYSTERESIS*

Figure 14-13 shows the hysteresis loop for a nickel-iron alloy called Deltamax.

What is the approximate value of the power dissipation per cubic meter and per cycle when the material is driven to saturation both ways?

14-19 *THE FLUXGATE MAGNETOMETER AND THE PEAKING STRIP*

A wire is would around a strip of Deltamax (Fig. 14-13), forming a solenoid. A few turns of wire are then wound over the solenoid, and connected to an oscilloscope.

An alternating current is applied to the solenoid, driving the Deltamax to saturation in both directions.

a) What is the shape of the waveform observed on the oscilloscope?

b) What happens to the pattern on the oscilloscope if a steady magnetic field is now applied parallel to the strip?

It can be shown that, upon application of the steady field, the output voltage contains a component at double the frequency of the applied alternating current and proportional to the magnitude of the steady field. This is the principle of operation of the *fluxgate magnetometer*. Fluxgate magnetometers are often used on board satellites for measuring magnetic fields in outer space. They are useful from about 10^{-10} to 10^{-7} tesla.

The *peaking strip* also utilizes a solenoid wound on a material with a square hysteresis loop like Deltamax (Fig. 14-13), and a secondary winding. It is used differently, however. The peaking strip is placed inside a solenoid whose current is adjusted so as to cancel the ambient field, making the pattern on the oscilloscope symmetrical. The peaking strip is used to measure magnetic inductions in the range of about 10^{-6} to 10^{-2} tesla.

CHAPTER 15

MAGNETIC FIELDS: VIII

Magnetic Circuits

In general, it is not possible to calculate magnetic fields accurately, when magnetic materials are involved. There are several reasons for this. (a) The relation between **H** and **B** for ferromagnetic materials is non-linear. It even depends on the previous history of the material, as we saw in Sec. 14.9. Also, ferromagnetic materials are often non-isotropic. (b) The iron cores that are used to confine and guide the magnetic flux are often quite inefficient. As we shall see, a large part of the flux can be situated outside the core. (c) Permanent magnets are not as simple as one would like them to be: their magnetization **M** is non-uniform and depends on the presence of neighboring magnetic materials.

It is nonetheless necessary to be able to make approximate calculations. The calculation serves to design a model on which magnetic fields or magnetic forces can be measured, and which can then be modified to give the final design.

15.1 MAGNETIC CIRCUITS

Figure 15-1 shows a ferromagnetic core around which is wound a short coil of N turns carrying a current I. We wish to calculate the magnetic flux Φ through the core.

In the absence of ferromagnetic material, the lines of **B** are as shown in the figure. At first sight, one expects the **B** inside the core to be much larger close to the winding than on the opposite side. This is not the case, however, and **B** is of the same order of magnitude at all points within the ferromagnetic material.

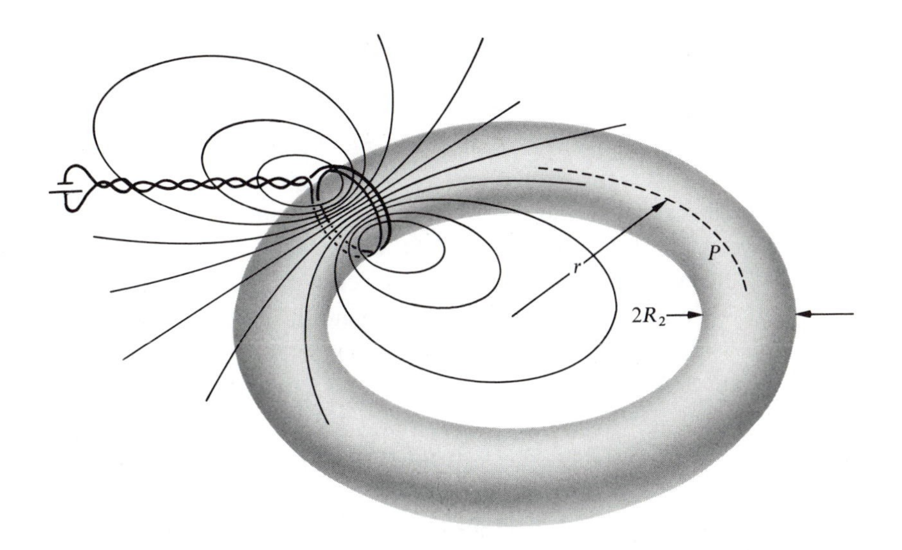

Figure 15-1 Ferromagnetic toroid with concentrated winding. The lines of force shown are in the plane of the toroid and are similar to those of Fig. 8-6. They apply only when there is no iron present.

This can be understood as follows. The magnetic induction of the current I magnetizes the core in the region near the coil, and this magnetization gives equivalent currents that both increase **B** and extend it along the core. This further increases and extends the magnetization, and hence **B**, until the lines of **B** extend all around the core.

Of course some of the lines of **B** escape into the air and then return to the core to pass again through the coil. This constitutes the *leakage flux* that may, or may not, be negligible. For example, if the toroid is made up of a long thin wire, most of the flux leaks across from one side of the ring to the other and the flux at P is negligible compared to that near the coil.

Let us assume that the cross-section of the toroid is large enough to render the leakage flux negligible. Then, applying Ampère's circuital law, Eq. 14-18, to a circular path of radius r going all around inside the toroid,

$$\oint \mathbf{H} \cdot \mathbf{dl} = NI, \tag{15-1}$$

$$2\pi r B/\mu_r \mu_0 = NI, \tag{15-2}$$

$$B = \mu_r \mu_0 NI/2\pi r. \tag{15-3}$$

Taking R_1 to be the radius corresponding to the average value of B, and R_2 to be the minor radius of the toroid,

$$\Phi = \mu_r \mu_0 \pi R_2^2 NI / 2\pi R_1. \tag{15-4}$$

The flux through the core is therefore the same as if we had a toroidal coil (Sec. 9.1.2) of the same size and if the number of ampere-turns were increased by a factor of μ_r. In other words, for each ampere-turn in the coil there are $\mu_r - 1$ ampere-turns in the core. The amplification can be as high as 10^5.

This equation shows that the magnetic flux is given by the magnetomotance NI multiplied by the factor

$$\mu_r \mu_0 \pi R_2^2 / 2\pi R_1,$$

which is called the *permeance* of the magnetic circuit. The inverse of the permeance is called the *reluctance*. Thus

$$\Phi = \frac{NI}{\mathscr{R}} = \frac{NI}{2\pi R_1 / \mu_r \mu_0 \pi R_2^2}, \tag{15-5}$$

and the reluctance is

$$\mathscr{R} = 2\pi R_1 / \mu_r \mu_0 \pi R_2^2. \tag{15-6}$$

Reluctance is expressed in henrys^{-1}.

The analogy with Ohm's law is obvious: if an electromotance $\mathscr{V}$ were induced in the core, the current would be

$$I = \frac{\mathscr{V}}{2\pi R_1 / \sigma \pi R_2^2}, \tag{15-7}$$

where $2\pi R_1 / \sigma \pi R_2^2$ is the resistance of the core.

Thus the corresponding quantities in electric and magnetic circuits are as follows:

Current I	Magnetic flux Φ
Current density $\mathbf{J}$	Magnetic induction $\mathbf{B}$
Conductivity σ	Permeability $\mu = \mu_r \mu_0$

Electromotance $\mathscr{V}$	Magnetomotance NI
Electric field intensity $\mathbf{E}$	Magnetic field intensity $\mathbf{H}$
Conductance $G = 1/R$	Permeance $1/\mathscr{R}$
Resistance R	Reluctance $\mathscr{R}$

There is one important difference between electric and magnetic circuits: the magnetic flux cannot be made to follow a magnetic circuit in the manner that an electric current follows a conducting path. Indeed, a magnetic circuit behaves much as an electric circuit would if it were submerged in tap water: part of the current would flow through the components, and the rest would flow through the water.

If a magnetic circuit is not properly designed, the leakage flux can easily be an order of magnitude *larger* than that flowing around the circuit.

15.2 MAGNETIC CIRCUIT WITH AN AIR GAP

Figure 15-2 shows a circuit with an air gap whose cross-section is different from that of the soft-iron yoke. Each winding provides $NI/2$ ampere-turns. We wish to calculate the magnetic induction in the air gap.

We assume that the leakage flux is negligible. As we shall see, this assumption will result in quite a large error.

Applying Ampère's circuital law to the circuit, we see that

$$NI = H_i l_i + H_g l_g, \tag{15-8}$$

where the subscript i refers to the iron yoke and g to the air gap; l_i and l_g are the path lengths. The path length l_i in the iron can be taken to be the length measured along the center of the cross-section of the yoke.

If we neglect leakage flux, the flux of $\mathbf{B}$ must be the same over any cross-section of the magnetic circuit, so

$$B_i A_i = B_g A_g, \tag{15-9}$$

where A_i and A_g are, respectively, the cross-sections of the iron yoke and of the air gap.

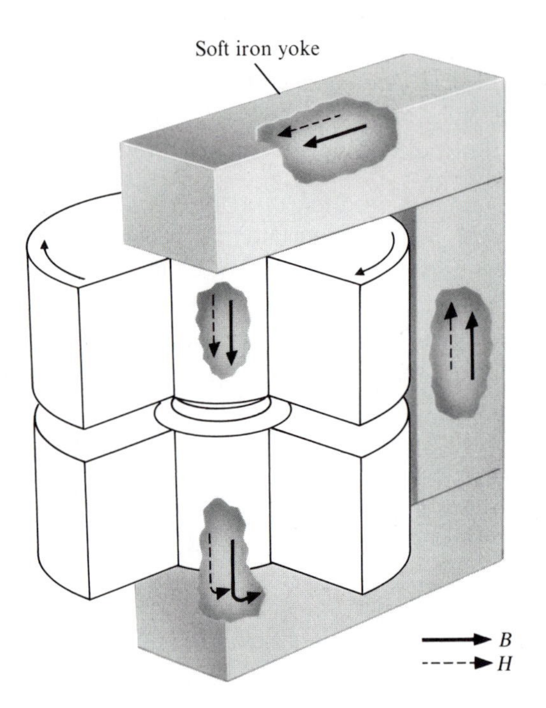

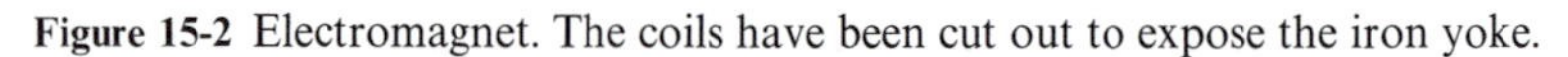
Figure 15-2 Electromagnet. The coils have been cut out to expose the iron yoke.

Combining these two equations,

$$B_g A_g \left[\frac{l_i}{\mu_r \mu_0 A_i} + \frac{l_g}{\mu_0 A_g} \right] = NI, \qquad (15\text{-}10)$$

and the magnetic flux is

$$\Phi = B_g A_g = \frac{NI}{\dfrac{l_i}{\mu_r \mu_0 A_i} + \dfrac{l_g}{\mu_0 A_g}}. \qquad (15\text{-}11)$$

The magnetic flux is therefore equal to the magnetomotance divided by the sum of the reluctances of the iron and of the air gap.

This is a general law: reluctances in series in a magnetic circuit add in the same way as resistances in series in an electric circuit; permeances in parallel add like conductances in parallel.

Since we have neglected leakage flux, the above equation can only provide an upper limit for Φ.

Now let $\mathscr{R}_i = l_i/\mu_r\mu_0 A_i$ be the reluctance of the iron, and $\mathscr{R}_g = l_g/\mu_0 A_g$ be the reluctance of the air gap. Note that

$$H_i l_i = \mathscr{R}_i \Phi, \tag{15-12}$$

$$H_g l_g = \mathscr{R}_g \Phi. \tag{15-13}$$

If $\mathscr{R}_i \ll \mathscr{R}_g$, as is usually the case, since $\mu_r \gg 1$, Eqs. 15-8 and 15-11 become

$$NI \approx H_g l_g, \tag{15-14}$$

$$\Phi \approx NI\mu_0 A_g/l_g. \tag{15-15}$$

15.2.1 *EXAMPLE: ELECTROMAGNET*

Let us calculate B and the stored energy in the electromagnet of Fig. 15-2. If there is a total of 10,000 turns in the two windings, and setting $I = 1.00$ ampere, $A_i = 10^4$ square millimeters, $A_g = 5 \times 10^3$ square millimeters, $\mu_r = 1{,}000$, $l_i = 900$ millimeters, and $l_g = 10$ millimeters, then

$$\Phi = \frac{10^4}{\dfrac{0.9}{10^3 \times 4\pi \times 10^{-7} \times 10^{-2}} + \dfrac{10^{-2}}{4\pi \times 10^{-7} \times 5 \times 10^{-3}}}, \tag{15-16}$$

$$= 6.0 \times 10^{-3} \text{ weber}, \tag{15-17}$$

$$B_g = 6.0 \times 10^{-3}/5 \times 10^{-3} = 1.2 \text{ teslas}. \tag{15-18}$$

The self-inductance is

$$L = N\Phi/I = 10^4 \times 6.0 \times 10^{-3}/1.00 = 60 \text{ henrys}. \tag{15-19}$$

The length l_i is measured along the middle of the cross-section of the yoke. The stored energy is

$$(1/2)LI^2 = (1/2) \times 60 \times 1.00 = 30 \text{ joules}. \tag{15-20}$$

In this particular case the leakage flux is 70% of the flux in the gap. In other words, the magnetic induction in the gap is not 1.2 teslas, but only $1.2/1.7 = 0.71$ tesla.
There exist empirical formulae for estimating leakage flux for simple geometries.

15.3 MAGNETIC FORCES EXERTED BY ELECTROMAGNETS

Electromagnets are commonly used to actuate various mechanisms such as switches, valves, and so forth. A switch activated by an electromagnet is called a *relay*. These electromagnets comprise a coil and an iron core that is made in two parts, one fixed and one movable, called the *armature*, with an air gap between the two. The coil and its core together are usually termed a *solenoid*. The magnetic force is *attractive*.

In the simplest cases, one has a flat air gap, perpendicular to **B**. Then the magnetic force of attraction is $B^2/2\mu_0$ times the cross-section of the gap (Sec. 13.2) where B is calculated as in Sec. 15.2. If the solenoid is energized with direct current, then B is approximately proportional to the inverse of the gap length, and since the force is proportional to B^2, the force increases rapidly with decreasing gap length.

The air gap is often designed so that the attractive force will vary with the gap length in some prescribed way. The design of the magnetic circuit is largely empirical.

15.4 SUMMARY

The concept of *magnetic circuit* is widely used whenever the magnetic flux is guided mostly through magnetic material. We then have a magnetic equivalent of Ohm's law:

$$\Phi = NI/\mathscr{R}, \qquad (15\text{-}5)$$

where NI is the *magnetomotance* of the energizing coil and $\mathscr{R}$ is the *reluctance* of the magnetic circuit.

The correspondence between electric and magnetic circuits is as follows:

Current I	Magnetic Flux Φ
Current density $\mathbf{J}$	Magnetic induction $\mathbf{B}$
Conductivity σ	Permeability $\mu_r\mu_0$
Electromotance $\mathscr{V}$	Magnetomotance NI
Electric field intensity $\mathbf{E}$	Magnetic field intensity $\mathbf{H}$
Conductance	Permeance
Resistance R	Reluctance $\mathscr{R}$

The reluctance of a bar of magnetic material of permeability $\mu_r\mu_0$, cross-section A, and length l is $l/\mu_r\mu_0 A$. Reluctances in series and in parallel are treated like resistances in series and in parallel.

The magnetic force of attraction exerted by an electromagnet is $B^2/2\mu_0$ times the cross-section of the gap.

PROBLEMS

15-1E *RELUCTANCE*

Show that the energy stored in a magnetic circuit is

$$W = (1/2)\Phi^2\mathscr{R},$$

where Φ is the magnetic flux and $\mathscr{R}$ is the reluctance.

15-2E *RELUCTANCE*

Show that the inductance L of a coil of N turns is given by

$$L = N^2/\mathscr{R},$$

where $\mathscr{R}$ is the reluctance of the magnetic circuit.

15-3E *CLIP-ON AMMETER*

It is often useful to be able to measure the current flowing in a wire without disturbing the circuit. This can be done with a clip-on ammeter. As a rule, clip-on ammeters are transformers, as in Prob. 18-16, and can therefore be used only with alternating currents.

It is also possible to make a clip-on ammeter that will measure direct currents as in Fig. 15-3. In this instrument the magnetic flux through the yoke is measured with

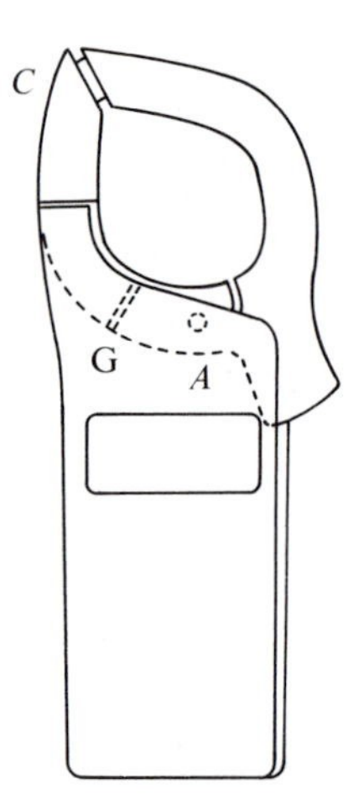

Figure 15-3 Clip-on ammeter for direct currents. A hinge A permits the iron yoke to be opened at C so that it can be clipped around the current-carrying wire. A gap G in the yoke contains either a Hall generator or a magnetoresistor. The magnetic induction in the gap is a measure of the current I. See Prob. 15-3.

a Hall generator (Sec. 10.8.1), or with a magnetoresistor (Prob. 10-15) situated in the gap of length l_g.

a) Show that, if $l_g \gg l_i/\mu_r$, the magnetic induction in the gap is $\mu_0 I/l_g$. What is the advantage of using an iron core?

b) Is this magnetic induction affected by the position of the wire inside the ring?

15-4 *MAGNETIC CIRCUIT*

The ductile cast-iron ring of Prob. 14-14 is cut, leaving an air gap 1 millimeter long. The current in the toroidal coil is again 2.4 amperes.

Calculate B in the air gap, using the relative permeability curve of Fig. 15-4.

You will have to solve this problem by successive approximations. Start by assuming a reasonable value for μ_r, say 500. Find the corresponding B. This will prob-

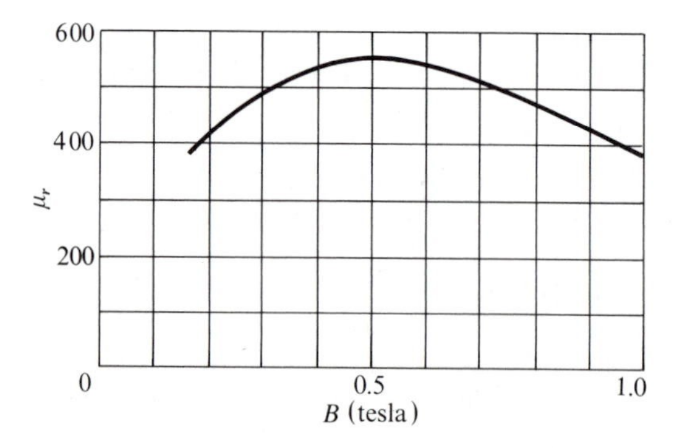

Figure 15-4 Relative permeability as a function of magnetic induction for ductile cast iron. See Prob. 15-4.

ably not give the correct B on the curve. Then try another value of μ_r, etc. Draw a table of μ_r and of the calculated B as a help in selecting your next approximation.

The calculated value of B need not agree to better than 10% with the B on the curve.

15-5E *MAGNETORESISTANCE MULTIPLIER*

The magnetic circuit shown in Fig. 15-5 gives an output voltage V that is proportional to the *product* $I_a I_b$. The voltmeter draws a negligible current.

The air gaps contain magnetoresistances (Prob. 10-15) R_1 and R_2, connected as in Fig. 15-5b. This circuit is identical to that of Prob. 5-7.

From Prob. 10-15,

$$R_1 = R_0[1 + \mathscr{M}^2(B_a + B_b)^2],$$

$$R_2 = R_0[1 + \mathscr{M}^2(B_a - B_b)^2],$$

where $\mathscr{M}$ is the mobility of the charge carriers.

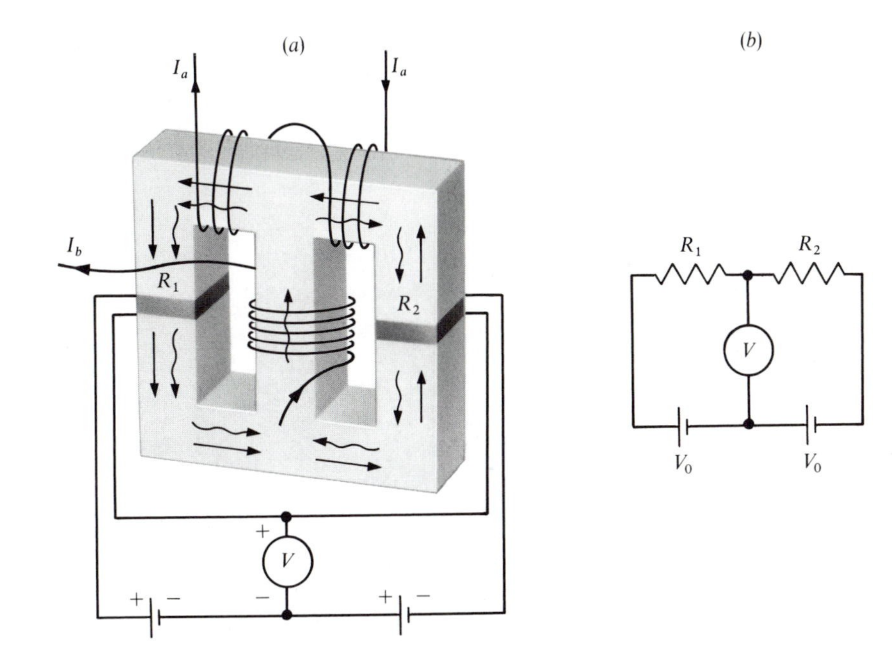

Figure 15-5 (*a*) Magnetic circuit with a pair of magnetoresistances R_1 and R_2 for obtaining a voltage V proportional to the product of I_a and I_b. The straight arrows show the flux through the iron core due to I_a. The wavy arrows show the flux due to I_b. In the left-hand gap, $B = B_a + B_b$, while on the right $B = B_a - B_b$. (*b*) The electric circuit. See Prob. 15-5.

Show that

$$V = -2\mathcal{M}^2 B_a B_b V_0,$$

as long as $\mathcal{M}^2 B_a^2 \ll 1$, $\mathcal{M}^2 B_b^2 \ll 1$.

15-6 *MAGNETIC FORCE ON THE ARMATURE OF AN ELECTROMAGNET*

Figure 15-6 shows a small electromagnet. All dimensions are in millimeters. Each coil has 500 turns and carries a current of 2 amperes.

a) Calculate B in the air gaps as a function of the gap length x. Set $\mu_r = 1000$, and neglect the leakage flux.

Solution: The magnetomotance is

$$NI = 2 \times 2 \times 500 = 2000 \text{ ampere turns.} \tag{1}$$

The iron yoke has a cross-section of 10×10 square millimeters, and a mean length of 2×25 millimeters for the vertical parts, plus 23 millimeters for the horizontal part. Hence its reluctance is

$$\mathcal{R}_y = 73 \times 10^{-3}/(1000 \times 4\pi \times 10^{-7} \times 10^{-4}) = 5.8 \times 10^5 \text{ henry}^{-1}. \tag{2}$$

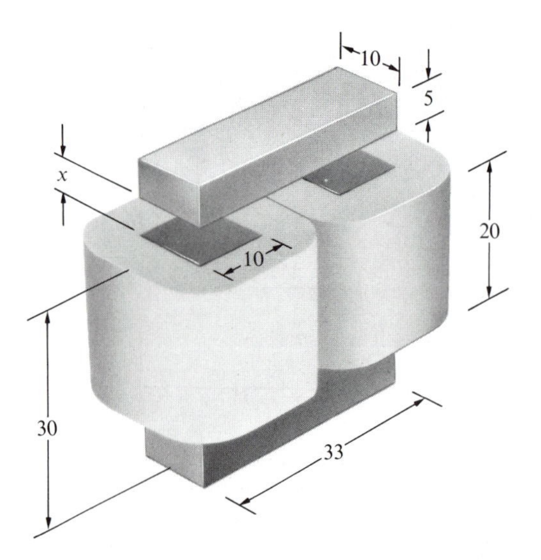

Figure 15-6 Electromagnet. See Prob. 15-6.

The armature has a cross-section of 50 square millimeters. The mean path length of the flux is about 23 millimeters. Hence

$$\mathscr{R}_a = 23 \times 10^{-3}/(1000 \times 4\pi \times 10^{-7} \times 50 \times 10^{-6}) = 3.7 \times 10^5 \text{ henry}^{-1}. \quad (3)$$

Finally, the two air gaps have a reluctance

$$\mathscr{R}_g = 2x/(4\pi \times 10^{-7} \times 10^{-4}) = 1.6 \times 10^{10} x \text{ henry}^{-1}. \quad (4)$$

Thus, in the gaps,

$$B = NI/\mathscr{R}A = \frac{2000}{(5.8 \times 10^5 + 3.7 \times 10^5 + 1.6 \times 10^{10}x)10^{-4}}, \quad (5)$$

$$= \frac{200}{9.5 + 1.6 \times 10^5 x} \text{ teslas.} \quad (6)$$

b) Calculate the force of attraction exerted on the armature when $x =$ 5 millimeters.

Solution: The force is equal to the product of the energy density in the gap, by its cross-section A:

$$F = 2B^2 A/2\mu_0, \quad (7)$$

$$= \frac{4 \times 10^4 \times 10^{-4}}{(9.3 + 1.6 \times 10^5 \times 5 \times 10^{-3})^2 4\pi \times 10^{-7}} \approx 5 \text{ newtons.} \quad (8)$$

c) Would it make sense to use Eq. 6 to calculate the force (i) at $x = 0.1$ millimeter, (ii) at $x = 20$ millimeters?

Solution: At $x = 0.1$ millimeter, Eq. 6 gives a magnetic induction of about 12 teslas. This is much more than the saturation field of one or two teslas for iron (Sec. 14.8). Equation 6 is not valid at $x = 10^{-4}$.

At $x = 20$ millimeters, the reluctance of the air gap is large and the leakage flux is much larger than the flux in the gaps. The actual force is much smaller than that calculated from Eq. 7. Equation 6 is again not valid.

So the magnetic induction found in Eq. 6 is valid only over a limited range of x.

15-7 *RELAY*

Calculate the force of attraction on the armature of a relay such as the one shown in Fig. 15-7 whose magnetic circuit has the following characteristics:

Coil, 10,000 turns; resistance, 1000 ohms; rated voltage, 10 volts.

Gap length, 2 millimeters; gap cross-section, 1 square centimeter.

For simplicity, we assume that the gap has a uniform length, that the reluctance of the iron is negligible when the gap is open, and that there is zero leakage flux.

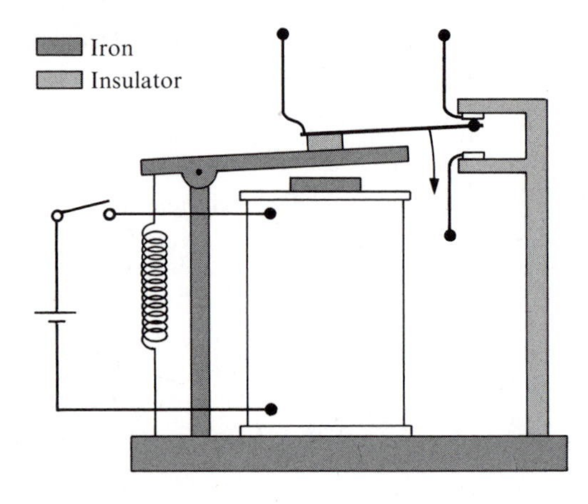

Figure 15-7. Relay. When the coil is energized by closing the switch, the armature falls, opens the upper contact, and closes the lower one. Tens of contacts can be actuated simultaneously in this way. See Prob. 15-7.

15-8 *MAGNETIC FLUIDS*

There exist magnetic fluids consisting of fine particles (≈ 10 nanometers in diameter) of iron oxides suspended in various fluids.†

a) If the magnetic fluid is placed in a non-magnetic container on a non-magnetic support, a permanent magnet placed in it will remain suspended near the bottom, but without touching the container, either at the bottom or at the sides. Why is this?

b) Can you find applications for magnetic fluids?

---

† Manufactured by the Ferrofluidics Corporation, 144 Middlesex Turnpike, Burlington, Massachusetts 02103.

CHAPTER 16

ALTERNATING CURRENTS: I

Complex Numbers and Phasors

Most electric and magnetic devices use alternating or, at least, fluctuating currents. There are many reasons for this, but the two major ones have to do with power technology and with the transmission and processing of information.

First, with *alternating* currents, the electric power supplied by a source at a given voltage can be made available at almost any other convenient voltage by means of transformers. This makes electric power adaptable to a broad variety of uses.

The power supplied by a *direct*-current source can also be changed from one voltage to another. But this is done by first switching the current periodically, to obtain an alternating current, then feeding this to a transformer, and then rectifying and filtering the output. The operation is relatively costly and inefficient.

The second reason for using alternating or fluctuating currents is that they can serve to transmit information. For example, a microphone transforms the information contained in a spoken word into a complex fluctuating current. This current then serves to modulate a radio-frequency current that is fed to an antenna. The antenna launches an electromagnetic wave, and so forth.

A good part of this chapter will be devoted to the method used for solving the differential equations associated with circuits carrying alternating currents. This method requires the use of complex numbers. No previous knowledge of the subject is required.

We shall assume that the voltages, the currents, and the charges are all cosine functions of time, with appropriate phases. This is not always the case. The current through a microphone is not normally sinusoidal. Or one might have only the positive part of the cosine function, or a cosine function whose period is a function of the time. However, any *periodic* function can be analyzed into an infinite series of sine and cosine terms, forming what is called a *Fourier series.*

16.1 ALTERNATING VOLTAGES AND CURRENTS

Voltages are often of the form

$$V = V_0 \cos \omega t, \tag{16-1}$$

where V_0 is the *peak voltage*, $\omega = 2\pi f$ is the *circular frequency*, expressed in radians per second, and f is the *frequency*, expressed in *hertz*. Such voltages are said to be *alternating* (Sec. 11.2.1). We have arbitrarily set the *phase* ωt of V equal to zero at $t = 0$.

If such a voltage is applied to a circuit, then the branch currents are of the form

$$I = I_0 \cos (\omega t + \varphi), \tag{16-2}$$

where I_0 is the peak *current*, $\omega t + \varphi$ is the phase, and φ is the phase of the branch current I with respect to the source voltage V.

It is unfortunately the custom to use the expressions AC *current* and AC *voltage*, where AC stands for Alternating Current.

16.2 RMS, *OR EFFECTIVE VALUES*

If an alternating voltage $V_0 \cos \omega t$ is applied across a resistor R as in Fig. 16-1a,

$$I = \frac{V_0 \cos \omega t}{R} = I_0 \cos \omega t, \tag{16-3}$$

as in Fig. 16-1b. In that case the current is in phase with the voltage, and the peak current I_0 is V_0/R.

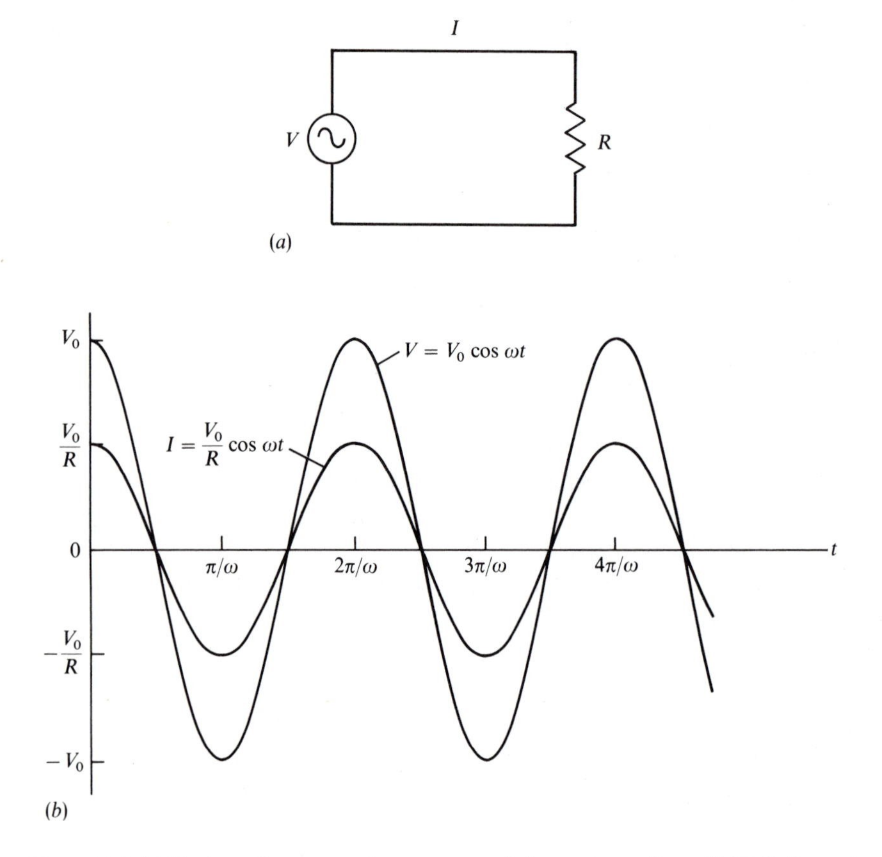

Figure 16-1 (*a*) Resistor R connected to a source of alternating voltage V. (*b*) Voltage V and current I as functions of the time for a resistance of 2 ohms.

The instantaneous power dissipated in the resistor is

$$P_{\text{inst}} = VI = \frac{V_0^2}{R} \cos^2 \omega t = I_0^2 R \cos^2 \omega t. \tag{16-4}$$

Over one cycle, the average power dissipation is

$$P_{\text{av}} = \frac{1}{2} \frac{V_0^2}{R} = \frac{1}{2} I_0^2 R, \tag{16-5}$$

as in Fig. 16-2.

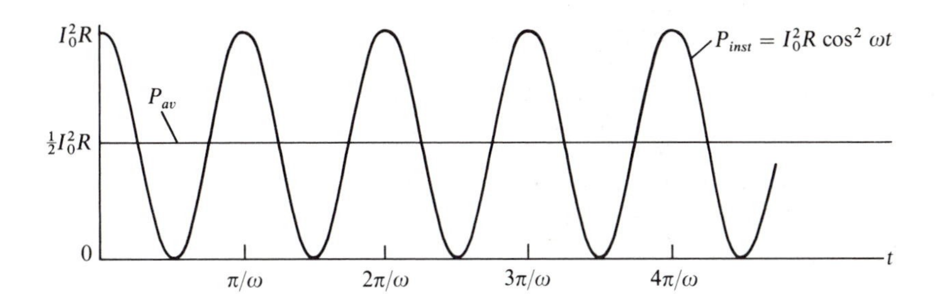

Figure 16-2 Instantaneous and average power dissipation in a resistor.

Unless specified otherwise, one always avoids the factor of $\frac{1}{2}$ by using the *rms*, or *effective* voltage and current,

$$V_{\text{rms}} = V_0/2^{1/2}, \qquad I_{\text{rms}} = I_0/2^{1/2}, \tag{16-6}$$

instead of V_0 and I_0. Here "rms" stands for *root mean square*, or the square root of the mean value of the square of the function. See Prob. 16-2. Then

$$P_{\text{av}} = V_{\text{rms}}^2/R = I_{\text{rms}}^2 R. \tag{16-7}$$

The peak values V_0 and I_0 are $2^{1/2}$ times larger than the rms values, as in Fig. 16-3, as long as V and I are sinusoidal functions of the time.

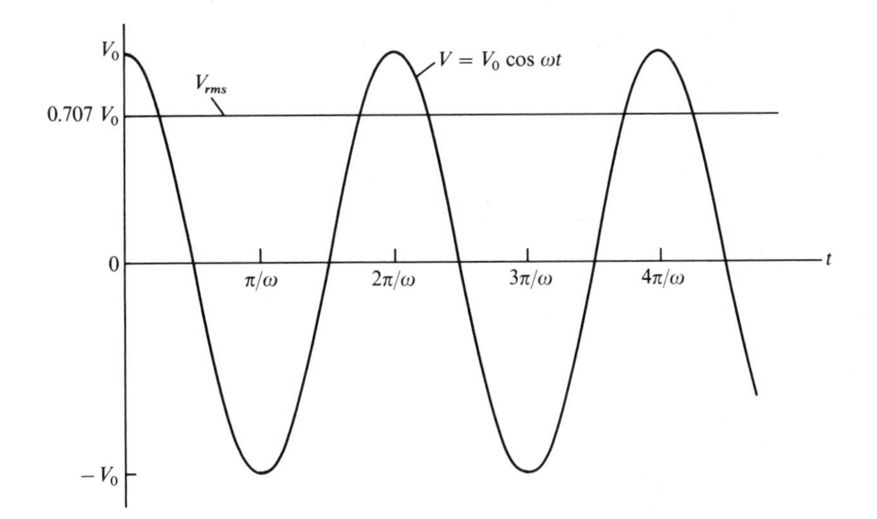

Figure 16-3 Relation between peak and *rms* values for a cosine function.

It is useful to remember that

$$2^{1/2} = 1.414 \text{ and } 1/2^{1/2} = 2^{1/2}/2 = 0.707. \tag{16-8}$$

Unless stated otherwise, numerical values for voltages, currents, and charges are always rms values.

16.2.1 *EXAMPLE: THE OUTPUT VOLTAGE OF AN OSCILLATOR*

An oscillator supplies 10 volts at a pair of terminals, one of which is grounded. Then the potential at the other terminal is $1.414 \times 10 \cos \omega t$. The potential at that terminal varies between -14.14 and $+14.14$ volts with respect to ground.

16.3 *THREE-WIRE SINGLE-PHASE ALTERNATING CURRENT*

As a rule, low-power circuits, such as those used in electronics, utilize alternating voltages and currents as in Sec. 16.1. For higher powers, such as those used in a house, one can proceed differently.

Figure 16-4a shows two alternating current sources connected to two resistors R_1 and R_2 by means of four wires. The signs and the arrows show the polarities and the current directions at a particular instant.

The two center wires of Fig. 16-4a can of course be replaced by a single wire as in Fig. 16-4b, where the center wire carries only the *difference* between the load currents. One then has *three-wire single-phase current.* Thus three wires do the work of four and, moreover, the I^2R loss in the center wire is low, whenever the loads are reasonably well balanced.

In North America, electric utilities maintain 120 volts between A and B, 120 volts between C and B, thus 240 volts between A and C. The potential of point D with respect to ground is not quite zero because of the current flowing through the center wire.

It is the custom to operate low-power devices, such as light bulbs, at 120 volts, but high-power devices, such as baseboard heaters, are operated at 240 volts. For a given power consumption, a high-power device thus draws only half the current it would draw at 120 volts. This permits the use of lighter wire, and reduces the cost of the wiring.

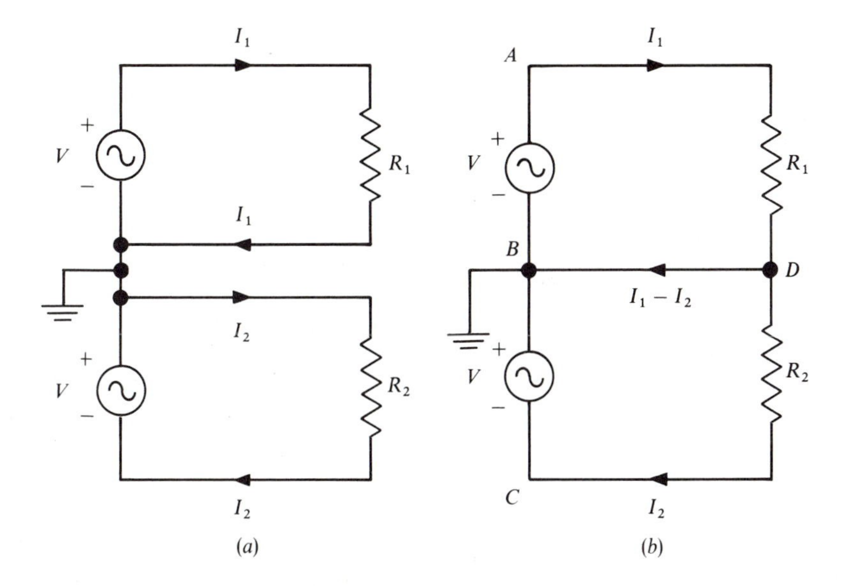

Figure 16-4 (*a*) Two similar sources with opposite phases connected to two resistors R_1 and R_2. (*b*) Three-wire single-phase supply. If the two resistances are equal, the connection *BD* is unnecessary.

16.4 *THREE-PHASE ALTERNATING CURRENT*

Figure 16-5a shows three sources of alternating current feeding resistors R_1, R_2, R_3. The sources are oriented 120° apart, on the figure, so as to indicate their relative phases. For example, if the phase at A is zero at $t = 0$, then

$$V_A = V_0 \cos \omega t, \tag{16-9}$$

$$V_B = V_0 \cos (\omega t + 2\pi/3), \tag{16-10}$$

$$V_C = V_0 \cos (\omega t + 4\pi/3). \tag{16-11}$$

If the three resistances are equal, the total current flowing in the three grounded wires is

$$(V_0/R)[\cos \omega t + \cos (\omega t + 2\pi/3) + \cos (\omega t + 4\pi/3)].$$

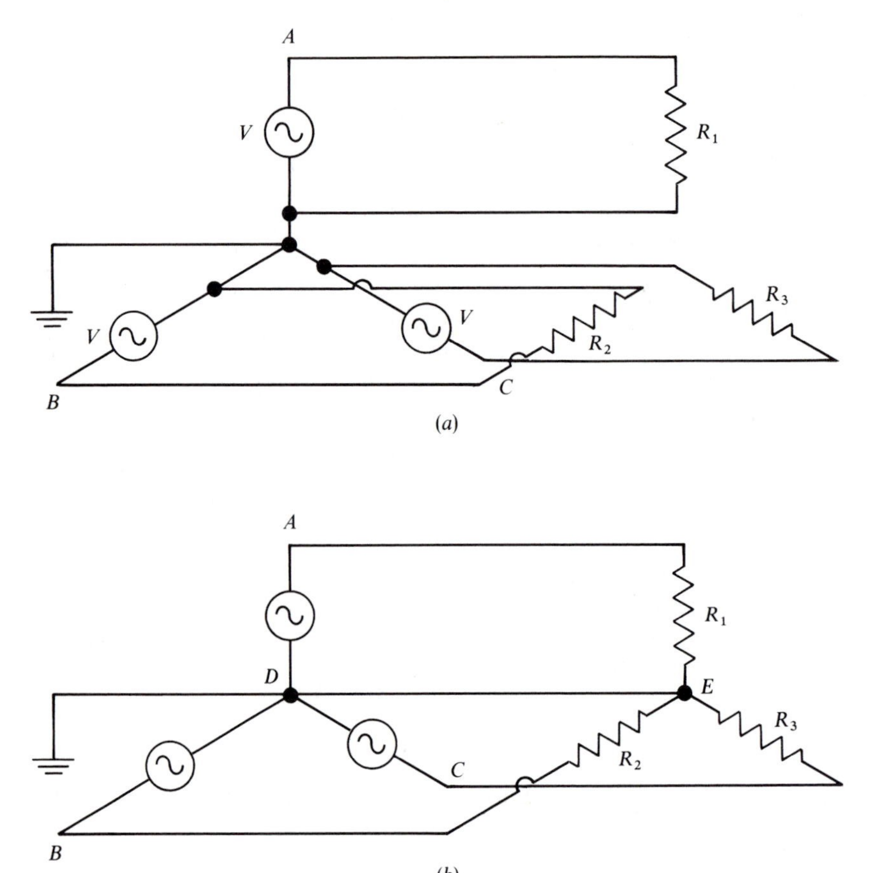

Figure 16-5 (*a*) Three similar sources of alternating current, with phase differences of 120 degrees, connected to three resistors R_1, R_2, R_3. (*b*) Three-phase supply. If the resistances are equal, one may omit the connection *DE*.

It will be shown in Prob. 16-6 that the bracket is always equal to zero. (At $t = 0$, it is equal to $1 - 0.5 - 0.5$.) Then the sources can be connected to the resistances as in Fig. 16-5b and the wire *DE* can be dispensed with.

If the resistances are unequal, the currents do not completely cancel in *DE*. We now have four wires doing the work of six, with low I^2R losses in the wire *DE*.

A set of three sources, star-connected and phased as in Fig. 16-5b, supplies *three-phase alternating current*. The main advantage of three-phase current is that it can be used to produce the revolving magnetic fields that are required for large electric motors. See Prob. 16-7.

Electric power stations generate three-phase currents. This can be seen from the fact that high-voltage transmission lines have either three or six wires, plus one or two light wires.

Except for a few problems at the end of this chapter, and except for the examples in Secs. 16.7.1 and 16.7.3, we shall be concerned henceforth solely with two-wire single-phase currents, as in Fig. 14-4.

16.5 ALTERNATING CURRENTS IN CAPACITORS AND INDUCTORS

Figure 16-6a shows a source V connected to a capacitor C. Then

$$Q/C = V = V_0 \cos \omega t, \tag{16-12}$$

$$I = \frac{dQ}{dt} = C\frac{dV}{dt} = -\omega C V_0 \sin \omega t, \tag{16-13}$$

$$= \omega C V_0 \cos (\omega t + \pi/2). \tag{16-14}$$

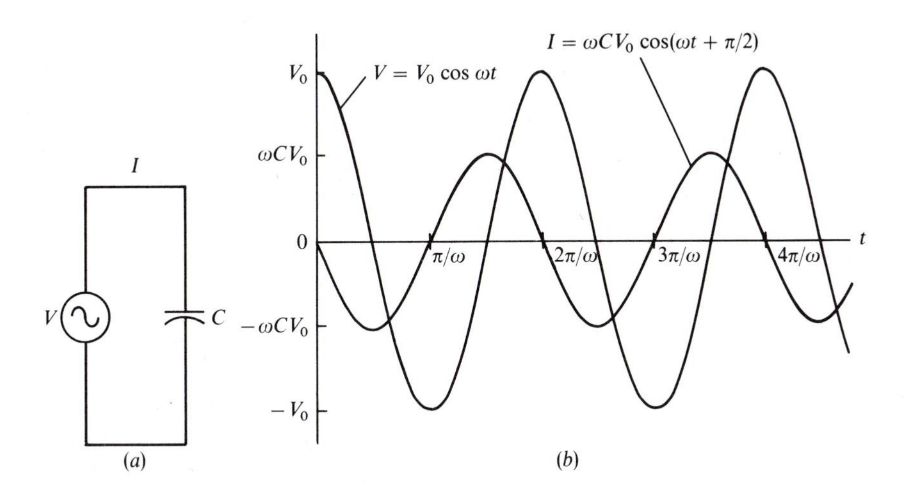

Figure 16-6 (*a*) Capacitor C connected to a source of alternating voltage V. (*b*) Voltage V and current I as functions of the time for a capacitor, with $1/\omega C = 2$ ohms.

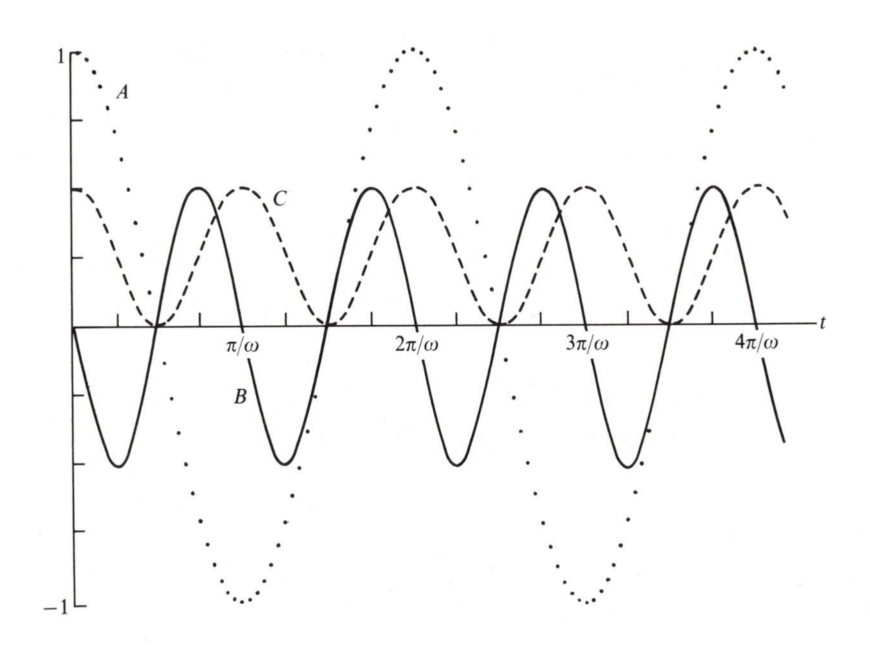

Figure 16-7 Power transfer between a source and a capacitor, as in Fig. 16-6*a*, for $C = \omega = V_0 = 1$. A: applied voltage $V_0 \cos \omega t$ as a function of t. B: $-C\omega V_0^2 \sin \omega t \cdot \cos \omega t$, the power absorbed by the capacitor. C: $(\frac{1}{2})\, CV_0^2 \cos^2 \omega t$, the energy stored in the capacitor.

The current is proportional to the frequency and to C. It *leads* the voltage by $\pi/2$ as in Fig. 16-6b. The current is said to be *in quadrature* with the voltage. This expression is used whenever the phase difference between two quantities is $\pi/2$, leading or lagging.

The instantaneous power absorbed by the capacitor is

$$VI = -\omega C V_0^2 \cos \omega t \sin \omega t. \tag{16-15}$$

This quantity is shown as curve B in Fig. 16-7. It will be observed that power flows alternately into and out of the capacitor. Curve C of the same figure shows the energy $\frac{1}{2}CV^2$ stored in the capacitor as a function of the time. The stored energy oscillates between zero and $\frac{1}{2}CV_0^2$.

Figure 16-8a shows an ideal inductor L with zero resistance connected

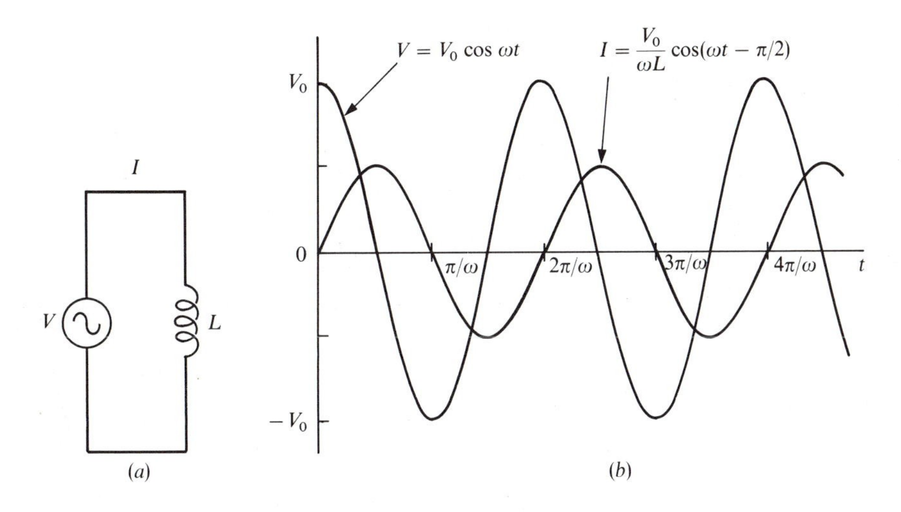

Figure 16-8 (*a*) Inductor L connected to a source V. (*b*) Voltage V and current I as functions of the time for an inductor, with $\omega L = 2$ ohms.

to a source V. From Sec. 12.3,

$$L \frac{dI}{dt} = V = V_0 \cos \omega t, \tag{16-16}$$

$$I = \frac{V_0}{\omega L} \sin \omega t = \frac{V_0}{\omega L} \cos (\omega t - \pi/2). \tag{16-17}$$

We have neglected the constant of integration because we are only interested in the steady-state sine and cosine terms. The current is inversely proportional to the frequency and to L. Figure 16-8b shows V and I as functions of t. The current is again in quadrature with V. It *lags* the voltage by $\pi/2$ radians.

As in the case of the capacitor, the average energy dissipation in the inductor is zero:

$$VI = \frac{V_0^2}{\omega L} \cos \omega t \sin \omega t, \tag{16-18}$$

as in curve B of Fig. 16-9. The stored energy oscillates between zero and $\frac{1}{2}L(V_0/\omega L)^2$, or $\frac{1}{2}LI_0^2$ (Sec. 13.5).

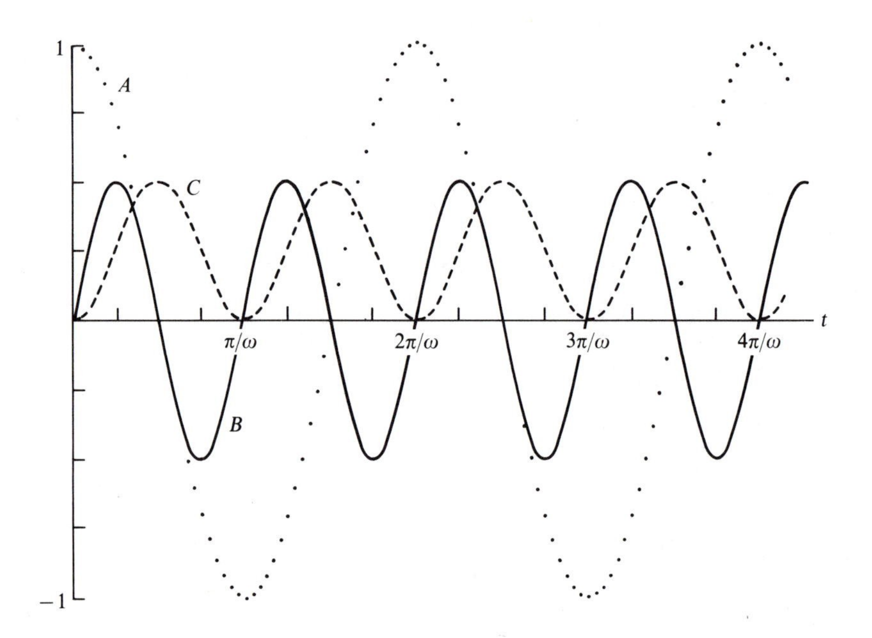

Figure 16-9 Power transfer between a source and an inductor as in Fig. 16-8*a*, for $L = \omega = V_0 = 1$. *A*: applied voltage $V_0 \cos \omega t$ as a function of t. *B*: $(V_0^2/\omega L) \sin \omega t \cos \omega t$, the power absorbed by the inductor. *C*: $(\frac{1}{2})\, L(V_0/\omega L)^2 \sin^2 \omega t$, the energy stored in the inductor.

16.6 COMPLEX NUMBERS

Complex numbers are extensively used for the analysis of electric circuits.

Ordinarily, numbers are used to express quantity, size, etc. For example, one says that there are 20 chairs in a room, or that a building is 30.5 meters high. Such numbers are said to be *real numbers.*

Imaginary numbers are real numbers multiplied by the square root of minus one.

There are two symbols used for $(-1)^{1/2}$. In general, one uses the symbol i. However, whenever one deals with electric and magnetic phenomena, the custom is to use j instead of i because, in the past, the symbol for current was i. The SI symbol for current is now I. Thus $3i$, or $3j$, is an imaginary number.

A *complex number* is the sum of a real number and an imaginary number, like $2 + 3j$, for example. It is useful to represent complex numbers

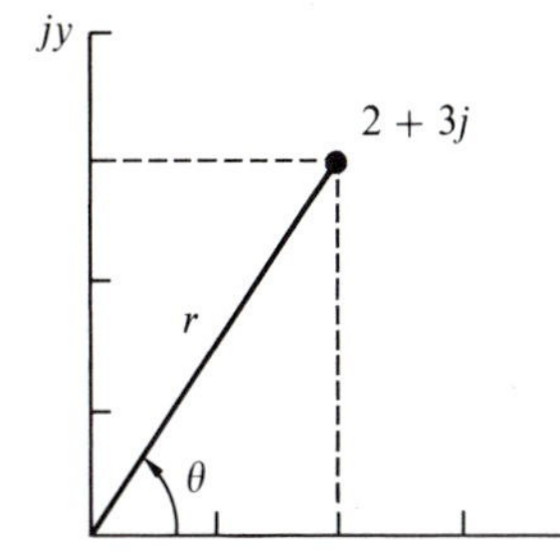

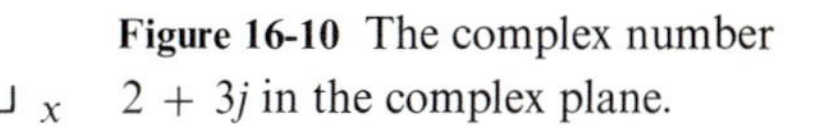

Figure 16-10 The complex number $2 + 3j$ in the complex plane.

in the *complex plane*, as in Fig. 16-10, with the real part (2) plotted horizontally, and the imaginary part ($3j$) vertically.

In Cartesian coordinates, the complex number z is thus written

$$z = x + jy. \tag{16-19}$$

In polar coordinates, from Fig. 16-10,

$$z = r\cos\theta + jr\sin\theta = r(\cos\theta + j\sin\theta), \tag{16-20}$$

where r is the *modulus* of z, and θ is its *argument*. Here

$$r = (x^2 + y^2)^{1/2}, \tag{16-21}$$

and θ is the angle between the radius vector and the x-axis, as in the figure.

If x is positive, z is either in the first or in the fourth quadrant and

$$\theta = \arctan\,(y/x). \tag{16-22}$$

If x is negative, z is in the second or in the third quadrant and

$$\theta = \arctan\,(y/x) + \pi, \tag{16-23}$$

where the first term on the right is assumed to be either in the first or in the fourth quadrant.

Now

$$e^{j\theta} = \cos\theta + j\sin\theta, \tag{16-24}$$

if θ is expressed in *radians*. This rather surprising relation, known as *Euler's formula*, can be checked by writing down the series for $e^{j\theta}$, for $\cos\theta$, and for

$\sin\theta$. This will be done in Prob. 16-12. Therefore, from Eq. 16-20,

$$z = x + jy = re^{j\theta}, \tag{16-25}$$

where x, y, r, θ are related as in Eqs. 16-21, 16-22, and 16-23.

Complex numbers in Cartesian form may be added or subtracted by adding or subtracting the real and imaginary parts separately.

To add or subtract complex numbers in polar form, one first transforms the numbers into Cartesian form.

Complex numbers in Cartesian form may be multiplied in the usual way, remembering that $j^2 = -1$:

$$(a + bj)(c + dj) = (ac - bd) + j(ad + bc). \tag{16-26}$$

In polar form,

$$(r_1 e^{j\theta_1})(r_2 e^{j\theta_2}) = r_1 r_2 e^{j(\theta_1 + \theta_2)}. \tag{16-27}$$

Note that, in the product, the modulus is equal to the product of the moduli, while the argument is the sum of the arguments.

To divide one complex number by another, in Cartesian coordinates, one proceeds as follows:

$$\frac{a + bj}{c + dj} = \frac{(a + bj)(c - dj)}{(c + dj)(c - dj)} = \frac{(ac + bd) + j(bc - ad)}{c^2 + d^2}. \tag{16-28}$$

Division in polar coordinates is simple:

$$\frac{r_1 e^{j\theta_1}}{r_2 e^{j\theta_2}} = \frac{r_1}{r_2} e^{j(\theta_1 - \theta_2)}. \tag{16-29}$$

In the quotient, the modulus is equal to the quotient of the moduli, and the argument is that of the numerator, minus that of the denominator.

16.6.1 *EXAMPLES: USING COMPLEX NUMBERS*

a) From Eq. 16-24,

$$e^{j\pi/2} = j, \qquad e^{j\pi} = -1, \qquad e^{-j\pi/2} = -j. \tag{16-30}$$

b) Let us express the complex number $2 + 3j$ in polar form:

$$2 + 3j = (2^2 + 3^2)^{1/2} \exp j\left(\arctan \frac{3}{2}\right) = 3.61 \exp(0.983j). \tag{16-31}$$

Remember that the angle θ must be expressed in *radians*.

c) The square of $2 + 3j$ is

$$(2 + 3j)^2 = (2 + 3j)(2 + 3j) = 4 - 9 + 12j = -5 + 12j. \tag{16-32}$$

In polar form,

$$(2 + 3j)^2 = 3.61^2 \exp(2 \times 0.983j) = 13 \exp(1.966j). \tag{16-33}$$

d) Finally, let us calculate $(2 + 3j)/(1 + j)$, first in Cartesian and then in polar coordinates.

$$\frac{2 + 3j}{1 + j} = \frac{(2 + 3j)(1 - j)}{(1 + j)(1 - j)} = \frac{(2 + 3) + (3 - 2)j}{2}, \tag{16-34}$$

$$= 2.5 + 0.5j, \tag{16-35}$$

or

$$\frac{2 + 3j}{1 + j} = \frac{3.61 \exp(0.983j)}{1.414 \exp[(\pi/4)j]} = 2.55 \exp(0.197j). \tag{16-36}$$

16.7 PHASORS

When dealing with alternating currents and voltages, one must differentiate sine and cosine functions with respect to time over and over again. Now the usual procedure for differentiating these functions is inconvenient. For example, to differentiate $\cos \omega t$, the cosine function is changed to a sine, the result is multiplied by ω, and the sign is changed. To find the second derivative, the sine is changed back to a cosine, and the result is again multiplied by ω, but this time without changing sign, and so on.

It is possible to simplify these operations in the following way.

First, remember that

$$e^{j\omega t} = \cos \omega t + j \sin \omega t, \tag{16-37}$$

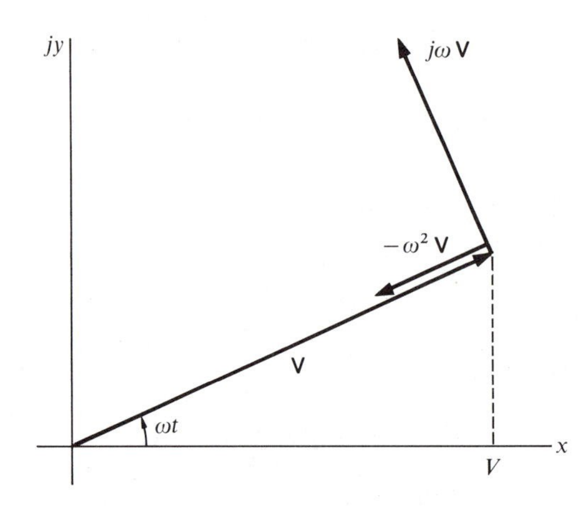

Figure 16-11 The phasors $\mathbf{V}$, $j\omega\mathbf{V}$, and $-\omega^2\mathbf{V}$, and the voltage V in the complex plane. Note that the phasor rotates about the origin at the angular velocity ω. We have made ω equal to 1/2. The quantities $d\mathbf{V}/dt$ and $d^2\mathbf{V}/dt^2$ are the projections of $j\omega\mathbf{V}$ and $-\omega^2\mathbf{V}$ on the real axis.

from Eq. 16-24. Then

$$V = V_0 \cos \omega t = \mathrm{Re}\,(V_0 e^{j\omega t}), \tag{16-38}$$

where the operator Re means "real part of" what follows.

Remember also that the time derivative of the exponential function reduces to a simple multiplication:

$$\frac{d}{dt}(V_0 e^{j\omega t}) = j\omega(V_0 e^{j\omega t}). \tag{16-39}$$

Consider now the quantity

$$\mathbf{V} = V_0 e^{j\omega t} = V_0 \cos \omega t + jV_0 \sin \omega t. \tag{16-40}$$

This quantity, which is called a *phasor*, is a vector that rotates at the angular velocity ω in the complex plane, and its projection on the real axis is V, as in Fig. 16-11. In differentiating $\mathbf{V}$, the real and the imaginary parts remain separate:

$$\frac{d\mathbf{V}}{dt} = -\omega V_0 \sin \omega t + j\omega V_0 \cos \omega t, \tag{16-41}$$

$$\frac{d^2\mathbf{V}}{dt^2} = -\omega^2 V_0 \cos \omega t - j\omega^2 V_0 \sin \omega t. \tag{16-42}$$

The same applies to all the derivatives. Thus

$$\frac{dV}{dt} = \mathrm{Re}\,\frac{d\mathbf{V}}{dt}, \qquad \frac{d^2V}{dt^2} = \mathrm{Re}\,\frac{d^2\mathbf{V}}{dt^2}, \cdots. \tag{16-43}$$

One therefore replaces the real voltage

$$V = V_0 \cos \omega t \tag{16-44}$$

by the phasor

$$\mathbf{V} = V_0 e^{j\omega t}. \tag{16-45}$$

This then adds to V the parasitic term $jV_0 \sin \omega t$. From then on, the *operator* d/dt can be replaced by the *factor* $j\omega$. In the end, one recovers the real part. In this way, differential equations become simple algebraic equations.

Inversely, an integration with respect to time is replaced by a division by $j\omega$.

Figure 16-11 also shows $j\omega\mathbf{V}$ and $-\omega^2\mathbf{V}$. Note that multiplying a phasor by $j\omega$ increases its argument by $\pi/2$ radians and multiplies its modulus by ω.

Phasors are used for V, I, and Q. They are also widely used to represent all kinds of sinusoidally varying quantities.

We use bold-faced sans serif letters for phasors (for example, $\mathbf{V}$), but it is more common to use ordinary light-faced italic letters. For example, the letter V usually means either $V_0 \cos \omega t$ or $V_0 \exp j\omega t$, depending on the context. Moreover V often stands for $V_0/2^{1/2}$! Surprisingly enough, this seldom leads to confusion.†

16.7.1 *EXAMPLE: THE PHASORS IN A THREE-PHASE SUPPLY*

In Fig. 16-5, the voltages at A, B, C can be written either as in Eqs. 16-9 to 16-11, or as follows.

$$\mathbf{V}_A = V_0 \exp j\omega t, \tag{16-46}$$

$$\mathbf{V}_B = V_0 \exp j(\omega t + 2\pi/3), \tag{16-47}$$

† In a handwritten text, one can identify a phasor by means of a wavy line under the symbol, for example, $\underset{\sim}{\mathbf{A}}$.

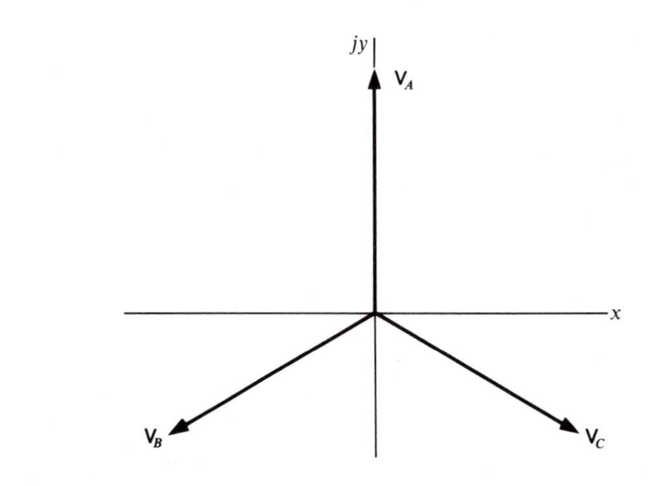

Figure 16-12 The phasors $\mathbf{V}_A$, $\mathbf{V}_B$, $\mathbf{V}_C$ for the voltages at *A*, *B*, *C* in Fig. 16-5, at $\omega t = \pi/2$.

$$\mathbf{V}_C = V_0 \exp j(\omega t + 4\pi/3). \quad (16\text{-}48)$$

These three phasors are shown in Fig. 16-12 at $\omega t = \pi/2$.

16.7.2 *ADDITION AND SUBTRACTION OF PHASORS*

The real quantity V, or I, or Q, that is represented by a phasor is given by the projection of the phasor on the x-axis. Thus, two voltages, or two currents, or two charges, can be added by first adding their phasors vectorially, and then taking the projection of the sum on the x-axis. This is because the projection of the sum of two vectors is equal to the sum of the projections. Similarly, the difference between two voltages, or two currents, or two charges, can be obtained from the difference between the corresponding phasors.

16.7.3 *EXAMPLE: THE PHASORS IN A THREE-PHASE SUPPLY*

Figure 16-13 shows the three phasors $\mathbf{V}_A$, $\mathbf{V}_B$, $\mathbf{V}_C$ of Fig. 16-12 at $t = 0$. We wish to find the voltage of A with respect to B, or $V_A - V_B$, as a function of t, knowing that V_A is $V_0 \cos \omega t$.

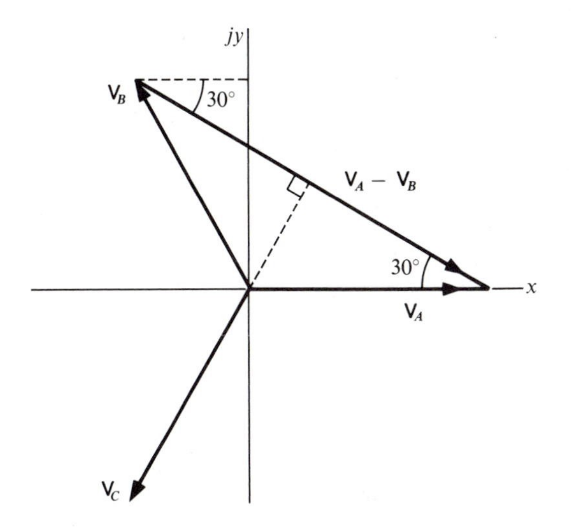

Figure 16-13 Phasor diagram for calculating $\mathbf{V}_A - \mathbf{V}_B$ for the star-connected sources of Fig. 16-5.

We could, of course, find $V_A - V_B$ by trigonometry, since

$$V_A - V_B = V_0 \cos \omega t - V_0 \cos (\omega t + 2\pi/3). \qquad (16\text{-}49)$$

Is it easier and more instructive to use phasors. One can see immediately, from Fig. 16-13, that the phasor $\mathbf{V}_A - \mathbf{V}_B$ has a modulus of $2(3^{1/2}/2)V_0$ and an argument of -30 degrees, or $-\pi/6$ radian at $t = 0$. Thus

$$V_A - V_B = 3^{1/2} V_0 \cos (\omega t - \pi/6). \qquad (16\text{-}50)$$

If the rms voltage at A, B, C is 120 volts, the rms voltage between A and B is $3^{1/2} \times 120$, or 208 volts.

Figures such as 16-13 that serve to perform calculations on phasors are known as *phasor diagrams.*

16.7.4 WHEN NOT TO USE PHASORS

a) One can use phasors *only* if the time dependence is strictly of the form $\cos (\omega t + \varphi)$, where φ is a constant. Phasors must *not* be used if one has, say, a square wave for V, or even a damped sine wave. They must *not* be used if the components are non-linear, for example, if one has non-linear resistors as in Sec. 5.3.

b) Whenever products of phasors, or products of $e^{j\omega t}$ terms are encountered, one reverts to the sine or cosine functions.

For example, suppose one wishes to calculate the power dissipated in a resistor. The instantaneous power is VI, with

$$V = V_0 \cos \omega t, \qquad I = I_0 \cos \omega t. \tag{16-51}$$

If one uses the phasors

$$\mathbf{V} = V_0 e^{j\omega t}, \qquad \mathbf{I} = I_0 e^{j\omega t}, \tag{16-52}$$

one is tempted to conclude that

$$\mathbf{VI} = V_0 I_0 e^{2j\omega t} \quad \text{and that} \quad VI = V_0 I_0 \cos 2\omega t. \tag{16-53}$$

This is *wrong*, as one can see from Sec. 16.2.

The error, here, comes from the fact that

$$\mathrm{Re}\,(e^{j\omega t})\,\mathrm{Re}\,(e^{j\omega t}) \neq \mathrm{Re}\,(e^{2j\omega t}). \tag{16-54}$$

One can divide a phasor by another phasor or by a complex number, however, as we shall see in the next chapter. In those cases one has straightforward operations with complex numbers.

16.8 *SUMMARY*

Alternating voltages and currents are of the form

$$V = V_0 \cos \omega t, \tag{16-1}$$

$$I = I_0 \cos (\omega t + \varphi). \tag{16-2}$$

We have assumed here that the phase of V is zero at $t = 0$. In general, the current I is not in phase with the applied voltage. Unless specified otherwise, one does not use V_0, but rather

$$V_{rms} = V_0/2^{1/2} \tag{16-6}$$

as a measure of the magnitude of a voltage. The same applies to a current: one uses

$$I_{\text{rms}} = I_0/2^{1/2} \tag{16-6}$$

to specify the magnitude of a current.

In *three-wire single-phase current*, one uses two sources, back to back, with a common center wire. With *three-phase current*, one has three sources with phase differences of 120 degrees, connected to their loads with three wires, plus a common center wire.

In a capacitor C, the current is proportional to ωC and *leads* the applied voltage by $\pi/2$ radians. In an inductor L, the current is inversely proportional to ωL and *lags* the voltage by $\pi/2$ radians. In both cases energy flows alternately in and out of the element and there is zero average energy transfer.

Complex numbers are of the form $a + bj$, where a and b are real, and $j = (-1)^{1/2}$. To plot complex numbers in the complex plane, one writes

$$z = x + jy, \tag{16-19}$$

or

$$z = r(\cos\theta + j\sin\theta), \tag{16-20}$$

where r is the *modulus*, and θ is the *argument* of z, as in Fig. 16-10, and

$$r = (x^2 + y^2)^{1/2}, \tag{16-21}$$

$$\theta = \arc\tan\frac{y}{x} \quad \text{or} \quad \theta = \arc\tan\frac{y}{x} + \pi. \qquad (16\text{-}22, 16\text{-}23)$$

Also,

$$e^{j\theta} = \cos\theta + j\sin\theta, \tag{16-24}$$

so that

$$z = re^{j\theta}. \tag{16-25}$$

Complex numbers may be added, subtracted, multiplied, and divided.

A *phasor* is a complex number whose real part is a cosine function of the time. For example,

$$\mathbf{V} = V_0 e^{j\omega t} = V_0 \cos \omega t + jV_0 \sin \omega t. \tag{16-40}$$

The addition of the parasitic term $jV_0 \sin \omega t$ to the real voltage $V_0 \cos \omega t$ transforms the cosine function into an exponential function. Differentiation with respect to time can then be replaced by a multiplication by the factor $j\omega$. Inversely, an integration with respect to time becomes a division by $j\omega$.

PROBLEMS

16-1E *RECTIFIER CIRCUITS COMPARED*

Figures 16-14a and b show, respectively, *half-wave* and *full-wave* rectifier circuits. What are the waveforms across the resistances?

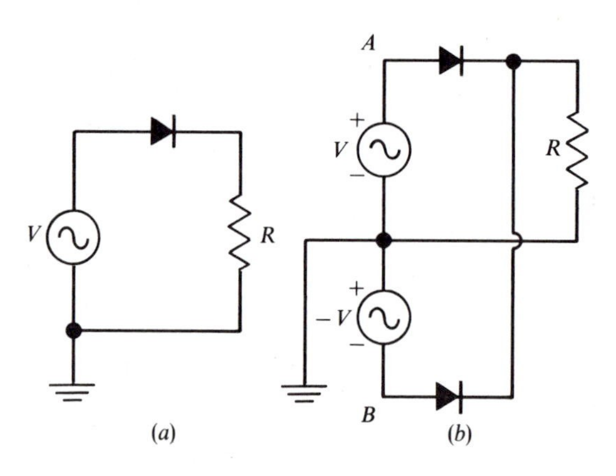

Figure 16-14 (*a*) Half-wave rectifier circuit. The triangular figure at the top represents a rectifier, or diode. Current can pass through it only in the direction of the arrow head. (*b*) Full-wave rectifier circuit. The signs show the polarities at a given instant: points *A* and *B* have opposite phases. See Prob. 16-1.

16-2E RMS *VALUE OF A SINE WAVE*

Show that the root mean square value of $V_0 \cos \omega t$ is $V_0/2^{1/2}$.

16-3

BRIDGE RECTIFIER CIRCUIT

The voltage $V_A - V_B$ applied to the circuit of Fig. 16-15a is a symmetrical saw-tooth wave as in Fig. 16-15b, with a peak voltage of 100 volts. The diodes in AC, AD, etc., pass current only in the direction of the arrow head.

Battery chargers use either this circuit or those of Fig. 16-14. The resistance shown in the figures then represents the battery. The source for a battery charger is a transformer connected to AC power line, and V_0 is about 20 volts for a 12-volt battery.

a) Calculate the average and rms values of $V_A - V_B$ in Fig. 16-15b.

Solution: The average value of $V_A - V_B$ is zero.
The rms value is given by

$$(V_A - V_B)^2_{\text{rms}} = \frac{1}{2}\int_0^2 (V_A - V_B)^2\, dt. \qquad (1)$$

The limits of integration cover one period. To evaluate this integral, we split the time interval into four equal parts of 0.5 second, giving four integrals:

$$\frac{1}{2}\int_0^{0.5} (V_A - V_B)^2\, dt + \frac{1}{2}\int_{0.5}^{1} (V_A - V_B)^2\, dt + \cdots .$$

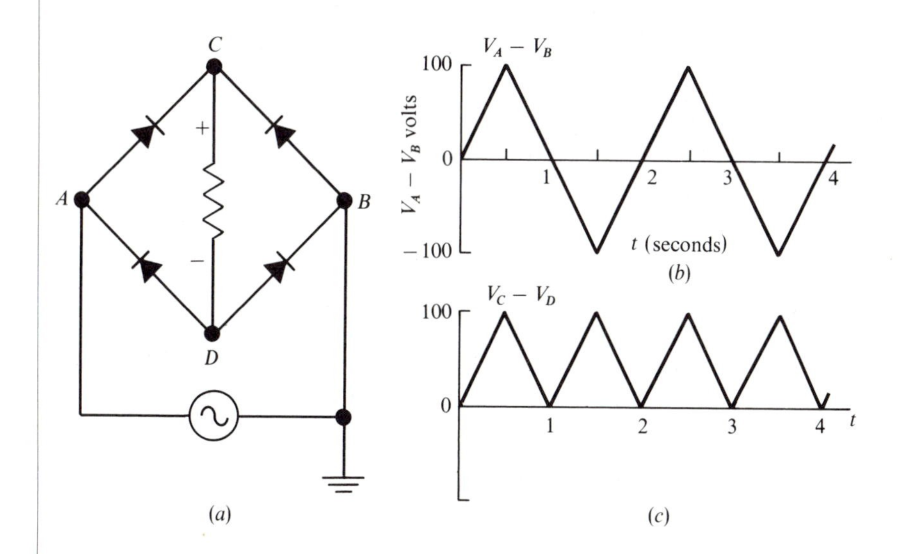

Figure 16-15 (*a*) Bridge rectifier circuit. (*b*) Saw-tooth voltage applied between points A and B. (*c*) Voltage between points C and D. See Prob. 16-3.

Now these integrals are all equal and

$$(V_A - V_B)^2_{\text{rms}} = \frac{4}{2}\int_0^{0.5}\left(\frac{100}{0.5}\,t\right)^2 dt, \tag{2}$$

$$= 2 \times 4 \times 10^4\left[\frac{t^3}{3}\right]_0^{0.5} = \frac{10^4}{3}, \tag{3}$$

$$(V_A - V_B)_{\text{rms}} = 100/3^{1/2} = 57.7 \text{ volts.} \tag{4}$$

b) Draw the curve of $V_C - V_D$ as a function of the time.

Solution: Consider the interval 0–1 second in Fig. 16-15b, where A is positive with respect to B. During this time the current flows along the path $ACDB$ and $V_C - V_D$ is positive as in Fig. 16-15c.

During the interval 1–2 seconds, A is negative with respect to B and the current flows along $BCDA$. Thus $V_C - V_D$ is again positive, as in Fig. 16-15c, and so on.

c) What are the average and rms values of $V_C - V_D$?

Solution: The average value is 50 volts.
The rms value is the same as for $V_A - V_B$, or 57.7 volts.

16-4 RMS *VALUE*

Find the rms values for the waveforms shown in Fig. 16-16.

16-5E *THREE-WIRE SINGLE-PHASE CURRENT*

What is the maximum value of the voltage difference between points A and C in Fig. 16-4b, in the 120/240 volt system commonly used for interior wiring?

16-6E *THREE-PHASE CURRENT*

Show that, for any time t,

$$\cos \omega t + \cos(\omega t + 2\pi/3) + \cos(\omega t + 4\pi/3) = 0.$$

16-7 *ROTATING MAGNETIC FIELD*

Three identical coils, oriented as in Fig. 16-17, produce magnetic fields as follows:

$$B_a = B_0 \cos \omega t, \qquad B_b = B_0 \cos(\omega t + 2\pi/3), \qquad B_c = B_0 \cos(\omega t + 4\pi/3).$$

Show that the resulting field has a magnitude of $1.5B_0$ and rotates at an angular velocity ω.

This is the principle utilized to obtain rotating magnetic fields in large electric motors.

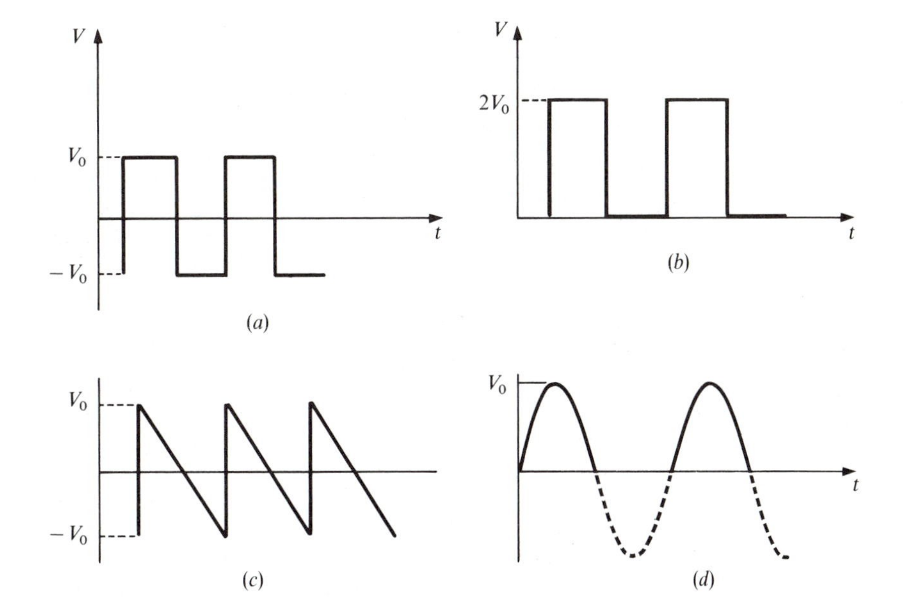

Figure 16-16 See Prob. 16-4.

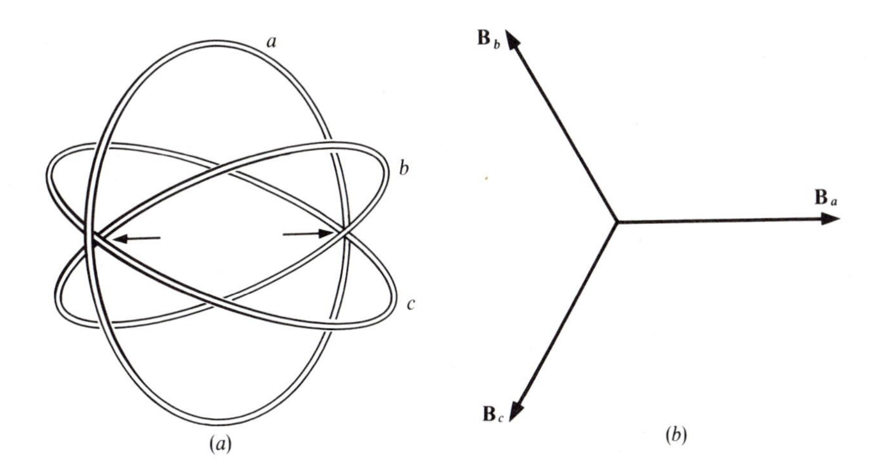

Figure 16-17 (*a*) Three coils, fed with alternating currents of the same amplitude and same frequency, produce magnetic fields of equal magnitudes and oriented as in (*b*). The two arrows in (*a*) show where the coils touch. If the currents are phased as in Prob. 16-7, the direction of the resulting magnetic field rotates at the frequency f.

16-8E *DIRECT-CURRENT HIGH-VOLTAGE TRANSMISSION LINES*

From many points of view, alternating current is much preferable to direct current for power distribution. However, as we shall see, line losses are lower with direct current.

Consider a long high-voltage overhead transmission line. Its maximum operating voltage depends on several factors, such as corona losses (Prob. 7-14), the size of the insulators, etc. So, for a given line, the instantaneous voltage between one conductor and ground must never exceed a certain value, say V_0. Otherwise, the down time and the cost of maintenance become excessive. The cost of a line increases rapidly with its voltage rating.

The current in the line can be made nearly as large as one likes without damaging it, since the conductors are well cooled by the ambient air. However, the power loss increases as the square of the current. So, the lower the current the better.

With direct current, one has two conductors operating at $+V_0$ and $-V_0$ with respect to ground. The power delivered to the load is $2V_0I_{DC}$.

a) With single phase alternating current, we have two wires at $+V_0 \cos \omega t$ and $-V_0 \cos \omega t$.

Show that, for the same power at the load, the rms current I_{sp} is $2^{1/2}I_{DC}$.

The I^2R losses in the line are twice as large as those with direct current.

b) With three-phase alternating current, we have three wires at $V_0 \cos \omega t$, $V_0 \cos (\omega t + 2\pi/3)$, $V_0 \cos (\omega t + 4\pi/3)$. We assume that the three load resistances connected between these wires and ground are equal. Then the current in the ground wire is zero.

Show that, for the same total power delivered to the three load resistances, the rms currents I_{tp} are $(2/3)2^{1/2}I_{DC} \approx I_{DC}$.

With three-phase alternating current, the rms currents are about the same as with direct current, but we have three current-carrying wires instead of two, so that the losses are 50% larger than with direct current.

16-9 *ELECTROMAGNET OPERATING ON ALTERNATING CURRENT*

An electromagnet with a variable gap length is operated on alternating current.

Show that the rms value of the magnetic flux is independent of the gap length, for a given applied alternating voltage, and neglecting leakage flux.

16-10E *COMPLEX NUMBERS*

a) Express the complex numbers $1 + 2j$, $-1 + 2j$, $-1 - 2j$, $1 - 2j$ in polar form.

b) Simplify the following expressions, leaving them in Cartesian coordinates:

$$(1 + 2j)(1 - 2j), \qquad (1 + 2j)^2,$$

$$1/(1 + 2j)^2, \qquad (1 + 2j)/(1 - 2j).$$

16-11E *COMPLEX NUMBERS*

Complex numbers in polar form are often written as $r\angle\theta$, where r is the modulus and θ is the argument.

Express $1 + 2j$ in this way.

16-12E *COMPLEX NUMBERS*

Use the series

$$e^x = 1 + x + (x^2/2!) + (x^3/3!) + \cdots$$

$$\sin x = x - (x^3/3!) + (x^5/5!) - (x^7/7!) + \cdots$$

$$\cos x = 1 - (x^2/2!) + (x^4/4!) - (x^6/6!) + \cdots$$

to show that

$$e^{jx} = \cos x + j \sin x.$$

16-13E *COMPLEX NUMBERS*

Show that $\exp j\pi = -1$.

16-14E *COMPLEX NUMBERS*

What happens to a complex number in the complex plane when it is (a) multiplied by j, (b) multiplied by j^2, (c) divided by j?

16-15 *THREE-PHASE ALTERNATING CURRENT*

Show, by trigonometry, that the expressions on the right-hand side of Eqs. 16-49 and 16-50 are equal.

16-16 *CALCULATING AN AVERAGE POWER WITH PHASORS*

In Sec. 16.7.4 we saw that, with alternating currents, one must *not* calculate a power by using the product $\mathbf{VI}$.

a) Show that, for the general case where

$$V = V_0 \cos \omega t, \qquad I = I_0 \cos(\omega t + \varphi)$$

in a load, the average power is

$$P_{av} = (1/2) V_0 I_0 \cos \varphi = V_{rms} I_{rms} \cos \varphi.$$

The term $\cos \varphi$ is called the *power factor* of the load (Sec. 18.2).

b) Show that

$$P_{av} = (1/2) Re \mathbf{VI}^*,$$

where $\mathbf{I}^*$ is the *complex conjugate* of $\mathbf{I}$, obtained by changing the sign of its imaginary part.

CHAPTER 17

ALTERNATING CURRENTS: II

Impedance, Kirchoff's Laws, Transformations

This chapter roughly parallels Chapter 5, most of which dealt with direct currents flowing through resistive circuits. Instead of direct currents, we now have alternating currents and, instead of resistances, we have impedances.

As we shall see, both Kirchoff's laws and the star-delta transformations remain valid. We shall also learn how to transform a mutual inductance into a star; this transformation does not have a direct-current equivalent.

17.1 *IMPEDANCE*

Let us return to the simple circuits of Figs. 16-1, 16-6, and 16-8, where an alternating voltage is applied successively to a resistor, to an inductor, and to a capacitor.

If an alternating voltage $\mathbf{V}$ is applied to a resistor,

$$\mathbf{I} = \mathbf{V}/R. \tag{17-1}$$

If $\mathbf{V}$ is applied to an inductor, then, rewriting Eq. 16-16,

$$\mathbf{V} = j\omega L\mathbf{I} \tag{17-2}$$

and

$$\mathbf{I} = \mathbf{V}/j\omega L. \tag{17-3}$$

The quantity $j\omega L$ plays the same role as R and is called the *impedance* of the inductor. The symbol for impedance is Z. Impedances are measured in ohms, like resistances.

When $\mathbf{V}$ is applied to a capacitor,

$$\mathbf{I} = Cj\omega\mathbf{V} = \frac{\mathbf{V}}{1/j\omega C}, \tag{17-4}$$

from Eq. 16-13. Here $1/j\omega C = -j/\omega C$ is the impedance of the capacitor.

More generally, an impedance is complex and is written as

$$Z = R + jX = |Z|e^{j\theta}, \tag{17-5}$$

where R is the *resistance*, X is the *reactance*, $|Z|$ is the *modulus* or the *magnitude* of the impedance, and θ is its *phase angle*. A positive reactance is said to be *inductive*, while a negative reactance is said to be *capacitive*.

Thus, in general, the voltage $\mathbf{V}$ across an impedance Z carrying a current $\mathbf{I}$ is

$$\mathbf{V} = Z\mathbf{I}, \tag{17-6}$$

and

$$\mathbf{I} = \mathbf{V}/Z, \qquad Z = \mathbf{V}/\mathbf{I}. \tag{17-7}$$

The inverse of an impedance is an *admittance*:

$$Y = 1/Z. \tag{17-8}$$

Impedances in series and in parallel are calculated in the same way as resistances in series and in parallel (Secs. 5.4 and 5.5).

17.1.1 *EXAMPLE: THE IMPEDANCE AND THE ADMITTANCE OF A COIL*

A coil has a resistance of 1000 ohms and an inductance of 3.000 henrys. Its impedance at a frequency of 100.0 hertz is

$$R + j\omega L = 1000 + 2\pi \times 100 \times 3j = 1000 + 1885j \text{ ohms}, \tag{17-9}$$

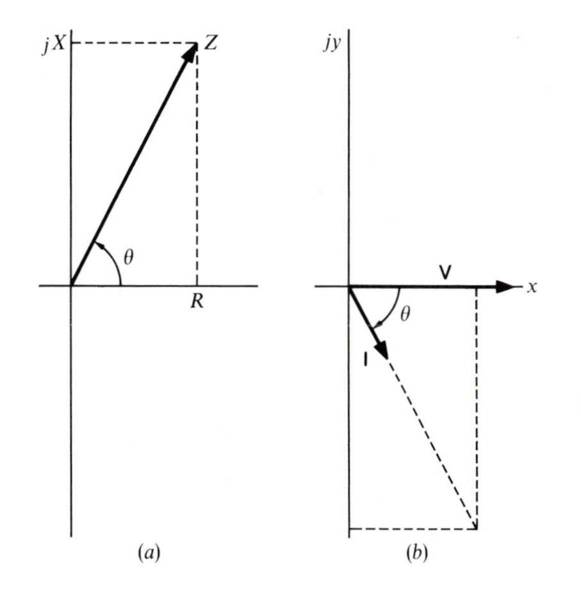

Figure 17-1 (*a*) The impedance of the coil of Sec. 17.1.1 in the complex plane. (*b*) **V** and $\mathbf{I} = \mathbf{V}/Z$ for this coil at $t = 0$. The moduli of **V** and **I** are not to scale.

or

$$Z = (1000^2 + 1885^2)^{1/2} \exp\left(j \arc\tan \frac{1885}{1000}\right),$$

$$= 2134 \exp 1.083j \text{ ohms.} \tag{17-10}$$

Figure 17-1 shows Z in the complex plane.

If one applies across this coil a voltage $V_0 \cos \omega t$ with an rms value of 10 volts, the current is

$$\mathbf{I} = \frac{\mathbf{V}}{2134 \exp 1.083j} = \frac{V_0}{2134} \exp j(\omega t - 1.083), \tag{17-11}$$

$$I = \frac{2^{1/2} \times 10}{2134} \cos(\omega t - 1.083). \tag{17-12}$$

Then the current I has an rms value of 10/2134, or 4.686 milliamperes. The current lags the voltage by 1.083 radians.

The admittance of the coil is

$$Y = \frac{1}{Z} = \frac{1}{1000 + 1885j} = \frac{1000 - 1885j}{1000^2 + 1885^2} = \frac{1000 - 1885j}{(2134)^2}, \tag{17-13}$$

$$= (2.196 - 4.139j)10^{-4} \text{ siemens}, \tag{17-14}$$

or

$$Y = \frac{1}{2134} \exp(-1.083j), \tag{17-15}$$

$$= 4.686 \times 10^{-4} \exp(-1.083j) \text{ siemens}. \tag{17-16}$$

17.2 THE EQUATION $\mathbf{I} = \mathbf{V}/Z$ IN THE COMPLEX PLANE

It is instructive to consider the equation $\mathbf{I} = \mathbf{V}/Z$ in the complex plane. Writing

$$Z = |Z|e^{j\theta}, \tag{17-17}$$

then

$$\mathbf{I} = \frac{\mathbf{V}}{|Z|e^{j\theta}} = \frac{V_0}{|Z|} e^{j(\omega t - \theta)}. \tag{17-18}$$

Thus the modulus of the phasor $\mathbf{I}$ is the modulus of $\mathbf{V}$ divided by $|Z|$, while its argument is the argument of $\mathbf{V}$ minus that of Z, as in Fig. 17-1.

17.3 KIRCHOFF'S LAWS

Kirchoff's laws of Sec. 5.7 apply to alternating currents just as well as to direct currents.

The current law states that the sum of the currents entering a node is zero. As we saw in Sec. 5.9, one normally uses mesh currents, instead of branch currents, so as to avoid using this law explicitly.

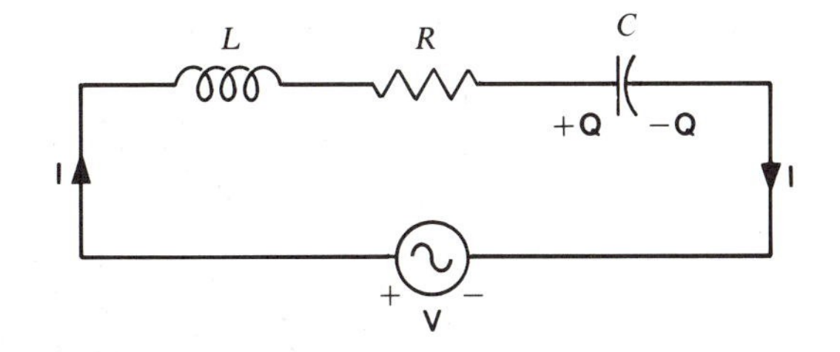

Figure 17-2 Inductor L, resistor R, and capacitor C, connected in series across a source **V**.

The voltage law states that the sum of the voltage drops around a mesh is zero. To use this law, one shows plus and minus signs for the sources and for the charges, and arrows for the currents, as in Fig. 17-2.

The signs and the directions of the arrows are arbitary, save for one exception. If a current **I** flows into a capacitor electrode carrying a charge **Q**, and if one wishes to write that $\mathbf{I} = d\mathbf{Q}/dt$, then one must have the current flowing into the positive **Q**, as in Fig. 17-2. With the arrows in the opposite direction, one would have to write $\mathbf{I} = -d\mathbf{Q}/dt$.

Note that these signs and arrows are *not* meant to indicate the polarities at a particular instant, as they were in Fig. 16-4. They are simply used as an aid in writing down the mesh equations correctly.

17.3.1 *EXAMPLE: THE SERIES LRC CIRCUIT*

Figure 17-2 shows a circuit with an inductor L, a resistor R, and a capacitor C, connected in series across a source **V**.

We shall (a) write down the differential equation for the mesh, (b) rewrite it in phasor notation to find **Q** and **I**, (c) write down the value of **I** directly from the relation $\mathbf{I} = \mathbf{V}/Z$, (d) deduce I as a function of the time, and (e) discuss the value of Z as a function of the frequency.

a) From Kirchoff's voltage law, starting at the lower left-hand corner,

$$-L\frac{dI}{dt} - RI - \frac{Q}{C} + V = 0, \tag{17-19}$$

or

$$L\frac{d^2Q}{dt^2} + R\frac{dQ}{dt} + \frac{Q}{C} = V. \tag{17-20}$$

b) In phasor notation, this equation becomes

$$\left[(j\omega)^2 L + j\omega R + \frac{1}{C}\right]\mathbf{Q} = \mathbf{V}, \tag{17-21}$$

so that

$$\mathbf{Q} = \frac{\mathbf{V}}{(j\omega)^2 L + j\omega R + (1/C)}. \tag{17-22}$$

Multiplying both sides by $j\omega$,

$$j\omega\mathbf{Q} = \mathbf{I} = \frac{\mathbf{V}}{R + j\omega L + (1/j\omega C)}. \tag{17-23}$$

c) Since we have impedances in series, the total impedance is the sum of the impedances:

$$Z = R + j\omega L + \frac{1}{j\omega C}. \tag{17-24}$$

Then

$$\mathbf{I} = \frac{\mathbf{V}}{Z} = \frac{\mathbf{V}}{R + j\omega L + (1/j\omega C)} \tag{17-25}$$

as above.

d) To find I as a function of the time, we first express the impedance in the polar form $|Z| \exp(j\theta)$:

$$Z = R + j[\omega L - (1/\omega C)], \tag{17-26}$$

$$= \{R^2 + [\omega L - (1/\omega C)]^2\}^{1/2} \exp\left[j \arctan \frac{\omega L - (1/\omega C)}{R}\right]. \tag{17-27}$$

Thus

$$|Z| = \{R^2 + [\omega L - (1/\omega C)]^2\}^{1/2}, \tag{17-28}$$

$$\theta = \arctan \frac{\omega L - (1/\omega C)}{R}, \tag{17-29}$$

and

$$\mathbf{I} = \frac{V_0 e^{j\omega t}}{|Z| e^{j\theta}}, \tag{17-30}$$

$$= \frac{V_0}{|Z|} e^{j(\omega t - \theta)}. \tag{17-31}$$

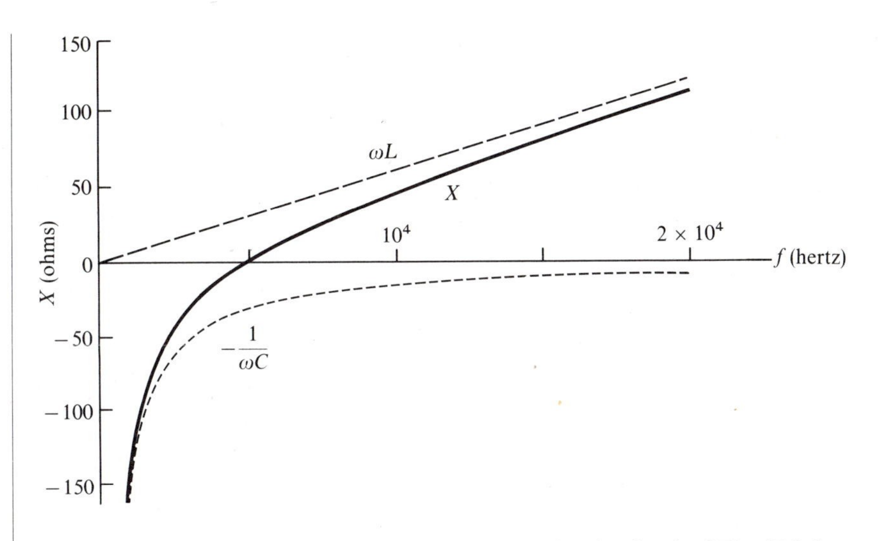

Figure 17-3 Reactance as a function of the frequency for the circuit of Fig. 17-2, for $L = 1$ millihenry and $C = 1$ microfarad. The reactance X is $\omega L - 1/\omega C$.

The current I is the real part of $\mathbf{I}$:

$$I = \frac{V_0}{|Z|} \cos(\omega t - \theta). \tag{17-32}$$

e) Note that $|Z|$ and θ are both functions of the frequency.
Since

$$Z = R + jX = R + j\omega L\left(1 - \frac{1}{\omega^2 LC}\right), \tag{17-33}$$

the reactance X is zero when $\omega^2 LC = 1$. At that frequency the voltage drop on L exactly cancels that on C and the circuit is said to be *resonant*. With series resonance, the *voltages* across L and across C can be *larger* than the applied voltage.

Figure 17-3 shows the reactance as a function of the frequency for $L = 1$ millihenry and $C = 1$ microfarad.

This is a case of *series resonance* because the capacitor and the inductor are connected in series with the source. Problem 17-15 concerns parallel resonance.

In the case of parallel resonance, the *currents* through the capacitor and through the inductor partly cancel, and each one can be *larger* than the total current.

17.4 STAR-DELTA TRANSFORMATIONS

In Sec. 5.10 we saw how one can transform a resistive star circuit into an equivalent delta, and vice versa.

If one has complex impedances, as in Fig. 17-4a, instead of resistances, the transformation rules are similar. This comes from the fact that the reasoning we used in Sec. 5.10 is in no way limited to real impedances. Thus

$$Z_A = \frac{Z_b Z_c}{Z_a + Z_b + Z_c}, \tag{17-34}$$

$$Z_B = \frac{Z_c Z_a}{Z_a + Z_b + Z_c}, \tag{17-35}$$

$$Z_C = \frac{Z_a Z_b}{Z_a + Z_b + Z_c}, \tag{17-36}$$

$$Y_a = \frac{Y_B Y_C}{Y_A + Y_B + Y_C}, \tag{17-37}$$

$$Y_b = \frac{Y_C Y_A}{Y_A + Y_B + Y_C}, \tag{17-38}$$

$$Y_c = \frac{Y_A Y_B}{Y_A + Y_B + Y_C}. \tag{17-39}$$

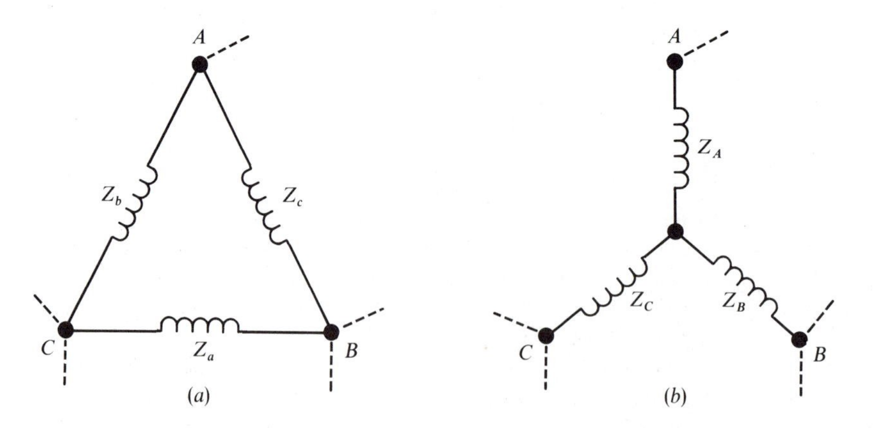

Figure 17-4 Three nodes *A*, *B*, *C*. (*a*) Delta-connected. (*b*) Star-connected. The wavy lines represent impedances.

17.4.1 *EXAMPLE: TRANSFORMING A DELTA INTO A STAR*

As an example, let us transform the delta of Fig. 17-5a into a star at a frequency of one kilohertz. In this case,

$$Z_a = Z_b = 1/j\omega C = -10^3 j/2\pi \text{ ohms,} \tag{17-40}$$

$$Z_c = 100 \text{ ohms.} \tag{17-41}$$

From Eqs. 17-34 to 17-36, and remembering that $Z = 1/Y$,

$$Z_A = Z_B = 45.51 - 14.30j \text{ ohms,} \tag{17-42}$$

$$Z_C = -22.75 - 72.43j \text{ ohms.} \tag{17-43}$$

Thus Z_A and Z_B are each composed of a resistance of 45.51 ohms, in series with a capacitor whose impedance is $-j/\omega C$, where

$$\omega C = 1/14.30, \tag{17-44}$$

$$C = 1/(14.30 \times 2\pi \times 1000) = 11.13 \text{ microfarads.} \tag{17-45}$$

Similarly, Z_C is composed of a resistance of -22.75 ohms, in series with a capacitor of 2.197 microfarads.

The equivalent star is shown in Fig. 17-5b.

Note that the resistances of the equivalent circuits can have negative values. A real circuit can therefore be equivalent to a fictitious one that is impossible to build.

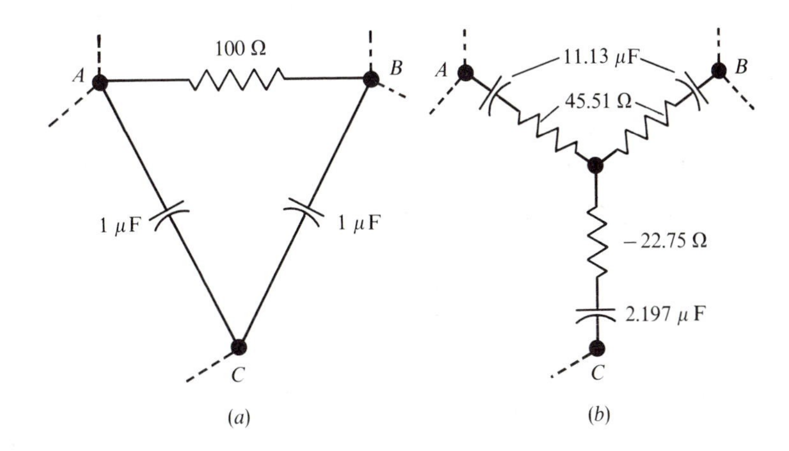

Figure 17-5 (*a*) Delta-connected circuit. (*b*) Its equivalent star at 1 kilohertz.

Note also that the resistances and capacitances in the star depend on the operating frequency. For example, in Prob. 17-16, it is shown that, at one megahertz, the delta of Fig. 17-5a is equivalent to a completely different star.

17.5 TRANSFORMATION OF A MUTUAL INDUCTANCE

Figure 17-6a shows a mutual inductance M with its two coils connected at C. We can transform this mutual inductance into the star of Fig. 17-6b in the following way.

The mesh currents $\mathbf{p}$, $\mathbf{q}$, $\mathbf{r}$ must be the same in the two circuits. For simplicity, let us assume that C is at ground potential. This will not restrict the generality of the transformation.

In Fig. 17-6a,

$$\mathbf{V}_A = (\mathbf{p} - \mathbf{q})Z_a + j\omega M(\mathbf{r} - \mathbf{q}). \tag{17-46}$$

The second term on the right is the voltage induced on the left-hand side, or M times the rate of change of the current on the right (Sec. 12.1). Here, M is positive if the two coils are wound in such a way that a current in the positive direction in one coil produces in the other coil a flux linkage that is in the same direction as that due to a positive current in that coil (Sec. 12.1). Otherwise, M is negative. We have assumed here that clockwise currents are positive.

In 17-6b,

$$\mathbf{V}_A = (\mathbf{p} - \mathbf{q})Z_A + (\mathbf{p} - \mathbf{r})Z_C. \tag{17-47}$$

Similarly, in 17-6a,

$$\mathbf{V}_B = (\mathbf{q} - \mathbf{r})Z_b - j\omega M(\mathbf{p} - \mathbf{q}), \tag{17-48}$$

while, in 17-6b,

$$\mathbf{V}_B = (\mathbf{q} - \mathbf{r})Z_B + (\mathbf{p} - \mathbf{r})Z_C. \tag{17-49}$$

Equating now the two values of $\mathbf{V}_A$, and then the two values of $\mathbf{V}_B$ and simplifying,

$$\mathbf{p}(Z_a - Z_A - Z_C) + \mathbf{q}(Z_A - Z_a - j\omega M) + \mathbf{r}(Z_C + j\omega M) = 0, \tag{17-50}$$

$$\mathbf{p}(Z_C + j\omega M) + \mathbf{q}(Z_B - Z_b - j\omega M) + \mathbf{r}(Z_b - Z_B - Z_C) = 0. \tag{17-51}$$

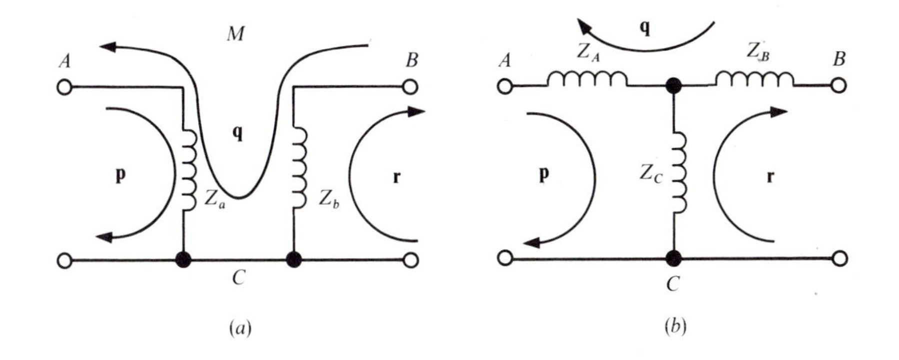

Figure 17-6 (*a*) Mutual inductance, and (*b*) its equivalent star circuit, with the same mesh currents.

Since these two equations are valid, whatever the values of **p**, **q**, **r**, the six coefficients between parentheses must be zero, and

$$Z_A = Z_a + j\omega M, \tag{17-52}$$

$$Z_B = Z_b + j\omega M, \tag{17-53}$$

$$Z_C = -j\omega M. \tag{17-54}$$

Note that, if M is positive, the reactance $-j\omega M$ is capacitive. On the other hand, if M is negative, then $-j\omega M$ is the reactance of a *pure* inductance $|M|$. Again, we have a circuit that is mathematically equivalent to one that cannot be realized in practice, since any inductor, unless it is superconducting, possesses a resistance.

17.5.1 *EXAMPLE: TWO SUPERPOSED COILS*

Figure 17-7a shows *two superposed coils* wound in the same direction, one over the other. Each coil has a resistance R and an inductance L. Since the two coils are superposed, $|M| = L$.

In this example, with the current directions shown in the figure, M is negative and the equivalent circuit is shown in Fig. 17-7c; the impedance is $R/2 + j\omega L$. If we changed the direction of one of the currents, we would still arrive at the same result.

The impedance $R/2 + j\omega L$ is easily explained: in effect, we have one coil made up of larger wire, so that the inductance is L, and the resistance is $R/2$.

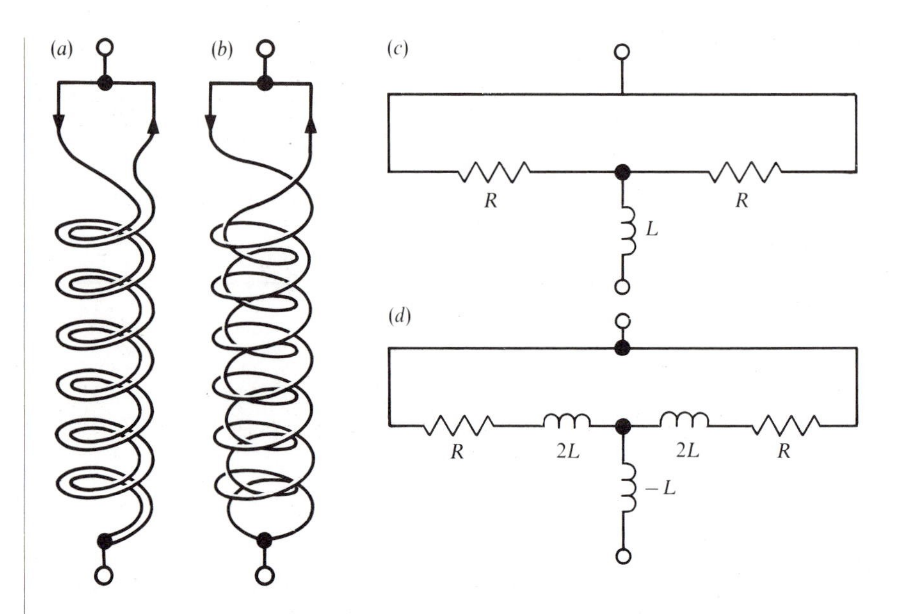

Figure 17-7 Two superposed coils, (*a*) wound in the same direction, and (*b*) wound in opposite directions; (*c*) and (*d*), equivalent circuits for (*a*) and (*b*), respectively.

Figure 17-7b shows a similar pair of coils, except that now the coils are wound in opposite directions. For the current directions shown, M is positive, the equivalent circuit is shown in 17-7d, and the total impedance is $R/2$.

This is correct since, in this example, the net magnetic field is zero, and the two resistances are in parallel.

17.6 SUMMARY

The current in a circuit element is given by the voltage across it, divided by the *impedance* of the element:

$$\mathbf{I} = \mathbf{V}/Z. \tag{17-1}$$

The impedance of a resistor is R, that of an inductor is $j\omega L$, and that of a capacitor $1/j\omega C$. An impedance is expressed as a complex number, either in Cartesian or in polar form:

$$Z = R + jX = |Z|e^{j\theta}, \tag{17-5}$$

where R is its *resistance*, X is its *reactance*, $|Z|$ is the *modulus* or *magnitude* of the impedance, and θ is its *phase angle*.

The inverse of an impedance is an *admittance*:

$$Y = 1/Z. \tag{17-8}$$

Series and parallel combinations of impedances are calculated in the same way as with resistances.

Kirchoff's laws apply to alternating currents as well as to direct currents.

The *star-delta transformations* of Chapter 5 apply to impedances.

A mutual inductance can be transformed into a star circuit as in Fig. 17-6 and Eqs. 17-52 to 17-54.

PROBLEMS

17-1 *IMPEDANCE*

a) Calculate the impedance Z of the circuit shown in Fig. 17-8. What is the value of Z (i) for $f = 0$, (ii) for $f \to \infty$?

b) What is the magnitude and the phase angle of Z at 1 kilohertz?

c) What is the magnitude and the phase angle of $Y = 1/Z$ at that frequency?

d) Calculate the power dissipation when the current is 100 milliamperes, again at 1 kilohertz.

e) Can the real part of the impedance become negative?

f) For what frequency range is the circuit equivalent to (i) a resistor in series with an inductor, (ii) a resistor in series with a capacitor?

g) At what frequency is the circuit equivalent to a pure resistance?

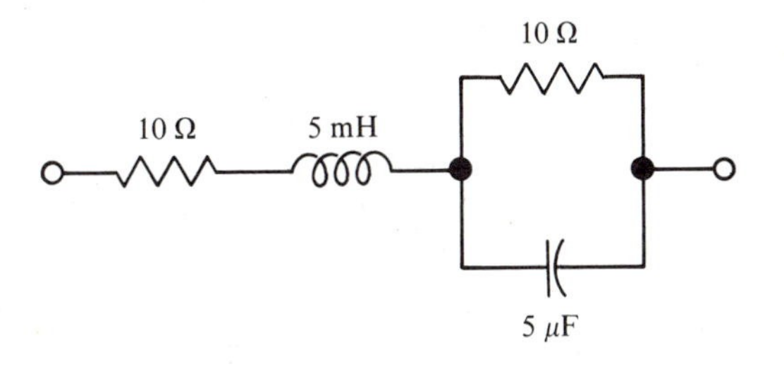

Figure 17-8 See Prob. 17-1.

17-2 *REAL INDUCTORS*

A real inductor has not only an inductance but also a resistance and a capacitance. Its equivalent circuit is shown in Fig. 17-9.

Show that the impedance between the terminals is

$$Z = \frac{R + j\omega[L(1 - \omega^2 LC) - R^2 C]}{(1 - \omega^2 LC)^2 + R^2 \omega^2 C^2}.$$

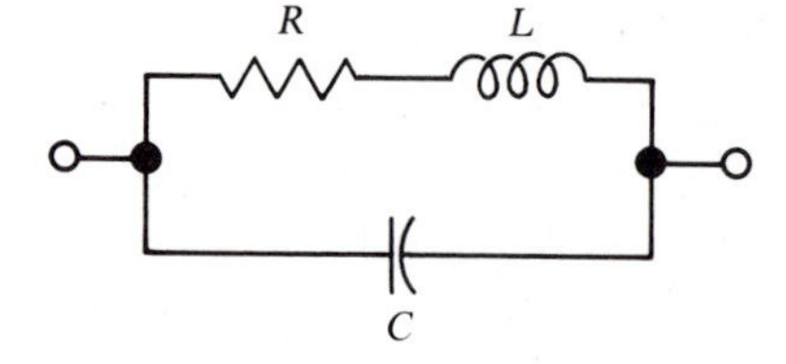

Figure 17-9 Equivalent circuit of a real inductor. The stray capacitance is C. See Prob. 17-2.

17-3E *COMPENSATED VOLTAGE DIVIDER*

The voltage divider of Prob. 5-5 cannot be used as such at high frequencies for the following reason. There are stray capacitances due to the wiring, etc., in parallel with R_1 and R_2. If the frequency is high enough, these stray capacitances carry an appreciable current and $\mathbf{V}_o/\mathbf{V}_i$ is a function of the frequency.

Show that, at any frequency

$$\frac{\mathbf{V}_o}{\mathbf{V}_i} = \frac{R_2}{R_1 + R_2}$$

if $R_1 C_1 = R_2 C_2$ in Fig. 17-10. Capacitances C_1 and C_2 are made large compared to the stray capacitances. The voltage divider is then said to be *compensated*.

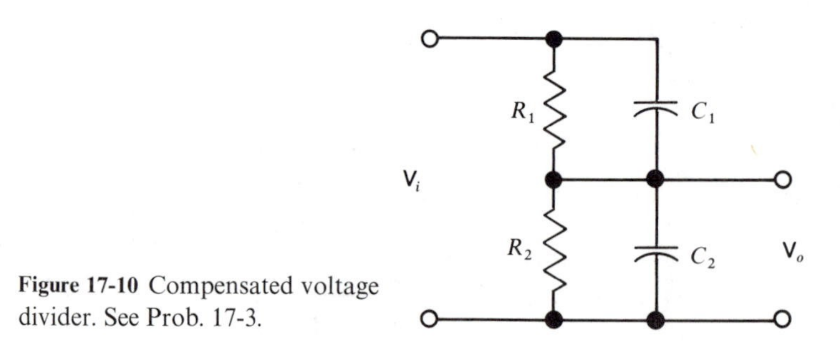

Figure 17-10 Compensated voltage divider. See Prob. 17-3.

17-4 *RC FILTER*

Four impedances are connected as in Fig. 17-11.

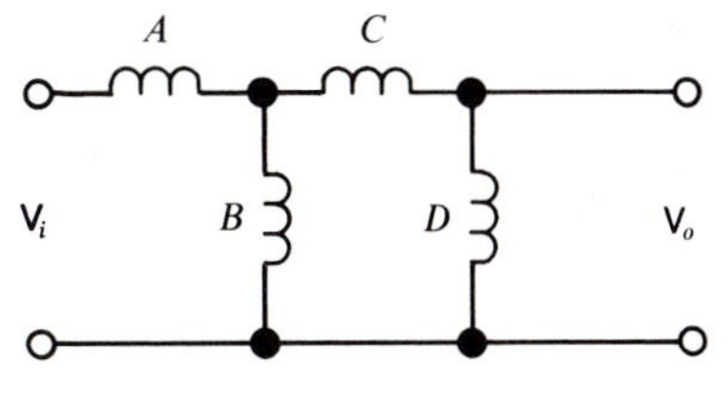

Figure 17-11 See Prob. 17-4.

a) Show that

$$\frac{\mathbf{V}_o}{\mathbf{V}_i} = \frac{BD}{AB + (A + B)(C + D)}.$$

b) Draw the curve of $|\mathbf{V}_o/\mathbf{V}_i|$ as a function of $R\omega C$ for the circuit of Fig. 17-12 for $R\omega C = 0.1$ to 10. Use a log scale for $R\omega C$.

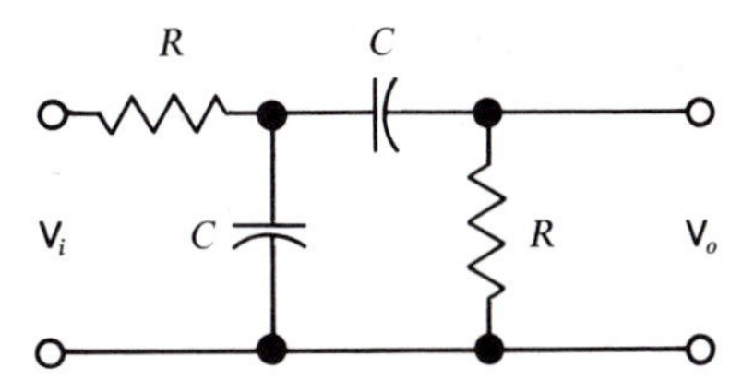

Figure 17-12 See Prob. 17-4.

17-5 *MEASURING AN IMPEDANCE WITH A PHASE-SENSITIVE VOLTMETER*

An unknown impedance Z is connected in series with a known resistance R' as in Fig. 17-13.

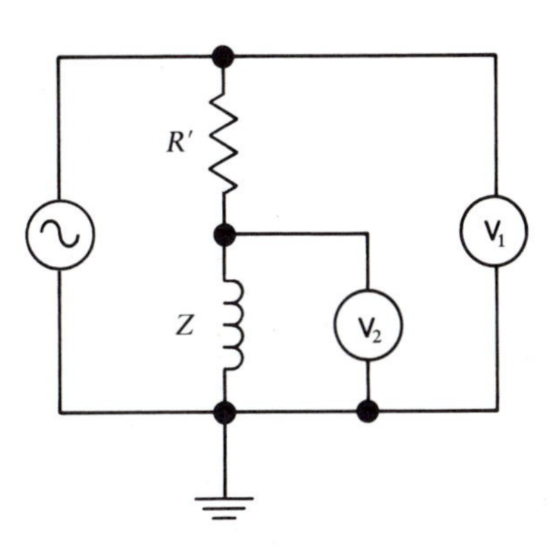

Figure 17-13 See Prob. 17-5.

Show that the impedance Z can be found by measuring the voltage ratio

$$r = a + bj = \mathbf{V}_2/\mathbf{V}_1,$$

in other words, by measuring the in-phase and the quadrature components of $\mathbf{V}_2$, $\mathbf{V}_1$ being known.

17-6E *IMPEDANCE BRIDGES. THE WIEN BRIDGE*

Figure 17-14 shows an impedance bridge. It is the alternating-current equivalent of the Wheatstone bridge of Prob. 5-8. In this case, $V = 0$ when

$$Z_1/Z_2 = Z_3/Z_4.$$

In alternating-current bridges the components must satisfy *two* independent equations to satisfy both the real and the imaginary parts of this equation.

There exist many types of impedance bridge.† One common type is the Wien bridge shown in Fig. 17-15. As a rule, one sets

$$R_1 = R_2/2, \qquad R_3 = R_4, \qquad C_3 = C_4.$$

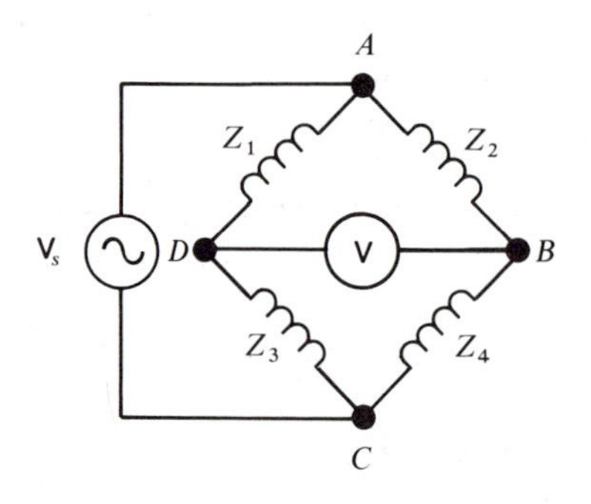

Figure 17-14 Impedance bridge. See Prob. 17-6.

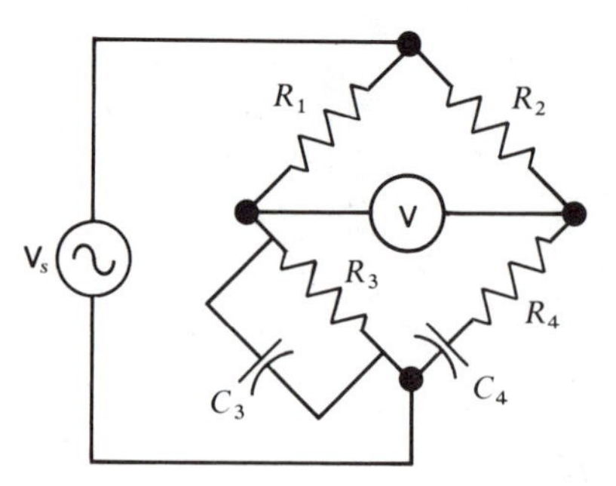

Figure 17-15 Wien bridge. See Prob. 17-6.

† See, for example, *Reference Data for Radio Engineers*, Fifth Edition, Chapter 11 (Howard W. Sams and Co., Inc., Indianapolis, 1968).

Show that, under these conditions, $R_3\omega C_3 = 1$.

The Wien bridge is used in tuned amplifiers and in oscillators, as well as for measuring or monitoring a frequency. To measure a frequency, one changes R_3 and R_4 simultaneously until $R_3\omega C_3$ is equal to unity.

See also Prob. 17-18.

17-7E *LOW-PASS RC FILTER*

The circuit of Fig. 17-16 is commonly used to reduce the AC component at the output of a power supply.

As a rule, the term *power supply* applies to a circuit that transforms 120-volt 60-hertz alternating current into direct current. One simple type is the battery charger of Prob. 16-3.

A power supply usually comprises three parts: a transformer, a rectifying circuit, and a filter. The output of the rectifying circuit is a pulsating direct current that can be considered to be a steady direct current, plus an alternating current. The function of the filter is to attenuate the alternating part. Filters are usually dispensed with in battery chargers.

The filter is often followed, or even replaced, by an electronic circuit that maintains either the output voltage or the output current at a set value (Sec. 5.11).

The circuit shown is a simple form of low-pass filter. The load resistance R_L is assumed to be large compared to $1/\omega C$.

a) Show that the alternating component is attenuated by the factor

$$|\mathbf{V}_o/\mathbf{V}_i| \approx 1/R\omega C$$

if $R\omega C \gg 1$. The voltage drop across the resistance R should be small compared to that across R_L, so

$$R_L \gg R \gg 1/\omega C.$$

b) Plot a curve of $|\mathbf{V}_o/\mathbf{V}_i|$ as a function of frequency from 0 to 150 hertz for $R = 10^4$ ohms and $C = 100$ microfarads.

c) Plot the corresponding curve of the attenuation in *decibels* as a function of frequency, where the number of decibels is $20 \log |\mathbf{V}_o/\mathbf{V}_i|$.

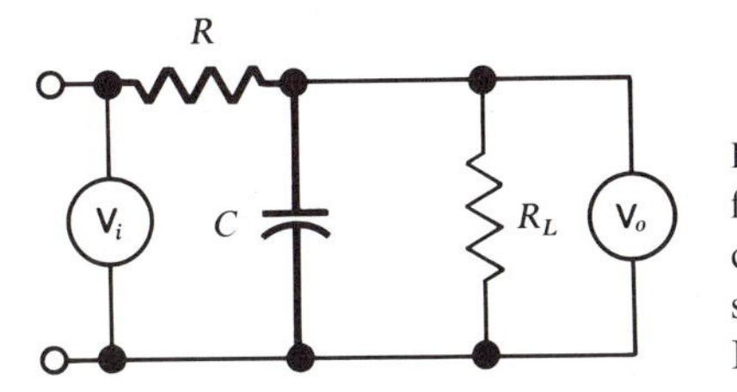

Figure 17-16 Simple low-pass RC filter for attenuating the alternating component at the output of a power supply, with $R_L \gg R \gg 1/\omega C$. See Prob. 17-7.

17-8 *PHASE SHIFTER*

It is often necessary to shift the phase of a signal. Figure 17-17 shows a simple circuit for doing this. The resistances are adjustable, but are kept equal.

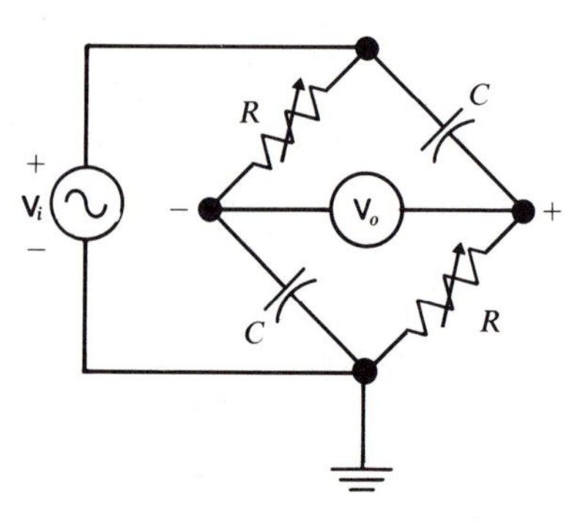

Figure 17-17 Phase shifter. See Prob. 17-8.

Use the polarities shown. These mean that $\mathbf{V}_i$ is the voltage of the top terminal with respect to the bottom one, and $\mathbf{V}_o$ is the voltage of the right-hand terminal with respect to the left-hand one.

a) Show that

$$\mathbf{V}_o/\mathbf{V}_i = \exp\left[2j \arctan\left(1/R\omega C\right)\right].$$

b) Draw a graph of the phase of $\mathbf{V}_o$ with respect to $\mathbf{V}_i$ in the range $R\omega C = 0.1$ to 10. Use a log scale for $R\omega C$.

17-9 *MEASURING SURFACE POTENTIALS AND SURFACE CHARGE DENSITIES ON DIELECTRICS WITH A GENERATING VOLTMETER, OR FIELD MILL*

In Prob. 6-11, we saw how one can measure the surface potential V and the surface charge density σ on a dielectric, with a probe connected to an electrometer. Here is a similar method, shown in Fig. 17-18a, in which the charge induced on the probe is modulated to produce an alternating current. The alternating current passes through a resistance, giving an alternating voltage that is then amplified to give an output voltage proportional to V, or to σ. Instruments that measure electric field intensities in this way are called *generating voltmeters*, or *field mills*.

One such instrument can measure potentials up to 3000 volts and has a field of view only about 2 millimeters in diameter. It serves to test the photoconductive materials used for xerography.

Figure 17-19 shows another one that measures the terminal voltage on a Van de Graaff high-voltage generator.

a) Use Fig. 17-18b to find E_a, as a function of σ, d_a, d_D, ϵ_r, without the vane of Fig. 17-18a. The current through R is then zero.

b) Now what is the value of σ_i?

Note that σ_i depends on both d_D and on ϵ_r.

c) What is the value of V in terms of σ, d_a, d_D, ϵ_r?

Note that V depends on d_a, and is therefore affected by the presence of the instrument.

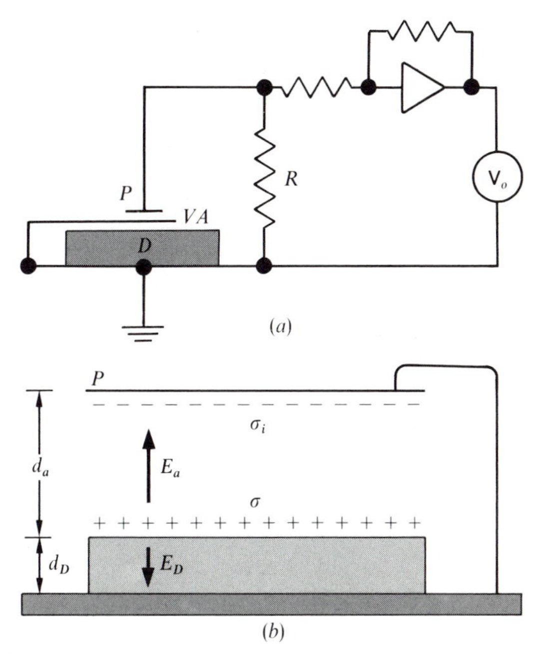

Figure 17-18 (*a*) Apparatus for measuring the potential at the surface of a dielectric slab *D*. The probe *P* is grounded through the resistance *R*, and the grounded vane *VA* is made to pass between *D* and *P* in such a way that *P* is alternately screened and exposed to *D*. An alternating current flows to *P* and produces a voltage drop on *R*. The voltage across *R* is amplified (Prob. 5-9), giving a voltage $\mathbf{V}_o$ proportional to the surface charge density. The voltage across *R* is orders of magnitude smaller than the voltage at the surface of *D*. (*b*) Probe *P* near the dielectric carrying a surface charge density σ.

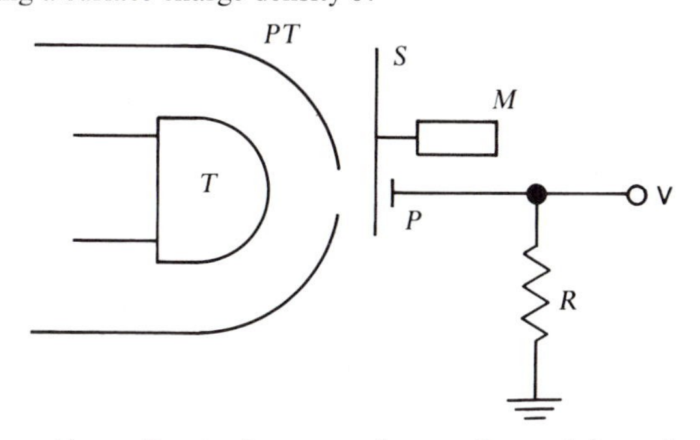

Figure 17-19 Generating voltmeter for measuring a voltage of the order of millions of volts, from a distance. A motor *M* rotates a shield *S* in front of a probe *P* in view of the high-voltage terminal *T* enclosed in a pressure tank *PT*. Sectors are cut out of the shield. An alternating current flows to *P* through *R*. The voltage **V** is proportional to the voltage on the terminal. See Prob. 17-9.

d) Now calculate the voltage across R when $R = 10$ megohms and $V = 1000$ volts, and when the capacitance C between P and the dielectric surface varies sinusoidally between 0 and 0.1 picofarad. The vane oscillates at 100 hertz and screens P once per cycle.

17-10E *REFERENCE TEMPERATURES NEAR ABSOLUTE ZERO*

The circuit shown in Fig. 17-20 is used to define fixed points on the temperature scale near absolute zero. It utilizes the fact that several elements exhibit sharp and reproducible transitions to and from the superconducting state at definite temperatures.

A slug of one of these elements, S, is inserted in the mutual inductance M_2. Then R_1/R_2 is adjusted to make $V = 0$. When S becomes superconducting, large eddy currents flow through it, canceling part of the magnetic field, M_2 is decreased, and $V \neq 0$.

Show that, when $V = 0$,

$$M_1/M_2 = R_1/R_2.$$

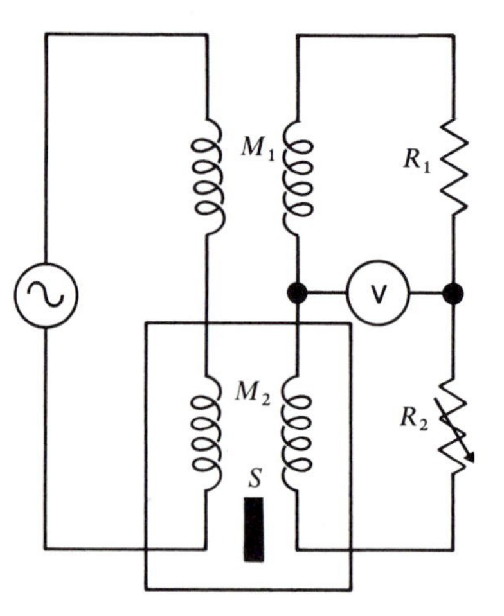

Figure 17-20 See Prob. 17-10.

17-11 *REMOTE-READING MERCURY THERMOMETER*

A good mercury thermometer is a simple, inexpensive device for making accurate measurements of temperature. There are many circumstances, however, where ordinary mercury thermometers are impractical. For example, one may wish to monitor the temperature at many points near the bottom of a river, every minute, 24 hours a day. Or one may wish to use a temperature reading to control a manufacturing process automatically.

It is possible to read a mercury thermometer electronically, and hence automatically and remotely, with the circuit shown in Fig. 17-21.

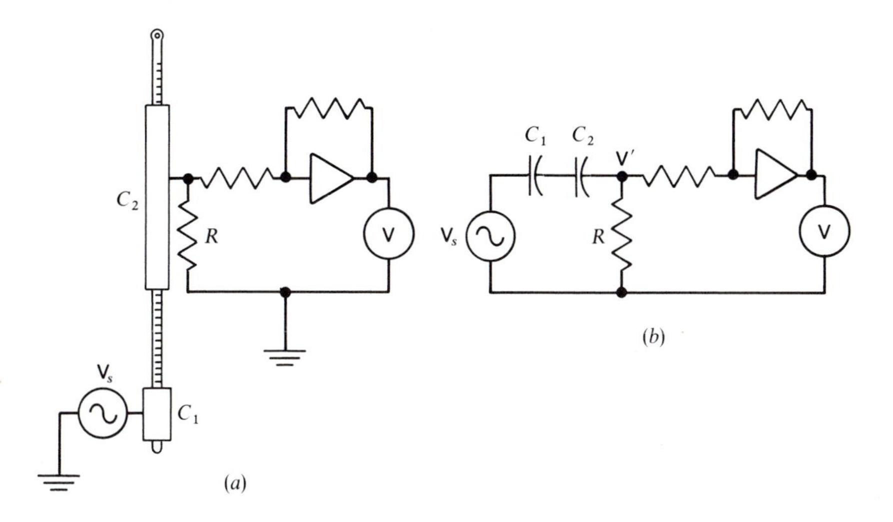

Figure 17-21 (*a*) Mercury thermometer adapted for remote reading. An oscillator applies about 10 volts at 10 megahertz to an electrode surrounding the bulb and forming a capacitance C_1. The capacitance C_2 between the mercury column and a coaxial tube varies linearly with temperature. The voltage on R is amplified (Prob. 5-9) and measured at **V**. (*b*) Equivalent electric circuit. See Prob. 17-11.

a) Find the voltage $\mathbf{V}'$ as a function of ω, the source voltage $\mathbf{V}_s$, C_2, and R, with $C_1 \gg C_2$.

b) Under what condition will V' vary linearly with temperature?

c) Estimate the value of C_2, and suggest a value for R.

d) Then what is the order of magnitude of V' if V_s is 10 volts?

17-12 WATTMETER

Figure 17-22 shows a Hall generator (Sec. 10.8.1) used as a wattmeter to measure the power supplied by a source to a load Z.

The coil of a small electromagnet E has an inductance L, with $\omega L \ll |Z|$. It supplies the magnetic field for the Hall generator. The resistances R_1 and R_2 are chosen so that $R_1 + R_2 \gg |Z|$. In this way, the power supplied by the source is approximately equal to that dissipated in the load.

The voltage and the current at the source are

$$V = V_0 \cos \omega t,$$

$$I = I_0 \cos (\omega t + \varphi).$$

Utilize the result of Sec. 10.8.1 to show that the average value of the Hall generator output voltage V' is proportional to $V_{\text{rms}} I_{\text{rms}} \cos \varphi$.

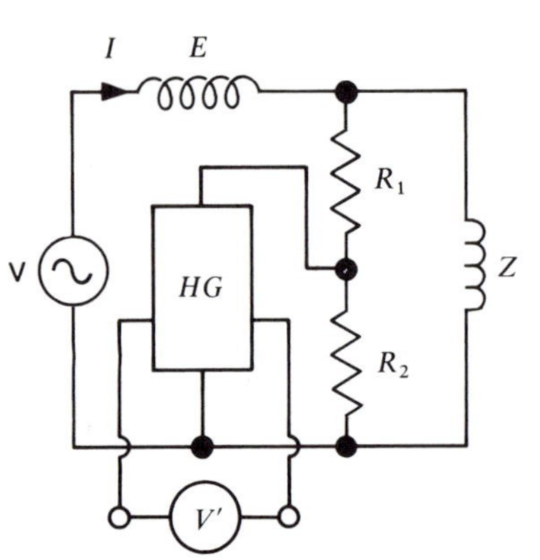

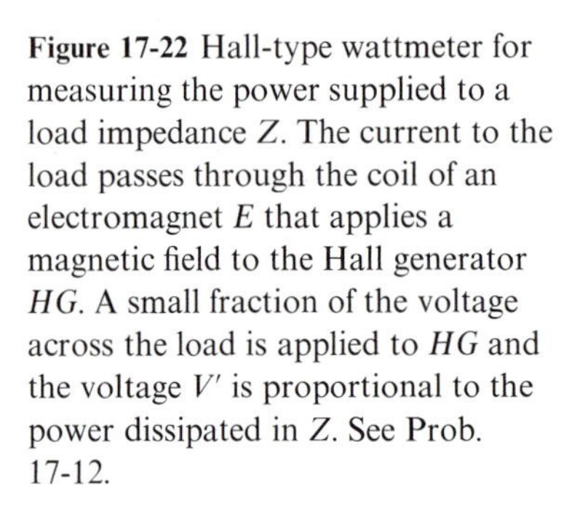

Figure 17-22 Hall-type wattmeter for measuring the power supplied to a load impedance Z. The current to the load passes through the coil of an electromagnet E that applies a magnetic field to the Hall generator HG. A small fraction of the voltage across the load is applied to HG and the voltage V' is proportional to the power dissipated in Z. See Prob. 17-12.

17-13 *TRANSIENT SUPPRESSOR FOR INDUCTOR*

In Prob. 12-13 we saw that, if one disconnects a large current-carrying inductor from a direct-current source, high transient voltages develop and the inductor can be destroyed. We also saw that the overvoltages can be eliminated by connecting a diode in parallel with the inductor.

If the inductor is fed by an alternating-current source, one can use the protective circuit shown in Figure 17-23. Here, R is the resistance and L is the inductance of the inductor, and $C = L/R^2$.

Show that the reactance between A and B is zero. Then the circuit between A and B must act like a pure resistance, and the voltage across the inductor must become zero when the switch is opened.

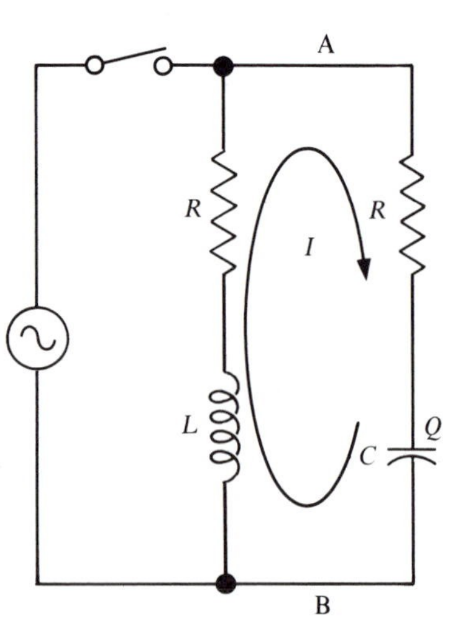

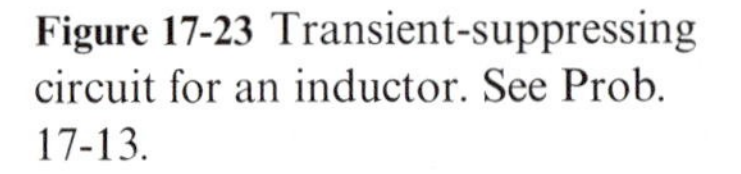

Figure 17-23 Transient-suppressing circuit for an inductor. See Prob. 17-13.

Solution:

$$Z = \frac{(R + j\omega L)[R + (1/j\omega C)]}{2R + j\omega L + (1/j\omega C)}, \tag{1}$$

$$= \frac{R^2 + (L/C) + j[\omega LR - (R/\omega C)]}{2R + j[\omega L - (1/\omega C)]}, \tag{2}$$

$$= \frac{\{R^2 + (L/C) + j[\omega LR - (R/\omega C)]\}\{2R - j[\omega L - (1/\omega C)]\}}{4R^2 + [\omega L - (1/\omega C)]^2}. \tag{3}$$

The reactance is the imaginary part of Z:

$$X = \frac{-[R^2 + (L/C)][\omega L - (1/\omega C)] + 2R[\omega LR - (R/\omega C)]}{4R^2 + [\omega L - (1/\omega C)]^2}. \tag{4}$$

The numerator of this expression is zero, Hence the reactance between A and B is zero at all frequencies!

17-14E *SERIES RESONANCE*

What is the locus of the impedance Z in the complex plane when the frequency applied to the series-resonant circuit of Fig. 17-2 changes from zero to infinity?

17-15 *PARALLEL RESONANCE*

An inductor of inductance L and resistance R is connected in parallel with a capacitor C.

Set $R = 10$ ohms, $L = 1.0$ millihenry, and $C = 1.0$ microfarad.

a) Plot curves of the real part, the imaginary part, the magnitude, and the phase of the impedance as functions of the frequency, from zero to 10 kilohertz.

Note that, when Z is maximum, $\omega^2 LC \approx 1$.

b) At what frequency is the reactance zero?

c) Find the impedance at 8 kilohertz.

d) At that frequency, the circuit has the same impedance as a capacitor C' and a resistor R' in series. What are the values of R' and of C'?

This is a case of *parallel resonance*, L and C being in parallel.

The *bandwidth* is defined as the difference between the frequencies where the magnitude of the impedance is $1/2^{1/2}$, or 0.707 times its maximum value.

The Q (for Quality) of the circuit is defined to be the resonance frequency, divided by the bandwidth. The higher the Q, the sharper the resonance peak. It can be shown that

$$Q = \frac{1}{R}\left(\frac{L}{C}\right)^{1/2}.$$

The Q of the above circuit is 3.2.

Parallel resonance has innumerable applications. Here is one example. For high-speed, high-precision coil winding, it is important to monitor the wire diameter. For several reasons, mechanical devices are unacceptable. The monitoring can be done in the following way: a small coil of a few turns is wound around a quartz tube a few millimeters in diameter through which the wire passes. The coil of course has an inductance. It also has a capacitance; as in any coil, there is a voltage difference, and hence a capacitance, between turns. This *stray capacitance* is increased by the presence of the wire. So the coil acts as a parallel-resonant circuit, whose resonance frequency depends on the diameter of the wire. If the wire diameter *in*creases, the resonant frequency *de*creases and vice versa. The curve of Z as a function of frequency is like that calculated under part *a* above, the peak moving to the *right* when the wire diameter *de*creases, and to the *left* when the diameter *in*creases.

The coil is designed so that its resonance frequency is always slightly above 300 megahertz, and an oscillator operating at 300 megahertz is connected to the coil through a resistance. If the wire diameter *in*creases, the resonance peak moves to the left, Z *in*creases, and the voltage on the coil *in*creases. Measurements are accurate within 0.5%.

17-16E *STAR-DELTA TRANSFORMATION*

Transform the delta of Fig. 17-5a into a star at one megahertz.

17-17E *STAR-DELTA TRANSFORMATION*

Transform the star of Fig. 17-24 into a delta at one kilohertz.

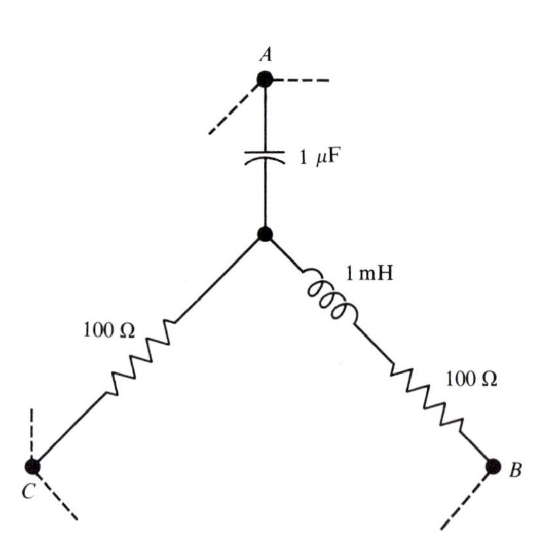

Figure 17-24 See Prob. 17-17.

17-18

T-*CIRCUITS*

Figures 17-25a and b show two types of T-circuits, the twin (or parallel) T and the bridged T. The components are selected in such a way that the output voltage V_o becomes zero at a specified frequency. Circuits that have this characteristic are called *notch filters*. T-circuits are used in oscillators and in tuned amplifiers. Some are also used for measuring impedances, in much the same way as the impedance bridges of Prob. 17-6. Note that the input and the output have a common terminal, which is grounded. This is a great advantage, since most oscillators, voltmeters, and oscilloscopes have one terminal grounded.

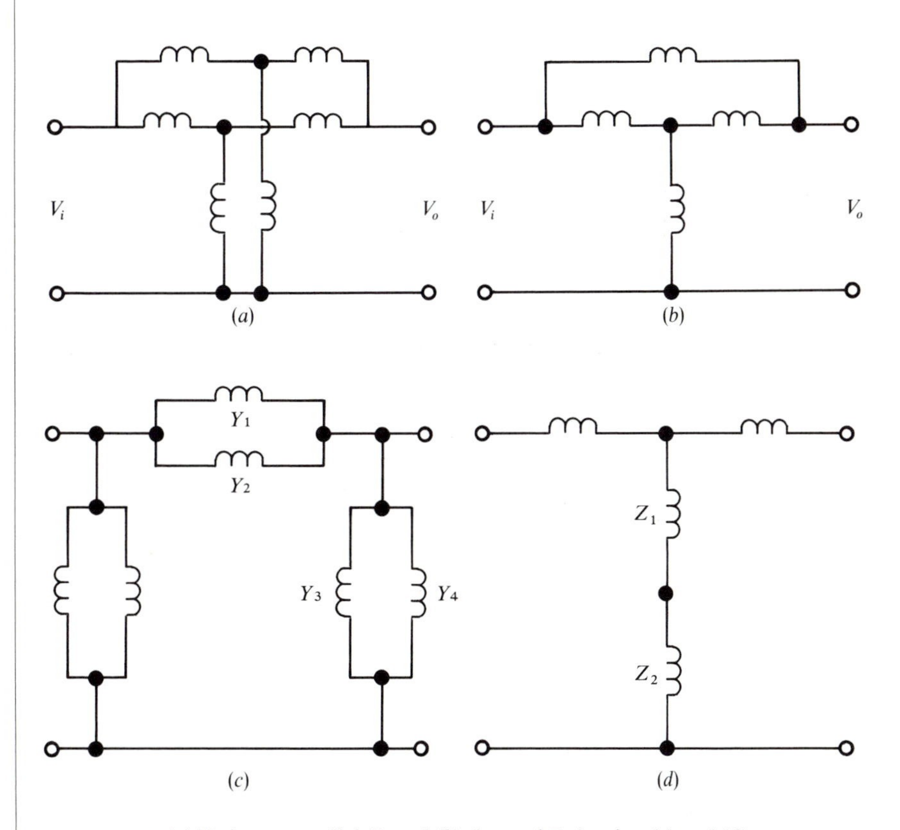

Figure 17-25 (*a*) Twin, or parallel, T, and (*b*) shunted T circuits; (*c*) and (*d*), equivalent circuits. See Prob. 17-18.

To find the conditions under which the output voltage of the twin T is zero, we transform the star circuits into deltas, giving the circuit of Fig. 17-25c, and we set $Y_1 + Y_2 = 0$.

For the bridged T, we transform the delta formed by the three impedances at the top into a star. The bridged T then becomes a simple T as in Fig. 17-25d, and the output voltage is zero when $Z_1 + Z_2 = 0$.

Figure 17-26 shows one common type of twin T.

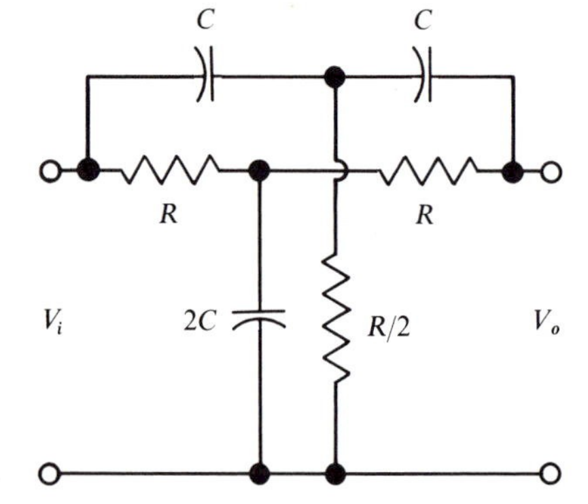

Figure 17-26 See Prob. 17-18.

a) Find the Y_1 and Y_2 of Fig. 17-25c for the twin T of Fig. 17-26, setting $R\omega C = x$, and show that $\mathbf{V}_o = 0$ when $x = 1$.

Solution: In the circuit of Fig. 17-25c, let us call Y_1 the admittance obtained by transforming the R-R-$2C$ star of Fig. 17-26. Then

$$Y_1 = \frac{(1/R)(1/R)}{(1/R) + (1/R) + 2j\omega C} = \frac{1}{2R + 2j\omega CR^2}. \tag{1}$$

Similarly, for the C-C-$R/2$ star,

$$Y_2 = \frac{j\omega C j\omega C}{j\omega C + j\omega C + (2/R)} = \frac{j\omega C j\omega CR}{2j\omega CR + 2}. \tag{2}$$

Then $\mathbf{V}_o = 0$ when

$$Y_1 + Y_2 = \frac{1}{2R(1 + jx)} + \frac{j\omega C jx}{2(jx + 1)} = \frac{1 - x^2}{2R(1 + jx)} = 0, \tag{3}$$

or when

$$1 - x^2 = 0, \tag{4}$$

$$x = 1. \tag{5}$$

b) Find $\mathbf{V}_o/\mathbf{V}_i$ as a function of x.

Assume that the load impedance connected to the right-hand terminals is infinite.

Solution: Considering first the star C-C-$R/2$, Y_3 of Fig. 17-25c is given by

$$Y_3 = \frac{j\omega C(2/R)}{j\omega C + (2/R) + j\omega C} = \frac{j\omega C}{1 + jx}. \tag{6}$$

For the star R-R-$2C$,

$$Y_4 = \frac{(1/R)2j\omega C}{(1/R) + 2j\omega C + (1/R)} = \frac{j\omega C}{1 + jx}. \tag{7}$$

Now

$$\frac{\mathbf{V}_o}{\mathbf{V}_i} = \frac{\dfrac{1}{Y_3 + Y_4}}{\dfrac{1}{Y_1 + Y_2} + \dfrac{1}{Y_3 + Y_4}} = \frac{Y_1 + Y_2}{Y_1 + Y_2 + Y_3 + Y_4}, \tag{8}$$

where

$$Y_1 + Y_2 = \frac{1}{2R(1 + jx)}(1 - x^2), \tag{9}$$

$$Y_3 + Y_4 = \frac{2j\omega C}{1 + jx}. \tag{10}$$

Thus

$$\frac{\mathbf{V}_o}{\mathbf{V}_i} = \frac{(1 - x^2)/(2R)}{[(1 - x^2)/2R] + 2j\omega C} = \frac{1 - x^2}{1 - x^2 + 4jx}. \tag{11}$$

c) Draw a curve of $|\mathbf{V}_o/\mathbf{V}_i|$ as a function of x between $x = 0.1$ and $x = 10$. Use a logarithmic scale for x.

Solution: See Fig. 17-27.

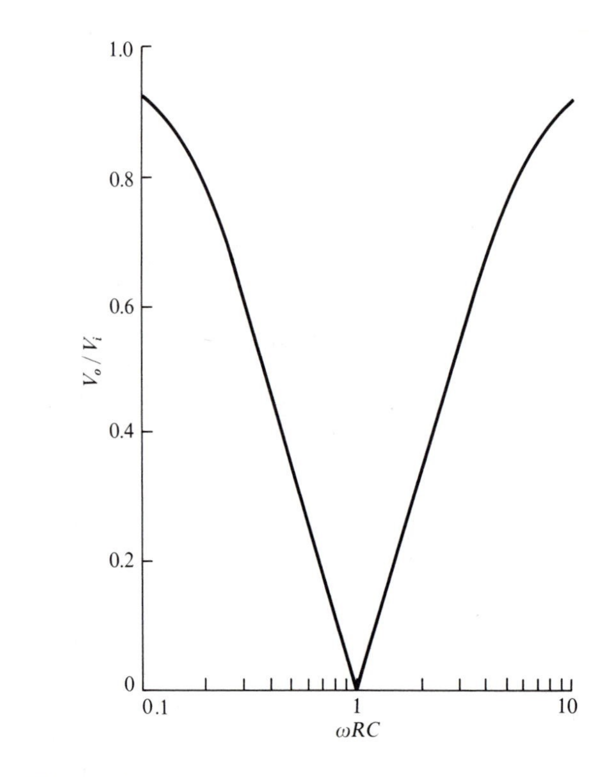

Figure 17-27 See Prob. 17-18.

17-19 *BRIDGED* T

Show that, for the bridged T of Fig. 17-28, $V_o = 0$ when

$$\omega^2 C^2 rR = 1, \qquad \omega^2 LC = 2.$$

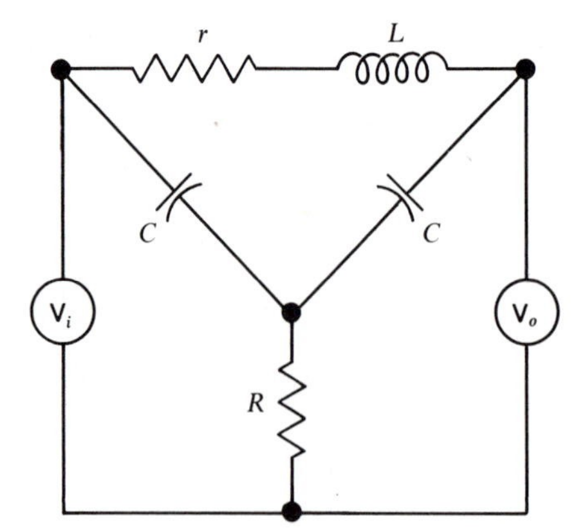

Figure 17-28 See Prob. 17-19.

17-20 *MUTUAL INDUCTANCE*

Find the rms values of the currents $\mathbf{I}_1$ and $\mathbf{I}_2$ in the inductances of the circuit of Fig. 17-29 when

$$V = 10 \text{ volts}, \qquad R = 5 \text{ ohms}, \qquad R_1 = R_2 = 10 \text{ ohms},$$

$$L_1 = L_2 = 1 \text{ henry}, \qquad M = +0.9 \text{ henry}, \qquad f = 1 \text{ kilohertz}.$$

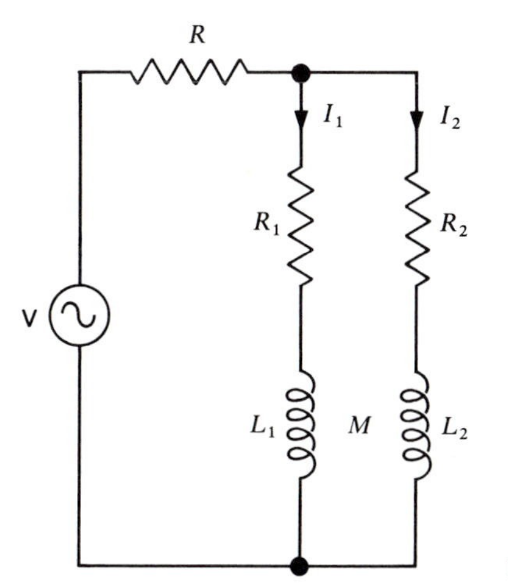

Figure 17-29 See Prob. 17-20.

CHAPTER 18

ALTERNATING CURRENTS: III

Power Transfer and Transformers

In this last chapter on alternating currents we shall be concerned with the transfer of power from a source to a load, first directly, and then through a transformer.

Transmission lines that carry large amounts of power over large distances must operate, for reasons of efficiency, at voltages of a few hundred thousand volts; some even operate at close to one million volts. But insulation problems prevent generators in power plants from being operated above about 20,000 volts. At the other end of the lines, consumers require electric power at voltages ranging from a fraction of a volt for soldering guns to 6000 volts for electric motors of a few hundred kilowatts.

It is the function of transformers to change the ratio V/I with little loss of power. In this way, electric power can be produced, transmitted, and utilized at the most convenient voltages, within limits. Transformers are also used for impedance matching, as we shall see in Sec. 18.6.

Transformers can act only on alternating currents.

18.1 POWER DISSIPATION

Let us consider the processes whereby the electric and magnetic energy flowing in a circuit is dissipated. Until now, we have considered only the I^2R loss in a resistance R carrying a current I. Of course, many other processes are possible. For example, electric motors produce mechanical energy, storage batteries can produce chemical energy, light bulbs and antennas produce electromagnetic waves, loudspeakers produce acoustic waves, and so forth.

Any one of these devices, when connected to a source, draws a current, and the ratio **V**/**I** is its impedance. How is this impedance related to the power that is withdrawn from the circuit?

Let us think of a current **I** flowing through an impedance $R + jX$. If the applied voltage **V** is $V_0 \exp j\omega t$,

$$\mathbf{I} = \frac{\mathbf{V}}{R + jX} = \frac{\mathbf{V}}{R^2 + X^2}(R - jX) = \frac{\mathbf{V}}{|Z|^2}(R - jX), \tag{18-1}$$

$$= \frac{V_0 \exp j\omega t}{|Z|^2}[R + X \exp(-j\pi/2)] \tag{18-2}$$

and, since I is the real part of **I**,

$$I = \frac{V_0}{|Z|^2}[R \cos \omega t + X \cos(\omega t - \pi/2)]. \tag{18-3}$$

The instantaneous power absorbed by the impedance is the product of this quantity by $V_0 \cos \omega t$:

$$P_{\text{inst}} = \frac{V_0^2}{|Z|^2}[R \cos^2 \omega t + X \cos(\omega t - \pi/2) \cos \omega t]. \tag{18-4}$$

Now the average of $\cos^2 \omega t$ over one cycle is $\frac{1}{2}$, while the average value of the second term between the brackets is zero. Thus the average power absorbed by the impedance $R + jX$ is

$$P_{\text{av}} = \frac{1}{2}\frac{V_0^2}{|Z|^2}R = \frac{V_{\text{rms}}^2}{|Z|^2}R = I_{\text{rms}}^2 R. \tag{18-5}$$

Note that P_{av} is $I_{\text{rms}}^2 R$, as for direct currents, but that it is V_{rms}^2/R *only if* $X = 0$.

If the impedance is that of some device that draws power from a circuit, then $I_{\text{rms}}^2 R$ must account for *all* the electric power that is absorbed.

18.1.1 *EXAMPLES: A LIGHT BULB, A RADIO-FREQUENCY ANTENNA, AN ELECTRIC MOTOR*

In a light bulb, $I_{\text{rms}}^2 R$ is equal to the heat energy that is lost by conduction and convection, plus the electromagnetic energy that is radiated in one second.

A radio-frequency antenna is usually tuned to have a purely resistive impedance R, which is called its *radiation resistance*. If the current fed to the antenna is I, then $I^2_{\text{rms}}R$ is the radiated power.

In an alternating-current electric motor, $I^2_{\text{rms}}R$ includes: Joule losses (Sec. 5.2.1) in the copper conductors; hysteresis losses (Sec. 14.9) in the iron; eddy-current losses (Sec. 11.3) both in the copper and in the iron; friction losses in the bearings; windage losses in the air; and, finally, useful mechanical power. The resistance R of an electric motor in operation is thus the sum of many terms. It is much larger than the resistance measured with an ohmmeter when the motor is stopped. See Prob. 18-1.

18.2 THE POWER FACTOR

The ratio $R/|Z|$ of an impedance is called its *power factor* and is usually expressed as a percentage. If the impedance Z is represented in the complex plane, as in Fig. 18-1, the power factor is $\cos \varphi$, φ being the phase angle (Sec. 17.1).

Whenever large amounts of power must be transmitted from a source to a load over a line of appreciable resistance, care must be taken to adjust the power factor close to unity. The reason for this is that the component of **I** that is in quadrature with **V** produces no useful power in the load, but nonetheless gives rise to a power loss and to a voltage drop in the line connecting the source to the load.

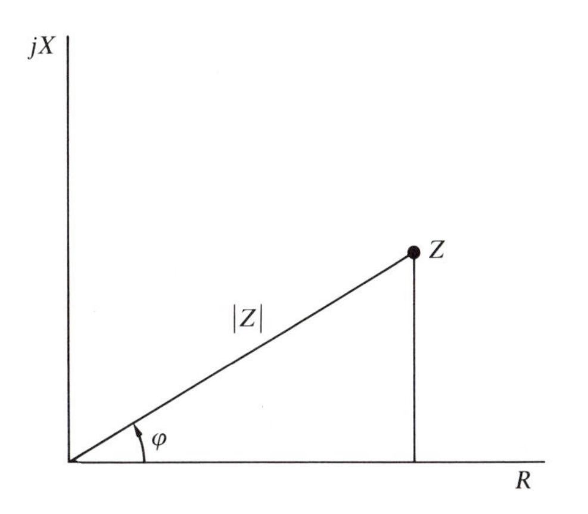

Figure 18-1 Impedance Z represented in the complex plane. The cosine of the phase angle φ is called the *power factor*.

Electric motors are inductive loads, and, in large installations, the power factor is corrected by connecting capacitors across the line, as close as possible to the motors. See Probs. 18-2 and 18-3.

18.2.1 *EXAMPLES: RESISTORS, CAPACITORS, AND INDUCTORS*

The power factor of a resistor is 100%, and that of a capacitor is 0%.

Real inductors have both resistance and inductance.† If $R = \omega L$, then $\varphi =$ 45 degrees, and the power factor is 70.7%.

18.3 POWER TRANSFER FROM SOURCE TO LOAD

In Fig. 18-2, we have replaced the source by an ideal voltage source $\mathbf{V}$, in series with an impedance $R_s + jX_s$. This is the general form of Thévenin's theorem (Sec. 5.12).

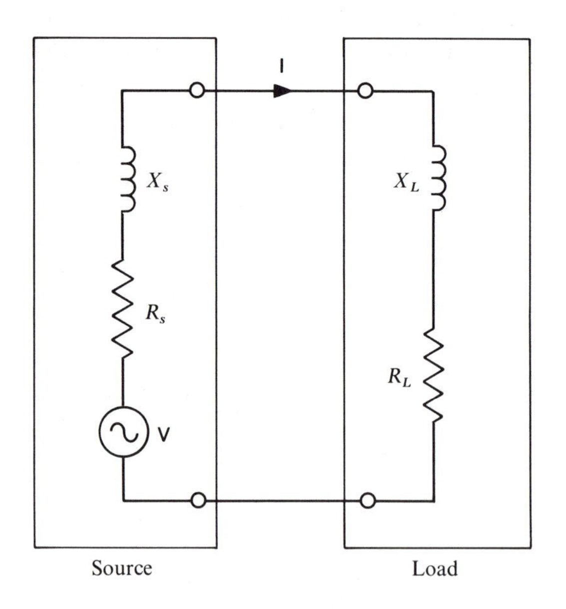

Figure 18-2 Source feeding an impedance $Z_L = R_L + jX_L$.

† If the frequency is quite high they become capacitive. See Prob. 17-2.

The *average* power dissipated in the load is

$$P_L = I_{\text{rms}}^2 R_L = \frac{V_{\text{rms}}^2}{(R_L + R_s)^2 + (X_L + X_s)^2} R_L. \tag{18-6}$$

Under what conditions is P_L maximum, for a given V_{rms}? First, $X_L + X_s$ should be zero:

$$X_L = -X_s. \tag{18-7}$$

In other words, the reactance of the load should be of the same magnitude as the reactance of the source, but of the opposite sign.

If that condition is satisfied, or if the reactances are both zero,

$$P_L = \frac{R_L V_{\text{rms}}^2}{(R_L + R_s)^2}. \tag{18-8}$$

We can now find the optimum value of R_L by setting

$$\frac{dP_L}{dR_L} = \frac{(R_L + R_s)^2 - R_L \times 2(R_L + R_s)}{(R_L + R_s)^4} V_{\text{rms}}^2 = 0. \tag{18-9}$$

Then

$$R_L = R_s. \tag{18-10}$$

Thus, for maximum power transfer to the load,

$$X_L = -X_s, \qquad R_L = R_s. \tag{18-11}$$

The load impedance is then said to be *matched* to that of the source.

The maximum power that can be dissipated in the load, for a given open-circuit voltage V_{rms} is

$$P_{L\max} = I_{\text{rms}}^2 R_L = \left(\frac{V_{\text{rms}}}{2R_s}\right)^2 R_s = \frac{V_{\text{rms}}^2}{4R_s}. \tag{18-12}$$

Also, from Eqs. 18-8 and 18-12,

$$\frac{P_L}{P_{L\max}} = \frac{R_L V_{\text{rms}}^2}{(R_L + R_s)^2} \frac{4R_s}{V_{\text{rms}}^2} = \frac{4(R_L/R_s)}{[(R_L/R_s) + 1]^2}. \tag{18-13}$$

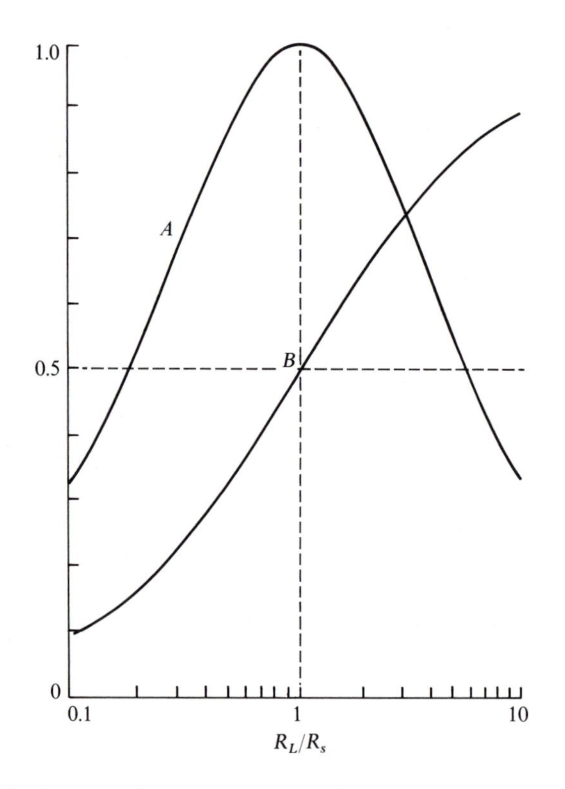

Figure 18-3 A: $P_L/P_{L\max}$ as a function of R_L/R_s. The power dissipated in the load is maximum when the load resistance is equal to the resistance of the source. B: Efficiency as a function of R/R_s. The efficiency tends to unity as $R \to \infty$. It is only 50% at maximum power transfer.

Figure 18-3 shows $P_L/P_{L\max}$ as a function of R_L/R_s. It will be observed that the condition $R_L = R_s$ for maximum power transfer is not critical. For example, with $R_L = 2R_s$, the power P_L dissipated in the load is still 89% of $P_{L\max}$.

The *efficiency* may be defined as the ratio $P_L/(P_L + P_s)$, where P_s is the power dissipated in the source:

$$\frac{P_L}{P_L + P_s} = \frac{I_{\text{rms}}^2 R_L}{I_{\text{rms}}^2 (R_L + R_s)} = \frac{R_L}{R_L + R_s}, \tag{18-14}$$

$$= \frac{R_L/R_s}{(R_L/R_s) + 1}. \tag{18-15}$$

Figure 18-3 also shows how the efficiency varies with the ratio R_L/R_s. The efficiency is equal to unity when $R_L/R_s \to \infty$, but then the power dissipated in the load tends to zero. For $R_L = R_s$, the power dissipated in the load is maximum and the efficiency is only 50%.

18.3.1 *EXAMPLE: WHIP ANTENNA*

A radio-frequency transmitter has an output resistance of 50 ohms. It is to be connected to the whip-antenna of Fig. 18-4, which has a radiation resistance of 37 ohms. What will be the efficiency of power transfer?

In this case,

$$R_s = 50, \qquad X_s = 0, \qquad R_L = 37, \qquad X_L = 0. \tag{18-16}$$

The efficiency will be 37/(50 + 37), or about 43%. This is satisfactory, compared with the optimum value of 50%.†

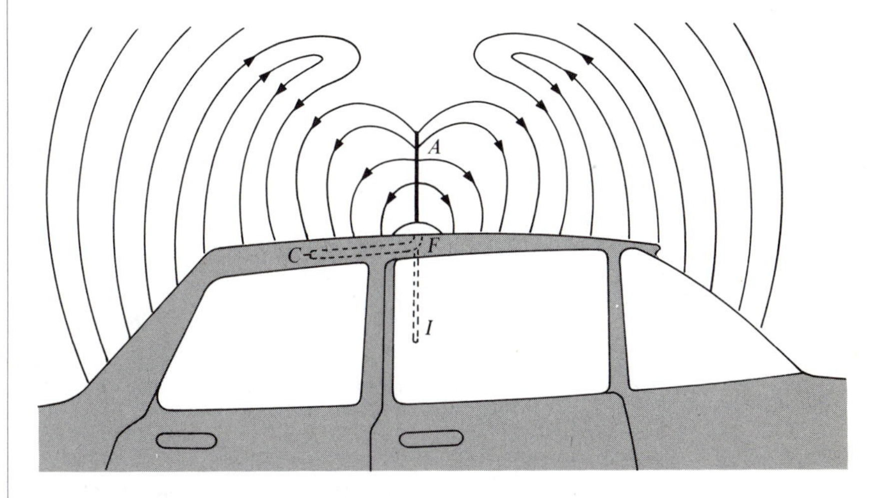

Figure 18-4 Quarter-wave whip antenna A mounted on a metal car top. The whip is one-quarter wavelength long and is supported by a feed-through insulator F. It is fed by a transmitter situated at the back of the car through a coaxial line C. The whip, together with its image I (Prob. 3-15) forms a half-wave antenna. Charges flowing in the antenna form lines of force that detach themselves and escape into space at the velocity of light. This type of antenna is used at frequencies ranging from 150 to 450 megahertz.

† One could achieve 50% efficiency by using a matching circuit. See, for example, *The Radio Amateur's Handbook*, American Radio Relay League, Newington, Conn., U.S.A. (published every other year).

18.4 TRANSFORMERS

A transformer is a mutual inductor that is used to change the voltage level of an electric current, as explained briefly in the introduction to this chapter.

The mutual inductor can have either a magnetic or a non-magnetic core. The former type is commonly referred to as a *magnetic-core transformer*, and the latter as an *air-core transformer*. Air-core transformers are used at high frequencies where eddy-current and hysteresis losses in a magnetic core would be excessive (Secs. 11.3 and 14.9).

The equivalent star circuit (Sec. 17.5) will give us some general results that are exact for air cores, and only approximate for magnetic cores.

Figure 18-5a shows a transformer inserted between a source supplying a voltage $\mathbf{V}_1$, and a load impedance Z_L. The impedance $R_1 + j\omega L_1$ of the primary winding is Z_1, and the impedance $R_2 + j\omega L_2$ of the secondary winding plus Z_L, is Z_2.

The mutual inductance is

$$M = k(L_1 L_2)^{1/2}. \tag{18-17}$$

Since the circuit of Fig. 18-5b is equivalent to that of Fig. 18-5a, from Sec. 17.5, the impedance "seen" by the source, or the *input impedance* Z_{in} of the transformer is

$$Z_{\text{in}} = Z_1 + j\omega M + \frac{(Z_2 + j\omega M)(-j\omega M)}{Z_2 + j\omega M - j\omega M}, \tag{18-18}$$

$$= Z_1 + \frac{\omega^2 M^2}{Z_2}. \tag{18-19}$$

The input impedance is therefore Z_1, plus the term $\omega^2 M^2/Z_2$, which is called the *reflected impedance* of the secondary.

The current $\mathbf{I}_1$ in the primary is of course

$$\mathbf{I}_1 = \mathbf{V}_1/Z_{\text{in}}. \tag{18-20}$$

The current in the secondary can be found by applying Kirchoff's voltage law to the circuit of Fig. 18-5b, using the value of $\mathbf{I}$ given in Eqs. 18-19 and 18-20:

$$\mathbf{I}_2 = -\frac{j\omega M}{Z_1 Z_2 + \omega^2 M^2}\mathbf{V}_1. \tag{18-21}$$

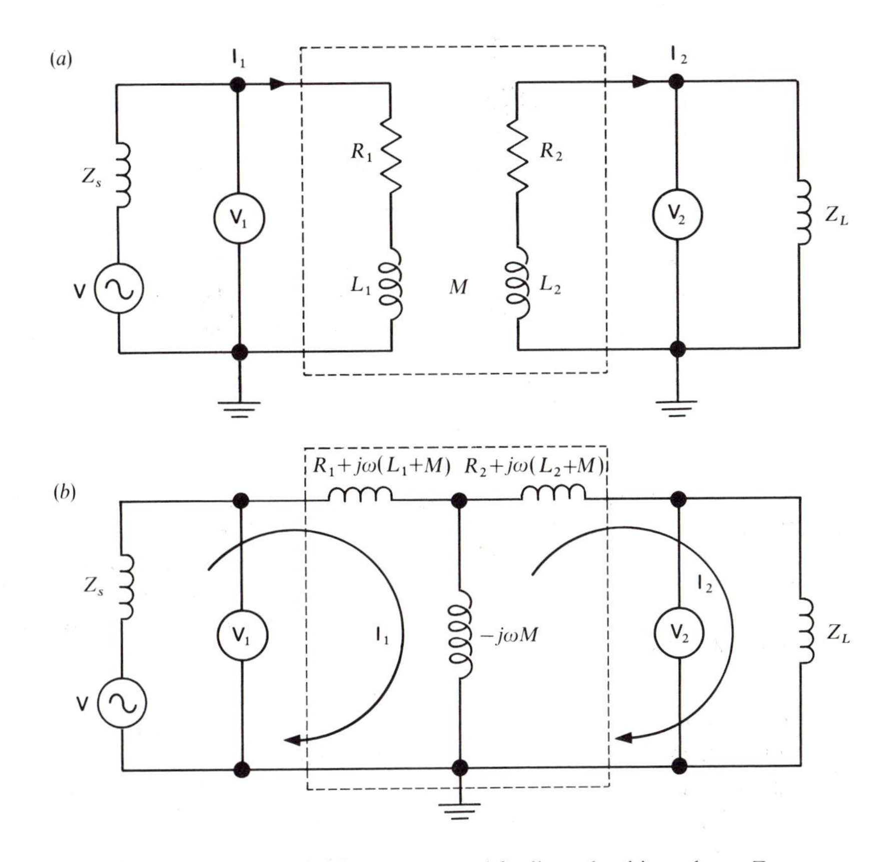

Figure 18-5 (*a*) Transformer fed by a source and feeding a load impedance Z_L. (*b*) Equivalent circuit.

18.5 MAGNETIC-CORE TRANSFORMERS

Magnetic-core transformers as in Fig. 18-6 are used at frequencies ranging from a few hertz to about one megahertz.

A magnetic-core transformer has two essential features. First, for a given core cross-section, the magnetic flux per ampere-turn in the primary is larger than with an air core by several orders of magnitude. Second, the magnetic flux through the secondary is nearly equal to that through the primary.

Thus, the magnetic core permits the design of transformers that are smaller and have fewer turns, both in the primary and in the secondary.

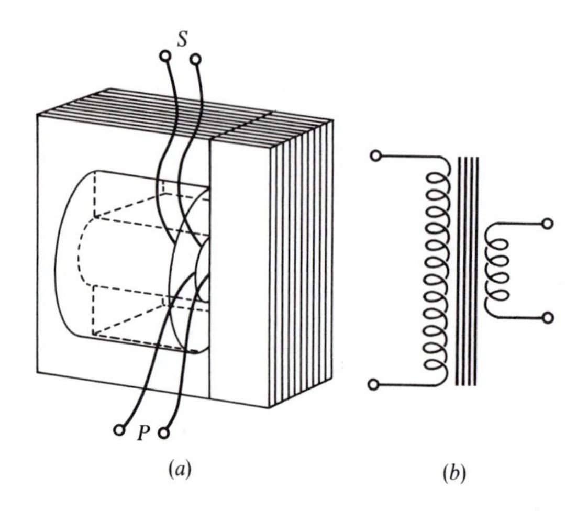

Figure 18-6 (*a*) Common type of magnetic-core transformer. The core has the shape of a figure eight. The primary and secondary windings *P* and *S* are wound on the center leg. The core is made of two types of lamination, one having the shape of an *E*, and the other the shape of an *I*, placed one next to the other. The laminations are assembled after the coil has been wound. (*b*) Schematic diagram of a magnetic-core transformer.

Also, with a magnetic core, there is very little stray magnetic field. For example, one can operate two such transformers side by side with negligible interaction.

Cores made of iron alloys are usually formed of parallel insulated sheets called *laminations*. This reduces the losses due to the eddy currents induced by the changing magnetic field. The laminations have a thickness of a fraction of a millimeter. At audio frequencies the alloy is sometimes in the form of a fine powder molded in a bonding agent. One then has a *powdered iron* core.

Ferrites are used at audio frequencies and above. These are ceramic-type materials that are molded from oxides of iron and various other metals. Their main advantage is that their conductivities are low, of the order of 1 siemens per meter, which is about 10^{-6} times that of transformer iron. The eddy current losses in ferrites are thus low.

The analysis of magnetic-core transformers is rendered difficult by several factors. First, the relationship between **B** and **H** in ferromagnetic materials is not linear (Sec. 14.9). For example, if the flux linkage in a circuit is Λ when it carries a current I, then the self-inductance L is Λ/I, as in Sec. 12.3. If there are no magnetic materials in the field, Λ is proportional to I and Λ/I depends solely on the geometry of the circuit. However, in the presence of magnetic materials, Λ is not proportional to I and the self-inductance L required for the calculations can only be approximate.

Moreover, the losses in a magnetic-core transformer are complex: there are eddy-current losses (Sec. 11.3) both in the iron core and in the copper windings, hysteresis losses in the iron core (Sec. 14.9), and Joule losses resulting from the currents flowing in the windings. All these losses can be expressed as an I^2R loss in the primary, but, for a given transformer, R depends on the voltage applied to the primary, on the current drawn from the secondary, and on the frequency.

18.5.1 *THE IDEAL TRANSFORMER*

We shall make the following three assumptions. (a) There are no Joule or eddy-current losses. (b) The hysteresis loop for the core is a straight line through the origin. Then $\mathbf{B}$ is proportional to $\mathbf{H}$ and there are no hysteresis losses either. (c) There is zero leakage flux. Then the coefficient of coupling k is equal to unity, and the flux through the primary is equal to that through the secondary.

As a consequence of the first two assumptions, the transformer is lossless, and its efficiency is 100%.

A fictitious transformer having these characteristics is called an *ideal transformer.*

The assumption that B is proportional to H, and hence to I, is most undesirable, but difficult to avoid.

As to the efficiency, it is close to 100% for large industrial transformers, but only about 70 or 80% for a small power transformer supplying, say, 10 watts.

Finally, we shall consider the usual situation, in which the load impedance is real and much smaller than the reactance of the secondary winding:

$$Z_L = R_L \ll \omega L_2. \tag{18-22}$$

18.5.2 *THE RATIO* $\mathbf{V}_1/\mathbf{V}_2$

With these assumptions, the voltage applied to the primary is

$$\mathbf{V}_1 = N_1 \frac{d\boldsymbol{\Phi}}{dt} = N_1 j\omega\boldsymbol{\Phi}, \tag{18-23}$$

and the voltage across the secondary is

$$\mathbf{V}_2 = N_2 \frac{d\boldsymbol{\Phi}}{dt} = N_2 j\omega\boldsymbol{\Phi} = \mathbf{I}_2 R_L, \tag{18-24}$$

where N_1 and N_2 are the numbers of turns in the primary and in the secondary windings, respectively, and where $\boldsymbol{\Phi}$ is the magnetic flux in the core:

$$\boldsymbol{\Phi} = \boldsymbol{\Phi}_1 + \boldsymbol{\Phi}_2, \tag{18-25}$$

$$= \frac{N_1 \mathbf{I}_1}{\mathscr{R}} + \frac{N_2 \mathbf{I}_2}{\mathscr{R}}, \tag{18-26}$$

where $\mathscr{R}$ is the reluctance of the core, from Sec. 15.1.

Thus

$$\mathbf{V}_1/\mathbf{V}_2 = N_1/N_2 \tag{18-27}$$

and, for an ideal transformer and for a given $\mathbf{V}_1$, $\mathbf{V}_2$ is independent of the load current. Also, $\mathbf{V}_2$ is either in phase with $\mathbf{V}_1$, or π radians out of phase. Of course, the phase of $\mathbf{V}_2$ can be changed by π radians at will by interchanging the wires coming out of the secondary winding.

Note that $\mathbf{V}_1$ is equal to $N_1 j\omega\boldsymbol{\Phi}$. Thus, for a given applied voltage, the total flux $\boldsymbol{\Phi}_1 + \boldsymbol{\Phi}_2$ is approximately independent of the current drawn from the secondary.

Also, Φ is B times the cross-section of the core. Since the maximum possible value for B depends on the core material, Eq. 18-23 shows that an increase in frequency permits the use of a core having a smaller cross-section.

18.5.3 THE RATIO L_1/L_2

Note that

$$L_1 = \Lambda_1/\mathbf{I}_1 = N_1\boldsymbol{\Phi}_1/\mathbf{I}_1 = N_1^2/\mathscr{R}. \tag{18-28}$$

Similarly,

$$L_2 = N_2^2/\mathscr{R}, \tag{18-29}$$

so that

$$L_1/L_2 = N_1^2/N_2^2. \tag{18-30}$$

18.5.4 THE INPUT IMPEDANCE Z_{in}

Since the coupling coefficient k is equal to unity, then, from Sec. 12.4,

$$M = (L_1 L_2)^{1/2}. \tag{18-31}$$

Also,

$$Z_1 = j\omega L_1, \qquad Z_2 = R_L + j\omega L_2, \tag{18-32}$$

where R_L is the load resistance connected between the terminals of the secondary.

Thus, from Eq. 18-19,

$$Z_{in} = j\omega L_1 + \frac{\omega^2 L_1 L_2}{R_L + j\omega L_2} = j\omega L_1 \left(1 - \frac{j\omega L_2}{R_L + j\omega L_2}\right), \tag{18-33}$$

$$= \frac{R_L j\omega L_1}{R_L + j\omega L_2}, \tag{18-34}$$

$$= \frac{L_1}{L_2} R_L = \left(\frac{N_1}{N_2}\right)^2 R_L. \qquad (R_L \ll \omega L_2) \tag{18-35}$$

18.5.5 THE RATIO $\mathbf{I}_1/\mathbf{I}_2$

From Sec. 18.4,

$$\mathbf{I}_1 = \frac{\mathbf{V}_1}{Z_{in}} = \frac{R_L + j\omega L_2}{R_L j\omega L_1} \mathbf{V}_1, \tag{18-36}$$

$$\mathbf{I}_2 = -\frac{j\omega M}{Z_1 Z_2 + \omega^2 M^2} \mathbf{V}_1 = -\frac{j\omega M}{j\omega L_1 (R_L + j\omega L_2) + \omega^2 L_1 L_2} \mathbf{V}_1, \tag{18-37}$$

$$= -\frac{j\omega M}{R_L j\omega L_1} \mathbf{V}_1, \tag{18-38}$$

$$\frac{\mathbf{I}_1}{\mathbf{I}_2} = -\frac{R_L + j\omega L_2}{j\omega M}. \tag{18-39}$$

Disregarding the sign,

$$\mathbf{I}_1/\mathbf{I}_2 = L_2/M = (L_2/L_1)^{1/2} = N_2/N_1. \qquad (R_L \ll \omega L_2) \tag{18-40}$$

18.6 POWER TRANSFER FROM SOURCE TO LOAD THROUGH A TRANSFORMER

As we saw in Sec. 18.3, the power dissipated in a load is maximum when its resistance R_L is equal to the internal resistance of the source R_s, and when $X_L = -X_s$. As a rule, the reactances are zero. If it is impossible to vary either R_L or R_s, then one can still achieve optimum power transfer by inserting a transformer between source and load as in Fig. 18-7a. Let us see how this comes about.

Let us assume that the transformer has a magnetic core, that the approximations used in Sec. 18.5.1 are satisfactory, and that $X_s = X_L = 0$. Then, from Eq. 18-35, the transformer has an input impedance of $(N_1/N_2)^2 R_L$ ohms. In other words, the current and the power supplied by the source are precisely the same as if the transformer and its load resistance R_L were replaced by the resistance $(N_1/N_2)^2 R_L$, as in Fig. 18-7b. Remember that, with our approximation, the transformer has an efficiency of 100%.

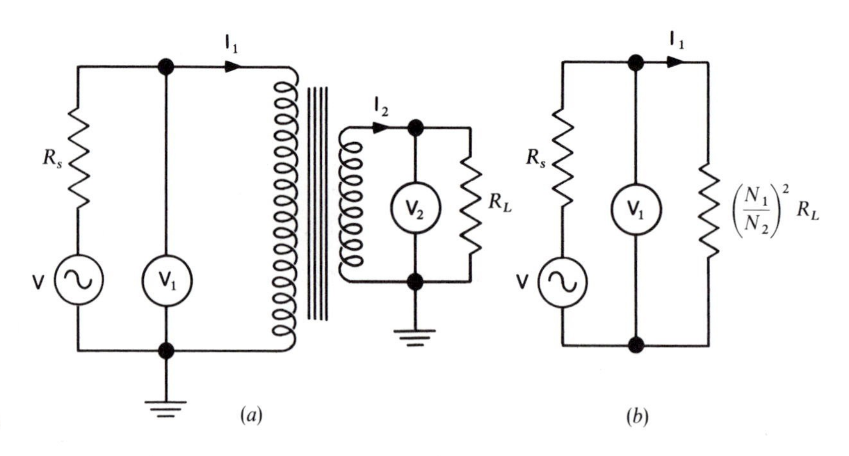

Figure 18-7 (*a*) Source feeding a load resistance R_L through a magnetic-core transformer. (*b*) Equivalent circuit.

Then, with a transformer as in Fig. 18-7a, the power transfer is optimum when

$$R_s = \frac{N_1^2}{N_2^2} R_L, \tag{18-41}$$

or when the turns ratio is

$$N_1/N_2 = (R_s/R_L)^{1/2}. \tag{18-42}$$

When this condition is satisfied, the power dissipation in the load is maximum, but the efficiency is again only 50%. The transformer is then said to be used for *impedance matching*.

18.7 SUMMARY

Any device that transforms the energy flowing in an electric circuit into some other form has a resistance R such that I^2R accounts for *all* the energy withdrawn from the circuit.

The *power factor* of the load is the ratio $R/|Z|$, or the cosine of the phase of $\mathbf{I}$ with respect to $\mathbf{V}$, and is usually expressed as a percentage.

For maximum power dissipation in a load, the following two conditions should be satisfied:

$$X_L = -X_s, \qquad R_L = R_s. \tag{18-11}$$

where the subscripts L and s refer, respectively, to the load and to the source.

The impedances are then said to be *matched*, and the efficiency is 50%. The *efficiency* is defined as the power delivered to the load, divided by the power produced by the source.

The *input impedance of a transformer* is equal to the impedance Z_1 of the primary winding, plus the *reflected impedance* of the secondary circuit, which is ω^2M^2/Z_2, where Z_2 is the impedance of the secondary winding, plus the load impedance Z_L.

For an *ideal magnetic-core transformer*,

$$\mathbf{V}_2/\mathbf{V}_1 = \mathbf{I}_1/\mathbf{I}_2 = N_2/N_1 \qquad (18\text{-}27,\ 18\text{-}40)$$

and the input impedance is

$$Z_{\text{in}} = \frac{L_1}{L_2} R_L = \frac{N_1^2}{N_2^2} R_L. \tag{18-35}$$

A transformer can serve to match the impedance of a load to that of a source, by making

$$N_1/N_2 = (R_s/R_L)^{1/2}. \tag{18-42}$$

The efficiency under these conditions is again 50%.

PROBLEMS

18-1E *DIRECT-CURRENT MOTORS*

Let us see how a direct-current motor operates.

Figure 18-8 shows a highly schematic diagram of a *series motor*. The term "series" refers to the fact that the field winding and the rotating coil, called the *armature*, are

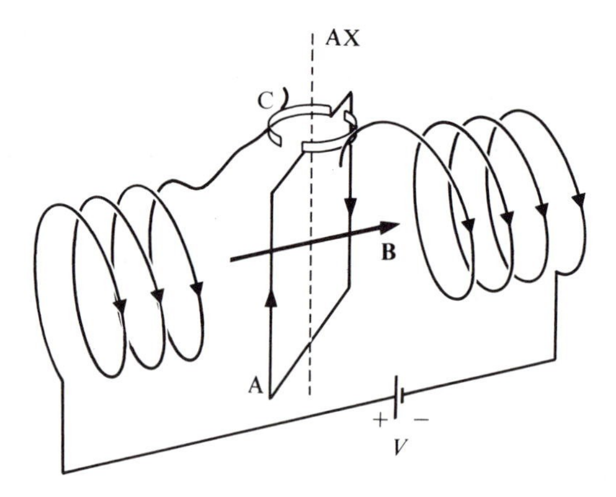

Figure 18-8 Schematic diagram of a direct-current series motor. The armature A rotates about the vertical axis AX. It is connected in series with the coils producing the magnetic field **B** through a commutator C. The commutator is a split copper ring. Contact to the ring is made with carbon brushes. The direction of the current through the armature is inverted twice every turn. The magnetic core is not shown. See Prob. 18-1.

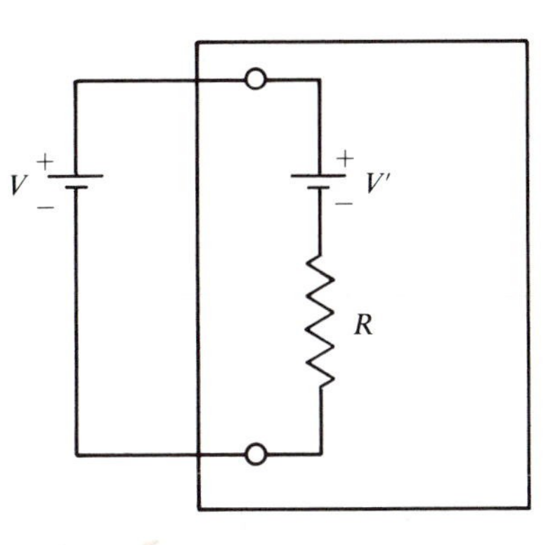

Figure 18-9 Equivalent circuit V', R of the electric motor of Fig. 18-8.

in series. In a *shunt motor* they are in parallel. *Compound motors* obtain part of their field with a series winding, and part with a shunt winding.

When the motor is stopped, it acts as a resistance. The resistance of the armature is much smaller than that of the field windings.

When the motor is in operation, the motion of the armature in the magnetic field generates a *counterelectromotive force* V' that opposes the flow of current as in Fig. 18-9. Here R is the resistance associated with the various losses enumerated in Sec. 18.1.1.

a) Use the substitution theorem to show that the mechanical power is equal to IV'.

b) How would you define the efficiency?

c) If a direct-current series motor is connected to a source of power and run without a load, it will gain more and more speed and, very probably, the armature will burst. A direct-current *series* motor that is not connected directly to its load is therefore *dangerous*.

Why is this?

Remember that V' is proportional to the speed of the motor, multiplied by the magnetic induction at the armature winding.

d) What happens when the mechanical power is increased?

18-2 *POWER-FACTOR CORRECTION*

A load has a power factor of 65% and draws a current of 100 amperes at 600 volts.

a) Calculate the magnitude of Z, its phase angle, and its real and imaginary parts.

b) Calculate the in-phase and quadrature components of the current.

c) What size capacitor should be placed in parallel with the load to cancel the reactive current at 60 hertz?

18-3 *POWER-FACTOR CORRECTION WITH FLUORESCENT LAMPS*

A *fluorescent lamp* consists of an evacuated tube containing mercury vapor and coated on the inside with a fluorescent powder. A discharge is established inside the tube, between electrodes situated at each end. The discharge emits most of its energy at 253.7 nanometers, in the ultraviolet. The fluorescent coating absorbs this radiation and re-emits visible light.

The discharge will operate correctly only when connected in series with an impedance. A resistor could be used, but would dissipate an excessive amount of energy, so it is the custom to use a series inductor. There exist many types of circuit.

A particular fluorescent fixture operating at 120 volts has a power dissipation of 80 watts. Its power factor is 50%.

a) Find the reactive current.

b) What is the size of the capacitor connected in parallel with the discharge tube and its inductor that will make the power factor equal to 100%?

18-4 *ENERGY TRANSFER TO A LOAD*

Figure 18-10 shows a circuit that has been used to measure the energy absorbed by a load Z when the source V produces a pulse of current. The currents I' and I'' are negligible compared to I, and the voltage drop across r is negligible compared to V. As we shall see, this energy is proportional to the area enclosed by the curve on the oscilloscope screen. The method is valid is valid even if Z is non-linear.

We shall demonstrate this for one cycle of an alternating current, but the same applies to an individual current pulse of any shape.

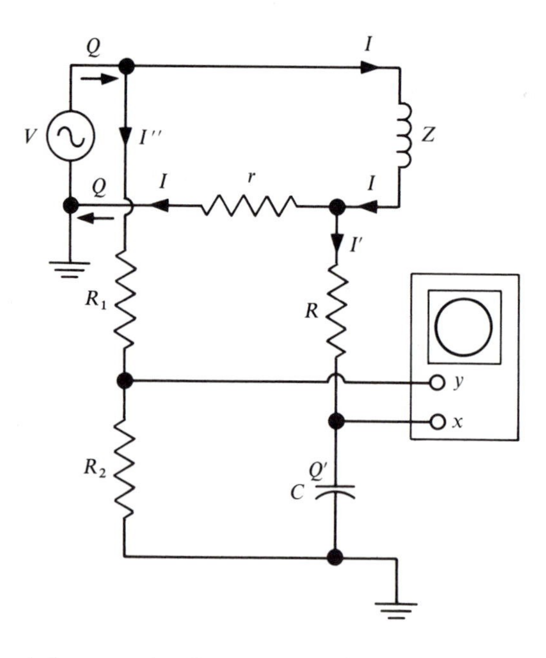

Figure 18-10 Circuit for measuring the energy transfer to a load Z during a current pulse; $r \ll Z$, $r \ll R$, $R \gg 1/\omega C$. The charge Q flows out of the top terminal of the source and into the lower one. The terminals x and y are the inputs to the horizontal and vertical amplifiers of the oscilloscope. See Prob. 18-4.

On the screen,

$$x \propto \frac{Q'}{C}, \qquad y \propto \frac{R_2}{R_1 + R_2} V.$$

Set $V = V_0 \cos \omega t$.

a) Show that, if $R \gg 1/\omega C$,

$$x \propto \frac{r}{RC} Q.$$

b) Show that the energy transferred per cycle is proportional to the area enclosed by the curve on the oscilloscope screen.

c) What is the shape of the curve when the load is (i) a resistance, (ii) a capacitance, (iii) a pure inductance, (iv) a resistance in series with an inductance, with $R = \omega L$?

d) What is the shape of the curve if the load draws current only when V is close to its maximum value?

18-5E *ENERGY TRANSFER TO A LOAD*

Figure 18-11 shows another circuit for measuring the energy dissipated when a pulse of current originating in G passes through a load Z. The resistor R' serves to discharge the capacitor C slowly between pulses; a negligible amount of charge flows through it during a pulse. Also, C is large and the voltage drop across it is negligible compared to that across Z.

Show that the area under the curve observed on the oscilloscope is proportional to the energy dissipated in Z, whether Z is linear or not.

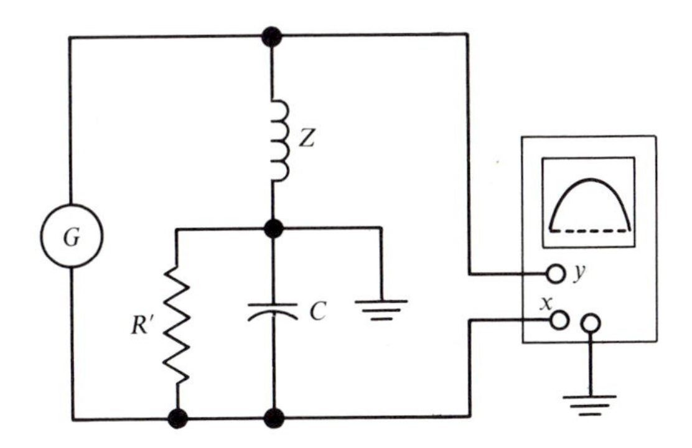

Figure 18-11 See Prob. 18-5.

18-6E *REFLECTED IMPEDANCE*

Show that a positive (inductive) reactance in the secondary of a transformer is equivalent to a negative (capacitive) reactance in the primary, and vice versa.

18-7E *MEASUREMENT OF THE COEFFICIENT OF COUPLING k*

A transformer has a primary inductance L_1, a secondary inductance L_2, and a mutual inductance M. The winding resistances are negligible.

Show that

$$Z_0/Z_\infty = 1 - k^2,$$

where Z_0 and Z_∞ are the impedances measured at the terminals of the primary, when the secondary is short-circuited and when it is open-circuited.

18-8E *REFLECTED IMPEDANCE*

How does the impedance at the terminals of the primary of a transformer vary with the resistance R_2 in the secondary circuit?

Consider a simple case where $\omega = L_1 = L_2 = k = 1$, $R_1 = 0$.

a) Show that

$$Z = \frac{R_2}{R_2^2 + 1} + j\frac{R_2^2}{R_2^2 + 1}.$$

(Note that, dimensionally, this equation is absurd. The reason is that we have assigned numerical values to some of the parameters. For example, the 1 in the denominators is really $\omega^2 L_2^2$.)

b) Draw curves of R, X, $|Z|$ as functions of R_2, for values of R_2 ranging from 0.1 to 10. Use a log scale for R_2.

18-9E *MEASURING THE AREA UNDER A CURVE*

You are asked to measure the area under a curve drawn on a sheet of paper.

You are given a pair of Helmholtz coils, as in Prob. 8-5, that give a uniform magnetic field over a sufficiently large area, also a source of alternating current, an electronic voltmeter, conducting ink, etc.

How will you proceed?

18-10E *SOLDERING GUNS*

A soldering gun consists in a step-down transformer that passes a large current through a length of copper wire.

One soldering gun dissipates 100 watts in a piece of copper wire ($\sigma = 5.8 \times 10^7$ siemens per meter) having a cross-section of 4 square millimeters and a length of 100 millimeters.

a) Find V and I in the secondary.

b) Find the current in the primary if it is fed at 120 volts.

18-11E *CURRENT TRANSFORMER*

Figure 18-12 shows a side-looking current transformer that is used for measuring large current pulses. It is more convenient than the Rogowski coil of Prob. 12-7 in that the current-carrying wire need not be threaded through the measuring coil.

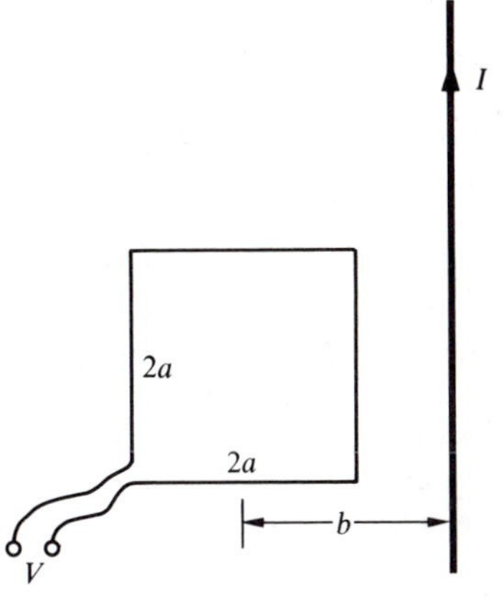

Figure 18-12 Side-looking transformer. See Prob. 18-11.

Show that, for a single-turn coil,

$$V = \frac{\mu_0 a}{\pi} \ln\left(\frac{b + a}{b - a}\right)\frac{dI}{dt}.$$

18-12E *INDUCED CURRENTS*

A thin-walled conducting tube has a length l, a radius a, a wall thickness b, and a conductivity σ. When placed in a uniform axial magnetic field of the form

$$B = B_0 \cos \omega t,$$

an induced current flows in the azimuthal direction.

Alternating currents tend to flow near the outer surface of a conductor. This phenomenon is called the *skin effect*. See Prob. 11-7. The *skin depth* δ is $(2/\mu_r\mu_0\omega\sigma)^{1/2}$.†

We are concerned here with a brass tube having a wall thickness of 0.5 millimeter and with a frequency of 60 hertz. Since the skin depth is 16.3 millimeters under those conditions, we may disregard the skin effect and assume that the induced current is uniformly distributed throughout the thickness of the tube.

Show that, for currents flowing in the azimuthal direction, the resistance and self-inductance of the tube are given by

$$R_t = 2\pi a/\sigma l b, \qquad L_t = \mu_0 \pi a^2/l,$$

so that

$$\omega L_t/R_t = \mu_0 \sigma \omega a b/2.$$

† *Electromagnetic Fields and Waves*, Sec. 11.5.

For a brass tube with $l = 200$ millimeters, $a = 5$ millimeters, $b = 0.5$ millimeter, and $\sigma = 1.60 \times 10^7$ siemens per meter, this ratio is only about 1% at 60 hertz.

18-13 *EDDY-CURRENT LOSSES*

Why do eddy-current losses increase as the square of the frequency?

18-14 *EDDY-CURRENT LOSSES*

Eddy-current losses in magnetic cores are minimized by assembling them from laminations.

Consider a core of rectangular cross-section as in Fig. 18-13. The eddy-current loss is proportional to V^2/R, where V is the electromotance induced around a typical current path such as the one shown by a dashed curve. The resistance is also difficult to define, but it is of the order of twice the resistance of the upper half, or

$$2\frac{a}{\sigma(b/2)L}.$$

Show that, if the core is split into n laminations, the loss will be reduced by a factor of n^2.

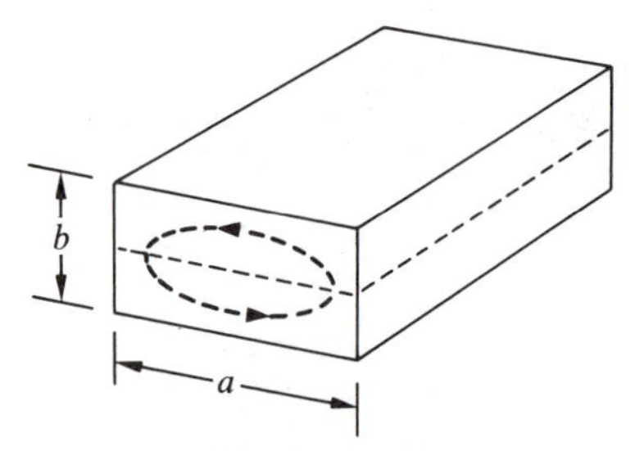

Figure 18-13 See Prob. 18-14.

18-15 *HYSTERESIS LOSSES*

Hysteresis losses are proportional to the operating frequency f, while eddy-current losses increase as f^2 (Prob. 18-13).

You are given a number of transformer laminations. Can you devise an experiment that will permit you to evaluate the relative importance of the two types of loss?

18-16 *CLIP-ON AMMETER*

Figure 18-14 shows a type of clip-on ammeter that is more common than that described in Prob. 15-3. The one shown here is simply an iron-core transformer whose primary is the current-carrying wire. It is similar to the Rogowski coil of Prob. 12-7, except for the iron core, and except for the fact that the secondary does not surround the wire. This type of transformer is often called a *current transformer*.

Instruments of this type are available for measuring alternating or pulsed currents ranging from milliamperes to hundreds of amperes.

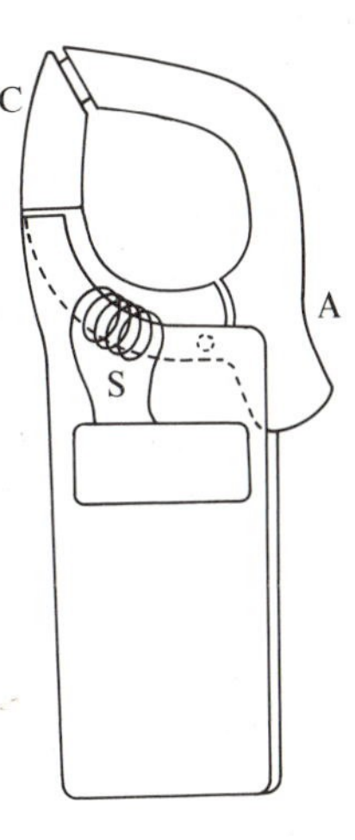

Figure 18-14 Clip-on ammeter. The ring-shaped magnetic yoke is made in two halves that can rotate about the hinge A. The yoke can be clipped around the wire by opening the jaws at C. The voltage V induced in winding S is proportional to the current flowing through the wire. See Prob. 18-16.

a) Calculate V when the measured current is one ampere at 60 hertz. The core has a mean diameter of 30 millimeters, a cross-section of 64 square millimeters, and a permeability of 10,000. The secondary has 1000 turns.

b) How could you increase the voltage V for a given current I and for a given instrument?

18-17 *AUTO-TRANSFORMERS*

Figure 18-15a shows a schematic diagram of an auto-transformer. Its single winding is tapped and serves both as a primary and as a secondary. Auto-transformers are widely used, particularly for obtaining an adjustable voltage at 60 hertz. The left-hand terminals are connected to the 120-volt line, and the right-hand

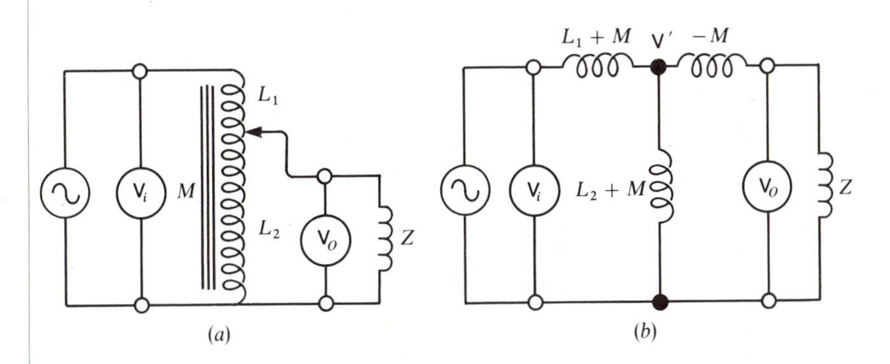

Figure 18-15 (*a*) Auto-transformer. (*b*) Equivalent circuit. See Prob. 18-17.

terminals are connected to the load. The tap can slide along the winding. The power ratings of commercial models vary from a few tens of watts to several kilowatts.

Calculate the ratio of output voltage to input voltage V_o/V_i. The self-inductances L_1 and L_2 are those of the winding, on either side of the tap. Set

$$L_1 = AN_1^2, \qquad L_2 = AN_2^2, \qquad M = (L_1L_2)^{1/2} = AN_1N_2, \tag{1}$$

where M is the mutual inductance between L_1 and L_2, and neglect the resistance of the winding. These assumptions are reasonable, in practice.

You will find that $\mathbf{V}_o$ is directly proportional to the number of turns in L_2, and is independent of Z, under the above assumptions.

Solution: First let us transform the circuit of Fig. 18-15a into that of Fig. 18-15b. Then

$$\frac{\mathbf{V}_o}{\mathbf{V}'} = \frac{Z}{Z - j\omega M}. \tag{2}$$

Similarly,

$$\frac{\mathbf{V}'}{\mathbf{V}_i} = \frac{\dfrac{(Z - j\omega M)j\omega(L_2 + M)}{Z - j\omega M + j\omega(L_2 + M)}}{\dfrac{(Z - j\omega M)j\omega(L_2 + M)}{Z - j\omega M + j\omega(L_2 + M)} + j\omega(L_1 + M)}. \tag{3}$$

Then

$$\frac{\mathbf{V}_o}{\mathbf{V}_i} = \frac{\mathbf{V}_o}{\mathbf{V}'}\frac{\mathbf{V}'}{\mathbf{V}_i} = \frac{Zj\omega(L_2 + M)}{Z[j\omega(L_2 + M) + j\omega(L_1 + M)] + \omega^2 M(L_2 + M) - \omega^2 L_2(L_1 + M)}. \tag{4}$$

Substituting the values of L_1, L_2, M,

$$\frac{\mathbf{V}_o}{\mathbf{V}_i} = \frac{Zj\omega A(N_2^2 + N_1N_2)}{Zj\omega A(N_1^2 + N_2^2 + 2N_1N_2) + \omega^2 A^2(N_1^2N_2^2 - N_1^2N_2^2)}, \tag{5}$$

$$= \frac{N_2}{N_1 + N_2}. \tag{6}$$

CHAPTER 19

MAXWELL'S EQUATIONS

At this stage we have found only three of Maxwell's four equations. These were Eqs. 6-12, 8-12, and 11-14. Our first objective in this chapter is to deduce the fourth one, namely the equation for the curl of **B**. Then we shall re-examine all four equations as a group.

19.1 THE TOTAL CURRENT DENSITY $\mathbf{J}_t$

There are several types of electric current.

a) There are first the usual *conduction currents through good conductors* such as copper, which are associated with the drift of conduction electrons (Sec. 5.1).

b) There are also *conduction currents in semiconductors* in which both conduction electrons and holes (Sec. 5.1) can drift.

c) *Electrolytic currents* are due to the motion of ions through liquids.

d) The motion of ions or electrons in a vacuum—for example, in an ion beam or in a vacuum diode—gives *convection currents.*

e) The *motion of macroscopic charged bodies* also produces electric currents. A good example is the belt of a Van de Graaff high-voltage generator.

All these currents are associated with the motion of free charges. We shall use the symbol $\mathbf{J}_f$ for the current density associated with free charges.

f) The *displacement current.* The displacement current density $\partial\mathbf{D}/\partial t$ has two components, the *polarization current* density $\partial\mathbf{P}/\partial t$, and $\epsilon_0\partial\mathbf{E}/\partial t$ (Sec. 7.6).

g) The *equivalent currents* in magnetic materials, with a volume density $\nabla \times \mathbf{M}$ and a surface density $\mathbf{M} \times \mathbf{n}_1$ (Secs. 14.2 and 14.3).

Thus, in the general case, the *total volume current density* is

$$\mathbf{J}_t = \mathbf{J}_f + \frac{\partial \mathbf{D}}{\partial t} + \nabla \times \mathbf{M}, \tag{19-1}$$

$$= \mathbf{J}_f + \epsilon_0 \frac{\partial \mathbf{E}}{\partial t} + \frac{\partial \mathbf{P}}{\partial t} + \nabla \times \mathbf{M}, \tag{19-2}$$

$$= \mathbf{J}_m + \epsilon_0 \frac{\partial \mathbf{E}}{\partial t}, \tag{19-3}$$

where $\mathbf{J}_m$ is the *volume current density in matter*, with

$$\frac{\partial \mathbf{D}}{\partial t} = \epsilon_0 \frac{\partial \mathbf{E}}{\partial t} + \frac{\partial \mathbf{P}}{\partial t}, \tag{19-4}$$

$$\mathbf{J}_m = \mathbf{J}_f + \frac{\partial \mathbf{P}}{\partial t} + \nabla \times \mathbf{M}. \tag{19-5}$$

19.2 *THE CURL OF* B

In Sec. 9.2 we found that, for static fields,

$$\nabla \times \mathbf{B} = \mu_0(\mathbf{J}_f + \nabla \times \mathbf{M}). \tag{19-6}$$

For time-dependent fields, we must replace the parenthesis by the total current density. Then

$$\nabla \times \mathbf{B} = \mu_0 \left(\mathbf{J}_f + \epsilon_0 \frac{\partial \mathbf{E}}{\partial t} + \frac{\partial \mathbf{P}}{\partial t} + \nabla \times \mathbf{M} \right), \tag{19-7}$$

or

$$\nabla \times \mathbf{B} - \epsilon_0 \mu_0 \frac{\partial \mathbf{E}}{\partial t} = \mu_0 \left(\mathbf{J}_f + \frac{\partial \mathbf{P}}{\partial t} + \nabla \times \mathbf{M} \right). \tag{19-8}$$

This is one of Maxwell's equations. In more concise form,

$$\boxed{\nabla \times \mathbf{B} - \epsilon_0 \mu_0 \frac{\partial \mathbf{E}}{\partial t} = \mu_0 \mathbf{J}_m.} \tag{19-9}$$

Since

$$\mathbf{B} = \mu_0(\mathbf{H} + \mathbf{M}), \tag{19-10}$$

we also have that

$$\nabla \times \mathbf{H} = \mathbf{J}_f + \frac{\partial \mathbf{D}}{\partial t}. \tag{19-11}$$

19.2.1 *EXAMPLE: DIELECTRIC-FILLED PARALLEL-PLATE CAPACITOR*

Figure 19-1 shows a parallel-plate capacitor connected across a source of alternating voltage $\mathbf{V}$. The capacitor contains a slightly conducting dielectric that has a permittivity $\epsilon_r \epsilon_0$ and a conductivity σ. We neglect edge effects. Then the field inside the capacitor is uniform and both $\mathbf{J}_f$ and $\partial \mathbf{D}/\partial t$ are independent of r.

We can find the magnetic induction inside the dielectric by transforming Eq. 19-8 into its integral form, setting $\mathbf{M} = 0$. Integrating over an area S,

$$\int_S \nabla \times \mathbf{B} \cdot \mathbf{da} = \mu_0 \int_S \left(\mathbf{J}_f + \frac{\partial \mathbf{D}}{\partial t} \right) \cdot \mathbf{da}, \tag{19-12}$$

and, using Stokes's theorem on the left,

$$\oint_C \mathbf{B} \cdot \mathbf{dl} = \mu_0 \int_S \left(\mathbf{J}_f + \frac{\partial \mathbf{D}}{\partial t} \right) \cdot \mathbf{da}, \tag{19-13}$$

where C is the curve bounding the surface S. By symmetry, the vector $\mathbf{B}$ is azimuthal.

Note that $\mathbf{B}$, $\mathbf{J}_f$, and $\mathbf{D}$ are at the same time vectors and phasors, here.

If C is now a circle of radius r inside the capacitor and parallel to the plates, as in the figure, we have the following phasor equation:

$$2\pi r \mathbf{B} = \mu_0 \left(\mathbf{J}_f + \frac{\partial \mathbf{D}}{\partial t} \right) \pi r^2, \tag{19-14}$$

$$\mathbf{B} = \frac{\mu_0}{2} \left(\mathbf{J}_f + \frac{\partial \mathbf{D}}{\partial t} \right) r. \tag{19-15}$$

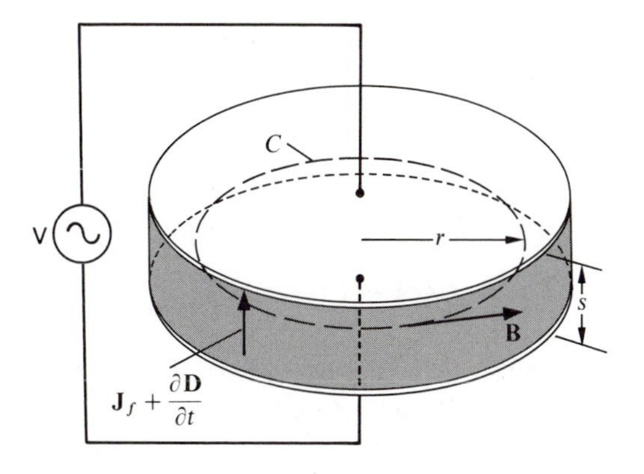

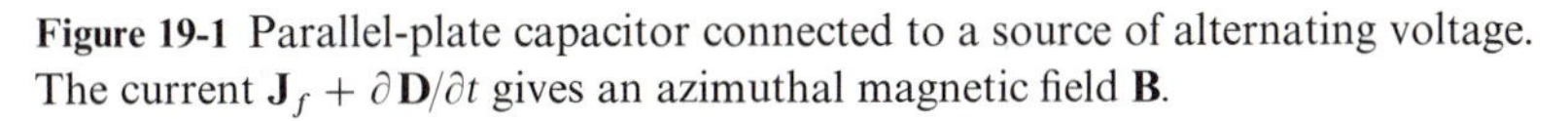

Figure 19-1 Parallel-plate capacitor connected to a source of alternating voltage. The current $\mathbf{J}_f + \partial \mathbf{D}/\partial t$ gives an azimuthal magnetic field **B**.

Now,

$$\mathbf{J}_f = \sigma \mathbf{E} = \sigma \mathbf{V}/s, \tag{19-16}$$

$$\frac{\partial \mathbf{D}}{\partial t} = \epsilon_r \epsilon_0 \frac{\partial \mathbf{E}}{\partial t} = j\omega \epsilon_r \epsilon_0 \frac{\mathbf{V}}{s}, \tag{19-17}$$

and

$$\mathbf{B} = \frac{\mu_0}{2s}(\sigma + j\omega \epsilon_r \epsilon_0)\mathbf{V}r. \tag{19-18}$$

Thus the vector **B** is azimuthal and its magnitude is proportional to r. Figure 19-2 shows lines of **B** in a plane parallel to the plates.

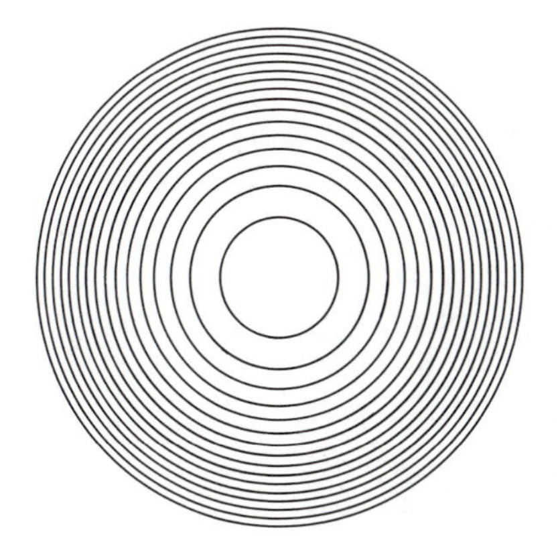

Figure 19-2 Lines of **B** in a plane parallel to the plates inside the capacitor of Fig. 19-1.

The magnetic induction is the sum of two terms. The one that is proportional to σ is associated with the conduction current. That **B** is in phase with **V**, like the conduction current $\sigma\mathbf{E}$. The other term, which is proportional to $\epsilon_r\epsilon_0$, is associated with $\partial\mathbf{D}/\partial t$ and is thus proportional to the time rate of change of **V**. It is in quadrature with **V**.

19.3 MAXWELL'S EQUATIONS

Let us group together the four Maxwell equations, which we found successively as Eqs. 6-12, 8-12, 11-14, 19-9:

$$\nabla \cdot \mathbf{E} = \rho_t/\epsilon_0, \tag{19-19}$$

$$\nabla \cdot \mathbf{B} = 0, \tag{19-20}$$

$$\nabla \times \mathbf{E} + \frac{\partial \mathbf{B}}{\partial t} = 0, \tag{19-21}$$

$$\nabla \times \mathbf{B} - \epsilon_0\mu_0 \frac{\partial \mathbf{E}}{\partial t} = \mu_0 \mathbf{J}_m. \tag{19-22}$$

As usual,

E is the electric field intensity in volts per meter;
$\rho_t = \rho_f + \rho_b$ is the total electric charge density in coulombs per cubic meter;
ρ_f is the free charge density;
ρ_b is the bound charge density $-\nabla \cdot \mathbf{P}$, where **P** is the electric polarization in coulombs per square meter;
B is the magnetic induction in teslas;
$\mathbf{J}_m = \mathbf{J}_f + \partial\mathbf{P}/\partial t + \nabla \times \mathbf{M}$ is the current density resulting from the flow of charges in matter, in amperes per square meter;
$\mathbf{J}_f$ is the current density of free charges;
$\partial\mathbf{P}/\partial t$ is the polarization current density;
$\nabla \times \mathbf{M}$ is the equivalent current density in magnetized matter;
M is the magnetization in amperes per meter;
ϵ_0 is the permittivity of free space, $8.854\,187\,82 \times 10^{-12}$ farad per meter;
μ_0 is the permeability of free space, $4\pi \times 10^{-7}$ henry per meter.

In conductors obeying Ohm's law (Sec. 5.2),

$$\mathbf{J}_f = \sigma \mathbf{E}, \tag{19-23}$$

where σ is the conductivity in siemens per meter.

Maxwell's equations are partial differential equations involving space and time derivatives of the field vectors **E** and **B**, the total charge density ρ_t, and the current density $\mathbf{J}_m$. They do *not* yield the values of **E** and **B** directly, but only after integration and after taking into account the proper boundary conditions.

These are the four fundamental equations of electromagnetism. They apply to all electromagnetic phenomena in media that are at rest with respect to the coordinate system used for the del operators.

19.4 MAXWELL'S EQUATIONS IN INTEGRAL FORM

We have stated Maxwell's equations in differential form; let us now state them in integral form, in order to arrive at a better understanding of their physical meaning.

The integral form is really more interesting and general. It is more interesting because one can visualize integrals more easily than derivatives. It is also more general because it is applicable even when the derivatives do not exist, for example at the interface between two media.

In Sec. 6.4, we deduced Eq. 19-19 from Gauss's law in integral form,

$$\int_S \mathbf{E} \cdot \mathbf{da} = \frac{1}{\epsilon_0} \int_{\tau'} \rho_t \, d\tau' = \frac{Q_t}{\epsilon_0}, \tag{19-24}$$

where S is the surface bounding the volume τ', ρ_t is the total charge density $\rho_f + \rho_b$, and Q_t is the total net charge $Q_f + Q_b$ contained within τ'. The outward flux of **E** through any closed surface S is therefore equal to $1/\epsilon_0$ times the total net charge inside, as illustrated in Fig. 19-3.

We proceeded similarly in Sec. 8.2, where we deduced Eq. 19-20 from the fact that

$$\int_S \mathbf{B} \cdot \mathbf{da} = 0, \tag{19-25}$$

where S is any closed surface. Thus the net outgoing flux of **B** through any closed surface S is zero. This is shown in Fig. 19-4.

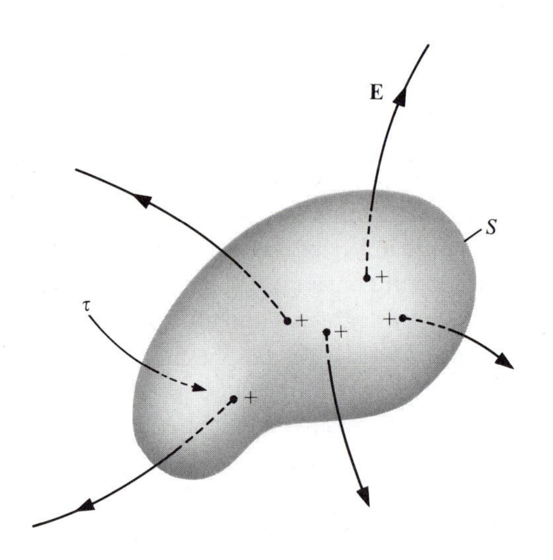

Figure 19-3 Lines of **E** emerging from a volume τ containing a net charge Q_t. The outward flux of **E** is equal to Q_t/ϵ_0.

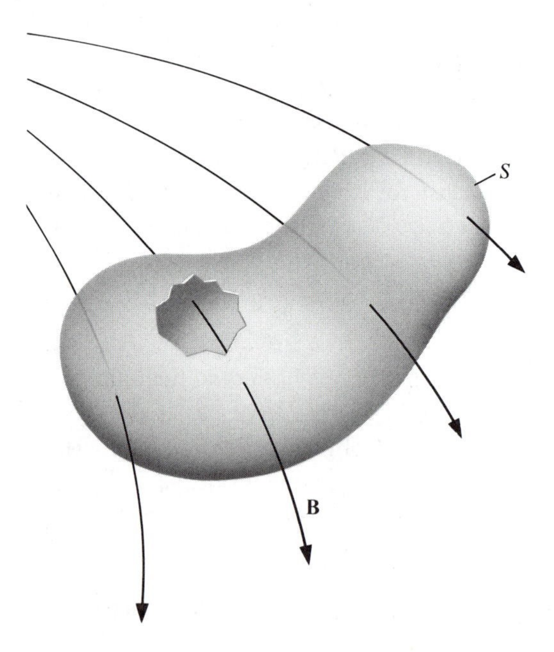

Figure 19-4 Lines of **B** through a closed surface S. The net outward flux of **B** is equal to zero.

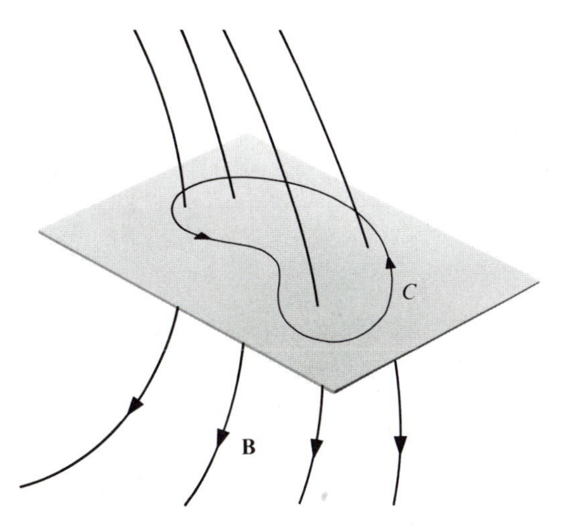

Figure 19-5 The direction of the electromotance induced around C is indicated by an arrow for the case where the magnetic induction **B** is in the direction shown and *in*creases. The electromotance is in the same direction if **B** is upward and decreases.

If we now integrate Eq. 19-21 over a surface S bounded by a curve C,

$$\int_S \nabla \times \mathbf{E} \cdot \mathbf{da} = -\int_S \frac{\partial \mathbf{B}}{\partial t} \cdot \mathbf{da}. \tag{19-26}$$

Or, if we use Stokes's theorem on the left and invert the operations on the right, the surface S being fixed in space,

$$\oint_C \mathbf{E} \cdot \mathbf{dl} = -\frac{\partial}{\partial t} \int_S \mathbf{B} \cdot \mathbf{da} = -\frac{d\Phi}{dt}. \tag{19-27}$$

Then the electromotance induced around a closed curve C is equal to minus the rate of change of the magnetic flux Φ linking C, as in Fig. 19-5. The positive directions for **B** and around C are related according to the right-hand screw rule (Sec. 1.12). If the curve C has more than one turn, then the surface S becomes complicated and the surface integral is called the flux linkage.

Finally, if we also integrate Eq. 19-22 over an area S bounded by a curve C,

$$\oint_C \mathbf{B} \cdot \mathbf{dl} = \mu_0 \int_S \left(\mathbf{J}_m + \epsilon_0 \frac{\partial \mathbf{E}}{dt} \right) \cdot \mathbf{da} = \mu_0 I_t. \tag{19-28}$$

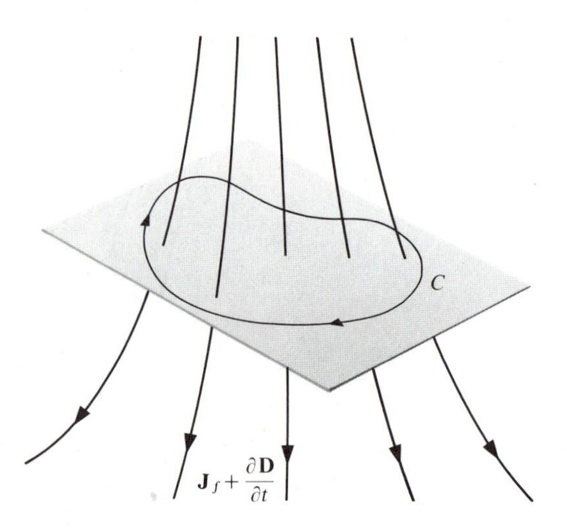

Figure 19-6 The direction of the magnetomotance around C is indicated by an arrow for the case where the current $\mathbf{J}_f + (\partial \mathbf{D}/\partial t)$ is in the direction shown. The displacement current is downward if $\mathbf{D}$ is downward and increases, or if it is upward and decreases.

If we perform the same operation on the corresponding equation for $\mathbf{H}$, as we did in Sec. 19.2,

$$\oint_C \mathbf{H} \cdot d\mathbf{l} = \int_S \left(\mathbf{J}_f + \frac{\partial \mathbf{D}}{\partial t} \right) \cdot d\mathbf{a}. \tag{19-29}$$

The magnetomotance around C is equal to the sum of the free current plus the displacement current linking C. This is illustrated in Fig. 19-6. The positive directions are again related by the right-hand screw rule.

19.5 SUMMARY

The *total volume current density* is the sum of four terms:

$$\mathbf{J}_t = \mathbf{J}_f + \epsilon_0 \frac{\partial \mathbf{E}}{\partial t} + \frac{\partial \mathbf{P}}{\partial t} + \nabla \times \mathbf{M}, \tag{19-2}$$

where the first term is the *free current density*, the second and third together form the *displacement current density* $\partial \mathbf{D}/\partial t$, and the fourth term is the *equivalent volume current density* in magnetic materials. Thus

$$\frac{\partial \mathbf{D}}{\partial t} = \epsilon_0 \frac{\partial \mathbf{E}}{\partial t} + \frac{\partial \mathbf{P}}{\partial t}. \tag{19-4}$$

Also,

$$\mathbf{J}_m = \mathbf{J}_f + \frac{\partial \mathbf{P}}{\partial t} + \nabla \times \mathbf{M} \tag{19-5}$$

is the *current density in matter.*

The following four equations are known as *Maxwell's equations*:

$$\nabla \cdot \mathbf{E} = \rho_t/\epsilon_0, \tag{19-19}$$

$$\nabla \cdot \mathbf{B} = 0, \tag{19-20}$$

$$\nabla \times \mathbf{E} + \frac{\partial \mathbf{B}}{\partial t} = 0, \tag{19-21}$$

$$\nabla \times \mathbf{B} - \epsilon_0 \mu_0 \frac{\partial \mathbf{E}}{\partial t} = \mu_0 \mathbf{J}_m. \tag{19-22}$$

In conductors obeying *Ohm's law*,

$$\mathbf{J}_f = \sigma \mathbf{E}. \tag{19-23}$$

In integral form, Maxwell's equations become

$$\int_S \mathbf{E} \cdot \mathbf{da} = Q_t/\epsilon_0, \tag{19-24}$$

$$\int_S \mathbf{B} \cdot \mathbf{da} = 0, \tag{19-25}$$

$$\oint_C \mathbf{E} \cdot \mathbf{dl} = -\frac{\partial}{\partial t} \int_S \mathbf{B} \cdot \mathbf{da} = -\frac{d\Phi}{dt}, \tag{19-27}$$

$$\oint_C \mathbf{B} \cdot \mathbf{dl} = \mu_0 \int_S \left(\mathbf{J}_m + \epsilon_0 \frac{\partial \mathbf{E}}{\partial t} \right) \cdot \mathbf{da} = \mu_0 I_t. \tag{19-28}$$

PROBLEMS

19-1

PARALLEL-PLATE CAPACITOR SUBMERGED IN AN INFINITE MEDIUM

Consider a parallel-plate capacitor situated in an infinite medium as in Fig. 19-7. For simplicity, we assume that the plates are circular. At $t = 0$, the plates carry charges $+Q$ and $-Q$. The capacitance between the plates is C. The medium is slightly conducting and the resistance between the plates is R.

From Sec. 5.14, the voltage difference V between the plates decreases exponentially with t:

$$V = V_0 e^{-t/RC}. \tag{1}$$

Therefore, at any point in the dielectric, the electric field intensity decreases in the same manner:

$$E = E_0 e^{-t/RC}, \tag{2}$$

where E_0 varies from one point to another.

a) Use the result of Prob. 6-17 to show that the total current density $J_f + \partial D/\partial t$ in the dielectric is zero. The dielectric has a conductivity σ and a relative permittivity ϵ_r.

Solution: At any point, the total current density is

$$J_f + \frac{\partial D}{\partial t} = \left(\sigma + \epsilon_r \epsilon_0 \frac{\partial}{\partial t}\right) E, \tag{3}$$

$$= \left(\sigma - \frac{\epsilon_r \epsilon_0}{RC}\right) E = 0, \tag{4}$$

since $RC = \epsilon_r \epsilon_0 / \sigma$, from Prob. 6-17.

b) What does this result tell you about the magnetic field?

Solution: According to Eq. 19-13, for any closed curve C,

$$\oint \mathbf{B} \cdot d\mathbf{l} = 0. \tag{5}$$

Now, by symmetry, if there is a magnetic field, it must be azimuthal. So consider a circle C as in Fig. 19-7. By symmetry again, since the plates are circular by hypothesis, B must be the same all around the circle. Because of the above equation, B is zero everywhere on the circle. But the circle can be situated anywhere, as long as it is parallel to the plates and centered on the axis of symmetry, so $\mathbf{B}$ is zero everywhere.

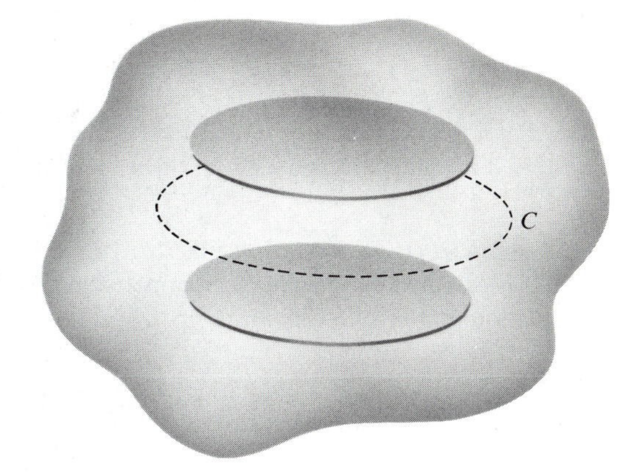

Figure 19-7 See Prob. 19-1.

19-2E *MAXWELL'S EQUATIONS*

Show that Maxwell's equations can also be written under the form

$$\nabla \cdot \mathbf{D} = \rho_f,$$

$$\nabla \cdot \mathbf{B} = 0,$$

$$\nabla \times \mathbf{E} + \frac{\partial \mathbf{B}}{\partial t} = 0,$$

$$\nabla \times \mathbf{H} - \frac{\partial \mathbf{D}}{\partial t} = \mathbf{J}_f.$$

19-3E *MAXWELL'S EQUATIONS*

Show that, for linear and isotropic media, and for fields that are sinusoidal functions of the time.

$$\nabla \cdot \epsilon_r \epsilon_0 \mathbf{E} = \rho_f,$$

$$\nabla \cdot \mu_r \mu_0 \mathbf{H} = 0,$$

$$\nabla \times \mathbf{E} + j\omega\mu_r\mu_0 \mathbf{H} = 0,$$

$$\nabla \times \mathbf{H} - j\omega\mu_r\mu_0 \mathbf{E} = \mathbf{J}_f.$$

Here again, we have quantities that are both vectors and phasors.

19-4E *MAXWELL'S EQUATIONS*

The equation for the conservation of charge of Sec. 5.1.1 is built into Maxwell's equations.

Show that it follows from Eq. 19-22.

The **J** used in Chapter 5 is the current density of free charges $\mathbf{J}_f$.

19-5E *MAGNETIC MONOPOLES AND MAXWELL'S EQUATIONS*

When it is assumed that magnetic monopoles do exist, it is the custom to write Maxwell's equations in the following form:

$$\nabla \cdot \mathbf{E} = \rho_t/\epsilon_0, \qquad \nabla \times \mathbf{E} = -\frac{\partial \mathbf{B}}{\partial t} - \mathbf{J}^*,$$

$$\nabla \cdot \mathbf{B} = \rho^*, \qquad \nabla \times \mathbf{B} = \mu_0 \left(\epsilon_0 \frac{\partial \mathbf{E}}{\partial t} + \mathbf{J}_m \right),$$

where ρ^* is the volume density of magnetic charge in webers per cubic meter, and $\mathbf{J}^*$ is the magnetic current density, expressed in webers per second per square meter.

a) Show that

$$\nabla \cdot \mathbf{J}^* = -\frac{\partial \rho^*}{\partial t}.$$

The modified Maxwell equations therefore postulate the conservation of magnetic charge.

By analogy with Coulomb's law, near a magnetic charge,

$$\mathbf{B} = \frac{Q^*}{4\pi r^2}\mathbf{r}_1.$$

In a vacuum,

$$\mathbf{H} = \frac{Q^*}{4\pi\mu_0 r^2}\mathbf{r}_1,$$

and the force between two magnetic charges is

$$\mathbf{F}_{ab} = Q_a^*\mathbf{H}_b = \frac{Q_a^* Q_b^*}{4\pi\mu_0 r^2}\mathbf{r}_1.$$

b) Show that, for any closed path C, and if $\partial\mathbf{B}/\partial t$ is zero,

$$\oint_C \mathbf{E} \cdot d\mathbf{l} = -I^*,$$

where I^* is the magnetic current. Note the negative sign; the induced electromotance in a coil is equal to minus the magnetic current linking it. See Prob. 12-7.

19-6 Express the following quantities in terms of kilograms, meters, seconds, and amperes:

joule, watt,
coulomb, volt, ohm, siemens, farad,
weber, tesla, henry.

CHAPTER 20

ELECTROMAGNETIC WAVES

In this last chapter we shall have no more than a glimpse of electromagnetic waves. These waves are everywhere—light waves are electromagnetic—and their applications are innumerable, some of the best known being radio, television, and radar. We shall start with a brief review of plane waves, and then discuss plane electromagnetic waves, first in free space, and then in dielectrics. This will require all four of Maxwell's equations.

Unfortunately, the methods that are used for launching or for detecting these waves are beyond the scope of this book. However, if you wish to go further, you will find many books on electromagnetic waves.†

20.1 WAVES

The essential nature of waves is illustrated by a stretched string, fixed at one end and moved rapidly in a vertical plane in some arbitrary fashion at the other end, as in Fig. 20-1.

If we call $y(t)$ the height of the moving end, then the height y at a distance z along the string is $y[t - (z/v)]$, assuming no losses, a perfectly flexible string, and no reflected wave. The height at z is thus equal to the height of the moving end at a previous time $t - (z/v)$, the quantity z/v being the time required for the disturbance to travel through the distance z at the velocity v.

---

† *Almost All About Waves* by John R. Pierce, MIT Press, 1974, is a delightful little book by an authority on the subject. See also *Electromagnetic Fields and Waves*, Chapters 11 to 14.

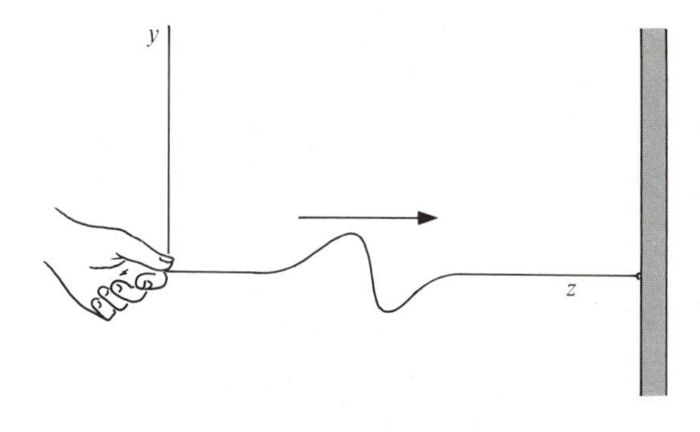

Figure 20-1 Wave on a stretched string.

We can generalize from this simple example and consider waves propagating in an extended region, such as acoustic waves in air or light waves in space.

The quantity propagated can be either a scalar or a vector quantity. For example, in an acoustic wave, we have the propagation of a pressure, which is a scalar. In an electromagnetic wave, we have the propagation of a vector **E** and of a vector **H**.

20.1.1 *PLANE SINUSOIDAL WAVES. PROPAGATION OF A SCALAR QUANTITY*

If a certain scalar quantity α propagating with a velocity v is defined at $z = 0$ by

$$\alpha = \alpha_0 \cos \omega t, \tag{20-1}$$

then, for any position z in the direction of the propagation,

$$\alpha = \alpha_0 \cos \left[\omega \left(t - \frac{z}{v} \right) \right]. \tag{20-2}$$

This expression describes a *plane wave* traveling along the z-axis, since α is independent of x and y. The wave is also *unattenuated*, since the amplitude α_0 is constant. The quantity between brackets is the *phase* of the wave. The *frequency* f is $\omega/2\pi$, and the *period* T is $1/f$.

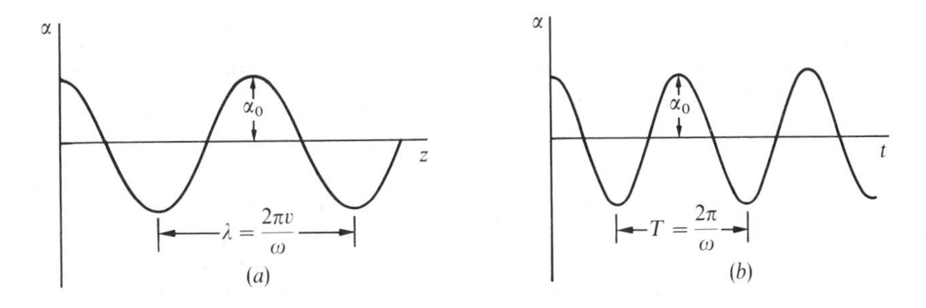

Figure 20-2 The quantity $\alpha = \alpha_0 \cos \omega[t - (z/v)]$ as a function of z and as a function of t.

For a given position z, we have a sinusoidal variation of α with time (Fig. 20-2a); for a given time t, we have a sinusoidal variation of α with z (Fig. 20-2b).

The surfaces of constant phase, called *wave-fronts*, are normal to the z-axis.

The quantity v is called the *phase velocity* of the wave, since it is the velocity with which a wavefront is propagated in space. For example, if the bracket is zero, the position of the wavefront at the time t is given by

$$z = vt. \tag{20-3}$$

The *wavelength* λ is the distance over which $\omega z/v$ changes by 2π radians, as shown in Fig. 20-2:

$$\omega\lambda/v = 2\pi, \tag{20-4}$$

$$\lambda = 2\pi v/\omega = v/f. \tag{20-5}$$

Also,

$$\bar{\lambda} = \lambda/2\pi = v/\omega, \tag{20-6}$$

where $\bar{\lambda}$ (read "lambda bar") is the *radian length.* It is the distance over which the phase of the wave changes by one radian; this is about $\lambda/6$. It turns out that the quantity that appears in nearly all wave calculations is $\bar{\lambda}$, and not λ. We are already familiar with the fact that in nearly all oscillation calculations it is the circular frequency ω that is used instead of the frequency f.

The intuitively simple quantity, namely the wavelength or the frequency, is not the one that is "natural" from a mathematical standpoint.

In phasor notation (Sec. 16.7), a sinusoidal wave traveling in the positive direction along the z-axis is described by

$$\alpha = \alpha_0 \exp j[\omega t - (z/\lambda\!\!\!^{-})].^\dagger \qquad (20\text{-}7)$$

Similarly, a wave traveling in the negative direction along the z-axis is described by

$$\alpha = \alpha_0 \exp j[\omega t + (z/\lambda\!\!\!^{-})]. \qquad (20\text{-}8)$$

20.1.2 *PLANE SINUSOIDAL WAVES. PROPAGATION OF A VECTOR QUANTITY*

In the above equations, α is a scalar. Now, as we noted at the beginning of this chapter, the quantity that is propagated can also be a vector, say the electric field intensity **E**. Then, the above equations apply to each one of the components E_x, E_y, E_z. Or, if we multiply the equation for E_x by **i**, that for E_y by **j**, and that for E_z by **k**, and take the sum, we have an equation like 20-7 with α replaced by **E**.

Note that the quantity **E** is then *both* a phasor and a vector. It is a vector because it is oriented in space, with the usual three components, and it is also a phasor because its time dependence is written in the form

$$\exp j(\omega t + \varphi).$$

20.2 *THE WAVE EQUATION*

Let us calculate the second derivatives of α (Eq. 20-7) with respect to t and to z:

$$\frac{\partial^2 \alpha}{\partial t^2} = -\omega^2 \alpha, \qquad \frac{\partial^2 \alpha}{\partial z^2} = -\frac{1}{\lambda\!\!\!^{-2}} \alpha. \qquad (20\text{-}9)$$

† In the last chapter, we shall follow the usual custom of *not* using a special notation for phasors: vector phasors will be written like vectors, and scalar phasors like scalars.

Thus

$$\frac{\partial^2 \alpha}{\partial z^2} = \frac{1}{\omega^2 \lambda^2} \frac{\partial^2 \alpha}{\partial t^2}, \tag{20-10}$$

$$= \frac{1}{v^2} \frac{\partial^2 \alpha}{\partial t^2}, \tag{20-11}$$

where v is the phase velocity. This is the *wave equation* for a plane wave propagating along the z-axis. This equation is valid even for non-sinusoidal waves.

More generally, for an unattenuated wave in space,

$$\nabla^2 \alpha = \frac{1}{v^2} \frac{\partial^2 \alpha}{\partial t^2}. \tag{20-12}$$

If we had a vector quantity instead of the scalar α, we would have the same equation with α replaced by, say, **E**. For an unattenuated sinusoidal wave,

$$\nabla^2 \alpha = -\frac{\omega^2}{v^2} \alpha. \tag{20-13}$$

20.3 THE ELECTROMAGNETIC SPECTRUM

We are now ready to study the propagation of electromagnetic waves in free space and in dielectrics.

Maxwell's equations impose no limit on the frequency of electromagnetic waves. To date, the spectrum that has been investigated experimentally is shown in Fig. 20-3. It extends continuously from the long radio waves to the very high energy gamma rays observed in cosmic radiation. In the former, the frequencies are about 10 hertz, and the wavelengths are about 3×10^7 meters; in the latter, the frequencies are of the order of 10^{24} hertz, and the wavelengths of the order of 3×10^{-16} meter. The known spectrum thus covers a range of 23 orders of magnitude. Radio, light, and heat waves, X-rays and gamma rays—all are electromagnetic, although the sources and the detectors, as well as the modes of interaction with matter, vary widely as the frequency changes by orders of magnitude. The nomenclature of radio waves is given in Table 20-1.

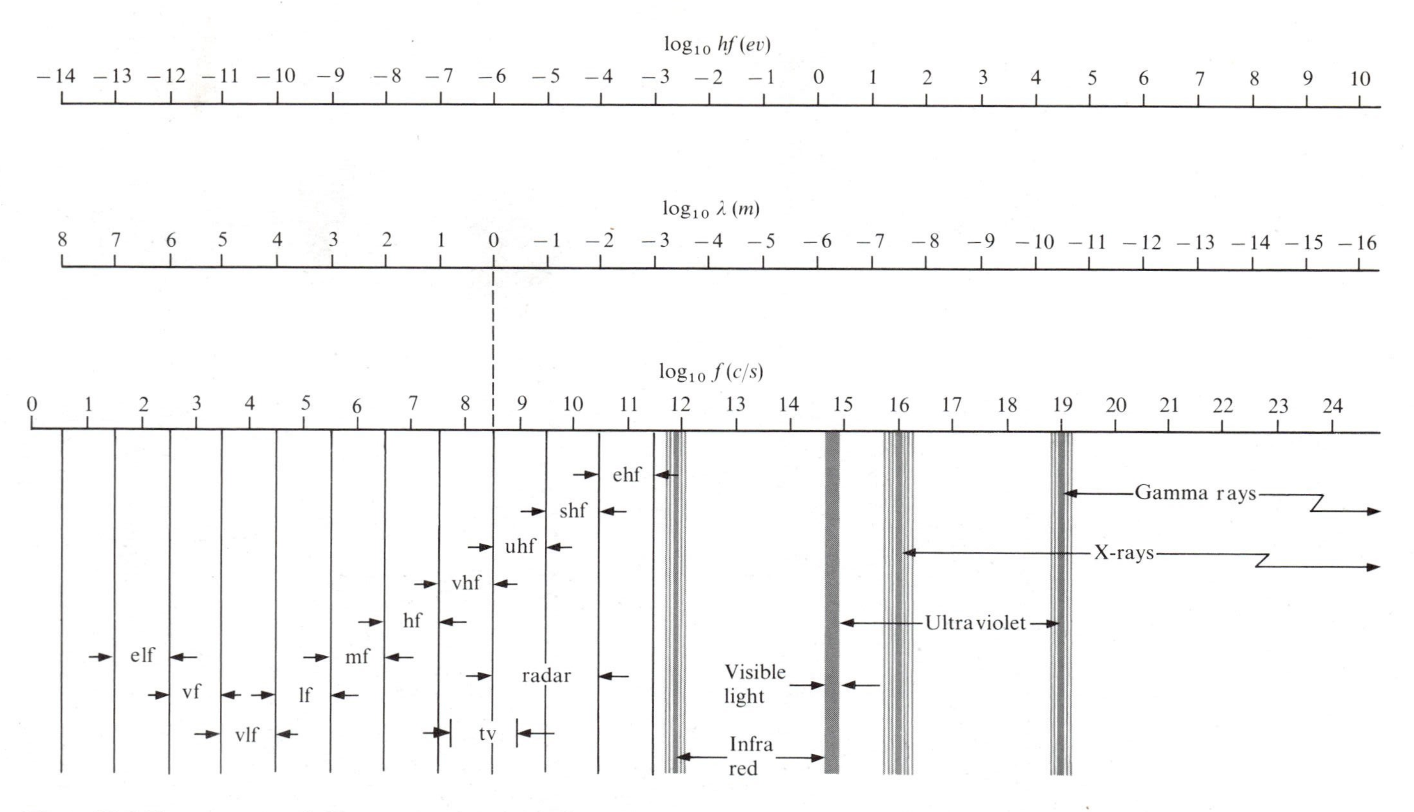

Figure 20-3 The spectrum of electromagnetic waves. The abbreviations elf, vf, vlf, . . . mean, respectively, extremely low frequency, voice frequency, very low frequency, low frequency, medium frequency, high frequency, very high frequency, ultrahigh frequency, super high frequency, and extremely high frequency. The limits indicated by the shaded regions are approximate. The energy hf, where h is Planck's constant (6.63×10^{-34} joule-second) and f is the frequency, is that of a photon or quantum of radiation.

Table 20-1

Band Numbers*	Frequency Range	Metric Subdivision	Abbreviation
1	3–30 Hz		
2	30–300 Hz	Megametric waves	elf
3	300–3000 Hz		vf
4	3–30 kHz	Myriametric waves	vlf
5	30–300 kHz	Kilometric waves	lf
6	300–3000 kHz	Hectometric waves	mf
7	3–30 MHz	Decametric waves	hf
8	30–300 MHz	Metric waves	vhf
9	300–3000 MHz	Decimetric waves	uhf
10	3–30 GHz	Centimetric waves	shf
11	30–300 GHz	Millimetric waves	ehf
12	300–3000 GHz	Decimillimetric waves	

* Band number N extends from 0.3×10^N to 3×10^N hertz.

The fundamental identity of all these types of waves is demonstrated by many experiments covering overlapping parts of the spectrum. It is also shown by the fact that in free space they are all transverse waves with a common velocity of propagation. For example, simultaneous radio and optical observations of flare stars have shown that the velocity of propagation is the same, within experimental error, for wavelengths differing by more than six orders of magnitude.

We shall follow the custom of using **H** rather than **B** in discussing electromagnetic waves, in spite of the fact that, until now, we have used **H** only in relation with magnetic materials. The main reason for using **H** instead of **B** in dealing with electromagnetic waves is that $\mathbf{E} \times \mathbf{H}$ gives the energy flux density, as we shall see.

20.4 PLANE ELECTROMAGNETIC WAVES IN FREE SPACE: VELOCITY OF PROPAGATION c

Let us start with the relatively simple case of a plane sinusoidal wave propagating in a vacuum in a region infinitely remote from matter. Then Maxwell's equations 19-19 to 19-22 become

$$\nabla \cdot \mathbf{E} = 0, \tag{20-14}$$

$$\nabla \cdot \mathbf{H} = 0, \tag{20-15}$$

$$\nabla \times \mathbf{E} + j\omega\mu_0 \mathbf{H} = 0, \tag{20-16}$$

$$\nabla \times \mathbf{H} - j\omega\epsilon_0 \mathbf{E} = 0. \tag{20-17}$$

Taking the curls of Eqs. 20-16 and 20-17,

$$\nabla \times \nabla \times \mathbf{E} + j\omega\mu_0 \, \nabla \times \mathbf{H} = 0, \tag{20-18}$$

$$\nabla \times \nabla \times \mathbf{H} - j\omega\epsilon_0 \nabla \times \mathbf{E} = 0, \tag{20-19}$$

or, using the identity of Prob. 1-32,

$$\nabla(\nabla \cdot \mathbf{E}) - \nabla^2 \mathbf{E} + j\omega\mu_0 \, \nabla \times \mathbf{H} = 0, \tag{20-20}$$

$$\nabla(\nabla \cdot \mathbf{H}) - \nabla^2 \mathbf{H} - j\omega\epsilon_0 \, \nabla \times \mathbf{E} = 0. \tag{20-21}$$

Finally, substituting Maxwell's equations,

$$\nabla^2 \mathbf{E} = -\epsilon_0 \mu_0 \omega^2 \mathbf{E}, \tag{20-22}$$

$$\nabla^2 \mathbf{H} = -\epsilon_0 \mu_0 \omega^2 \mathbf{H}. \tag{20-23}$$

Comparing now with Eq. 20-13, we find that these differential equations are those of a wave. Also, it follows that the field vectors **E** and **H** can propagate as waves in free space at the velocity

$$c = 1/(\epsilon_0 \mu_0)^{1/2}. \tag{20-24}$$

Equation 20-24 links three basic constants of electromagnetism: the *velocity of an electromagnetic wave* c, the permittivity of free space ϵ_0, and the permeability of free space μ_0. It will be remembered from Sec. 8.1 that the constant μ_0 was defined arbitrarily to be *exactly* $4\pi \times 10^{-7}$ henry per meter. The constant ϵ_0 can thus be deduced from the measured value for the velocity of electromagnetic waves,

$$c = 2.997\,924\,58 \times 10^8 \text{ meters/second:} \tag{20-25}$$

$$\epsilon_0 = 1/c^2\mu_0 = 8.854\,187\,82 \times 10^{-12} \text{ farad/meter.} \tag{20-26}$$

We shall use the approximate value of c, namely 3×10^8 meters per second, which is accurate within 7 parts in 10,000.

The permittivity of free space ϵ_0 can also be determined directly from measurements involving electrostatic phenomena. The measurements lead to the above value within experimental error, thereby confirming the theory.

20.5 *PLANE ELECTROMAGNETIC WAVES IN FREE SPACE: THE* E *AND* H *VECTORS*

For a plane electromagnetic wave propagating in the positive direction of the z-axis, **E** is independent of x and y, and

$$\nabla \cdot \mathbf{E} = \frac{\partial E_z}{\partial z} = 0. \tag{20-27}$$

Therefore the z component of **E** cannot be a function of z. We shall set

$$E_z = 0, \tag{20-28}$$

since we are interested in waves, and not in uniform fields.

The same argument applies to the **H** vector, and we can set

$$H_z = 0. \tag{20-29}$$

A plane electromagnetic wave propagating in free space is therefore *transverse*, since it has no longitudinal components.

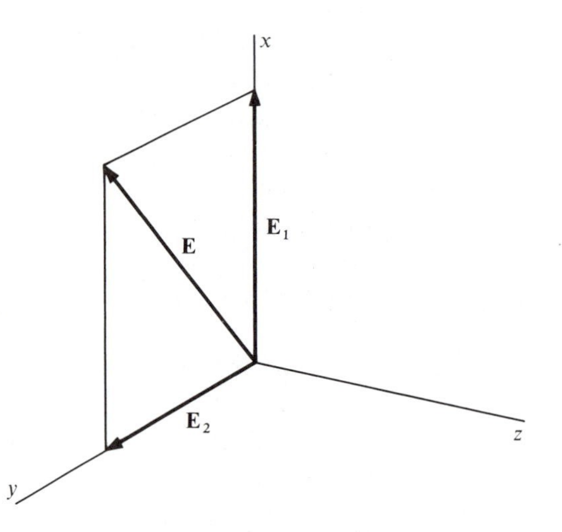

Figure 20-4 Decomposition of the vector **E** into two vectors $\mathbf{E}_1$ and $\mathbf{E}_2$.

We now assume that the wave is plane-polarized with its **E** vector always pointing in the direction of the x-axis. This does not involve any loss of generality since any plane-polarized wave can be considered to be the sum of two waves that are plane-polarized in perpendicular directions and in phase. For example, in Fig. 20-4, the vector **E** can be resolved into two mutually perpendicular vectors $\mathbf{E}_1$ and $\mathbf{E}_2$.

For a plane-polarized wave having its **E** vector in the direction of the x-axis (Fig. 20-5),

$$\mathbf{E} = E_0 \exp j\omega \left(t - \frac{z}{c} \right) \mathbf{i}. \tag{20-30}$$

Let us now substitute this value of **E** into Eq. 20-17:

$$\begin{vmatrix} \mathbf{i} & \mathbf{j} & \mathbf{k} \\ 0 & 0 & \dfrac{\partial}{\partial z} \\ H_x & H_y & 0 \end{vmatrix} = j\omega\epsilon_0 E_0 \exp j\omega \left(t - \frac{z}{c} \right) \mathbf{i}. \tag{20-31}$$

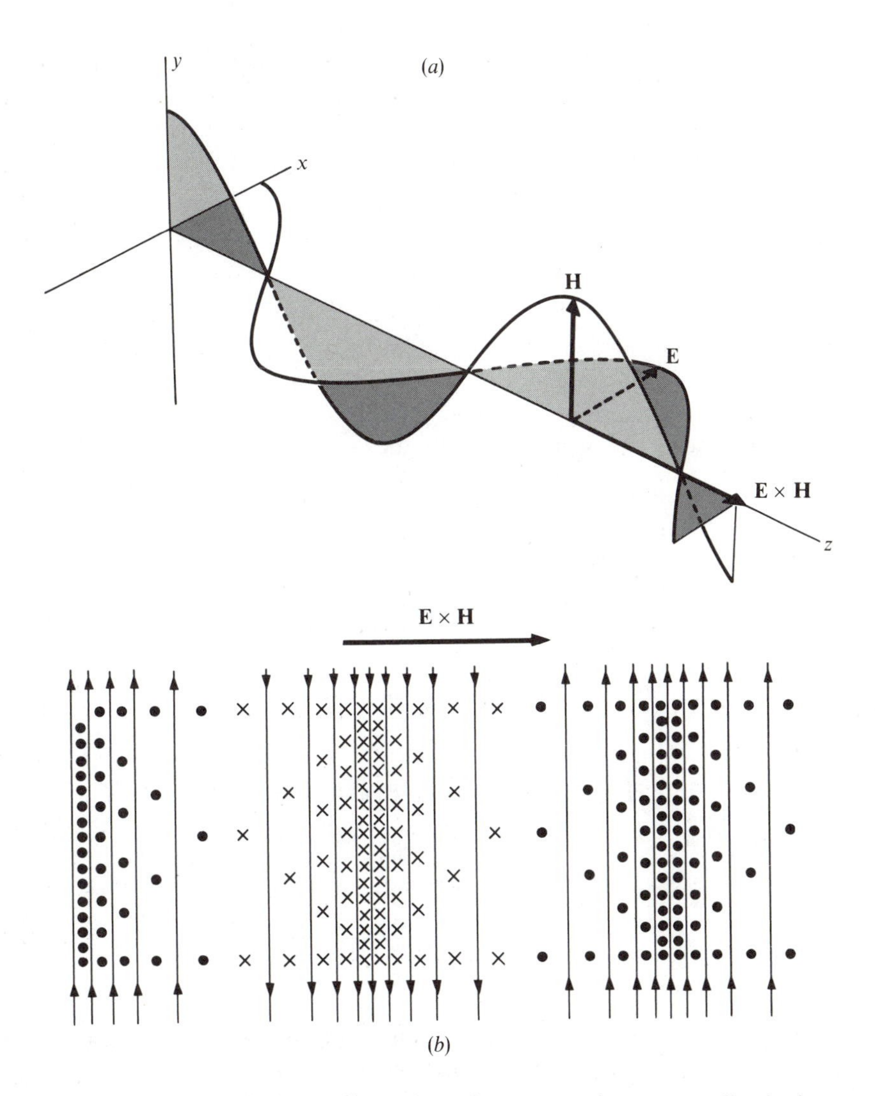

Figure 20-5 The **E** and **H** vectors for a plane electromagnetic wave traveling in the positive direction along the z-axis. (a) The variation of **E** and **H** with z at a particular moment. The two vectors are in phase, but perpendicular to each other. (b) The corresponding lines of force as seen when looking down on the xz-plane. The lines represent the electric field. The dots represent magnetic lines of force coming out of the paper, and the crosses represent magnetic lines of force going into the paper. The vector $\mathbf{E} \times \mathbf{H}$ gives the direction of propagation.

On the left-hand side we have set the derivatives with respect to x and to y equal to zero because we have a plane wave propagating along the z-axis. We have also set H_z equal to zero, from Eq. 20-29. Thus

$$-\frac{\partial}{\partial z} H_y = j\omega\epsilon_0 E_0 \exp j\omega\left(t - \frac{z}{c}\right), \tag{20-32}$$

$$\frac{\partial}{\partial z} H_x = 0. \tag{20-33}$$

Replacing the operator $\partial/\partial z$ by the factor $-j\omega/c$,

$$H_x = 0, \tag{20-34}$$

$$\mathbf{H} = c\epsilon_0 E_0 \exp j\omega\left(t - \frac{z}{c}\right)\mathbf{j}, \tag{20-35}$$

$$= (\epsilon_0/\mu_0)^{1/2} E\mathbf{j}. \tag{20-36}$$

Therefore **H** is perpendicular to **E**, and

$$E/H = 1/\epsilon_0 c = \mu_0 c = (\mu_0/\epsilon_0)^{1/2} = 377 \text{ ohms}, \tag{20-37}$$

$$E/B = c = 3.00 \times 10^8 \text{ meters/second}. \tag{20-38}$$

The **E** and **H** vectors of the wave are perpendicular and oriented in such a way that their vector product **E** × **H** points in the direction of propagation, as in Fig. 20-5. The **E** and **H** vectors are in phase, since E/H is real, and they have the same relative magnitudes at all points at all times.

The electric and magnetic energy densities are equal and in phase, since

$$(1/2)\epsilon_0 E^2 = (1/2)\mu_0 H^2. \tag{20-39}$$

At any instant, the total energy density is therefore distributed as in Fig. 20-6.

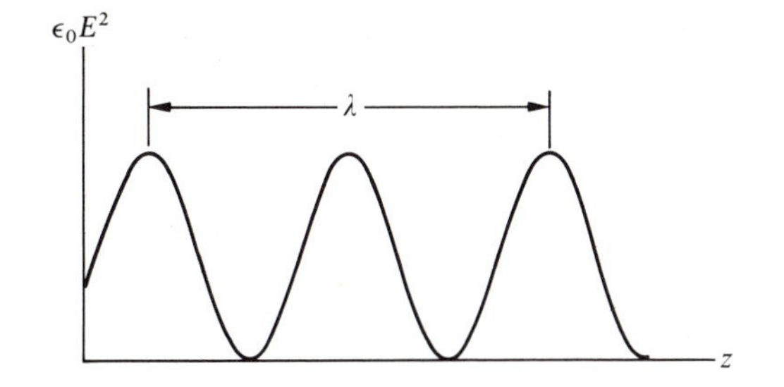

Figure 20-6 The energy density $\epsilon_0 E^2 = \mu_0 H^2$ as a function of z for a plane electromagnetic wave traveling along the z-axis in free space.

20.6 *THE POYNTING VECTOR $\mathscr{S}$ IN FREE SPACE*

We have found that a plane electromagnetic wave in free space propagates in the direction of the vector $\mathbf{E} \times \mathbf{H}$. Let us calculate the divergence of this vector for *any* electromagnetic field in free space. From Prob. 1-28,

$$\boldsymbol{\nabla} \cdot (\mathbf{E} \times \mathbf{H}) = -\mathbf{E} \cdot (\boldsymbol{\nabla} \times \mathbf{H}) + \mathbf{H} \cdot (\boldsymbol{\nabla} \times \mathbf{E}). \tag{20-40}$$

Now, using the Maxwell equations for $\boldsymbol{\nabla} \times \mathbf{H}$ and for $\boldsymbol{\nabla} \times \mathbf{E}$,

$$\boldsymbol{\nabla} \cdot (\mathbf{E} \times \mathbf{H}) = -\mathbf{E} \cdot \epsilon_0 \frac{\partial \mathbf{E}}{\partial t} - \mathbf{H} \cdot \mu_0 \frac{\partial \mathbf{H}}{\partial t}, \tag{20-41}$$

$$= -\frac{\partial}{\partial t}\left(\frac{1}{2}\epsilon_0 E^2 + \frac{1}{2}\mu_0 H^2\right). \tag{20-42}$$

Integrating over a volume τ bounded by a surface S, and using the divergence theorem on the left-hand side,

$$\int_S (\mathbf{E} \times \mathbf{H}) \cdot d\mathbf{a} = -\frac{\partial}{\partial t}\int_\tau \left(\frac{1}{2}\epsilon_0 E^2 + \frac{1}{2}\mu_0 H^2\right) d\tau. \tag{20-43}$$

The integral on the right is the sum of the electric and magnetic energies, according to Secs. 4.4 and 13.3. The right-hand side is thus the energy lost per unit time by the volume τ, and, since there must be conservation of energy, the left-hand side must be the total outward flux of energy through the surface S bounding τ.

The quantity

$$\mathscr{S} = \mathbf{E} \times \mathbf{H} \tag{20-44}$$

is called the *Poynting vector*. When integrated over a closed surface, it gives the total outward flow of energy per unit time.

In a plane electromagnetic wave, the Poynting vector points in the direction of propagation of the wave. Its instantaneous value at a given point in space is $\mathbf{E} \times \mathbf{H}$ and, according to Eq. 20-37,

$$\mathscr{S} = \frac{1}{\mu_0 c} E^2\mathbf{k} = c\epsilon_0 E^2\mathbf{k}. \tag{20-45}$$

The time average of $\mathscr{S}$ is

$$\mathscr{S}_{av} = (\epsilon_0 E_{rms}^2)c\mathbf{k} = [(1/2)\epsilon_0 E_0^2]c\mathbf{k} = (1/2)(\epsilon_0/\mu_0)^{1/2}E_0^2\mathbf{k}, \tag{20-46}$$

$$= 2.66 \times 10^{-3} E_{rms}^2\mathbf{k} \text{ watts/meter}^2. \tag{20-47}$$

The energy can therefore be considered to travel with an average density

$$\epsilon_0 E_{rms}^2 = (1/2)\epsilon_0 E_0^2 = (1/2)\epsilon_0 E_{rms}^2 + (1/2)\mu_0 H_{rms}^2 \tag{20-48}$$

at the velocity of propagation $c\mathbf{k}$.

20.6.1 *EXAMPLE: LASER BEAM*

A *laser beam* carries a power of 20 gigawatts and has a diameter of 2 millimeters. Let us calculate the peak values of E and of B.

From Eq. 20-47,

$$E_0 = 2^{1/2}E_{rms} = 2^{1/2}\left(\frac{10^3}{2.66} \times \frac{20 \times 10^9}{\pi \times 10^{-6}}\right)^{1/2}, \tag{20-49}$$

$$= 2.2 \times 10^9 \text{ volts/meter}. \tag{20-50}$$

This is an enormous electric field; it corresponds to a voltage difference of about a quarter of a volt over a distance of 10^{-10} meter, which is about the diameter of an atom.

We can now find B_0 from Eq. 20-38:

$$B_0 = \frac{E_0}{c} = \frac{2.2 \times 10^9}{3 \times 10^8} = 7.3 \text{ teslas}. \tag{20-51}$$

This is about ten times larger than the magnetic induction between the pole pieces of a powerful permanent magnet.

20.7 PLANE ELECTROMAGNETIC WAVES IN NON-CONDUCTORS

In a non-conductor with a permittivity $\epsilon_r\epsilon_0$ and a permeability $\mu_r\mu_0$, Maxwell's equations are similar to Eqs. 20-14 to 20-17, except that ϵ_0 and μ_0 are replaced by $\epsilon_r\epsilon_0$ and $\mu_r\mu_0$. The wave equations are therefore

$$\nabla^2\mathbf{E} = -\epsilon_r\epsilon_0\mu_r\mu_0\omega^2\mathbf{E}, \tag{20-52}$$

$$\nabla^2\mathbf{H} = -\epsilon_r\epsilon_0\mu_r\mu_0\omega^2\mathbf{H}, \tag{20-53}$$

and all the results of Secs. 20.4 to 20.6 apply, if one replaces ϵ_0 by $\epsilon_r\epsilon_0$ and μ_0 by $\mu_r\mu_0$.

The phase velocity is now

$$v = \frac{1}{(\epsilon_r\epsilon_0\mu_r\mu_0)^{1/2}} = \frac{c}{(\epsilon_r\mu_r)^{1/2}}. \tag{20-54}$$

The phase velocity in non-conductors is therefore less than in free space, and the *index of refraction* is

$$n \equiv c/v = (\epsilon_r\mu_r)^{1/2}. \tag{20-55}$$

In a non-magnetic medium, $\mu_r = 1$ and the index of refraction is equal to the square root of the relative permittivity:

$$n = \epsilon_r^{1/2}. \tag{20-56}$$

Note, however, that n and ϵ_r are both functions of the frequency. We have discussed briefly the variation of ϵ_r with frequency in Sec. 7.7. Since tables of n are usually compiled at optical frequencies, whereas ϵ_r is usually measured at much lower frequencies, such pairs of values cannot be expected to correspond.

Also, as in Sec. 20.5,

$$H_y = (\epsilon_r\epsilon_0/\mu_r\mu_0)^{1/2}E_x. \tag{20-57}$$

In non-conductors, the **E** and **H** vectors are in phase and the electric and magnetic energy densities are equal:

$$(1/2)\epsilon_r\epsilon_0E^2 = (1/2)\mu_r\mu_0H^2. \tag{20-58}$$

The total instantaneous energy density is thus $\epsilon_r\epsilon_0 E^2$ or $\mu_r\mu_0 H^2$, and the average total energy density is $\epsilon_r\epsilon_0 E_{\text{rms}}^2$ or $\mu_r\mu_0 H_{\text{rms}}^2$. The average value of the Poynting vector is

$$\mathscr{S}_{\text{av}} = \frac{1}{2}\left(\frac{\epsilon_r\epsilon_0}{\mu_r\mu_0}\right)^{1/2} E_0^2\mathbf{k} = \left(\frac{\epsilon_r\epsilon_0}{\mu_r\mu_0}\right)^{1/2} E_{\text{rms}}^2\mathbf{k}, \tag{20-59}$$

$$= (\epsilon_r\epsilon_0 E_{\text{rms}}^2)v\mathbf{k}, \tag{20-60}$$

$$= 2.66 \times 10^{-3}(\epsilon_r/\mu_r)^{1/2}E_{\text{rms}}^2\mathbf{k} \text{ watts/meter}^2. \tag{20-61}$$

The average value of the Poynting vector is again equal to the phase velocity multiplied by the average energy density.

20.7.1 *EXAMPLE: LASER BEAM*

If the 20 gigawatt *laser beam* of Sec. 20.6.1 is in a glass whose index of refraction is 1.6, then $\epsilon_r^{1/2}$ is 1.6, μ_r is unity, and E_0 is *smaller* than in air by a factor of $1.6^{1/2}$ or 1.26. Then E_0 in the glass is 1.7×10^9 volts per meter.

To calculate B_0 we can use the fact that the Poynting vector $(\frac{1}{2})E_0H_0$ is the same in the glass as in the air; since E_0 is smaller in the glass by a factor of 1.26, B_0 must be *larger* by the same factor.

You can check that this makes the electric energy density equal to the magnetic energy density (Eq. 20-58).

20.8 SUMMARY

For any plane wave propagating in the positive direction of the z-axis at a velocity v, the disturbance at z, at the time t, is the same as that at $z = 0$, at a previous time $t - (z/v)$. It is assumed that there is no attenuation. The quantity v is the *phase velocity*.

For a plane, unattenuated sinusoidal wave,

$$\alpha = \alpha_0 \cos\left[\omega\left(t - \frac{z}{v}\right)\right], \tag{20-2}$$

where α_0 is the *amplitude*, and the quantity between brackets is the *phase*. Surfaces of constant phase are *wavefronts*. The wavelength is

$$\lambda = v/f. \tag{20-5}$$

The *wave equation* is

$$\frac{\partial^2 \alpha}{\partial z^2} = \frac{1}{v^2} \frac{\partial^2 \alpha}{\partial t^2} \tag{20-11}$$

or, in three dimensions,

$$\nabla^2 \alpha = \frac{1}{v^2} \frac{\partial^2 \alpha}{\partial t^2}. \tag{20-12}$$

The known spectrum of electromagnetic waves extends from 10 to 10^{24} hertz.

In free space, the *wave equations for electromagnetic fields* are

$$\nabla^2 \mathbf{E} = -\epsilon_0 \mu_0 \omega^2 \mathbf{E}, \tag{20-22}$$

$$\nabla^2 \mathbf{H} = -\epsilon_0 \mu_0 \omega^2 \mathbf{H}. \tag{20-23}$$

The *phase velocity of electromagnetic waves in free space* is

$$c = 1/(\epsilon_0 \mu_0)^{1/2} = 2.997\,924\,58 \times 10^8 \text{ meters/second.} \tag{20-25}$$

Plane electromagnetic waves in free space are transverse. Their **E** and **H** vectors are orthogonal and oriented so that $\mathbf{E} \times \mathbf{H}$ points in the direction of propagation. The magnitudes of **E** and **H** are related to give equal electric and magnetic energy densities:

$$\frac{E}{H} = (\mu_0/\epsilon_0)^{1/2} = 377 \text{ ohms.} \tag{20-37}$$

The quantity

$$\mathscr{S} = \mathbf{E} \times \mathbf{H} \tag{20-44}$$

is called the *Poynting vector*. It is expressed in watts per square meter, and, when its normal outward component is integrated over a closed surface, it gives the electromagnetic power flowing out through the surface.

For a plane wave, $\mathscr{S}$ is equal to the energy density multiplied by the phase velocity.

In *non-conductors*, the phase velocity is

$$v = c/(\epsilon_r \mu_r)^{1/2} \tag{20-54}$$

and, in non-magnetic media ($\mu_r = 1$), the index of refraction n is related to the relative permittivity by the relation

$$n = \epsilon_r^{1/2}. \tag{20-56}$$

The vectors **E** and **H** are in phase, and the electric and magnetic energy densities are equal.

PROBLEMS

20-1 *PLANE WAVE IN FREE SPACE*

A plane electromagnetic wave of circular frequency ω propagates in free space in the direction of the z-axis.

Show that, from Maxwell's equations,

$$\mathbf{k} \times \mathbf{E} = \mu_0 c \mathbf{H},$$

$$\mathbf{k} \times \mathbf{H} = -\epsilon_0 c \mathbf{E},$$

where **k** is the unit vector in the direction of the z-axis, and c is the velocity of light in a vacuum.

20-2 *LOOP ANTENNA*

A 30-megahertz plane electromagnetic wave propagates in free space, and its peak electric field intensity is 100 millivolts per meter.

Calculate the peak voltage induced in a 1.00 square meter, 10-turn receiving loop oriented so that its plane contains the normal to a wave front and forms an angle of 30 degrees with the electric vector.

20-3E *POYNTING VECTOR*

a) Calculate the electric field intensity of the radiation at the surface of the sun from the following data: power radiated by the sun, 3.8×10^{26} watts; radius of the sun, 7.0×10^{8} meters.

b) What is the electric field intensity of solar radiation at the surface of the earth? The average distance between the sun and the earth is 1.5×10^{11} meters.

c) Calculate the value of $\mathscr{S}$ at the surface of the earth, neglecting absorption in the atmosphere.

20-4E *SOLAR ENERGY*

Calculate the area required for producing one megawatt of electric power from solar energy at the surface of the earth, assuming that the average efficiency over one year is 2%. The efficiency of solar cells is presently about 15% at normal incidence.

See Prob. 20-3.

20-5 *POYNTING VECTOR*

An electromagnetic wave in air has an electric field intensity of 20 volts rms per meter. It is absorbed by a sheet having a mass of 10^{-2} kilogram per square meter and a specific heat capacity of 400 joules per kilogram kelvin.

Assuming that no heat is lost, calculate the rate at which the temperature rises.

20-6 *POYNTING VECTOR*

A long superconducting solenoid carries a current that increases with time.

a) Draw a sketch showing, on a longitudinal cross-section of the solenoid, $\mathbf{A}$, $-\partial\mathbf{A}/\partial t$, $\mathbf{H}$, and $\mathbf{E} \times \mathbf{H}$.

b) Can you explain the existence and the orientation of the Poynting vector?

20-7D *POYNTING VECTOR*

A circularly polarized wave results from the superposition of two waves that are (a) of the same frequency and amplitude, (b) plane-polarized in perpendicular directions, and (c) 90° out of phase.

Show that the average value of the Poynting vector for such a wave is the sum of the average values of the Poynting vectors for the two plane-polarized waves.

Hint: Proceed as in Prob. 16-16b.

20-8E *POYNTING VECTOR*

A wire of radius a has a resistance of R' ohms per meter. It carries a current I.

Show that the Poynting vector at the surface is directed inward and that it gives the correct Joule power loss of I^2R' watts per meter.

20-9 *COAXIAL LINE*

A structure designed to guide a wave along a prescribed path is called a *waveguide*.

The simplest type is the coaxial line illustrated in Fig. 20-7. An electromagnetic wave propagates in the annular region between the two coaxial conductors, and there is zero field outside. The medium of propagation is a low-loss dielectric. (See Probs. 6-5 and 6-6.)

We assume that a wave propagates in the positive direction of the z-axis and that there is no reflected wave in the opposite direction. We also neglect the resistance of the conductors.

As in a cylindrical capacitor, $\mathbf{E}$ is radial and inversely proportional to the radius. Since we have a wave,

$$E = \frac{K}{r} \exp j\left(\omega t - \frac{z}{\lambda\!\!\!^{-}}\right),$$

where K is a constant. We shall consider an $\mathbf{E}$ pointing outward to be positive.

The radian length $\lambda\!\!\!^{-}$ is the same as for a plane wave in the same dielectric.

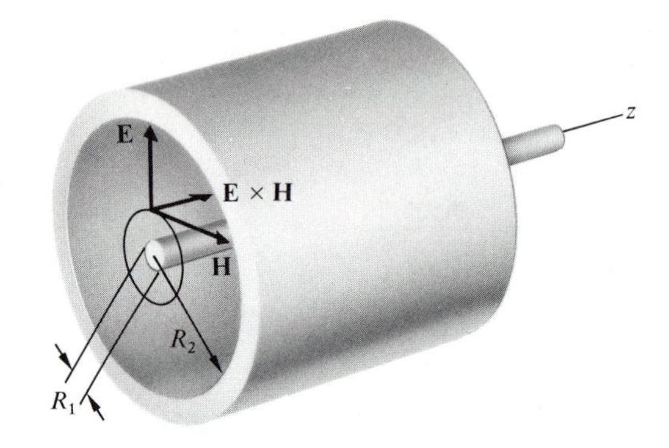

Figure 20-7 Coaxial line. See Prob. 20-9.

a) Draw a longitudinal section through the line and show vectors **E** at a given instant over at least one wavelength, varying the lengths of the arrows according to the strength of the field.

Solution: See Fig. 20-8.

b) Find the voltage of the inner conductor with respect to the outer one.

Solution:

$$V = \int_{R_1}^{R_2} \mathbf{E} \cdot \mathbf{dr} = K \exp j\left(\omega t - \frac{z}{\lambda}\right) \int_{R_1}^{R_2} \frac{dr}{r}, \tag{1}$$

$$= K \ln \frac{R_2}{R_1} \exp j\left(\omega t - \frac{z}{\lambda}\right). \tag{2}$$

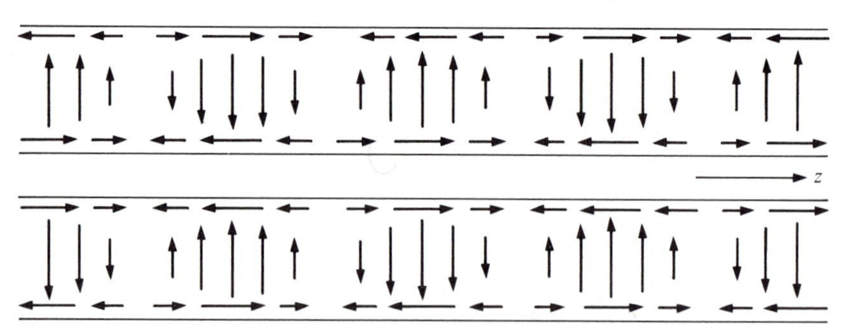

Figure 20-8 Section through a coaxial line showing the electric field intensity (vertical arrows) and the current (horizontal arrows) over 1.5 wavelengths at a given instant. The wave travels from left to right. The surface charge density is positive where the current arrows point right, and negative where they point left.

c) The vectors **E** and **H** are related as in a plane wave, so that

$$H = \left(\frac{\epsilon_r \epsilon_0}{\mu_0}\right)^{1/2} \frac{K}{r} \exp j\left(\omega t - \frac{z}{\lambda}\right). \tag{3}$$

The magnetic field is azimuthal.

Find the current on the inner conductor. An equal current flows in the opposite direction along the outer conductor.

Note that V and I are interrelated.

Solution: From Ampère's circuital law (Sec. 9.1),

$$I = 2\pi R_1 H = 2\pi R_1 \left(\frac{\epsilon_r \epsilon_0}{\mu_0}\right)^{1/2} \frac{K}{R_1} \exp j\left(\omega t - \frac{z}{\lambda}\right), \tag{4}$$

$$= \frac{K\epsilon_r^{1/2}}{60} \exp j\left(\omega t - \frac{z}{\lambda}\right). \tag{5}$$

Figure 20-8 shows how the current varies along the line. Note that E, H, and I are all in phase.

d) Calculate the time averaged transmitted power, first from the product VI, and then by integrating the Poynting vector over the annular region.

Solution:

(i)

$$VI = K \ln \frac{R_2}{R_1} \frac{K\epsilon_r^{1/2}}{60} \cos^2\left(\omega t - \frac{z}{\lambda}\right), \tag{6}$$

$$(VI)_{av} = \frac{K^2 \epsilon_r^{1/2}}{120} \ln \frac{R_2}{R_1} \text{ watts.} \tag{7}$$

(ii)

$$EH = \frac{K^2}{r^2}\left(\frac{\epsilon_r \epsilon_0}{\mu_0}\right)^{1/2} \cos^2\left(\omega t - \frac{z}{\lambda}\right), \tag{8}$$

$$(EH)_{av} = \frac{1}{2} K^2 \epsilon_r^{1/2} \int_{R_1}^{R_2} \frac{2\pi r dr}{r^2} = \frac{K^2 \epsilon_r^{1/2}}{120} \int_{R_1}^{R_2} \frac{dr}{r}, \tag{9}$$

$$= \frac{K^2 \epsilon_r^{1/2}}{120} \ln \frac{R_2}{R_1} \text{ watts} \tag{10}$$

20-10E COAXIAL LINE

A coaxial line has an inner diameter of 5 millimeters and an outer diameter of 20 millimeters. The outer conductor is grounded; the inner conductor is held at +220 volts. The current is 10.0 amperes.

Integrate the Poynting vector over the annular region between the conductors, and compare with the power 220 × 10.0 watts.

20-11D *REFLECTION AND REFRACTION. FRESNEL'S EQUATIONS.*

When light or, for that matter, any electromagnetic wave, passes from one medium to another, the original wave is separated into two parts, a reflected wave and a transmitted wave. There are two exceptions: in *total reflection* there is only a reflected wave, and a wave incident upon an interface at the *Brewster angle* gives only a transmitted wave.

The relative amplitudes of the incident, reflected, and transmitted waves are given by *Fresnel's equations*.

Let us find Fresnel's equations for an incident wave polarized with its **E** vector parallel to the plane of incidence as in Fig. 20-9. The *plane of incidence* is the plane containing the incident ray and the normal to the interface.

Of course the angle of incidence is equal to the angle of reflection and, according to *Snell's law*,

$$n_i \sin \theta_i = n_t \sin \theta_t.$$

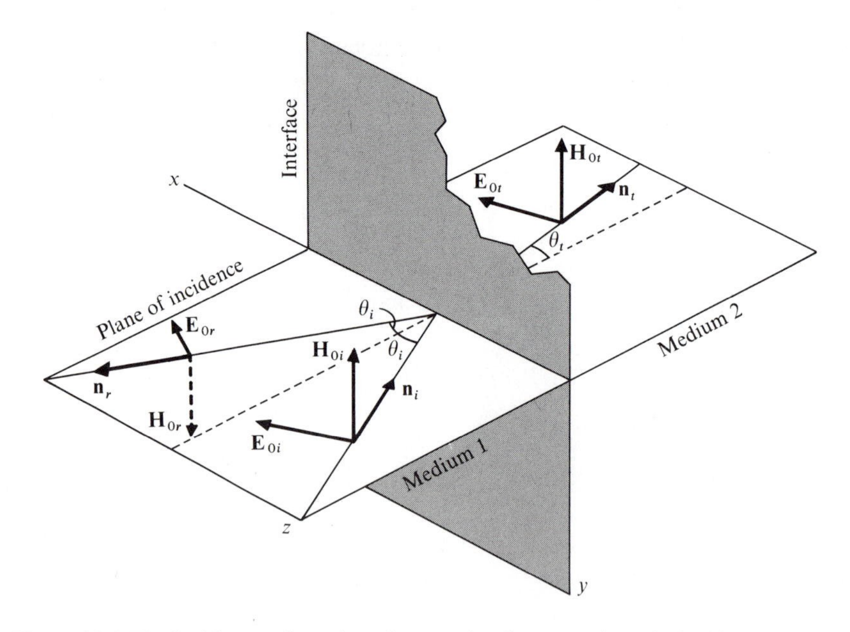

Figure 20-9 The incident, reflected, and transmitted waves when the incident wave is polarized with its **E** vector parallel to the plane of incidence. The arrows for **E** and **H** indicate the positive directions for the vectors at the interface. See Prob. 20-11.

The quantity $n \sin \theta$ is therefore conserved on crossing the interface. Total reflection occurs when this equation gives values of $\sin \theta_t$ that are larger than unity.

a) Find Fresnel's equations

$$\frac{E_{0r}}{E_{0i}} = \frac{-\cos \theta_i + \left(\dfrac{n_1}{n_2}\right) \cos \theta_t}{\cos \theta_i + \left(\dfrac{n_1}{n_2}\right) \cos \theta_t},$$

$$\frac{E_{0t}}{E_{0i}} = \frac{2\left(\dfrac{n_1}{n_2}\right) \cos \theta_i}{\cos \theta_i + \left(\dfrac{n_1}{n_2}\right) \cos \theta_t},$$

by using the conditions of continuity for the tangential components of **E** and of **H** at the interface (Secs. 7.1 and 14.10).

b) Draw curves of E_{0r}/E_{0i} and of E_{0t}/E_{0i} as functions of the angle of incidence for a wave in air incident on a glass surface with $n_2 = 1.5$.

c) Show that there is no reflected ray when

$$\theta_i + \theta_t = \pi/2.$$

The angle of incidence is then called Brewster's angle. The reflected ray disappears at Brewster's angle only if the incident ray is polarized with its **E** vector in the plane of incidence as in Figure 20-9.

d) Show that, for light propagating in glass with an index of refraction of 1.6, the Brewster angle is 32.0 degrees.

APPENDIX A

VECTOR DEFINITIONS, IDENTITIES, AND THEOREMS

DEFINITIONS

1) $\nabla f = \dfrac{\partial f}{\partial x}\mathbf{i} + \dfrac{\partial f}{\partial y}\mathbf{j} + \dfrac{\partial f}{\partial z}\mathbf{k}$

2) $\nabla \cdot \mathbf{A} = \dfrac{\partial A_x}{\partial x} + \dfrac{\partial A_y}{\partial y} + \dfrac{\partial A_z}{\partial z}$

3) $\nabla \times \mathbf{A} = \left(\dfrac{\partial A_z}{\partial y} - \dfrac{\partial A_y}{\partial z}\right)\mathbf{i} + \left(\dfrac{\partial A_x}{\partial z} - \dfrac{\partial A_z}{\partial x}\right)\mathbf{j} + \left(\dfrac{\partial A_y}{\partial x} - \dfrac{\partial A_x}{\partial y}\right)\mathbf{k}$

4) $\nabla^2 f = \dfrac{\partial^2 f}{\partial x^2} + \dfrac{\partial^2 f}{\partial y^2} + \dfrac{\partial^2 f}{\partial z^2}$

5) $\nabla^2 \mathbf{A} = \nabla^2 A_x \mathbf{i} + \nabla^2 A_y \mathbf{j} + \nabla^2 A_z \mathbf{k}$

IDENTITIES

6) $\mathbf{a} \times (\mathbf{b} \times \mathbf{c}) = \mathbf{b}(\mathbf{a} \cdot \mathbf{c}) - \mathbf{c}(\mathbf{a} \cdot \mathbf{b})$

7) $\nabla(fg) = f\,\nabla g + g\,\nabla f$

8) $\nabla \cdot (f\mathbf{A}) = (\nabla f) \cdot \mathbf{A} + f(\nabla \cdot \mathbf{A})$,

9) $\nabla \cdot (\mathbf{A} \times \mathbf{B}) = \mathbf{B} \cdot (\nabla \times \mathbf{A}) - \mathbf{A} \cdot (\nabla \times \mathbf{B})$,

10) $\nabla \times (f\mathbf{A}) = (\nabla f) \times \mathbf{A} + f(\nabla \times \mathbf{A})$,

11) $\nabla \times \nabla \times \mathbf{A} = \nabla(\nabla \cdot \mathbf{A}) - \nabla^2 \mathbf{A}$

12) $\nabla' \left(\dfrac{1}{r}\right) = \dfrac{\mathbf{r}_1}{r^2}$, where the gradient is calculated at the source point (x', y', z') and $\mathbf{r}_1$ is the unit vector *from* the source point (x', y', z') *to* the field point (x, y, z).

13) $\nabla\left(\frac{1}{r}\right) = -\frac{\mathbf{r}_1}{r^2}$, where the gradient is calculated at the field point, with the same unit vector.

THEOREMS

16) Divergence theorem: $\int_S \mathbf{A} \cdot \mathbf{da} = \int_\tau \nabla \cdot \mathbf{A}\, d\tau$,
where S is the closed surface that bounds the volume τ.

17) Stokes's theorem: $\oint_C \mathbf{A} \cdot \mathbf{dl} = \int_S (\nabla \times \mathbf{A}) \cdot \mathbf{da}$,
where C is the closed curve that bounds the surface S.

APPENDIX B

SI UNITS AND THEIR SYMBOLS

Quantity	Unit	Symbol	Dimensions
Length	meter	m	
Mass	kilogram	kg	
Time	second	s	
Temperature	kelvin	K	
Current	ampere	A	
Frequency	hertz	Hz	1/s
Force	newton	N	$kg \cdot m/s^2$
Pressure	pascal	Pa	N/m^2
Energy	joule	J	$N \cdot m$
Power	watt	W	J/s
Electric charge	coulomb	C	$A \cdot s$
Potential	volt	V	J/C
Conductance	siemens	S	A/V
Resistance	ohm	Ω	V/A
Capacitance	farad	F	C/V
Magnetic flux	weber	Wb	$V \cdot s$
Magnetic induction	tesla	T	Wb/m^2
Inductance	henry	H	Wb/A

APPENDIX C

SI PREFIXES AND THEIR SYMBOLS

Multiple	Prefix	Symbol
10^{18}	exa	E
10^{15}	peta	P
10^{12}	tera	T
10^{9}	giga	G
10^{6}	mega	M
10^{3}	kilo	k
10^{2}	hecto	h
10	deka†	da
10^{-1}	deci	d
10^{-2}	centi	c
10^{-3}	milli	m
10^{-6}	micro	μ
10^{-9}	nano	n
10^{-12}	pico	p
10^{-15}	femto	f
10^{-18}	atto	a

† This prefix is spelled "déca" in French.

CAUTION: The symbol for the prefix is written next to that for the unit *without* a dot. For example, mN stands for millinewton, and m·N is a meter newton, or a joule.

APPENDIX D

CONVERSION TABLE

Examples: One meter equals 100 centimeters. One volt equals 10^8 electromagnetic units of potential.

Quantity	SI	CGS Systems: esu	CGS Systems: emu
Length	meter	10^2 centimeters	10^2 centimeters
Mass	kilogram	10^3 grams	10^3 grams
Time	second	1 second	1 second
Force	newton	10^5 dynes	10^5 dynes
Pressure	pascal	10 dynes/centimeter^2	10 dynes/centimeter^2
Energy	joule	10^7 ergs	10^7 ergs
Power	watt	10^7 ergs/second	10^7 ergs/second
Charge	coulomb	3×10^9	10^{-1}
Electric potential	volt	$1/300$	10^8
Electric field intensity	volt/meter	$1/(3 \times 10^4)$	10^6
Electric displacement	coulomb/meter^2	$12\pi \times 10^5$	$4\pi \times 10^{-5}$
Displacement flux	coulomb	$12\pi \times 10^9$	$4\pi \times 10^{-1}$
Electric polarization	coulomb/meter^2	3×10^5	10^{-5}
Electric current	ampere	3×10^9	10^{-1}
Conductivity	siemens/meter	9×10^9	10^{-11}
Resistance	ohm	$1/(9 \times 10^{11})$	10^9
Conductance	siemens	9×10^{11}	10^{-9}
Capacitance	farad	9×10^{11}	10^{-9}
Magnetic flux	weber	$1/300$	10^8 maxwells
Magnetic induction	tesla	$1/(3 \times 10^6)$	10^4 gausses
Magnetic field intensity	ampere/meter	$12\pi \times 10^7$	$4\pi \times 10^{-3}$ oersted
Magnetomotance	ampere	$12\pi \times 10^9$	$(4\pi/10)$ gilberts
Magnetic polarization	ampere/meter	$1/(3 \times 10^{13})$	10^{-3}
Inductance	henry	$1/(9 \times 10^{11})$	10^9
Reluctance	ampere/weber	$36\pi \times 10^{11}$	$4\pi \times 10^{-9}$

NOTE: We have set $c = 3 \times 10^8$ meters/second.

APPENDIX E

PHYSICAL CONSTANTS

Constant	**Symbol**	**Value**
Gravitational constant	G	6.6720×10^{-11} Nm^2/kg^2
Avogrado's constant	N_A	6.022045×10^{23} mol^{-1}
Proton rest mass	m_p	$1.6726485 \times 10^{-27}$ kg
Electron rest mass	m_e	9.109534×10^{-31} kg
Elementary charge	e	$1.6021892 \times 10^{-19}$ C
Permittivity of vacuum	ϵ_0	$8.85418782 \times 10^{-12}$ F/m
Permeability of vacuum	μ_0	$4\pi \times 10^{-7}$ H/m
Speed of light in vacuum	c	2.99792458×10^8 m/s

APPENDIX F

GREEK ALPHABET

Letter	Lowercase	Uppercase
Alpha	α	A
Beta	β	B
Gamma	γ	Γ
Delta	δ	Δ
Epsilon	ϵ	E
Zeta	ζ	Z
Eta	η	H
Theta	θ	Θ
Iota	ι	I
Kappa	κ	K
Lambda	λ	Λ
Mu	μ	M
Nu	ν	N
Xi	ξ	Ξ
Omicron	o	O
Pi	π	Π
Rho	ρ	P
Sigma	σ	Σ
Tau	τ	T
Upsilon	υ	Υ
Phi	ϕ φ	Φ
Chi	χ	X
Psi	ψ	Ψ
Omega	ω	Ω

INDEX